RUTHERFORD

RUTHERFORD
Simple Genius

David Wilson

The MIT Press
Cambridge, Massachusetts

To
Clare and Andrew
Hatty and Jim

QC
16
·R8
W54
1983

First MIT Press edition, 1983

Printed and bound in Great Britain

Library of Congress Cataloging in Publication Data

Wilson, David, 1927–
 Rutherford, simple genius.

 Bibliography: p.
 Includes index.
 1. Rutherford, Ernest, 1871–1937. 2. Physicists—
Great Britain—Biography. 3. Nuclear physics—History.
I. Title.
QC16.R8W54 1983   539.7'092'4 [B]   83-9827
ISBN 0–262–23115–8

Introduction and Acknowledgments

The Introduction is almost invariably the last part of a book to be written. When this book was started there was no intention of coinciding with any anniversary or jubilee – yet, as I write this, it is fifty years exactly since the time of Rutherford's greatest public triumph. March and April 1932 saw first the discovery of the neutron by Chadwick, Rutherford's assistant director of the Cavendish Laboratory at Cambridge University, and then the announcement by Rutherford himself that two of his younger colleagues, J. D. Cockcroft and E. T. S. Walton, "had successfully disintegrated the nuclei of lithium and other light elements by protons entirely artificially generated". 1932 was the "annus mirabilis" not just for Cambridge physics but for physics in general and for Rutherford in particular.

Rutherford himself did not see these discoveries as the opening of the era of atomic energy, though he began to suspect this in the five years before his death in 1937, when the Cavendish Laboratory was the greatest single centre of experimental atomic physics in the world. Only ten years after his death the first atomic bombs had been exploded, the first plans for nuclear power stations were being made, but the Cavendish was shabby, run-down and unenthusiastic, no longer a place of importance in the world of atomic or nuclear physics, and already shifting its priorities to those other fields in which it has since proved pre-eminent. Under the first post-war Labour government, however, Rutherford's other and unacknowledged legacy, publicly-financed, or government, science, was flourishing mightily. It was largely his "boys" who had developed radar and other war-winning devices; it was his assistants Chadwick and Cockcroft, who built the Atomic Energy Authority; it was through "his" DSIR that funds flowed into scientific research in new and old laboratories all round the country; and above all this money was being applied largely on principles that he had enunciated.

Virtually everyone who ever worked with Rutherford has delivered a "Rutherford Memorial Lecture" or written a short memorial of him. These have almost all concentrated on his scientific work and the vast repertoire of "Rutherford stories". I have tried to show Rutherford outside the laboratory as well.

5

Writing a biography, I have discovered, is rather like being a detective. The trail starts from a set of bare facts, the bald entries in standard works of reference, and ploughs deep into the wealth of information contained in those books and memoirs written shortly after the subject's death. Those who wrote more or less immediately of the Rutherford they had known, loved and respected, could not know where his scientific discoveries would lead. Those who have written in the knowledge that Rutherford, more than any other single person, brought us into the era of atomic energy, have tended to concentrate on the history of the excitements of entering that age. So the trail of Rutherford has been obscured by more prominent personalities and events in the foreground. There are quite a number of people still surviving who knew Rutherford well, though their numbers are decreasing – sadly two who gave me great help at the start of my work, Lord Snow and Lord Ritchie Calder, have died very recently. The trail has taken me to great libraries and suburban houses, to Cotswold cottages and the shores of Loch Lomond. It has also taken me to previously unknown parts of places I thought I knew well – it took me to the Written Archives Department of the British Broadcasting Corporation, and to the Script Library, which I had never visited before although I have worked for the BBC for many years – and I have to thank both those sections of the Corporation for their help, as well as my own News Department for giving me a generous spell of leave.

The same trail took me, a journalist, into the world of libraries, and many kind people there taught me how to use their facilities, and helped whenever I got stuck – particularly Mr A. E. B. Owen and his staff in the Manuscript Room at Cambridge University Library and Mr Norman Robinson and the staff of the Royal Society Library. But I was also helped in the Library of the New Zealand High Commission in London, at the Royal Institution of London, at Trinity College, Cambridge, at the Bodleian Library in Oxford, at the UK Atomic Energy Authority, at the Medical Research Council, and by Dr Spencer Weart at the American Institute of Physics Center for the History of Physics, who provided invaluable material from the Niels Bohr Archive.

I suspect that some of the motive for writing this book comes from a mixture of resentment and disappointment at my own failure to enjoy, or be interested by, physics when I was an undergraduate at the Cavendish. There can have been few less enthusiastic students at Pembroke College, Cambridge, yet thirty years after my undistinguished sojourn there, the college unstintingly provided help and accommodation and hospitality when the time came for me to spend some months working in the

6

University Library. At Cambridge, too, I had help from the Cavendish Laboratory in a number of ways, including a special visit to their Museum, for which I must thank Professor Sir Brian Pippard, and the Secretary, Mr J. Deakin. Oxford has also contributed help, mostly through Professor Margaret Gowing and the Institute of Contemporary Scientific Archive, with advice from its Director, Mrs Jeannine Alton. Dr W. Hackman, of that University's Museum of the History of Science, also gave of his time and advice. Mr Don Cawthron drew my attention to the Rutherford material in the Medical Research Council records, an act of friendship to be put alongside many others in the many years I have known him, and I have to thank Lady Todd for permission to use the materials her father, Sir Henry Dale, left in the care of the Royal Society.

Professor E. T. S. Walton, Sir Mark Oliphant, Professor J. Allen, Professor S. Devons and Dr G. N. Burckhardt, all patiently answered questions in my letters, and Dr J. D. R. Thomas of UWIST, helped me to understand – I hope – some points of chemistry.

I was able to interview quite a number of those who knew and worked with Rutherford as scientists in the Cambridge days – so I owe thanks to Sir Fred Dainton, Professor Philip Dee, Professor T. E. Allibone, Dr A. Ratcliffe, Dr A. E. Kempton, Professor D. Shoenberg and Professor Sir Neville Mott. Several of these patient men also provided me with reprints of articles or memoirs they have written, as did Academician P. L. Kapitsa, to whom I am particularly grateful for permission to use freely the otherwise confidential files on the "Kapitsa Affair". Admiral Philip Burnett was most helpful on the technical problems of the use of sonar as it was developed from Rutherford's "ultrasonics". Professor Sir Harrie Massey not only let me grill him on his memories but most generously obtained access for me to the records of the Royal Society.

On a more personal level, Lady Alice Bragg, the late Mrs Gwendy Caroe and Mrs Phyllida Cooke mixed the provision of hospitality and the supplying of their memories of Rutherford in most gracious ways. Professor Peter Fowler, Rutherford's eldest grandchild, also provided hospitality, and added to it family papers, photographs, help, support and advice; his sister, Dr Ruth Edwards, also helped with family memories of life in Cambridge in the 1930s.

But the first great surprise and reward of following the Rutherford trail was to find a little treasure house of previously unsuspected material within a mile of my own home. This came from

Mrs Albert Wood, of Hampton, who found for me, in her late husband's study, dozens of Rutherford's own wartime letters, press cuttings, photographs, memoirs and other mementoes of Rutherford's Manchester days. To her I owe many thanks, but she was not the only one near at hand who provided help. Mr Frederick Sully, who for many years owned the grocery in our suburban high street, is a bibliophile who was most helpful in providing advice as to sources of the many out-of-print books I needed. Once again I have to thank Miss Marianne Stone for her quick and efficient typing, and I am also most grateful to Mrs Pat Rogers who managed to keep my desk tidy for four years without ever losing a piece of paper. My daughter, Dr Clare Poulter, who was doing her own research in the history of science field while I was writing this book, was an invaluable source of advice and discussion throughout.

Collecting material from the New Zealand and Canadian areas of Rutherford's life proved a great deal easier than I had expected through the kindness of the following: the McGill University archives under Dr Robert Michel, Dr A. W. Tickner, the archivist of the Canadian National Research Council, Professor W. B. Lewis of Queen's University, Kingston, and Dr Lewis Pyenson of the University of Montreal; and from New Zealand I received much help from Mr Neil C. Robinson, of Hodder and Stoughton's office in Auckland, from Mr Frank Allan, the local historian of Nelson, and above all from Dr John Campbell of the Physics Department of Rutherford's old college, the University of Canterbury at Christchurch, who is carrying out his own studies of Rutherford's earliest years.

Contents

Illustrations

Acknowledgments
1   The United Kingdom Atomic Energy Authority
2   Mrs A. B. Wood
3   Peter Fowler

1

New Zealand Education

A soft, wet, green version of Hollywood's Wild West describes New Zealand in the 1870s. There were the wooden buildings lining the solitary street of one-horse towns. There were occasional gold strikes, the pioneering spirit, and the first railways being driven through the wilderness. There were even some "Indians", in the shape of a fierce race of brown warriors recently responsible for various "massacres". The role of the US Cavalry, however, was rather woodenly performed by the red-coated files of the infantry of the Victorian British Army, brought over from the former penal settlements of Australia when occasion demanded.

The people, the white settlers, however, were different in culture and outlook from even the soberest citizens of America's western frontier. They were serious, self-conscious emigrants from "Home" in Britain, and there was no glamorous dash of the French voyageur or the hard-riding cowboy. They were people from almost every class of British society selected or self-selected for their skill, ability or even financial backing. They came to New Zealand in carefully planned and often well-financed parties, with their laws and their land distribution worked out before they sailed. They were examples of that Victorian "high-mindedness" which we, a century later, may find embarrassing, even hypocritical, but which we may regret losing.

Many of the settlers were Scottish, especially those who went to the South Island of New Zealand, and perhaps it was their influence that counted most. But right from the colony's earliest days education was of enormous importance, and publicly accepted to be of enormous importance, throughout the colony.

It was into this isolated, sober, almost crime-free society that Ernest Rutherford was born. Among his few known ancestors there is no trace of anything more than solid worth. Among his many collateral relatives still alive in New Zealand – cousins, nieces, nephews and their offspring by the dozen – there is none who has risen to fame. His only child was in no way remarkable for intellect, and though two of his four grandchildren are scientists in their own right, neither of them would claim to be their grand-father's equal.

Yet this Ernest Rutherford, born on August 30th, 1871, in a remote country area thirteen miles south of the pioneering town of Nelson in the South Island of New Zealand, the fourth child in a family of twelve, became, quite simply, the greatest experimental scientist of his age, "the father of nuclear energy". The ashes of Ernest, Baron Rutherford of Nelson were buried in Westminster Abbey, close to the tomb of Isaac Newton, sixty-six years later.

There can, therefore, be no accounting for Rutherford, no explanation of him, in terms of a sudden brilliant flowering of an unusual genetic stock. Still less was he the culminating bloom of some brilliant school or subculture. Writers of the 1920s and 1930s, including some of Rutherford's own colleagues acting as obitu-arists, were very prone to speak of the fine "stock" exported from Britain to New Zealand in the shape of the first organised colonists, and they seem to have felt, unaware of the Hitlerian overtones we now perceive, that such a stock transported to the fertile and empty spaces of New Zealand was sufficient to account for the appear-ance of a Rutherford. But there was no one to equal him among his own generation of New Zealanders, and it is not easy to find any subsequent New Zealander in any field who seems to be of his stature.

The explanation of Rutherford must therefore lie in the inter-action of his particular individual qualities and personality with the cultures and environments in which he found himself, and more especially in the precise stage of development of the areas of science and technology in which he chose (or was lucky enough) to work. In fact only a "Life and Times" of Rutherford can explain the man and his achievements. The picture that emerges is not that of a brilliant man; it is no portrait of a genius but a story of an exceptionally powerful man, a huge personality, a problem-solver, who provided humanity with a new explanation and description of the world in which we live and who provided compelling reasons for everyone to accept his view. It was his closest colleague, Sir James Chadwick, the man who worked longest with him, who triumphed most with him and who suffered most from him, who explained that Rutherford was not even a clever man, he simply was like an enormously powerful battleship ploughing majestically through the storms and seas.

The unusual thing about Rutherford is that despite this power, his outstanding characteristic was that, without any effort, he made himself loved; few men could claim to be admired and loved by such disparate persons as Marie Curie and Stanley Baldwin.

So Rutherford must be studied first in the environment of New Zealand, the country which can well be described as having the least history of any nation in the world. But it is a country with an exceptionally close relationship to the development of science, a country which has shown an exceptional interest in science.

New Zealand was discovered by Abel Tasman, the Dutch explorer, in 1642, and it did not provide much welcome. Attempting to establish the limits of "New Holland" – Australia – for the Dutch East India Company, he never discovered that there were two main islands; he never landed, for everywhere he was met by fierce warriors, who killed some of his men. It cannot be said that Tasman put New Zealand on the map in the ordinary sense, for the Dutch were so secretive as part of their commercial policy that his discoveries were not published for many years.

So for all practical purposes New Zealand was discovered by that great man, Captain James Cook. He first sighted it in 1769, and he paid the islands five visits during his explorations of the South Pacific. His first visit alone lasted six months, and at the end of his life the country was truly on the map, properly surveyed and well described. The main feature of the description was undoubtedly the natives, incredibly fierce and intelligent cannibals, brave to such a degree as to be undeterred by firearms either from attacking or trading with the Europeans. Lives were lost on both sides, brown and white.

Captain Cook could not possibly have realised the most striking natural features of the islands – they contained no animals and the flora included no major crop plants. Perhaps he may have suspected something of the kind, for he landed pigs – whose descendants still survive as wild-pig among the mountain forests – as well as seed potatoes, and the seeds of cabbage and turnips.

And so the Maori had not expanded to fill the Islands. They remained mostly confined to the coastal or riverside villages, living comfortably enough by fishing, by hunting birds in the vast and often gloomy forests, and by cultivating their comparatively limited vegetable crops in gardens beside their houses. They were a Stone Age people when Cook first brought metals into their ken. And there cannot have been more than two or three hundred thousand of them throughout New Zealand. Huge areas of the South Island, in particular, were uninhabited. All this became important only when large-scale white colonisation started, for it meant that the new settlers did not have to dispute every inch of land with the indigenous "natives", and though there were disputes and wars and skirmishes a-plenty, the relations between the two races have rarely been bitter.

Although Captain Cook claimed New Zealand on behalf of his King, that King's government rejected the acquisition of lands so far – twelve thousand miles – from home and took no interest in New Zealand for nearly eighty years. It was not until May 1840 that Britain formally annexed New Zealand, and then the government acted only to forestall the French, who had a colonising expedition underway, and because the first properly organised British attempts at settlement had already begun after years of official discouragement.

Before 1840, that is to say only thirty years before Rutherford was born, the white population of New Zealand was tiny, perhaps only two or three thousand, consisting mostly of whalers, sealers, traders and missionaries. Throughout the 1830s, however, the movement to colonise the country had been growing in strength in Britain – partly this was the first appearance of an "Imperial vision" of Britain's role in the world; partly it was a reaction against the growth of population and the living conditions which we associate with the early days of the Industrial Revolution in the home country.

Wellington, at the extreme south of the North Island, which was to become the capital city, Nelson, at the extreme north of the South Island, and on the opposite shore of the Cook Straits to Wellington, and Christchurch on the east coast of the South Island, which was to develop into the most English city of the British Empire, were the three major settlements of the colonisation of the 1840s. The expeditions organised by the New Zealand Company under its guiding spirit, Gibbon Wakefield, were financed by the sale of parcels of land before the emigrants left the shores of the home country. This in itself implied that many of the emigrants were people of some skill or standing. Care was taken to see that, as far as possible, all the trades and techniques that would be needed on the other side of the world were included in the make-up of the emigrant parties. Provision for religious and educational endowments was made before embarkation, and teachers were included in the earliest parties. Later, when secondary education became desirable, teachers and professors were recruited from English public schools and the English and Scottish universities. The settlement of Otago, with its chief city at Dunedin, was carried out on similar lines further south, with the difference that the majority of the settlers were dissidents from the Scottish Free Church, and there was therefore the possibility of some sectarian disagreements.

The North Island, where the majority of the Maoris lived, and where the greater part of the earliest, more free-wheeling, white settlement had taken place, also received its share of planned settlements.

Both Rutherford's parents came to New Zealand as children with their parents in the early waves of organised settlers. His grandfather, George Rutherford, a wheelwright from Dundee who worked for a Perth firm of coachbuilders, came to the Nelson colony in 1843. He had been specifically recruited by Captain Thomas Thoms to set up saw-milling equipment which had been sent out by an earlier boat. George Rutherford came on the ship *Phoebe*. He was then thirty-four, and he brought with him his thirty-one-year-old wife, Barbara, and their four sons, the third of whom, James, then aged five, was to be Ernest Rutherford's father.

George Rutherford duly set up the saw-milling equipment at Motueka, where it was driven by a twenty-horsepower water wheel. Then he moved his family to the nearby Wairoa river valley to the south of the chief city, Nelson, and set up in his old business of wheelwright. In the pioneering circumstances a wheelwright was unlikely to find much of a carriage-trade and therefore his work mostly concerned the local saw-mills and flour-mills – indeed he went into an unsuccessful flour-milling business himself. All his sons followed in his footsteps, working as wheelwrights, setting up small local mills and businesses in flour and flax, utilising the water power of the Wairoa river. The three daughters born in New Zealand all married local men and the various Rutherford and related establishments were clustered in the same area around the village of Brightwater, which was later renamed Spring Grove.

It is quite possible that in the more class-conscious society of England the Shuttleworths, a Sussex family, might not have spoken very much to a family of Scottish wheelwrights such as the Rutherfords. But one of their daughters, Caroline, who had married the son of a flour miller at Hornchurch in Essex, was tragically widowed when her young husband, who was reputed to be a clever mathematician, and who worked in his father's counting house, died young. So Caroline Thompson, with her daughter Martha and three sons, emigrated to New Zealand with her Shuttleworth parents in 1855. They sailed in the *Bank of England*, landed at Auckland and immediately went on by the brig *Ocean* to the Taranaki district of the North Island on the fertile and beautiful slopes of Mount Egmont. Caroline Thompson had a profession of her own as a teacher and the whole group settled down in the New Plymouth area.

New Plymouth and the Taranaki were among the New Zealand Company's early settlements, but, for reasons as much connected with the inter-tribal wars of the Maoris as with any fault of the white settlers, they were one of the few places where there were troublesome land disputes between the two races. The problems had not been cleared up under the Governorship of Admiral Robert Fitzroy, who had been told in no uncertain terms in the 1840s that the settlers there were faced with the ruin of their hopes. This Governor Fitzroy, as Captain Robert Fitzroy, had commanded the ship *Beagle* when it brought Charles Darwin for a few weeks' botanising on the North Island in 1838, during that famous trip around the world which is said to have provided the material for *The Origin of Species* and the Theory of Evolution. His Governorship, short though it was, had been a great deal less productive than his captaincy of the *Beagle*, for his autocratic methods had produced the first major clash with the Maoris, the *casus belli* being a flag-post.

So when the one and only serious war between settlers and Maoris broke out in 1860 it was over a land problem in Taranaki – the Waitara land purchase. The white settlers were nearly all driven to take refuge in New Plymouth. The Shuttleworths sent Caroline Thompson, her four children, and another of their women to the safety of Nelson and the South Island.

The Thompsons never returned to the North Island. Caroline got herself a job as schoolteacher at Spring Grove and there she stayed until she married William Jeffries, a man who had at one time been in partnership with George Rutherford. Her daughter, Martha, who had helped her mother in the school, had risen at four o'clock on many mornings to study in order to qualify as a teacher. Eventually she took over the teaching at Spring Grove school when her mother married. At Spring Grove Martha met and married James Rutherford on April 20th, 1866, and five years later their second son and fourth child, Ernest, was born at the same place.

By this time, 1871, the ten years of war with the Maoris had at last come to an end. There had been virtually no fighting in the South Island, in any case. Gold had been discovered in various places, notably around Otago, and a flood of population had poured in. The country as a whole was booming and the boom was led by the southern provinces of Canterbury and Otago, where it had been discovered that a land without animals or foodcrops of its own was ideal for sheep running in enormous numbers on the tussock grass of the bare downlands sloping up to the eastern flanks of the mountains. And so, at exactly the time Ernest Rutherford was born, the first generation of New Zealanders began to develop the future of their country on a firm basis, and being the people they were, they started with public works, mostly railways, and education.

The time of Rutherford's birth was also the time of the birth of many of the institutions that were to educate him and in which he was to spend his life; and the conclusion is unavoidable that the future achievements of the man must have been, to a large extent, the product of the system which was being created to nurture people such as he and that he arrived at the right time to benefit from it.

The New Zealand system of free, secular primary education was being founded in the first five years of his life and was officially blessed and brought into formal existence by an Act of the New Zealand Parliament in 1875. Balfour Stewart introduced laboratory experimental work into the teaching of physics at Manchester in the academic year 1870–1871. The Cavendish Laboratory at Cambridge, where Rutherford was first to make his mark on the scientific world and where as Professor and Director he was later to lead the world of physics, was created in 1870; its first Professor, Clerk Maxwell, was taking up his post as Rutherford was born and delivered his inaugural lecture in October 1871. Rutherford's own university undergraduate education and his first research was completed at Canterbury College, Christchurch, which was launched at a meeting on August 14th, 1871, when the Canterbury Collegiate Union was formed. The idea of a university in New Zealand had received official sanction in 1870, the year 1871 saw bitter battles over the shape of the new institution, but by 1874 the issues had been resolved and the new colony had its university barely thirty years after it had itself been brought into existence.

Many of these events which would affect Rutherford's development were connected among themselves as facets of the same movement for educational reform which also had a profound effect on British institutions. This reform movement reached its first flowering in the 1851 Great Exhibition in London, where Prince Albert was among its leaders. It was the profits from the Great Exhibition reappearing as scholarships forty years later that took Rutherford to Europe and Britain. But the movement reached its culmination in the decade 1867–77 when Gladstone suddenly reversed his earlier implacable opposition and himself carried measures for the reform and restructuring of the Universities of Oxford and Cambridge. On the face of it, these reforms were the removal of religious tests and qualifications for entrance to, and the holding of posts and fellowships in, the two universities. But in reality they meant the opening of the old universities as places of general education rather than their continuation as primarily training establishments for the Church of England.

A high proportion of the men called to New Zealand's young educational institutions as professors, headmasters or teachers were Oxford and Cambridge men, and many of them had been imbued with the spirit of reform which was then abroad in the home country. Some of those who were directly involved in the education movement around the time of Rutherford's birth were to be among the strongest intellectual influences on his life. Balfour Stewart, through a book, was to have a seminal effect. Clerk Maxwell not only set the Cavendish Laboratory tradition, but seems to have been one of Rutherford's exemplars in his method of work, starting, like Faraday, from a physical view or model of the phenomena of physics and translating this into geometrical and hence mathematical terms.

New Zealand's insistence on secular primary education does seem at first sight to be in contradiction to the religious motivation of much of its early settlement. Otago, as has been seen, was a settlement largely of Scottish dissidents from the established Presbyterian Church. Canterbury was looked upon by many of its founders as one of the great hopes of establishing the ideals of the Anglican Church, which could never be realised at home: "It is my only consolation that the Church of England is putting forth vigorous scions in foreign lands when the time of her glory, as far as human eyes can judge, is departing from her," wrote one correspondent on the subject in 1850. Perhaps the purpose of secularisation was to avoid the problems of sectarianism which had already reared their heads over the starting of university education in the 1860s.

Scientifically, however, New Zealand has been well on the map ever since its first discovery. Sir Joseph Banks botanised "with our usual good success" when he came with Captain Cook. Darwin explored the Bay of Islands area on the voyage of the *Beagle*. A country so cut off from the rest of the world, in the evolutionary sense, offered a very happy hunting ground – such as Rutherford himself would later have recognised as a "Tom Tiddler's ground" – for many men anxious to make names for themselves as observational and classifying botanists and geologists in the first half of the nineteenth century, when modern science was first becoming important. Sir Joseph Dalton Hooker, son of the great botanical pioneer Sir William Hooker, for instance, landed in New Zealand in 1841, at the time of the first colonisation, while he was botanist to the Ross expedition to the Antarctic. Twelve years later he published *Flora Novae-Zelandiae* as one of the official works of the expedition, but in the meanwhile he had paid William Colenso, the man who had acted as guide to his own first explorations, to make plant collections in various parts of the islands, and he had paid out of his own pocket.

Then there came a series of notable men who settled by choice in New Zealand because it offered such fine opportunities of practising science. Julius von Haast, the German geologist, building on the pioneering work of the Austrian Hochstetter, won a world reputation for his systematic study of New Zealand. His work was highly practical, too, for he discovered a coal seam near Nelson, and his collection of specimens, both geological and fossil, was the foundation of Canterbury Museum, which in turn became part of Canterbury College. James Hector, a Scot, was rather similarly an exploring geologist and map-maker, but more comfortable in association with official enterprises. By 1880 he had passed from exploring to leading the government's geological survey of New Zealand, and finally became the country's leading scientific official, organising the Meteorological office, the Government Laboratory and the botanical gardens. He was also Chancellor of the university. Frederick W. Hutton, a former Captain in the British Army who had fought in the Indian Mutiny, occupied a corresponding pioneering position in the biological sciences, and was the authority on the country's Mollusca. Both Hutton and Haast were among the early Professors at the infant University Colleges of Otago and Canterbury.

The importance of these men lies not only in the first-rate work they did, which earned them firm places in the (still largely European) world of science, but also in the value of their work to the early colonists and the fact that they showed the importance of science to New Zealanders. It can clearly be claimed that no other country has appreciated the value of science and placed public value on the work of scientists as much as New Zealand. An official historian of New Zealand science goes so far as to claim, "In proportion to their numbers, New Zealanders have done more for the progress of modern science than any other people."

Whether this claim can be substantiated or not is of little import. What matters is that New Zealand in the latter half of the nineteenth century, when Rutherford was growing up, was a place where science and scientists were valued, where they received official recognition, where they were among the chief public figures. Admittedly the approach to science was in expectation of practical benefits: scientists were valued for the practical value of their work in discovering mineral deposits, in improving fertility, in dealing with agricultural disease and other problems. And this was very different from the attitude of officialdom and the leaders of society, both political and educational, back in Britain – as the story of the foundation of the University of New Zealand shows.

The chapter on the history of the founding of this university in the government-sponsored book on *New Zealanders and Science* is entitled "The Establishment of a Scientific University". The good Scottish people of Otago, arriving with their national spirit of reverence for learning, had agreed that one-eighth of the price they paid for their land should be set aside for religious and educational purposes – and a university was one of the specified projects. Twenty years after their arrival, by 1869, they were hard at work setting up their university, which they planned to start with four chairs: one in philosophy, one in classics and literature, and two in science. What was unprecedented, shocking even, was that these chairs should all be equal in standing and stipend – never before in the English-speaking world had chairs of science been given foundation status and equal salaries with the humanities. The long battle over whether each province was to have a separate university, or whether they should join in a national university, and whether that university was simply to be an examining body need not concern us here. But the first Council of the University of New Zealand, rather wantonly against the wishes of the government that had founded it, refused to accept Otago's view that the curriculum should give equal status to science. Otago proceeded independently for several years with a university of its own.

When the province of Canterbury joined the battle with its own College at Christchurch, it also joined Otago in demanding that students should be allowed to take degrees in either arts or sciences and that they should be provided with equal facilities for both studies. Since the two provinces were then much the wealthiest and most prosperous in the country, they won their battle and the New Zealand University Act of 1874 provided that the national university should simply be an examining body to which independent provincial colleges should be affiliated. But when the new university applied to the British authorities for a Royal Charter, offering degrees in medicine, law, music, arts and science, the Crown would only grant letters patent if the original Act were amended to exclude any science degrees.

The colonial government decided that half a loaf was best, and actually accepted this extraordinarily conservative dictatorship by passing an Amending Act excluding science degrees in 1875, so that it got its charter a year later. One student did complete a science course in 1877 but had to be satisfied with a Bachelor of Arts degree. By 1884 however the colony got its way and was allowed to grant science degrees by a supplement to the charter. New Zealand's first science degree was granted, appropriately enough at Otago, in 1888, only three years before Rutherford started as an undergraduate.

For some reason biographers seem to like their scientific subjects to have a rural background – to be felt to have spent their formative years in close contact with nature whose secrets they are later to unravel, perhaps because they have learned the value of close and intimate observation of living things and the forces of weather, sun and water. The young Rutherford certainly grew up in the country, in pioneering, almost untamed, country at that. He could hardly have had an urban background since there was no truly urban environment within two thousand miles or so – and that is counting Sydney in Australia in the 1870s as truly urban. The irony of this is that Rutherford in his great scientific days was not an observer of nature in the usual sense; his fame was to come from his ability to manipulate and interpret the unobservable.

Yet stories of his childhood are easy enough to come by, as is only to be expected from a still largely rural community containing many dozens of his relatives and the immediate descendants of those who taught him and went to school with him.

There is no doubt that his mother was the dominating influence of Rutherford's early life. She set the tone of the home, and indeed it was reported just before she died at the age of ninety-two in 1935 that she "still ruled" the house. Since she was a schoolmistress it is hardly surprising that she valued learning and that the home was one in which intellectual endeavour was valued above what would normally be expected in a small wooden house near the edge of civilisation.

It can, indeed, be argued that his mother remained the dominating influence throughout his life. Certainly he wrote to her once a week or once a fortnight without fail for his entire life after he had left New Zealand. He placed all his triumphs before her rather than anyone else, and the story of his telegram to her when he was ennobled in 1930 is widely quoted: "Now Lord Rutherford, Honour more yours than mine," he cabled. He continued to say throughout his life that he owed everything he had gained to his parents and he frequently referred to the sacrifices they had made to procure his education. It is impossible to draw a line between proper filial duty and love of parents on the one hand, and that slightly unhealthy feeling of life-long psychological dominance over a child that one often meets. In Rutherford's case his unusual sexual diffidence, his hatred of physical contact with others, his delight in receiving overt honours, justifications of his worth to present to his mother as it were, leave a feeling that her influence was rather greater than perfect balance would demand. On the

23

other hand, he was always scrupulously fair to women professionally, and a pioneer in encouraging women to work in his laboratories and to become professional scientists, all of which must be accounted to the same influence of a schoolmistress mother.

His father is a more shadowy figure, at least in personal terms. He was plainly honest and upright, a pioneering "bush engineer", hardworking and inventive, able to turn his hand to most tasks of practical necessity in a new country, and perfectly capable of adapting existing machinery to the new tasks of a strange environment and even of inventing new devices to cope with new situations. He was firm and even daring in his decisions but he was not the man to carve an empire of his own out of the bush. He provided successfully for his family, avoiding both luxury and starvation, in a series of small, not always successful, business enterprises in logging, milling and flax cultivation. But he was often away from the family home for most of the working week, or even for longer spells, and the roles of disciplinarian and family treasurer both fell on his wife.

There were, however, so many children, twelve in all, that sibling relationships must have been as important as any others and must have tended to balance out any deficiencies or overemphases in parent-child relationships.

In social and economic terms the childhood of Ernest Rutherford occurred in a period of ever-increasing depression, the worst depression New Zealand has ever known. Nelson, the province in which he was born, had started as the most promising of all the private-enterprise colonies set up by Wakefield. The harbour of Nelson, almost concealed from the sea behind a shingle-bank, was a far finer anchorage than Astrolabe Bay used by earlier explorers such as d'Urville, and it had been revealed to the colonist by friendly Maoris. Nearby was a fine and fertile coastal plain, and superb deeply forested hills and mountains provided a backing. The climate was temperate, reasonable, and as good as any in New Zealand.

But the province proved deceptive. The coastal plain was too small for the number of new settlers – almost 2,000 in the first year and 4,000 by 1850. The steep valleys that run northwards to the sea were fertile enough in the bottom lands, but the hills separating them were steep and high and rose to the main mountain spine of the South Island inland. The superb timber of the slopes was cut and burnt unmercifully and wastefully, only to reveal a soil which was not fertile enough to support rich pastureland such as was found further south in Canterbury and Otago. Soon the mountain slopes reverted to a dull and uninspiring bush of gorse and man-uka. The province was cut off from the rest of New Zealand by the mountains to the south and by Cook Strait to the north. It was even cut up into parts itself by the steep parallel ranges of hills and much of the transport from one part to another was easier by sea than by land.

True, gold was found in the 1860s, but not in such quantities as in Otago to the south or in Auckland in North Island. By the time of Rutherford's birth the population of Nelson province had risen to nearly 15,000, and there it virtually stayed – there was an increase of only 3,000 in the next twenty years. A third of these people lived within the Nelson city boundaries. Nelson, in fact, with its sunny, mild climate, troubled only by the earthquakes of New Zealand's fault zone, was settling into its present reputation as "a quiet and contented backwater" due to the "unambitious outlook of its own people" – these quotations come from the New Zealand government's own survey of the country's geological and agricultural resources.

Neither the depression nor Nelson's somnolence had begun in 1871, though both were threatening. The Maori wars had just ended, and in any case only one per cent of Nelson's people were Maori. The first railway in the colony had been started there, pushing south-west from Nelson up the Wairoa valley. Under the Vogel administration, with its programme of public works, the railway was pushed further, with a view to connecting the province to the richer country of the south – a project which has never reached fulfilment. It was this railway development that formed the backcloth to Rutherford's earliest years – and has caused some problems to his biographers.

He was born in Spring Grove, which was a scattered community along the Wairoa river valley containing a church, a schoolhouse and a post office. As the railway advanced, two stations were built about a mile apart, the one further from Nelson being called Spring Grove. The original Rutherford wheelwright workshop and the churchyard containing the graves of his grandparents and several aunts and uncles, which were once in Spring Grove are now in Brightwater, the name of the other railway station.

The house in which Ernest Rutherford was born has now disappeared, though a photograph of it survives. It was a small and simple, though not crude, single-storey wooden dwelling with a roof of wooden shingles and a verandah outside. The photograph brings to mind a deserted Hollywood "Western" set, and then makes one pause to think how small it was for the rearing of twelve children.

It is not certain whether this was the only house the Rutherfords lived in during Ernest's first five years, but it certainly cannot have been the only house he knew, for there were a number of Rutherford and Thompson relatives living in the same area. There is some reason for believing that for a few weeks he may have attended the Spring Grove school where his mother and grandmother had been teachers, but for all practical purposes he began his education at Foxhill school a few miles further up the valley. As the railway advanced his family moved to Foxhill, to a house which still exists and which seems to be simply a slightly larger version of the house in which he was born – single-storey, woodbuilt, standing alone and surrounded by a simple verandah.

The reason for the move surely lay with the railway, for Rutherford's father made most of his income at this time with a government contract to provide railway sleepers, cut and shaped from brown or black beech. And it was at Foxhill school that Rutherford first came under an important influence outside the family. The schoolmaster was a Mr Henry Ladley, and under him the boy stayed for six years, working his way up to Standard IV and obviously getting a solid grounding.

Nevertheless the home influence must still have predominated and the picture provided by the brothers and sisters in their old age is one of an earnest, God-fearing, but happy and laughing family, grouped round the pioneer hearth. Mrs Rutherford had her Broadwood piano, her most prized possession. Father played his violin. Everyone read or listened to others reading aloud. Sunday meant, of course, church services, though not of the three-times-a-day routine. There were spelling bees and sing-songs and church hymns at home. Mrs Rutherford pinned up maps on the wall to illustrate current affairs. Some memorialists have even asserted that this life taught Ernest to have a deep love of music, though this is plainly pushing admiration too far, for a less musically-minded man can scarcely have existed. What it did impart to him were hymn tunes which he continued to reproduce in a mixture of grunting and humming throughout his life. What this life definitely taught him was his habit of voracious and continual reading. It is recorded that Rutherford found Pickwick particularly amusing and his relatives remember that his easy, booming and infectious laugh was already developing. But likewise he was already picking up his delight in what were then called "yellow-backs" – thrillers, shockers and crime stories, that he continued to consume at an enormous rate throughout his life.

The temptation to see the first signs of later genius in the young boy must be avoided. But there can be no doubt that at the age of ten Ernest Rutherford did possess a copy of Balfour Stewart's *Primer of Physics* because it has been preserved and shown in public exhibitions with his name inscribed and the date of July 31st, 1882.

This book was one of a series published in Britain, a venture which was in itself part of the educational reform movement of the times. The aim of the books was to impart the basic principles of the various sciences by encouraging the readers to perform simple experiments for themselves. Balfour Stewart's work asked the reader to provide only the simplest materials of weighing pans, wires, small coins, and the like. It is impossible to measure its influence on the boy of the 1880s. Sir Ernest Marsden, Rutherford's pupil and life-long friend, who eventually became the leading scientific official of his time in New Zealand and who devoted much effort to collecting information about Rutherford's childhood, points out that Balfour Stewart recommended and praised exactly the sort of methods that were to bring Rutherford the great results and insights of his adult life: "Simple direct experiments were outstandingly the feature of his methods of work as opposed to abstruse theoretical speculations removed from a directly visualised conception."

Balfour Stewart, it should also be noted, was the man who introduced experiments into the undergraduate physics course at Manchester, and was the most influential of the teachers of J. J. Thomson, who was in turn Rutherford's teacher.

All this makes Rutherford sound like a backwoods bookworm. It is quite certain that he was no such thing. All the children in the large family had their tasks to do around the house. To the very end of his life among the well-manicured fields and gardens of Southern England, Ernest chopped wood, cut down trees and hacked at overgrown bushes as a sort of hobby and as almost the only vigorous exercise he took; and it is known that this occupation began in his Foxhill days. He was fond in later life of recalling the rural tasks of his childhood. He told of the occasion when he was sent out to gather firewood and, with typical enthusiasm, gathered more than he could comfortably carry. His answer was to tie the bundle to the tail of the family cow and drive her home before him, which proved satisfactory until the cow came to a gap in the boundary-fence which was big enough for her but not big enough for the bundle of wood. In the ensuing struggle the cow lost the end of her tail but the firewood was brought safely home. When asked what he had done with the damning evidence of the remainder of the tail he replied that he had buried it, "because I had been taught that new things would grow from cuttings" which may not have been very funny but was typical of the way he would amuse children and which was also characteristic of his rather earthy sense of humour.

There is no doubt, too, because there were many alive in the days of his greatness who had been the companions of his youth, that he got out and about into the forests and streams of that beautiful and only half-tamed country. He caught fish, he nearly drowned in a river and had to be rescued by his brother when he got out of his depth; he shot birds with a rather primitive gun, and it is claimed he once shot sixteen of the elusive tree-pigeons in a single day by working out a technique of firing just at the moment when they spread their wings to land on the high branches. Similarly there is a story that, when his mother asked him to help with the education of his younger sisters by giving them a lesson, he willingly agreed as long as his mother would "muster" them. When this mustering was done the boy solved the problem of keeping the little girls still and in place by the simple expedient of tying their pigtails together. Another well-remembered occasion was when he had propelled a missile right over the roof of their small house and caused considerable hurt, though little injury, to a younger brother on the other side. Ernest himself recalled that he stifled the complaints of the younger child by "sitting on his head" until comfort, or submission, was secured.

Whether these stories are "true" in an academic or historical sense matters little. They show simply that Rutherford was brought up in a rough, homespun, if not rowdy manner, which made him a breezy and extrovert character all his life. In later life, at English universities he was, not unnaturally, found a little ungraceful by men who had been brought up in the comfort of the middle-class homes of English clergymen or professors. At the same time this upbringing presumably toughened him physically and endowed him with the endurance to work extremely hard when he found that desirable.

More seriously, he undoubtedly learned in his youth to be adept with his hands and inventive in a society where "toyshops" were not available. He took an early interest in photography – indeed it is recorded that hardly a square foot of the country round his home was left unphotographed when this enthusiasm swept him along. And there is no doubt that his mother kept him and his brothers and sisters "up to the collar" (the local phrase) with their school work.

Throughout this growing-up period there came a number of changes in the family circumstances. The railway up the Wairoa valley never broke through to the rich South Island sheep-pastures over the mountains. His father moved his interests to flax-milling at a place called Pukaka over the mountains to the east, where the native wild flax grew abundantly in a large swamp where the Pelorus River debouches into the large sea-inlet called Pelorus Sound. James Rutherford spent more and more time away from the Foxhill home, and was reduced to writing weekly letters to his wife – the road across from Nelson is still fairly precipitous. Eventually, in 1883 he moved his whole family and all his stock and possessions to his new base. The move was made by sea, the Rutherfords leaving Nelson on May 26th, on the paddle-steamer *Lady Barkly*, which was normally used on the service between Tasman Bay and Collingwood, but which had evidently been hired for this special run.

James Rutherford prospered at first. The family home was set up in Havelock. There was plenty of flax and he obtained more contracts for railway sleepers – 40,000 of them at two shillings and eightpence each. These sleepers were for the rapidly expanding railways around Christchurch, and were shipped from a jetty built by James Rutherford himself.

The New Zealand flax, *phormiuim tenax*, produces leaves about five feet long and five inches wide directly from the ground in the swampy area where it grows. The Maoris showed the early settlers how to cut the leaves and strip away the green matter to leave the strong fibres. The Maoris used these fibres for matting, clothing, making fishing lines and baskets. The Europeans used it basically for making ropes and other utilitarian purposes, rather as a substitute for jute – but the fibres were too coarse for processing into linen and textiles in the European fashion. And, of course, the Europeans invented various primitive machines to assist in the several processes needed to prepare the material. James Rutherford had his flax-mill near his jetty and exported most of his produce by sea in scows.

The eldest son of the family, George, had won a scholarship to Nelson College in 1881, shortly before the family moved to Havelock, so he was away for most of the time. And the period at Havelock was dreadfully marred by the tragic drowning of two of the younger brothers, Herbert and Charles, in a boating accident with another local boy out on the Pelorus Sound. Martha Rutherford was worst affected by the catastrophe; it seems almost to have broken her for some considerable time – the surviving children remembered that their mother stopped playing her piano for good and that all the light and laughter seemed to go out of the house. Ernest helped his father and remaining brothers in a fruitless search which lasted for months along the shores of the Sound in an attempt to find the bodies.

The stay at Havelock was not completely forlorn however. In particular, summer holidays were enjoyed and there were several trips back to the old Rutherford and Thompson country around Foxhill. They even stayed once in their old home, now deserted, when Ernest and the other boys were particularly attentive to their mother and strove to provide her with some amenities that they "cobbled up" from whatever materials they could find. The boys also earned pocket money on some of these trips by hop-picking in the fields around Nelson. And it is from one of these visits, when he was taken to his grandmother's home, that we hear of Ernest and the potato-masher. Apparently he noticed the old lady's masher was disintegrating so he promptly made her a new one of his own, improved design. This potato-masher is still preserved by the Royal Society in London, but there was another long-lasting relic of these summer holidays which was of slightly more value to Ernest himself.

It has been recorded locally that the three Rutherford brothers earned £13 for six weeks' work in the hop-fields one summer. Whether it was precisely after this season or on some other occasion, young Ernest certainly banked some of his hard-earned wages, for only a couple of years before his death Lord Rutherford of Nelson had to write to his sister, Mrs Strief, in New Zealand telling her that he had just been informed that he had a "windfall". In a letter dated May 24th, 1935 he says: "I have recently found that I have a Savings Bank Account in Nelson, New Zealand, and when I was nine years old, deposited £1.5s.0d. in it. This has been untouched and has now mounted up to £6.4s.6d. I have an idea that I must have earned this prodigious sum of money [at that time] by helping my Uncle Thompson on the farm." Without giving way unduly to the false wisdom of hindsight, it is worth remarking that, though he had to be very careful of the pennies in his young days, Rutherford's disregard of personal money-making remained one of his most pleasant characteristics throughout his life.

At this stage Ernest Rutherford came for the first time under the influence of an above-average teacher, Jacob H. Reynolds, the master and sole teacher at the Havelock school. Reynolds would give his brightest pupils an hour's extra lesson in Latin before the start of normal school hours. Ernest quickly showed that he was worthy of such extra tuition, and began to shine as a really clever boy. More accurately, we can now see, it was not so much that he was clever as that at this stage of his life there began to appear that power of concentration and that ability to master the subject in hand that was to distinguish him afterwards. It was always the power of his mind and personality that impressed acute observers rather than any cleverness. The evidence of a Miss Buckeridge, collected in the 1920s by Ernest Marsden, illuminates this emergence of intellectual power; she was employed by Martha Rutherford as a sort of governess and help with the younger children and she remembered: "Ernest never needed to study; having read a school book once he knew it." She added a very different sort of tribute, however: "He was a lively boy and I loved to see him and the other children at their mother's knee at prayers." The younger brother, James, recorded an interview when he was eighty years old and he probably meant much the same thing, the power of concentration, when he remembered an occasion when the children were watching a thunderstorm on the mountains but "Ern" was found to be "dreaming" – thinking about something to the exclusion of what was going on around him.

Havelock was in a different province from Nelson, and the local centre was the small town of Blenheim. Marlborough is one of the smallest of the New Zealand provinces, certainly in terms of population, and the opportunity for further education came almost solely through the offer of scholarships, few in number, to Nelson College. In a society that valued education so highly there was real competition between the small country schools for their pupils to win one of these scholarships, but local experience showed that candidates from the rural areas were disturbed and "put off their stroke" by having to travel to Blenheim and spend several days in strange surroundings, so that they rarely did themselves justice. In any case, Sir Ernest Marsden has pointed out, local research has shown that at the age of around fifteen, country boys tend to be about a year behind their urban contemporaries, especially in reading. The Havelock authorities had therefore campaigned to have the scholarship examinations held in their own local school for their local candidates. The point had been won and an inspector was appointed to travel to the school for this great occasion when Ernest Rutherford came to take his turn in 1886. He was then around fifteen, rather older than many of the other candidates, certainly no child prodigy.

Nearly fifty years later when Sir Ernest Rutherford, Professor of Physics at Cambridge, Nobel Prize winner and one of the world's best-known scientists, made a triumphal tour of his homeland in 1925, the occasion of the scholarship exam was still remembered. The retired local inspector of schools, a Mr Lambert, related how on that hot summer's day of 1886 various interested local residents came quietly to the door of the two-roomed school at Havelock to enquire anxiously and in whispers how the local pride was doing. The supervisor, who apparently collected each page as it was completed by the candidate, passed out the news including one comment on the arithmetic paper: "He is doing fine but has made a bloomer in question X." Gloom, it was reported, spread through the hamlet, but Ernest was a quick worker and had time in hand to check his answers at the end and put his errors right.

He won his scholarship with the previously unattained total of 580 marks out of a possible 600, including a maximum of 200 marks out of 200 in the arithmetic paper. The other subjects were geography, history and English.

The scholarships to Nelson College carried free tuition and a £40 grant for residential board. Ernest was therefore able to enter the school in February 1887, having been taken by his father across the mountain ranges on horseback.

Nelson College epitomised the high aims of the earliest New Zealand colonists in education. It was founded in 1856, just fifteen years after they had arrived in the new country, yet it was planned to be an imitation of Eton College both institutionally and architecturally. Though its first buildings were of wood, five of its first seven headmasters were graduates of Oxford or Cambridge and several of them had teaching experience at English public schools.

From the very start it had been forced to include a boarding element in its structure, drawing pupils from such a scattered pioneering community. The basic fees were £6 per annum, according to the first announcement of its existence in the local *Examiner* on April 7th, 1856, but the same announcement said there would be accommodation for boarders and day boarders and it promised that the college would be based on unsectarian principles.

Not only was the college one of New Zealand's earliest steps into secondary education, it was also one of the most successful (and remains so to the present). There was a steady increase in numbers throughout its early years. In the 1870s there came a setback when Mr McKay, described as "the strong disciplinary force in the school", moved off to become head of the new Wellington College on the other side of the Cook Straits in the North Island. He took with him fifteen of the senior pupils, some of the teachers and even some of the domestic staff. Nelson was left to "rebuild the boarding and teaching traditions".

There had been a long-running battle between the governors of the college and their successive headmasters over the relationship between the headmaster and the boarders at the school – it boiled down to whether the head should live in the same house as the boarders to exercise supervision like an English "housemaster" or whether the head could live in his own house, but this seems to have been resolved by the time Rutherford first went to Nelson. Nevertheless the college was changing rapidly during the time when he was a pupil. In 1885 the governors decided to discontinue keeping a vegetable garden and cows. The year before his arrival, 1886, the first telephone had been installed in the college. More important than these small physical signs of the times was the fact that in 1886 the "not very satisfactory" system allowing students to stay on at the college as undergraduates while taking degrees from the University of New Zealand, had been stopped. From then onwards university studies had to be carried out at one of the university colleges proper. Instead there came the system by which senior pupils could matriculate for the university entrance at the end of their careers at the secondary school, a system which was introduced in 1888, towards the end of Rutherford's stay at the college.

In physical terms, too, the college was changing, notably in the fact that there was a serious falling off in number of students as New Zealand battled with its long depression. Numbers had been up to 160 in the early 1880s, despite the secession to Wellington College, but there were only about eighty boys at college in Rutherford's time. Various additions had been made to the school buildings during the prosperity period. A reading room and library had been built in the early 1880s, and the college history records: "As early as 1881, Mr J. W. Morton had fitted up the old bathroom and boothouse as a Chemistry Laboratory . . . and had ordered a stock of chemicals." Some thirty pupils joined his first chemistry classes, and even further strides were taken in science teaching when W. S. Littlejohn came on to the staff in 1882. The college history describes the progress: "Apparatus for work in other branches of science was added . . . Complaints about the inadequacy of the damp, draughty, old shanty known as the Chemistry Room, plus a growing awareness in New Zealand of the importance of practical science . . ." led the governors to appoint Littlejohn into the official position of science master and to build a proper laboratory in 1890.

When Rutherford first arrived at the college he would undoubtedly have been classed as a "small potato" in the schoolboy jargon of that place and time. And this despite the fact that he was a rather large and healthy boy, a little older than the average new boy, and that he started up the school higher than was usual – in Standard V – presumably because of his outstanding scholarship results. It is characteristic of the man he was to become that he called important people "big pots" for many years – throughout his life his admirers noted a "boyish enthusiasm" – while the few who found his company uncomfortable were irritated precisely by his lack of adult phraseology and seriousness in his outward behaviour.

That Rutherford was treated from the start as a "clever" boy is demonstrated by his mother's memory (recorded many years later by Marsden) of the first interview between the Rutherford parents and the headmaster of Nelson College. The old lady described the conversation:

> After a short talk concerning Ernest and his abilities he said to us both, "I have a personal favour to ask of you and will tell you the reason why. Bishop X's son is learning Greek as well as Latin but where there is no competition there is no emulation, so I would esteem it a favour if you would allow your son to learn Greek as he has shown his knowledge of Latin in his exam papers." After thinking it over we consented and at the end of the year he won the Stafford and Senior Classical scholarships, so all was well.

A big, strong boy from a country background does not fall easily into the unpopular character of "a swot", and it is fairly clear that Rutherford had the physical size and the emotional fire to avoid what is probably the most unpopular of all schoolboy roles, while still retaining his power to work. A contemporary (C. E. Broad) recalled that "he had such powers of concentration that he would continue to read in the uproar of an unsupervised common-room, but when a missile landed on his head he would roar into activity with good-humoured rage". Eventually, in his last year at Nelson, he got into the rugger team as an efficient, if undistinguished, forward, but in New Zealand of all places, competence at the national sport (some say religion) of rugby football is a guarantee of popularity and high regard.

The headmaster who persuaded Rutherford's parents to let him learn Greek was W. J. Ford, a Cambridge M.A. who had taught for nine years at Marlborough, one of the better known, and more athletic of English public schools, and there he had become a housemaster. Ford was appointed to Nelson in 1886, just before Rutherford arrived, and he left early in 1889, while Rutherford was still a pupil. The official history of the college says that "a combination of circumstances led to his resignation", while a local history says he left "for financial reasons". He is described as "a very tall man with a black beard, a mighty hitter at cricket, capable of singing a lively song either solo or in a quartet with other members of the staff". All sources agree that he was a very good teacher, particularly of classics, and when Rutherford revisited the college in 1925 he spoke of Ford's "profound influence on me particularly in a literary direction".

But the man who really influenced Rutherford was the second master, William Still Littlejohn, an M.A. of the University of Aberdeen. He was in charge of the teaching of maths and science and in every respect was the opposite of Ford – small, wiry, bespectacled with jutting and unruly red hair, but with a personality that outshone any physical deficiency, so that Rutherford nearly fifty years later described him as "a fine figure of a man with his leonine head and ruddy hair and beard".

Throughout his life Rutherford's writing style was cool and lucid. This makes his books and papers exceptionally easy to read and understand. It makes his letters businesslike and impersonal. But when he came to write about himself he became unutterably boring. In the whole of his vast collection of papers there is only one in which fire and emotion breathe through. It is a description of a contest between Ford and Littlejohn which Rutherford wrote in 1934 in answer to a fairly routine request from a New Zealand writer, Alec Einar Pratt, for reminiscences of Littlejohn for biographical purposes.

The battle took place on the cricket field one weekend when Ford was entertaining a visitor from England who was also fancied as a cricketer. These two men apparently took on a team of Nelson College boys strengthened by Littlejohn. Rutherford wrote, "I was not a player but watched the game from the terrace overlooking the cricket field," and he describes how the match soon became a friendly competition between Ford and his guest to see who could make most runs – and in the course of this competition plainly forgot all obligations of sportsmanship towards their opponents. The two men hit everything the boys could bowl at them all over the field, wrote Rutherford, and he described how Littlejohn, playing wicket-keeper and quite without any chance of stumping the two great men, eventually took the ball "from the now weary boys". Although "not much of a bowler", he tried for over an hour using both overarm and underarm bowling; he never gave in and "attacked with the light of battle in his eyes". The interesting feature of this childhood reminiscence is that it did not end with victory for the boy's obvious hero – it ended with defeat – "in this case virtue was not rewarded. The team had been martyred to make a Roman holiday for the visitor." And Rutherford's final conclusion was that this had left him "with an enduring impression of high courage and resource under difficulties, and, though technically defeated, I thought Littlejohn was the true hero of the occasion".

Littlejohn was certainly the man who directed Rutherford towards science, although it was by no means clear at the beginning that this was the course he would take. The master was in essence a mathematician and Rutherford was honest enough to recognise this, however much he admired the man: "I thoroughly enjoyed the teaching in maths, but the science teaching did not attract me so much, for I imagine Littlejohn had not the same width of knowledge in the subject as in mathematics." Also the teaching of physics, chemistry and practical chemistry in a converted boot-room may be thought a rather tall order for a man who was basically a mathematical specialist.

Rutherford paid Littlejohn a compliment which probably sounds more valid in our egalitarian age than it did in his own time – he wrote: "He was a fine teacher of mathematics. The boys varied in ability but he grounded them all thoroughly in algebra, Euclid and mechanics." The teacher, however, was perfectly willing to devote his own time to the more able, and he gave Rutherford private coaching at his own home in maths and science subjects. Littlejohn's children have remembered being sent off to the back quarters with strict instructions to be quiet in their play when Rutherford came to their father for special tuition. And it is well remembered in the Nelson neighbourhood that Littlejohn and Rutherford went for long walks both in the hills and along the coastline at weekends and Littlejohn would often halt to draw diagrams with his stick in the sand or the earth to illustrate the points he wanted to make in their discussions. There are also tributes from the pupil to the master's good standing as both humorist and disciplinarian, including the memory of the occasion in a science class when a boy, who was not attending properly, casually, almost abstractedly, picked up a bottle of chemical reagent from a shelf; Littlejohn saw this out of the corner of a good schoolmaster's eye and sharply but quietly told him to "drop it": the boy obeyed him implicitly and the bottle smashed to the floor. Littlejohn also brought his enormous enthusiasm to the running of the college Officer Training Corps in which Rutherford came to shine as a sergeant – his first and last experience of any military drill or discipline.

Both Nelson College and Rutherford himself remembered with pride the exceptionally smart display of drill provided by the Corps when Lord Onslow paid an official visit to the school following his appointment to the post of Governor of New Zealand. This occurred only a few months before Rutherford came to the end of his career there, and he was "Dux" of the school. It fell to him, therefore, to deliver the address of welcome, which had been written by the headmaster, Joynt. For some reason he was shy, and this was remembered many years later when F. J. Mules wrote to him in 1937: "I sometimes recall your reluctance to discharge your responsibilities as head boy and resign the duty of reading Joynt's address of welcome to the Governor."

Joynt remained as headmaster of the college until 1898, when he became registrar of the University of New Zealand, and was immediately succeeded by Littlejohn as headmaster of Nelson College.

Rutherford's academic achievements at Nelson were impressive, though it must be remembered this was a provincial colonial college with only eighty pupils. He won prizes in every one of his three years there, prizes in English, French, history and Latin as well as in his strongest subject, maths, the most valuable being the Stafford and Simmons scholarships, each worth £100, which made a considerable difference at a time when the family fortunes over the hill at Havelock were undergoing considerable strain. In his last year he was top of the top class in all subjects. His report for June 1888 has been preserved and is of some interest: in classics "an occasional careless blunder" is his only fault; in English "has a retentive memory and a great power of reproduction"; in modern languages "a very careful scholar". No hint of genius or brilliance or imagination; that sort of note only comes in the report on maths: "Very quick, a very promising mathematician". The cricketing W. J. Ford recorded of him "Top in every class and his conduct irreproachable", but the succeeding headmaster, J. W. Joynt said, "Rutherford displayed some capacity for maths and physics but not to an abnormal degree and he was a keen footballer and popular boy."

Rutherford's earlier biographers have tended to leave his school career at that, implying that he carried all before him. But the records of Nelson College show that this was not so. At the time of Rutherford's schooling the question of examining the boys was a subject of controversy and change because the college had just been disaffiliated from the New Zealand University. Thus a visiting examiner in 1887, a Mr W. J. Kelly, a graduate of Dublin University, although recommending Rutherford for the classical scholarship, pointed out that another boy (C. E. Broad) was "slightly superior" in Greek. The mathematical scholarship, however, he awarded to Frederick Neve: "In Trigonometry and Arithmetic he was somewhat surpassed by Rutherford but this was more than counterbalanced by the other subjects." The same C. E. Broad also outclassed Rutherford the following year when the headmaster himself did the examining and gave Broad the first classical scholarship while Rutherford won the mathematical scholarship. And again F. Neve appeared in the honours list as well. These reports show that Rutherford did not study chemistry, but chose the alternative subject of French.

It is therefore worth examining some of Rutherford's contemporaries at Nelson, for plainly they were not completely outclassed by him, as we have been led to believe. They were mostly honourable men, distinguished in their own small arena, but none approaching Rutherford's international status and fame. Broad had become headmaster of the same Nelson College by the time Rutherford revisited the school in 1925. Neve also became a schoolmaster. Ten Rutherfords are recorded in the College roll, including Ernest's two brothers. Some of them were not close relatives: none of them achieved more than local distinction and some of them hardly even that. The other five boys who won Education Board Scholarships to the college at the same time as Rutherford, were C. J. McEachen, who became a stationmaster, the same eminence as achieved by J. P. Petrie. J. A. Cowles became a teacher and inspector of schools, while J. H. C. Bond became superintendent of the New Zealand Shipping Company. C. A. Craig sounds the most imaginative of the group; he became a Catholic priest in Dublin and later returned to teach at Nelson.

Rutherford was, then, the best scholar but by no means outstanding in a small collection of worthy, but not particularly distinguished boys in a small, worthy college on the outermost periphery of European culture. He had at this stage shown no particular interest in science, but had received a sound and very broad education. Clearly even the most perceptive and sympathetic of his teachers regarded him as a good worker rather than a genius; Littlejohn's final report says, "*Nunquam non paratus. Should give a good account of himself*".

The young man took the next step in his career by winning his university scholarship to Canterbury College at Christchurch.

It was as well that he won the scholarship, for certainly at this stage his family could not afford him any help. Catastrophe had struck his father's enterprises at Havelock. The Atkinson government stopped the further building of railways and cancelled the order for sleepers. Floods on the Pelorus River drowned and killed the flax in the swamps, and James Rutherford himself was seriously injured, breaking several ribs, when he slipped and fell during loading operations on his jetty.

So, in what was still pioneering country and society, James Rutherford decided on another wholesale uprooting of his family. Through a relative on his wife's side he learned that there was flax available near New Plymouth in Taranaki province, where New

Zealand's most famous landmark, the beautiful cone of Mount Egmont, descends into the sea. He was able to buy land for £3 an acre at Pungarehu, thirty miles south of New Plymouth, and later he was able to cut flax from other nearby swamps under "royalty". After he had reconnoitred the place he chartered the sailing ship *Murray* with her captain, Captain Vickerman, and put his entire family, his horses, his flax-milling machinery, some spare timber, all his furniture and three extra "hands" on board and sailed for the North Island. Three days later they landed at New Plymouth and then the whole cargo had to be moved by wagon and horse along the still rough roads to their new home. In the course of his stay at Pungarehu James Rutherford invented and developed several new devices for treating his flax. His younger daughters married in the North Island, and the majority of Rutherford relatives are today to be found in that Island rather than in the South.

It is worth noting that James Rutherford was also a pioneer in attempting to plant flax, rather than relying solely on the natural growth and propagation of the plant, and that he went to consider-able effort to find and experiment with specially selected varieties of the plant that the Maoris cultivated. But it was no way to make an easy living – a five-horse team could only draw three tons of flax on the two-day journey to the port of New Plymouth and the final product fetched only £13 a ton when it was sold in Melbourne.

So Ernest's family could hardly give him much financial support during his student days at Christchurch, and he could only go home during the long summer holidays – at other times he had to stay with his grandmother or other relatives still living near Nelson.

On at least one occasion when he did go home he landed at Wellington only to find that it was one of the days when no coach ran round the coast to Pungarehu. Without hesitation he walked the thirty-one miles.

He was still expected to take his share of the chores when he went home. Family memories assert that he repainted the house on one of his vacations and he also lent a strong hand in the various operations of treating and preparing the flax. But essentially, as with any other young undergraduate, this was the period when Ernest Rutherford cut loose from his family, and the break, in his case, must have been accentuated by the distance between home and university and the difficulty of the journey, including the sea crossing over a strait famed for its windiness.

Canterbury College was still in the first stage of its development. It was taking in just over 150 students a year, judging by the numbers matriculating, but there were only about 300 students actively working at any one time. The majority of the seven professors then in residence were those who had been appointed when the college was opened in the early 1870s and were, therefore, born and educated in Britain. The college's total income was less than £10,000 a year and only about £400 of this came from students' fees – the largest part came from the 300,000 acres of land with which the college had been endowed by the early government of the province.

From the fact that Ernest Rutherford is listed as Number 338 in the catalogue of graduates from the college (prepared in 1927), and from the discrepancy between the annual numbers matriculating and the numbers attending, we may deduce that there was a high proportion of "drop-outs", but doubtless many of these would be youngsters whose families moved in that highly mobile society, and others would be country boys for whom the financial strain of a college education in those depression-times had proved too much. Rutherford himself lived largely on scholarships – these awards appear now on his list of academic distinctions, but there can be no doubt but that at the time they represented his main financial resource.

His academic record at Canterbury College is certainly impressive. He arrived with a Junior University Scholarship in 1889 and his date of matriculation is given as March 23rd, 1891. He won maths exhibitions in 1890 and 1891. The following year he won exhibitions in experimental science and maths. In 1893 he was awarded his B.A. degree and won a senior scholarship in maths. In 1894 he achieved his M.A. with the rare distinction of a "double first" – First Class Honours in Physical Sciences and Mathematics.

There is some slight problem about the dates in this academic list, in that they do not correspond to the dates given in, for instance, the official testimonials drawn up for Rutherford by the university authorities when he applied for the "1851 Exhibition Scholarship" in 1895. Most of the dates given in these testimonials are one year earlier than in the Canterbury College history, but this can be easily explained by the fact that he would have taken his exams for, say, the M.A. and passed them at the end of the academic year 1893, yet he would only have been officially "gazetted" with the result in 1894. There is more difficulty over the question of whether he was officially awarded a BSc degree by Canterbury. The first, and major, biography by A. S. Eve, records him as having obtained this degree. But the official history of the college does not record it. His testimonial in 1895, provided by Arthur Dandy, Chairman of the Professorial Council, says simply that he passed "the University examination for BSc" in 1894, but does not say whether he was actually awarded, or took, the degree.

The academic record as it stands seems to present a clear and simple picture of a brilliant student sweeping all before him and taking all the necessary early steps on his way to becoming the great scientist of the future. A closer look at the details reveals something rather different. When passing the first section of the exam for B.A. in 1891, Rutherford took not only maths and mechanics, but also Latin and English. A letter from his headmaster in his final year at Nelson, J. W. Joynt, dated July 11th, 1890, that is after Rutherford had left college and before he went to Christchurch, expresses doubt "if it will be possible to procure Bradley's *Aids to Latin Prose* in New Zealand". But the headmaster will try to get it and will have it sent if it can be found and adds, "You are right to try and do some work waiting for term to open."

When passing the final section of the B.A. exam in 1892, Rutherford took French and physical sciences, and also won the Senior University Scholarship in maths. When obtaining his "double first" in the M.A. exams he took optics and astronomy in the mathematics part of the course and we hear of "electricity and magnetism", which was to be his eventual choice for research, for the first time only in the physical sciences part of the course. Yet in the following year, 1894, when taking the BSc exam, Rutherford took papers in maths and applied maths, in Latin, English, chemistry, physical science and geology. We know, further, from the earliest of his notebooks to be preserved, that in 1893 he was taking courses in botany and biology.

There is a tendency among those distinguished scientists who have given Rutherford Memorial Lectures to praise all this as an example of a "truly broad education", and the avoidance of early specialisation. It can equally well be interpreted as not knowing exactly where he was going – a young man who undoubtedly had the power to work hard so that he could pass any exam in any subject he cared to take, but who showed no particular flair for anything and who certainly showed no sign of youthful genius in the subject in which he was to dominate all his contemporaries.

This was obviously the opinion of the soundest of his teachers, C. H. H. Cook, the Professor of Mathematics who wrote in his testimonial, "He is possessed of considerable mathematical abilities, and is endowed with great power of work so that he has been a highly successful student," which is hardly what even the "soundest" of professors would write if he felt he had a "high flyer" to write about.

Charles Henry Herbert Cook, first Professor of Mathematics and Natural Philosophy at Canterbury College, was undoubtedly one of the major influences on Rutherford's intellectual development. He has been described as "solid, serious, dignified, greatly esteemed, a very corner stone of strength and stability", yet despite his large beard he could look smiling and cherubic. He loved choral and church music, yet he was a keen supporter of cricket and athletics at the university. He was a major proponent of university reform and played an important part in the national discussions about the shape of the University of New Zealand. But, above all, he was an excellent teacher and Rutherford acknowledged this throughout his life. And, most important for Rutherford at this crucial stage of development, Cook was completely sound and orthodox both in his teaching methods and in his own published works.

Cook had been born in London in 1844, and was therefore not yet fifty when Rutherford came under his influence. He had been taken to Australia when very young and graduated at Melbourne University. He returned to England, went to Cambridge and emerged as Sixth Wrangler in the mathematical finals of 1872. He then decided on a career as a lawyer and started reading for the Bar, but within less than a year he accepted the offer of the Professorship of Mathematics at Canterbury and emigrated finally to New Zealand.

It has often been said of Rutherford that even at the height of his power he was "no mathematician". In the sense that this phrase might be used by his great contemporaries among the European theoretical physicists this was true – he was never a man for Hamiltonians or quaternions. But it is only a half-truth. In all his greatest discoveries Rutherford was strictly quantitative. Measurement and mathematics were always present. This point will have to be emphasised again and again in examining Rutherford's scientific work – but though the mathematics he used as a scientist was simple to the mathematician, at all times the results observed in experimental situations were strictly compared with the results to be expected if such-and-such a law described the situation being examined – often it was a comparison with some simple law of scattering. The physical meaning – the model or vision of the Universe – was therefore either being confirmed or refuted.

Whitehead wrote of seventeenth-century science, "Mathematics supplied the background of imaginative thought with which the men of science approached the observation of nature" – and this is what Rutherford's maths was for him. And this is what he owed to Cook.

But it is also interesting that there is an eye-witness account of Cook discussing a problem in optics with a clever pupil, M. C.

Keane, a man some years junior to Rutherford, who despite his brilliance in mathematics at university subsequently became one of New Zealand's most respected journalists. The discussion took place at a point where the coach took over from the train on the journey to the West Coast. The story tells of Cook drawing diagrams on the dusty surface of the country road while discussing some problem that had been set in the exams. Keane, who was without coat, collar or tie, and in ragged trousers, was also without shoes, and he drew his diagrams with his bare big toe, holding up the coach while the argument went on and the professor happily missed his lunch. The episode is described as one "that could only have occurred in New Zealand" – but it is remarkably similar not only to the stories of Rutherford and Littlejohn on their Saturday walks in the Nelson area, but also the legends of Archimedes scrawling on the dusty floors of his villa in Syracuse. Plainly the New Zealanders (certainly of that time) liked to visualise themselves as simple, "pure", and poor classical scientists, and it will become clear that this is how Rutherford saw himself, too.

Cook, however, also had another role in Rutherford's development. He was counterweight and antidote to the other chief influence, Alexander William Bickerton, the first of all Canterbury's professors, the Professor of Chemistry. Bickerton was both Rutherford's inspiration and a perpetual "awful warning". He was heterodox, speculative, willing to dabble in almost any field. It was Bickerton who started Rutherford on research and appreciated and supported his first independent work. Yet forty years later Bickerton was still there in Rutherford's life, causing a struggle in the conscience of the President of the Royal Society, a struggle between Rutherford's famous loyalty and love on the one hand, and his scientific integrity on the other, as he despairingly tried to save the old man's "face" and finances whilst admitting that he had become an elderly scientific crank.

Bickerton was born in 1842 in Hampshire, trained as an engineer, and worked first in railway surveying. Soon he invented a new wood-working process and set up a factory in the Cotswolds, but also started teaching in a technical school in Birmingham. Thence by scholarship he progressed to the Royal School of Mines in London, and his experimental methods of teaching science led to his appointment to develop the science work at the Hartley Institute in Southampton and to an appointment at Winchester College as well as a job as county analyst. He was offered five professorships by the time he was thirty-two, and the one he accepted was Canterbury.

Bickerton came to Christchurch in 1874 as Professor of Chemistry – to which subject physics was later added. He had been selected by Lord Lyttelton who had been asked to find someone "young and with promise of future excellence", and his English thesis on the correlation between heat and electricity had been influential in obtaining his selection. He was offered a salary of £600 a year and he arrived in time to begin work in the first term in which Canterbury College officially began teaching. He was, therefore, the first professor of the university.

"He had a gift for public demonstration that was invaluable in those early days and an almost divine enthusiasm for science that at times carried him to excess," according to one author, Jenkinson, who is not prone to criticise. Another writes:

> Bickerton was nothing of a mathematician, otherwise, perhaps his physical ideas could not have been so heterodox, but also his originality was enhanced by his completely direct approach to the problems in which he was interested . . . There was, it seems, a strange mixture of the true scientist and the completely uninformed amateur in this remarkable individual, but as far as his influence on Rutherford is concerned two things at least may be said to his credit. He had boundless enthusiasm for research – and, in his more solid achievements he showed, by his very neglect of mathematical analysis, that the experimenter does well who keeps mathematics, not as a mistress, but as handmaid – and even then dispenses with her services at times.

This latter description was written by Norman Feather, one of Rutherford's own pupils who later became Professor of Physics at Edinburgh. It is therefore the more interesting that the less critical author also tells this story of Bickerton:

> . . . He was a poor mathematician with little faculty for exact arithmetic. His mistakes in the simplest problems of addition or subtraction were the standing joke of the back row in his classes. The professor, however, had an extraordinary faculty for a mental graphic arithmetic of his own. After looking at a long collection of complicated figures on the board, Bickerton would close his eyes for a few seconds and then dreamily announce that the final answer was about 430,000. No one in the class could tell offhand whether the answer would be closer to 0 or 40,000,000 but excited calculators would soon whisper some such figure as 437,618 round the amazed audience.

Rutherford obviously picked up Bickerton's public attitude to both mathematics and calculation, even though his serious commitment to both subjects was much greater than his master's.

Bickerton's many papers and publications varied from "On

Chlorine as a Cure for Consumption" to "On Molecular Attraction", from "On the Equilibrium of Gaseous Cosmic Spheres" to "On Hail". But his great work was his astrophysical "Theory of Partial Impact", an explanation of the origin of many astronomical phenomena such as double-stars, variable stars, novae and so on. This theory was put forward in a series of eight papers delivered to the Philosophical Institute of Canterbury between 1878 and 1880 – papers that even his kindest critics have been compelled to call "rambling, discursive and self-repeating". Some argue that Bickerton's work in this field has been wrongly neglected; it is only important here to wonder whether the concept of one body with great energy crashing into an isolated system of other bodies in space may have left any impression on the mind of the young Rutherford, so that when his turn came to consider the problem of energetic particles crashing into systems or bodies such as atoms he was freer to use his imagination than any of his contemporaries.

The principal forum for this remarkable if unreliable professor was the Philosophical Institute of Canterbury, the local branch of the eminently respectable New Zealand Institute, which has now become the Royal Society of New Zealand, and is the country's major learned society. It was, at the time when Rutherford was first elected a member on July 6th, 1892, rather typical of such institutions in a pioneering society – it was trying to build itself a tradition. In just the same way Canterbury College insisted on maintaining a regulation of 1878 that undergraduates must wear academical dress of gown and mortar-board at all times when on the college premises. Rutherford's first major research, published on November 7th, 1894 in a paper he read himself, was the first major research to be announced to the Institute, though Bickerton had read, typically, dozens of speculative papers to the same body over the years.

Yet it was undoubtedly Bickerton, and not Cook, the stern mathematician, who encouraged Rutherford to start research, although the young man was, at first sight, a mathematician. And it was Bickerton who appreciated, more than anyone, the importance of Rutherford's first researches. In his testimonials for the 1851 Scholarship it was Bickerton who wrote:

From the first he exhibited an unusual aptitude for experimental science and in research work showed originality and capacity of a high order .... Mr Rutherford conducted a long and important investigation in the time effects of electric and magnetic phenomena in rapidly alternating fields, and by means of an ingenious apparatus of his own design, was enabled to measure and observe phenomena occupying less than 1/100,000th of a second.

Bickerton also added, in his typically enthusiastic way, a comment that was completely inappropriate to a scholarship recommendation, yet is of great importance to the understanding of the young man he was recommending.

"Personally," he wrote, "Mr Rutherford is of so kindly a disposition and so willing to help other students over their difficulties, that he has endeared himself to all who have been brought into contact with him." The heterodox professor has a lesson for the modern biographer – under the blaze and weight of Rutherford's scientific discoveries and behind the supreme scientific politician he was to become, it will inevitably be hidden and forgotten that he was a man of exceptional personal kindness. Everyone who remembers Rutherford remembers this – that he was personally kind to them far and away beyond the normal behaviour of a pleasant human being. It is recorded by many memorialists that "Rutherford never made an enemy or lost a friend". But the man who was one of Marie Curie's great supports in her troubles is still remembered by elderly widows in England for his kindness to them when they were little girls or young wives at formal social occasions.

Rutherford threw himself into the life of the uncertain but enthusiastic young University College in Christchurch, in no way confining himself to science. He played rugby and became secretary of the young Science Society, a livelier congregation than the Philosophical Institute. The programme card for the 1894 session of this society, when Rutherford was only a committee member, shows him lecturing on "Electrical Waves and Oscillations" – he had begun his own research into this subject by this date. But his talk, on May 12th, is sandwiched between "Standards of Conduct: a Survey of Ethical Systems" and "The Subterranean Crustacea of New Zealand". The programme card states that the Science Society normally met in the Chemical Lecture Theatre, but on one occasion, when Rutherford himself was secretary he noted, "It being cold in the normal lecture room, the society adjourned to Mr Page's room." In some earlier biographies it has been said that Rutherford was only reluctantly persuaded to become secretary because the Science Society had come in for considerable criticism after a programme of talks on Evolution, including a lecture by Rutherford himself on "The Evolution of the Elements". There is no documented evidence of such a talk nor even of the society coming under public criticism, though this subject was a favourite Rutherford title some twenty years later. The subject seems inherently unlikely in 1895 although it became a reasonable subject for speculation after the great unravelling of the mystery of radioactivity during Rutherford's spell in Canada in the first decade of the twentieth century.

If he did, as a young man, come to any form of public attention it is far more likely to have been on account of his activities in the Dialectic Society – the general debating society of Canterbury College – for throughout his life he loved a good argument for its own sake. When the subject of the Dialectic Society for the evening was "Is sculpture or architecture the greater art", Rutherford introduced a new note by arguing that the architectural beauty of the new College Hall (a late Victorian Gothic structure) was spoilt by the intrusion of an ugly telegraph post laden with wires. He asserted that the day was not far distant when the wires and therefore the post as well, would be unnecessary since science was on the threshold of discoveries that would lead to other methods of communication. He was referring obviously to his own early work in the field of "wireless communication", and his intervention must certainly have produced an interesting variation on a rather worn set of arguments, though we do not know whether he voted in favour of architecture or sculpture.

The list of his academic achievements in terms of scholarships and exhibitions is, indeed, impressive, though it is rendered rather less formidable by the realisation that the scholarship system in New Zealand was more highly developed and more officially supported than in Britain, for it was an accepted mechanism to enable the young in this scattered and pioneering community to achieve the generally desired high standard of education. More important, it has to be realised that Rutherford did not sweep all before him. He was not the most brilliant of his contemporaries, though he was in the first rank.

Rutherford had not entered Canterbury College at the head of the Scholarship list, and though he invariably headed the table in mathematics, he had several rivals who seemed to outrank him. In particular there was W. S. Marris, later Sir William Marris, who eventually became Governor of the United Provinces as the climax to a fine career in the Indian Civil Service. Marris won just as many mathematics exhibitions and scholarships as Rutherford and added to these many in Latin, though he was two years younger. Marris recorded, when A. S. Eve was gathering material for his biography, how he often triumphed over Rutherford in maths exams, because the papers set by Cook demanded a detailed knowledge of the "bookwork" at which he excelled while Rutherford was essentially a "problem solver". This evaluation by a fellow student seems to have been particularly acute, for in some unsigned notes which are a mark list for the Honours maths and physical sciences paper, Rutherford is placed top in optics, geometry and algebra with the remark, "The work on the practical paper is greatly superior to anything I remember to have received in previous years". Marris also remembered that Rutherford was nervous in the examination hall.

But there were several others who entered Canterbury College at the same time as Marris and Rutherford who seemed, on their academic records, to be at least in the same class – Edgar S. Buchanan in modern languages, J. A. Erskine, another scientist specialising in physics, and James Hight, who made his name in the English department and became New Zealand's first Ll.D.

Indeed, when Rutherford finally left Canterbury at the end of the 1894 academic year his view of the future was by no means clear. He had, a year earlier, seriously considered going into medicine. This is proved by the existence of a letter from John Stevenson, a contemporary at the university (his date of matriculation is officially recorded as ten days after Rutherford's), who wrote from Edinburgh in September 1893. The letter is clearly in answer to one from Rutherford asking for advice about the possibilities of his joining Stevenson in medical training at Edinburgh and Stevenson advises the study of zoology before any move to Scotland. The description of student life in the Scottish capital, while revealing, is not encouraging – there is an Australian Club but it is "little less than a smoking, boozing, gambling den"; some lectures are attended by 300 students "all male and a rowdy lot they are" and "there are some of the most selfish, uncouth, brutal fellows attending some of the classes I ever saw".

It seems likely, from the censorious tone of this letter, that Rutherford had moved into, or been drawn into, a rather stern and puritanical social grouping. He is known to have taken lodgings quite early in his university career – possibly at the end of his first year – in the house of a widow, Mrs Arthur de Renzy Newton in North Belt, Christchurch. And it seems certain that this household, while not of the pious type, was strict and earnest. It was certainly anti-smoking and very strongly teetotal and even a centre of campaigning against drink. One of Rutherford's earliest friends, J. A. Erskine, who followed him, also with an 1851 Scholarship, to Europe, writes in his youthful letters most apologetically about his own passion for opera and excuses himself for his visits to the opera houses of Europe in the tones of one who is well aware that this will be regarded as frivolity and time-wasting.

Forty years later, one of Rutherford's oldest friends and protégés, Dr Clinton Coleridge Farr, apologises in a letter to Lady Rutherford that in a "primitive" pamphlet published at Havelock he has stated that Rutherford was not a "potted angel" and adds, jovially or jokingly – it is impossible to tell which – that he has "armed himself with irrefutable evidence of the fact". In another letter the same Dr Farr, who was Professor of Physics at Christchurch, describes the vigorous activities of Mrs Newton in a local by-election.

This Mrs Newton, the widowed landlady, had four children, and Ernest Rutherford fell in love with the eldest of them, Mary. In the manner of those days, this attachment was at first an "unofficial" engagement, known only to the two families. Later the engagement was made official, but the marriage could not take place until he could afford to support a wife, which only came about when he obtained his first professorship. The courtship therefore lasted some six years. And that, for all practical purposes, is the end of the story of Rutherford's romantic sex life. There is no trace in the records or in the memory of those who knew him of any other attraction or attachment to a woman throughout the rest of his life. To the morals and psychology of the 1980s this may seem evidence of the sexual diffidence found in him by C. P. Snow. To his own generation it was evidence of an ideally happy marriage, as all the biographers and memorialists who wrote immediately after his death record it.

It was certainly one of the very few firm guiding lines of his life as he ended his university career. Otherwise the future must have seemed very clouded. Two descriptions of him by his contemporaries emphasise the point: "Very modest, friendly but rather shy and rather vague – a man who had not yet found himself and was not then conscious of his extraordinary powers," according to Sir William Marris; "A boyish, frank, simple and likeable youth with no precocious genius; but when once he saw his goal he went straight for the central point," Sir Henry Dale recorded.

It is only with hindsight that we can see that the physical research he had started in primitive conditions in a college basement was to open for him the doors to a wider world, was to provide him with the life and lifestyle he relished, and was already forming the mental processes and methodology by which he would change man's view of the world in which we all live.

2

First Research

Rutherford's start in research is shrouded in confusion; there is no record of his ever describing it in scientific terms or reverting to it for material in later life, though he often described with relish the difficult physical conditions he had to overcome.

On the face of the record it appears that he moved into the front line of advanced electromagnetic research in his very first paper – a paper which also included the concept of what was later to become Marconi's first successful radio-receiver. "He seems to have stepped straight into research from his last two years' studies of physics for his honours degree," wrote J. G. Crowther; and Norman Feather speaks of his first research as that of "a mind which was already moving abreast of the development of thought in science and was soon to lead it".

Certainly his first published paper – "Magnetisation of Iron by High-Frequency Discharges", which was read to the Philosophical Institute of Canterbury on November 7th, 1894, and was published in *The Transactions of the New Zealand Institute* in the same year – is quite extraordinary. It starts off by stating the views of Sir Oliver Lodge, J. J. Thomson, then Professor of Physics at the Cavendish Laboratory in Cambridge, and H. R. Hertz (who had recently proved the existence of electromagnetic waves in space – what we now call radio-waves) who were all agreed that iron was not magnetic for very rapid frequencies. He then calmly announces that he has proved them all wrong. It is the equivalent, in modern terms, of a young student in a new university in one of the developing countries declaring that Einstein, Bohr, and Rutherford himself were all wrong on some major point of physics. This is just not possible, especially in the absence of a strong professor with a working physics laboratory. However biased a biographer may be towards his subject, he cannot expect him to step like Venus fully-grown from the sea.

All the evidence points clearly towards the fact that Rutherford's first research was the work which was published as his second paper – "Magnetic Viscosity" which he read before the Philosophical Institute in Canterbury on September 4th, 1895. This reading must have been his last scientific action before leaving New Zealand, and in the printed version he is described as "1851 Exhibition Science Scholar".

Rutherford, we have seen, was still regarded primarily as a mathematician when he obtained his B.A. in 1892, taking papers in a wide variety of subjects. His Senior University Scholarship in that year was awarded for mathematics, and the comparable award in physical science went to Chisholm of Otago. His Double First in his M.A. in 1893 was in mathematics and physical science. It was therefore only in this year, and not before, that he began a serious study of physics, and we must assume that he would have started on experimental work in the shape of some class experiments at this time.

It was apparently the tradition of Canterbury in those days that anyone who stayed on after taking the M.A. should take the BSc degree and also try to do some research. We have in the first of Rutherford's preserved notebooks a clear record of this development. One end of the notebook contains his botany notes and the other his biology notes, and we know that he took papers in these subjects in his BSc examinations at the end of 1894. The two sets of life-science notes do not meet in the middle; they end with cryptic remarks: "Origin of Living Matter: Unsettled Question" and "Living Bodies do not obey solely the law of Physics". Furthermore (as Norman Feather has pointed out) these notes do not seem to reflect a mind totally absorbed in the subject, for there are designs and "doodles" in the margin and the question, "What does a gown and hood cost?"

But suddenly at the end of the biology notes the subject changes. There is the date "5th December 1893" and the subject "Times of rise of current in a coil of wire", and "Particulars of steel wire ring" with a series of meticulous measurements and remarks: "The steel wire was the finest that could be obtained"; it was "wound on a bobbin" and "each layer was covered with shellac". This immediately develops, not into a series of laboratory notes but into something like the first draft of a paper, "Experiments on Secondary Circuits". It reads:

> After reading Lord Kelvin's article on "An accidental illustration of the shallowness of transient current in an iron bar", page 473, Vol. III of his collected works and Lord Rayleigh's article in the *Philosophical Magazine* for 1886 on "Resistance of conductors conveying alternating currents" it occurred to me that Faraday's statement of the absolute equality of the time integral of the induced current on making and breaking the circuit was only true in the case of very fine wires and that the thickness of the wire in the secondary circuit would have

the effect of making the current at make and break different since the resistance of conductors increases with the shortness of duration of a transient current. It was not observed till later that the same idea had occurred to Sir W. Thomson in a paper in *Philosoph. Magazine* for March, 1890, entitled "On the Time Integral of a transient electro-magnetically induced current".

The work continues and includes some sets of numerical results, carefully written down in columns and obviously not in the rough form of figures noted down during actual laboratory work.

And a few pages later we have "Magnetic Viscosity" and the explanation: "This research was carried out in order to see if there was any appreciable time taken by the molecules of iron to align in their final position. The times investigated are ten times shorter than Hopkinson's Experiments and get rid of many of his sources of error."

All this shows several important points. Firstly, as Feather has shown in his analysis of these early notebooks, Rutherford's very first work did not show the effect he had expected: "Faraday's result we now know is absolutely true theoretically." In other words, his first research was technically a failure, although we can see in it the germs of all the ideas that were to result in his first successful papers. However, the date shows that Rutherford had already completed the laboratory measurements before December 5th, 1893. Furthermore, it shows that at that time he was working with steel and iron rings, while there is no mention of "Leyden jar discharges", the mechanism by which he was to produce "Hertzian waves" – radio high-frequency oscillations. Yet it is his second paper "Magnetic Viscosity" which uses steel and iron rings, and his first paper that uses Leyden jar discharges.

If we now examine his second published paper (p. 58 CPR)[1] we read: "I had already designed the apparatus and the method of reducing the experiments before a copy of the *Proceedings of the Royal Society*, April 20th, 1893, reached New Zealand." This issue of PRS contained an account of experiments by Hopkinson, Wilson and Lydall on the same subject, according to Rutherford, but he adds, "I determined to continue my experiments on the subject especially as I was enabled to deal with intervals of time much shorter than those in Hopkinson's experiments." Communication with New Zealand from London was, of course, much more tedious and lengthy in those days than it is now, but we must accept that copies of the April 1893 issue of PRS would arrive some

---

[1] *The Collected Papers of Lord Rutherford* (*CPR*), p. 58.

time in that same year and not be delayed until 1894 or 1895. And obviously the remark in the 1895-published paper about the time interval in Hopkinson's experiments is simply a less boastful version of what was written in the laboratory notebook of December 1893.

It seems quite clear, then, that Rutherford was doing the work he published in 1895, the material of his second paper, in December 1893, and that he had started it even earlier than that – perhaps indeed before he had even taken his M.A. papers.

There is similarly evidence that the work which formed the subject of his first paper, in 1894, was actually performed in the middle of that year. (It is probably worth reminding Northern Hemisphere readers that December and Christmas-time are the height of the summer holiday period in Southern Hemisphere New Zealand. In the two-term system then applying in Canterbury, the working periods of the university were, roughly, March to July and August to November.)

In Rutherford's *third* notebook we have a brief section on experiments on the magnetisation of iron by Leyden jar discharge, with a date, Monday, March 16th – possibly 1894, though this is by no means certain as the following page contains some notes which are plainly of his work at Cambridge and there is considerable confusion.

But it was on May 12th, 1894 that he spoke to the Canterbury Science Society on "Electrical Waves and Oscillations" and it is coldly recorded in the society's minutes that the talk was "illustrated by experiments performed by Mr Rutherford with the assistance of Mr Page and Mr Erskine". Page was demonstrator in physics and assistant to Professor Bickerton; Erskine, we have seen, was Rutherford's friend and fellow physicist, one year his junior. The apparatus of his 1894 paper was therefore in existence and in action at this time and the paper was published a few months after this demonstration.

It was this apparatus – in principle his radio-receiver – that Rutherford took with him when he went to England in 1895, and it was with this apparatus that he started work when he began research in Cambridge. There is no evidence that he gave it up for a while to perform magnetic viscosity experiments. The conclusion must be that Rutherford's first work in the laboratory produced no results; his second project was the subject of his second paper and was published *after* the really impressive work of his third project. And this is altogether more compatible with what we know of normal human progress.

It is therefore appropriate first to look more closely at Rutherford's second paper – his work on magnetic viscosity. It is not, in truth, an earthshaking work. It begins well enough with a typically simple Rutherford question – "This research was undertaken to see if steel or soft iron exhibited any appreciable magnetic viscosity when under the influence of very rapidly changing magnetic fields" – the sort of simple question asked of nature which was to lead him so far, and which so delighted all his fellow scientists. But the five conclusions he gives at the end of the paper are on the whole rather woolly and indeterminate; they certainly do not answer the question he sets himself at the beginning.

There is, however, one good, clear sentence in the conclusions: "It was experimentally shown that the iron did not take more than one ten-thousandth of a second for the rearrangement of the molecules into their final position . . ." Here lies the strength of the work, the accurate determination of extremely small intervals of time, probably beyond the discrimination of any other scientists of the time and achieved by a simple direct experimental device that was a true harbinger of the genius Rutherford was later to show.

This timing device – more precisely a device capable of producing very short time intervals accurately and repeatably – consisted of a heavy weight falling down a vertical wire. It was allowed to strike two arms during its descent and these impacts broke two circuits. The two arms were so arranged, one positioned by a Vernier screw, that they could be made to take slight but measurable and controllable differences in level down the wire. Hence the contact made by the falling weight occurred with a very small time difference separating them. By a simple arrangement of the circuits, one of which included a small coil around the ring of metal wires that he was investigating, he was able to measure the inductance and hence the magnetisation of the iron or steel in the ring when current flowed for an extremely short period of time, and therefore increased and decreased very rapidly.

Two further points from this paper are interesting. Firstly, although the style gets nowhere near that "masculine economy" which made his later work so lucid even to the non-specialist, there are, nevertheless, the first signs of his (perfectly reasonable) cavalier disregard of mathematical niceties, or, more exactly, disregard of extreme numerical accuracies. Thus we have one quantity which he writes is "0.000192 nearly". This is a time interval and he goes on: "Now this gives the time interval as derived from theory. In practice the time intervals . . . must be slightly greater." He gives three causes for this difference and then dismisses them as "very small"; "cannot be very great" and "quite insignificant" and he continues, "Later experimental verification will be given that the calculated values are very nearly the same as the true values."

Here we approach the truth of Rutherford's attitude to mathematics and to numerical sizes. There are some fairly complicated equations in this, his first real work. There is nothing abstruse but he handles the equations with ease and manipulates them neatly. He uses mathematics to help him obtain numbers for physical quantities, but when it comes to these numbers he is interested only in their sizes compared to each other, not in their absolute accuracy.

More than eighty years after this paper was written, Vivian Bowden, a pupil and fellow researcher with Rutherford in his great Cambridge days, now Lord Bowden of Chesterfield, remembered how, in his *viva* examination to gain a place in the Cavendish, Rutherford asked him to work out in his head the speed of a one-volt electron. He recorded how another of Rutherford's favourite questions to young men in this position was to ask them what would be the inductance of a small loop of wire about the size of a wedding ring – in other words he was asking about the size of the numbers involved in his own first research.

Lord Bowden, lecturing at Canterbury, added:

> He insisted that staff and students should understand orders of magnitude. They must know about how big physical quantities are . . . Not only did he know instinctively how big things are, he was very good indeed at mental arithmetic. No one else I ever knew could copy a dozen numbers down wrongly, add them up wrongly, and then come up with the right answer. It wasn't really fair. He had furthermore an unrivalled ability to put himself in the place of an alpha-particle in a piece of apparatus and decide just what he would do in the circumstances.

A modern scholar, T. J. Trenn, has even shown how in his Canadian days when he was studying radioactivity, Rutherford formally reached a correct conclusion by two self-cancelling arithmetical mistakes. So here we have one of the world's greatest scientists, and accepted as such by all his contemporaries, who continually made silly arithmetical mistakes, yet was described as very good at mental arithmetic. The answer to the apparent paradox lies in Bowden's acknowledgment that Rutherford was interested in "orders of magnitude". His appreciation of the sizes of physical things was not however "instinctive" – it came instead from continuous deep thought about the subject in hand. The immersion of his mind in the landscape that he was himself revealing to the world in his experiments enabled Rutherford to "feel" what size things should be and so enabled him to see at once whether a result, a suggestion, a theory, was likely to be right or wrong. Bowden also explained this phenomenon by a reference to Sir Isaac Newton, one of the few scientists of equal stature to Rutherford. Newton was asked how he worked out the orbit of the moon and planets after he had observed the famous apple falling; and the great man replied: "By constantly thinking about it."

Only at one single point in this 1895 paper on magnetic viscosity did Rutherford refer to his 1894 paper. This in itself suggests that the second paper was not following the work of the first, and the words used emphasise this view for they are in the unusual form "In my paper published last year . . ." This paragraph is then full of embarrassed phrases – ". . . not yet known . . .", ". . . quite probable that . . .", ". . . interpretation of the results very difficult".

Sir Edward Appleton, in his short essay introducing Rutherford's early work in the *Collected Papers of Lord Rutherford*, has shrewdly pointed out that "he found the interpretation of some of his experiments in this research somewhat difficult, and not entirely in unqualified conformity with the conclusions of his previous paper".

But this previous paper describing his third research project is of a totally different standard, presenting much more important results in a much more lucid style. It has already been remarked that it begins by saying that Lodge, J. J. Thomson and Hertz are all wrong – a bold claim by a twenty-three-year-old from the uttermost part of the earth. It is not generally realised by non-scientists, and not often admitted by scientists themselves that the first paragraph of a scientific paper is usually a clear statement by the author of the "league" in which he is playing. The standard format is to start by referring to the results and papers of other people in the field on which the rest of the paper will build. This means that the opening paragraph may be saying in effect, "Here is a small piece of work which tidies up a little corner of the field my professor opened"; on the other hand, if it refers only to papers previously published by the author or his colleagues it may be saying something like, "Here is our next piece of work and nobody else in this area has got anything to compete." Finally there are a few papers which begin by saying, "Look here; pay attention. Here's something entirely new and very important"; these are the papers which begin by referring to the work over many years of major figures in science, and the author thus lays his claim to be playing "in the top league" at the frontiers of science. This is what Rutherford's first paper did.

What he succeeded in doing was to show that thin iron wires or steel needles could be magnetised by alternating currents acting at very high frequencies when these currents were generated in a small coil of conducting wire wound round a bundle of the wires or needles being studied. He likewise showed that if the needles – the effect was particularly strong in steel needles – had already been magnetised they could be demagnetised by such high-frequency currents and the process was very rapid.

But he went a great deal further than that. With that glorious simplicity combined with experimental clarity which was his greatest gift, he showed that the magnetisation of the wires was only skin deep – he simply dissolved away the outer layer of iron with hot nitric acid – measuring the magnetisation all the time – until he came to a layer of metal which was magnetised in the opposite direction. The experimental apparatus he used also showed considerable technical improvement – he used a Voss machine and a Ruhmkorff coil, which we should consider primitive, but which were reasonably up-to-date at the time – in the process of producing his high-frequency currents. Yet he was nonetheless limited in material – part of a steel knitting needle and piano wire were among the materials he studied. And in addition to his challenge to the great names in science in his opening paragraphs was the fact that he was experimenting in a field that had only just been opened – for Hertz had demonstrated the existence of high-frequency electromagnetic waves only seven years before, a major step for physics in that it showed the validity of the predictions of Clerk Maxwell's electromagnetic field theory. The young New Zealander used equipment very similar to that used by Hertz in his classical demonstration of the existence of radio-waves.

Rutherford had undoubtedly begun this work as a straightforward piece of "pure" research, and we can see now that it was a logical follow-up from his earlier work on magnetic viscosity – a further discussion of how rapidly or slowly iron and steel can be magnetised. But he was quick to recognise that the work on the very rapid demagnetisation of a bundle of steel needles gave him a means of detecting the invisible Hertzian waves, and, indeed, of receiving a simple signal, invisibly propagated by these waves.

To generate radio-waves he used in these experiments a device similar to that with which the discoverer of radio-waves had worked – Hertz's Dumb-bell oscillator.

In the experiments previously considered an ordinary short pianoforte wire, 0.032 inches in diameter, acted very well as a detector, but when we come to rates of oscillation of over 100,000,000 per second a more delicate detector is required. Some very fine steel wire was taken, glass hard, and cut up into lengths of 1 cm. Twenty-four of these little needles were then built up into one, each being first dipped in paraffin to prevent eddy currents passing from one wire to the other. This little collection of needles formed a compound magnet and offered considerable surface to the action of rapidly varying magnetising forces. The detector was fixed in the end of a thin glass tube for convenience of handling . . . When magnetised and placed in a solenoid of two or three turns it supplied an extremely sensitive means of detecting and measuring oscillatory currents of high frequency.[1]

This was the device and the experiment Rutherford's contemporaries most clearly remembered; it was his demonstration of the detection of radio-waves many years ahead of Marconi and the other pioneers of "wireless" that stuck most clearly in their minds and which they commemorate to this day. In 1937 R. M. Laing and S. Page (the demonstrator who helped Rutherford show his experiments to the Science Society) both recalled, in the New Zealand newspapers covering Rutherford's death, this early work on radio-waves. Mr Laing said, "I saw him send a message by Hertzian waves from one end to the other of Professor Bickerton's laboratory, the old tin shed as it was called, through various intermediate walls." "The Den" – the lower storey of the physics lab – has now been fitted out as a memorial to Rutherford, and it seems clear that he was broadcasting and receiving over a distance of about sixty feet.

Laing's memory of the "sending of a message" cannot have been strictly accurate however, for Rutherford's device could only detect and measure the presence of the radio-waves by showing a signal on a galvanometer. This effect was actually a demagnetisation of the steel needles, and a major development by way of a device to remagnetise the needles was necessary before a receiver could be developed which would detect a series of signals by radio.

The conditions in this "Den" remained as one of Rutherford's memories – he would often talk about it as a cold, damp and concrete-floored half cellar which was normally used by students for depositing their hats, gowns and cases. He very rarely referred to his radio work, however, seeming to forget his essay into this field. He more often referred to his work on magnetic viscosity, remembering particularly his struggles with the five "Grove cells"

[1] *CPR*, p. 38.

– primitive forms of chemical battery – which he used as a power supply in his first experiments. He built a special wooden case for these cells when he was on holiday at his father's farm at Pungarehu, and he frequently described the very considerable time he had to spend each day cleaning the electrodes and filling the cells with acid before he could begin any measurements. Although he found the Grove cells a satisfactory power supply at the start of a day's research, they soon ran down and he had only limited time available in which he could make satisfactory measurements of the effects he was studying. Plainly the experience stuck in his mind, for the power supplies of a laboratory were one of the very few technical and logistic aspects of his work in which he showed much interest when he was in his prime as director of laboratories.

The radio work was not only the most dramatic of his experiments, it was also a major part of his claim for the 1851 Exhibition Scholarship which was to be his passport to Europe and the centre of the scientific stage. This scholarship was derived from a fund, administered by Commissioners, from the profits of the International Exhibition of 1851 in London. It was remarkably easy to sneer at Albert, the Prince Consort, and his Great Exhibition in 1851, and it has been a popular occupation among the intelligentsia of Britain ever since. But the exhibition left us with a number of permanent memorials – the site of the museums and the Imperial College of Science and Technology in South Kensington, the Crystal Palace and the 1851 Exhibition Scholarships. The Exhibition was the outward and visible sign of a belief in the advantages of science and technology which has remained unpopular ever since; yet it was this same belief that led to the reform of education and the universities in Britain from the 1870s onwards; this same movement which provided in New Zealand the educational system from which Rutherford benefited; which provided the changes in university regulations in Britain that allowed him to start his scientific career at Cambridge; and which, financially, through the scholarships, gave him the wherewithal to enter the scientific world. And he remembered this. Many years later when he himself was one of the Commissioners administering the fund it was suggested that the scholarships might have outlived their usefulness and that the money might be spent in other ways. Sir Henry Dale recalled his brisk intervention "as free from anything like mock-modesty as it was from any hint of pretentiousness" as Rutherford declared, "You might remember that if there hadn't been any overseas scholarships you wouldn't have had any Rutherford here."

The scholarships enabled young science graduates from

"Dominions" universities – that is universities in the British Empire – to come to England to work in universities or other laboratories to widen their contacts and improve their research techniques. They were so arranged that each Dominion had the opportunity of nominating a candidate of its own every two or three years. 1895 was one of New Zealand's years to offer a candidate, and in late 1894 Rutherford decided to apply. He failed. The award went instead to the chemist, J. C. Maclaurin of Auckland, who had obtained a First Class Honours in chemistry in 1892, had been working a year longer than Rutherford at research, and had published a number of papers on his work on gold, and chemical methods of treating and extracting the metal.

Quite unbeknown to Rutherford and his supporters in New Zealand, the awarding of the scholarship to Maclaurin had been the result of a considerable struggle in England – essentially a struggle between chemists and physicists – which was won by the chemists, as was usual in those days. The two judges were Professor A. Gray, Professor of Physics at the University College of North Wales, and Professor T. E. Thorpe, who was the Government Chemist, a man of considerable influence. Even when the award had been given to Maclaurin, the physicist, Gray pressed Rutherford's claims on Lord Playfair, the Chairman of the Commissioners, and asked for a second scholarship to be awarded in the special circumstances. Writing of the importance of Rutherford's research, he said:

> The value of the results obtained, as bearing on the construction of alternating dynamos and transformers, as well as their theoretical interest, and the unmistakable evidence of capacity for original investigation which they display . . . It appears to me that the results obtained are many of them of both theoretical and practical importance. Some appear to be new, and others, which seem to have been independently arrived at, agree with the results obtained by different experimenters by other methods.

(Gray is here probably referring to work by the American, Joseph Henry, who in 1842 had shown that a discharging Leyden jar induced effects on secondary circuits some distance away, and had also shown that lightning flashes caused the demagnetisation of steel needles inside a secondary coil. Later in his life Rutherford averred that he had not heard of Henry's work when he started research into the same problem on an entirely different scale; and in any case Henry had no idea that he was working with invisible electromagnetic waves of very high frequencies.)

Gray's insistence was enough to bring the matter up at a special meeting of the 1851 Exhibition Fund Commissioners on March 9th, 1895. They would not accept the idea of giving a second scholarship that year for fear of the precedents it might create. But they pointed out that New Zealand had missed a turn in the round of scholarships the previous year, and so could reasonably be expected to have another nomination the following year. Mr Rutherford might be put forward for the award then, they suggested, "if still eligible". (In fact this missed-turn award was taken up the following year by Rutherford's friend, Erskine.)

It is impossible to determine with any precision what Rutherford himself was doing in New Zealand meanwhile. Possibly he did some more research at Canterbury, putting finishing touches to his magnetic viscosity project and writing it up. It is known that he taught in a boys' secondary school for a short time. Some early biographers give this as the Boys' High School in Christchurch, but Marsden, who had better access to New Zealand records, says the school was at New Plymouth, the largest town near his father's farm in the North Island. Marsden's version is much to be preferred because it is known that Rutherford spent a considerable time on his father's farm that year, painting the house and building a tennis court. What is not in dispute is that he was an extremely unsuccessful teacher of physics and that he found secondary school teaching a very miserable occupation. It seems that his basic fault was that he went too fast and too far for his pupils, carried away by his own enthusiasm for the latest development in his subject.

It is also agreed that the young teacher had no idea of keeping order in a class. There is a story of how he would pick out some particularly rowdy youth in the uproar and send him off to get the "Appearing Book" in which the culprit's name should be written down for detention after hours. But the boys soon discovered that once they were out of the classroom the master would forget them and they were able to sneak back undetected and go unpunished.

It was during this period of misery that matters took an unexpected turn for the better, albeit on the other side of the world. Shortly after the Commissioners of the 1851 Exhibition Fund had decided that they would not award a second scholarship to New Zealand, and to Rutherford, they were informed that J. C. Maclaurin had withdrawn his candidature for family reasons. In fact he had decided to get married and to stay in New Zealand – for the scholarship was only worth £150 a year and even in 1895 it was difficult to keep a wife in a foreign university on so little money. Maclaurin eventually became Government Analyst in New Zealand and did sterling work on the cyanide process for recovering gold. (His brother R. S. Maclaurin went to Cambridge immediately afterwards and worked there with Rutherford at the beginning of a distinguished career which he ended as President of MIT.) The Commissioners therefore transferred the award to Rutherford at another meeting on April 6th, 1895.

It is not known when Rutherford received the news – except in so far as he was in the garden at Pungarehu digging potatoes at the time when his mother brought out the telegram. He flung down his spade and declared, "That's the last potato I'll dig." And it seems that he lived up to his word, for he never went in for gardening, although it was later to be his wife's chief hobby, and he confined his activities in this sphere to cutting down trees and hacking away at unruly bushes – an occupation he plainly enjoyed.

The difficulty of tracing his movements continues through that last New Zealand winter. He must have returned from Pungarehu to Christchurch for a short period because he read his second paper to the Philosophical Institute. But even here there is a difficulty. The minutes of the society record that the first two items on the agenda for August 7th, 1895 were two papers by Rutherford. But the published version records his paper on magnetic viscosity as being read on September 4th. This cannot be taken at face value because Rutherford was in London by the middle of September. Certainly J. J. Thomson wrote from Cambridge to Rutherford in London on September 24th accepting him as a research worker at the Cavendish, which carries the clear implication that he must have left New Zealand very shortly after reading his paper in August. We do not even know why he decided to go to Cambridge – there was nothing in the scholarship to specify this as his destination and there are implications in Erskine's early letters that Rutherford had been seriously considering going to study in Berlin.

All we can be quite certain about is that one of his earliest notebooks carries on its cover: "E. Rutherford, Cavendish Laboratory, Oct 3, '95."

3

The Wide, Wide World

As soon as he arrived in London Ernest Rutherford slipped on a banana skin and wrenched his knee, an injury that was to plague him for thirty years. He was very lonely in London. He developed neuralgia and an appalling cold and took to his bed for three days. He had arrived on borrowed money, he had nowhere to work and he knew no one. All he had was a scholarship, a primitive radio-wave detector and a few letters of introduction.

Although he did go to Cambridge University, to Trinity College, to the Cavendish Laboratory, which makes such a satisfactory symmetry when his life is seen as a whole, it cannot be assumed that all this was fixed and preordained when he arrived in England. Indeed the evidence shows clearly that Cambridge was one choice among several. Cambridge's acceptance of Rutherford, and Rutherford's acceptance of Cambridge, was still in the balance in September 1895. That the two came to accept each other was largely the result of a coincidental and timely change in the institutional regulations of the university, which emphasises once again how much his career was shaped by the development of the academic and scientific institutions of his time. Rutherford was the first to step on to a new ladder which was, at the risk of mixed metaphors, being constructed as he climbed up it, and which he, a little later, was to help construct.

Similarly, because he was extremely successful scientifically in the three years after his arrival in England, it cannot be assumed that England or its scientific community welcomed him. Indeed, as Rutherford himself was soon forced to realise, it was only by scientific success that he would be able to make himself socially acceptable and financially viable. He was fortunate that the greatest pleasure he could find – that of doing science – was also his best weapon for carving out his social niche. Throughout the rest of his career this remained the case for him, and he saw to it that excellence in scientific work continued to be a force for social mobility for most of his close collaborators in the laboratory.

It is quite clear from Thomson's first letter to Rutherford that nothing had been fixed, and there were alternatives still open when the young man arrived in London. The letter was addressed to London and it read:

24 Sept. 1895 I shall be very glad for you to work at the Cavendish Laboratory and will give you all the assistance I can. Though it is not absolutely necessary, I think you will find it advantageous to become a member of the University. We have now instituted a degree for research so that anyone who resides for two years and does an original investigation which receives the approval of the examiners receives a degree . . .

   If you could spare the time to come to Cambridge for a few hours, I should be glad to talk matters over with you; so much depends on the requirements and intentions of a student that a personal interview is generally more efficacious than even a long correspondence. In case you decide to visit Cambridge, if you will let me know when to expect you, I will arrange to be at the Laboratory . . . I am much obliged to you for your paper. I hope to take an early opportunity of studying it.

Whether the Professor did study the youngster's paper we do not know. If it was Rutherford's first paper that was sent he would have found that his book, *Recent Researches in Electricity and Magnetism*, published in 1893, had provided some of the inspiration. But he would also have found "the experimental method pursued here is entirely different from Professor J. J. Thomson's, but the final results obtained are the same. The results are also quantitative while Thomson's method only admitted of qualitative results". In other words, he would have found that his prospective new research student claimed already to have outdone him on at least one point.

   At any rate Thomson plainly set out to attract the young man. Rutherford reported back to Mary Newton in his letter of October 3rd 1895:

I went to the Lab and saw Thomson and had a good long talk with him. He is very pleasant in conversation and is not fossilised at all. [Rutherford seems to have arrived at Cambridge with this fixation that most of the seniors were fossilised – perhaps he was justified.] As regards appearance, he is a medium sized man, dark and quite youthful still; shaves very badly and wears his hair rather long. His face is rather long and thin; has a good head and has a couple of vertical furrows just above his nose. We discussed matters in general and research work and he seemed pleased with what I was going to do. He asked me up to lunch to Scroope Terrace where I saw his wife, a tall dark woman, rather sallow in complexion, but very talkative and affable. Stayed an hour or so after dinner and then went back to town again . . . I like Mr

& Mrs both very much. She tries to make me feel at home as much as possible and he will talk about all sorts of subjects and not shop at all.

Back in London the young man had "an extra powerful attack of neuralgia". "On Monday, finding myself not much better, I went to a doctor and got a prescription and, whether due to nature or the medicine, I was getting well again Tuesday, so that I went up to Cambridge with all my belongings."

When he got there he found that the Thomsons had been at work on his behalf. "Mrs Thomson had been very kind and looked me out some lodgings with a widow and gave me directions where to find her. They also asked me up to dinner in the evening which was expressly not a dress affair." At this dinner he was introduced to one of the research workers, a Miss Martin, and to another newcomer like himself, J. S. Townsend – "a young fellow graduate, of Dublin University, who didn't get a fellowship he'd been working for and so came over to Cambridge to do some research work. As he has no friends here he and I knock about together a good deal."

This letter was written to New Zealand when he had been in Cambridge for just three days – and two of those days he had worked in the laboratory. So it was not surprising that "I cannot pass an opinion about Cambridge as I have not yet been about enough. The University term does not really begin till the 10th but there are a lot of freshmen up just in for their 'little go'. They are mostly very young and innocent looking and so a good many are attended by their mamas and papas the first few days." For Rutherford himself the die was not yet cast. He had still to decide whether he wanted to stay in Cambridge and he found it both strange and expensive. But

Thomson wants me to go into residence for the reason that by a new regulation research degrees are offered to students [resident] . . . Research students, if they reside, will be in the same position as the graduates of Cambridge and will dine with them. The expenses will however be rather high, I am afraid, unless they can be cut down considerably. I will know for certain in a week or so at any rate. For the present I will stay in these lodgings till I see my way clear.

It almost seems as if the decision to stick to Cambridge was an emotional one. One biographer, C. H. Boltz, has written that it appears that Mrs J. J. Thomson "had almost fallen in love with Rutherford in a maternal way". Crowther writes, "Thomson's consideration for the student Rutherford was one of the most admirable acts of his own very distinguished career. He had about twenty-five research workers and could scarcely be expected to attend to their personal details. Apart from Rutherford's obvious talent, Thomson seems to have been touched by his coming all the way from New Zealand to work with him." For Rutherford, it must have been rather like finding a substitute mother and father in this land of strangers. He ends his first letter from Cambridge with:

> It is of course a bit strange at first with all one's friends across the sea . . . My success here will probably depend entirely on the research work I do. If I manage to do some good things, Thomson would probably be able to do something for me. I am very glad I came to Cambridge. I admire Thomson quite as much as I thought I would, which is saying a good deal. They have both been very kind to me as you may judge from what I have written.

England came as a great surprise to Rutherford – he found himself in a truly strange land, although the country was still called Home by New Zealanders of his time. He never expressed this feeling of strangeness precisely, but it creeps through in small sentences in a number of his early letters. "It was lonely in London but I didn't mind much as long as I could keep moving around. I saw a good deal . . ." Of his lodgings in Cambridge he wrote, "The place is in a street full of houses and is inhabited by an old lady of about 55 . . . Everything is so different as regards arrangements that I will go into particulars. Rent of bedroom and sittingroom 15/6 a week, coal and firewood 1/6, use of crockery 1/6. Besides that I pay for oil and all my own tucker which she cooks." On his first long walks into the countryside around Cambridge he tells his mother, "One comes across some very old villages with mud and stone houses, thatched and very dilapidated, but picturesque. The great difference one observes is the amount of cultivation one sees around; turnips, etc. seem to be growing everywhere for the cattle and sheep." And in the same letter he describes the laboratory in which he works as compared to the conditions he had endured at Canterbury, "The place is heated throughout with hotwater pipes and is quite warm all day long. The more I see of the laboratory the better pleased I am with it, for although it is not as well fitted up as I expected, it has a fine collection of instruments."

His first important Cambridge dinner at King's College, when he was invited to the High Table, is an event we shall return to. He sent Mary Newton, his fiancée back in Christchurch, a long description of "the general impression I have received of the life of these friars of learning, for really the system of dining in hall is a relic of the time of the old monasteries which has been kept intact to the present day. Some of the men I meet at dinner strike one as very capable especially in their conversation, and it is a pity so many of them fossilise as it were in Cambridge and are not that use in the world they might be." That first impression was little changed forty years later though he loved the style of life. At the other end of the scale he observed the English countryside scientifically and went out on an ethnological expedition to the village of Barrington where the party measured the head size, and height, of every male they could persuade to volunteer, and recorded the country customs and games of the village children in photographs. Rutherford wrote back to New Zealand, "You can't imagine how slow-moving, slow-thinking, the English villager is. He is very different to anything one gets hold of in the colonies."

His early letters also contain many passing remarks which demonstrate how much social and professional opposition he faced at Cambridge; and he remembered these problems all his life and in his last years still took some delight in his early triumphs. He told his biographer, Eve, for instance about the small number of demonstrators who had "the ancient prejudice that no good things can come from the colonies", and how these men would pass his door at the laboratory with a snigger. He dealt with this problem by asking the men to come into his room one day, explaining that he was in some difficulty with his experiments and would be very grateful for their help in solving his problems. At the time he was working on his radio-wave detector, which was totally new to science. The offenders, of course, knew nothing about it and very little about the science behind its workings, and their ignorance was immediately shown up. Eve records that Rutherford ended this story by saying, "After that they gave me no more trouble; they had got on my nerves a bit." But these were an old man's memories, somewhat mellowing the incident. His letters tell of a more hostile relationship.

Three months after he had started work in Cambridge, when he and J. S. Townsend, the first two of the new research students had achieved their first successes, he wrote to Mary Newton on January 15th, 1896:

> The three demonstrators are all extremely friendly now they see we have made a strong position for ourselves and I grimly rejoice for they did not take any notice of us the first two months although they knew we were strangers and had no friends in Cambridge. My paper before the Physical Society was a heavy blow to their assumed superiority and now they all offer to help us in any way they can and tell me confidentially about their own little researches – so wags the world.

But a month later – February 21st, 1896 – he tells how after a lecture by J. J. Thomson at which the experiments did not go off very well:

> J.J.'s dander rose and he turned the laughter against them most skilfully – whereat I rejoiced for many of them are my enemies. I am of the opinion that the demonstrators regard the research students here with very little favour and try and put little obstacles in our path, but verily we rise superior to their machinations and they gnash their teeth with envy. There is one demonstrator on whose chest I should like to dance a Maori war-dance and which I will do in future if things don't mend. Adieu for the present to these warlike fancies, but the fact is my dander rose today and I have not yet quite recovered. The man from whom I was to get my cells for my midnight excursion [another experiment in radio transmission] failed me at the last moment and I, whose temper is on ordinary occasions most angelic, gave way to my feelings and did some tall talking for a few moments.

It is clear from this letter that, although most of the practical troubles came from the demonstrators in the Cavendish Laboratory, there were more than three men who were his "enemies". There are two accounts of this opposition which present fascinating contrasts. J. G. Crowther, the doyen of "science journalists" in Britain, a man who knew Rutherford well, but a man with no great love for the academic establishment of Oxbridge and who writes with Marxist "class consciousness", analyses the opposition. "When Rutherford and the new post-graduate students began work in the Cavendish Laboratory they were at first coldly ignored by some of the junior members of the staff who regarded them as barbarian interlopers who had illegitimately gained access to Cambridge without passing through the usual years of undergraduate initiation." This author regards Rutherford, with much justification, as essentially a non-European, an interloper from an agrarian non-capitalist society of the type described by Marx in *Das Kapital*, and he concludes, "If the bucolic New Zealander had not received so much help (from J. J. Thomson and others) he might not have succeeded in accomplishing the difficult change from a pioneering agrarian society to an ancient English University devoted for centuries to the training of the statesmen, officials, priests and technicians of an imperial governing class."

This gravely underestimates Rutherford's abilities, his innate self-confidence and his rather conservative views of society. It also underrates the adaptability of the imperial governing class then being formed at the university. Sir Henry Dale, a direct contemporary of Rutherford's at Cambridge, who was himself to become one of the great men of British Science, virtually admits that he was among those who did not at first appreciate the value of the research students.

> We ordinary Cambridge students, like not a few of our seniors, were inclined at first to look a little askance at these representatives of a new species, older than ourselves by several years, and neither just under-graduates nor proper dons, as we might have said. We soon found, however that most of them fitted into the picture remarkably well. Rutherford was probably the most brilliant of them all though we might not have recognised him as such for ourselves. He was open and friendly in his manner, simple and direct in his judgment of matters on which he thought himself entitled to an opinion; but he was free from any trace of those airs of portentous wisdom, of effortless brilliance, or artificial enthusiasm which a clever undergraduate from the English schools was often tempted to assume, and which his contemporaries were too ready to accept at face value. If we had been inclined to look critically at Rutherford, I think that we young cynics might have described his manner as rather hearty and even a little boisterous, but I can imagine us allowing, with pride in our tolerance, that a man who had been reared on a farm somewhere on the outer fringes of the British Empire, might naturally be like that, and might nevertheless have the remarkable ability which rumour was beginning to attribute to him.

So it was not among his contemporaries that the New Zealander could at first look for support and recognition. He was given his start by his seniors, notably by J. J. Thomson, Professor of Physics and Director of the Cavendish Laboratory.

J. J. Thomson's laboratory, the Cavendish Laboratory, was exactly as old as its newest recruit. The coincidence that both laboratory and Rutherford came into existence in 1871 is, in any rational sense, an irrelevancy. Yet it is striking, and it has more significance than might appear, for the foundation of the laboratory was an achievement of that same surge of thought, that same belief that science mattered and could help the world in the Victorian drive for progress, that gave Rutherford so many of his opportunities – from the 1851 Exhibition that provided his scholarship, to the opening up of the old universities to outside influences that was to allow him a place as the very first research student, the first man Cambridge had ever allowed to pursue postgraduate work who had not first graduated in the university. The Cavendish Laboratory was to provide him with his ethic, his standards, and at the end of his life his greatest triumphs. Rutherford was very much a man of the institutions in which he worked.

It seems extraordinary that Cambridge University, the home of

Sir Isaac Newton, should have waited so long to provide its first physics laboratory. The major German universities and Glasgow, for instance, had built such facilities many years before 1871; the first colleges of what was to become London University have had such laboratories more or less from their inception; even Oxford had a physics lab before Cambridge. But the Cambridge tradition was that those subjects which we now class as physics were taught as part of the mathematics course and were examined as part of the Mathematics Tripos. This was in some considerable measure due to the influence of Newton himself. He was a rather ill-tempered, quarrelsome recluse, who produced his great works almost entirely from the solitude of his college rooms in Trinity College or from his study in his Lincolnshire home. He had much influence on the drawing-up of the conditions for the Plumian Professorship of Mathematics when it was founded in 1704. The new professor had to provide not only an observatory with instrument and assistant to carry out observations, but he also had to give classes in his own residence on astronomy, optics, trigonometry, statics, hydrostatics and pneumatics. Students would have to pay for attending the course and for any experiments, thus providing the professor with a living. Payment by students for instruction in experimental physics was one of the ways in which Newton himself made his living and plainly he regarded the subject as a proper field of personal enrichment.

The academics of Cambridge taught Newton's "Principia" for the one hundred and fifty years following his death "rather as the theologians taught the Bible". And these academics were grouped in their colleges rather than in the university. Nearly all the teaching of undergraduates in 1850 was in the colleges, and both wealth and power were concentrated there rather than in the university. Furthermore, in the middle of the nineteenth century the whole university had become a very small, closed society, with far fewer students, for instance, than it had in the time of Queen Elizabeth I. The great developments of science and technology of the first half of the nineteenth century had no Cambridge connections at all – they occurred in France or Germany; with Faraday in London or with William Thomson (later Lord Kelvin and no relation at all to J. J. Thomson) in Glasgow.

The reform of Cambridge started from inside the university, when a group led by Whewell persuaded Albert, the Prince Consort, to become Chancellor in 1848. But this movement soon got out of the control of the reforming group and reforms were finally completed almost entirely by outside pressure. The Prince Consort took his task seriously and made enquiries about the state of the university which led to alarming discoveries, not the least of which was the extraordinary attitude towards science of even the keenest reformers. Whewell held that mathematical knowledge should have "paramount consideration because it is conversant with indisputable truths – that such departments of science as chemistry are not proper subjects of academical instruction". He believed that at least a century should be allowed before new scientific discoveries could be considered for inclusion in university courses – teaching was the imparting of knowledge and any uncertainty in the information offered to pupils would undermine the authority of the teacher. The statesman Sir Robert Peel, who was the Prince Consort's adviser, and was himself an example of the leaders of the Industrial and Agricultural revolutions, remarked, ". . . Are the students of Cambridge to hear nothing of electricity?"

The reforms introduced by the Prince Consort in 1848 were followed by the appointment of a Royal Commission in 1852. Its members suggested the introduction of engineering and modern languages into the university's teaching and even proposed the building of a chemistry laboratory, similar to those already provided at some of the new colleges in London. But they could not force through all the reforms they wished, and the conservatives – now including Whewell – kept such subjects as heat and electricity out of even the mathematics examinations.

Slowly, over the years, however, these, and similar subjects, were introduced into the syllabus, but only in the advanced papers of the Mathematical Tripos and still taught as branches of mathematics. As late as the second half of the 1880s J. J. Thomson organised "some demonstrations to enable mathematical students to get some realisation of the physical subjects, heat mechanics and hydrostatics included in the Mathematical Tripos", and found "many cases where men who could solve the most complicated problems about lenses . . . when given a lens and asked to find the image of a candle flame, would not know on which side of the lens to find the image. But perhaps the most interesting point was their intense surprise when any mathematical formula gave the right result. They did not seem to realise it was anything but something for which they had to write out proofs in examination papers."

There was, on the other hand, the problem of how to teach what was called "experimental physics". In various European centres there were different approaches to the problem of how to organise large classes of young men in such a way that they could all perform experiments for themselves – obviously large classrooms would be needed of a type that was unknown in the study of other subjects;

obviously, too, there was considerable expense involved in providing scientific instruments in large numbers at a time when such devices were not mass-produced but were the products of a very small number of craftsmen. In Glasgow, William Thomson maintained a master-apprentice, one-to-one relationship with his pupils. Balfour Stewart, whose elementary textbook had so much influenced the young Rutherford, only started applying his methods to large classes at Manchester in 1870.

Cambridge itself had first provided lecture-rooms for the teaching of science in 1786, but when the Natural Sciences Tripos was introduced in 1851 most of the science teaching was in botany, chemistry and anatomy, and took the form of lectures. The Jacksonian Professor could choose which of these subjects with some physics and applied mechanics he wished to teach. But it was not until the 1860s that subjects such as heat and electricity were introduced into the Mathematical Tripos, largely under the influence of James Clerk Maxwell, of whom very much more will be said shortly. Then in 1868, "after more than twenty years of agitation and deliberation" (according to the official *Centenary History of the Cavendish Laboratory*), Cambridge at last took a positive step to establish the teaching and study of "experimental physics" – they set up a committee, which in university terms is called a syndicate.

In February 1869 this syndicate reported. They admitted in their first sentence that the Royal Commission of 1850 – nineteen years previously – had urged "the importance of cultivating a knowledge of the great branches of experimental physics in the university". They had consulted all the professors involved in teaching mathematics and physical sciences and these gentlemen had agreed that they could not themselves meet the want of an extensive course of lectures on "Physics treated as such, and in great measure experimentally".

The syndicate found that the rules which regulated the Jacksonian Professorship of Natural Philosophy were "fanciful and obsolete", and that the rules governing the Plumian Professorship of Astronomy and Experimental Philosophy also needed revising, as the holder was asked to teach an enormously wide range of subjects. And after considering the possibility of waiting for a vacancy to occur in one of these professorships and then reassigning the work, the report came out firmly in favour of setting up a new Professorship of Experimental Physics, although this was only to continue, in the first instance, for the length of time the first holder should occupy the post. The new professor would teach primarily heat, magnetism and electricity at a salary of £500 a year.

However the Syndicate had done its job very thoroughly, if belatedly, and they insisted that such a professor must be provided with a new laboratory, properly equipped with apparatus, lecture-rooms and classrooms. There must also be a Demonstrator of Experimental Physics because "personal instruction is essential, and the Professor, who ought to be at least partly occupied in original research, and whose attention must at all events be sufficiently occupied in preparing himself to impart and in actually imparting in the most luminous manner the scientific principles of his subject, is not likely to have much time at his disposal for the instruction of tyros in the use of their tools" – wise words which obviously sprang from the working experience of the scientific members of the syndicate.

The financial estimates, with which the report concluded, set the likely cost of the new building at £5,000; allowed £1,300 for the purchase of apparatus and furniture; and allocated £660 a year for salaries of professor, demonstrator and a lecture room attendant who was to care for the apparatus. And it was on the financial rock that the plan struck and grounded.

It was not until 1882 that the Royal Commission on Universities forced the Cambridge colleges to contribute any major part of their wealth towards the university as a whole – and then the interesting solution of making the richer colleges pay a higher proportion of their income than the poorer ones was provided. But in 1869 the university income came from little more than capitation fees of less than £1 apiece from every student. In May 1869 another syndicate was set up to try to find ways of financing the proposed developments in physics teaching. A year later, in May 1870, it proposed a two-shilling increase in the capitation tax, but its report was not unanimous, and the tax proposal was defeated. Throughout the summer of 1870 it appeared that the new professorship and laboratory would not come into existence.

And then in October 1870 came the solution. The Chancellor of the University, the Duke of Devonshire, simply offered to pay the entire sum of £6,300 for the laboratory and its equipment out of his own pocket. William Cavendish, the seventh Duke of Devonshire, was not the sort of man one would normally think typical of the richest of Britain's landed aristocrats. He had had a distinguished academic career in his undergraduate days and had particularly excelled in mathematics, being Second Wrangler and First Smith's Prizeman. When he stood unsuccessfully for the Cambridge University seat in Parliament in 1829 his chief supporter was Charles Babbage, the man usually credited with conceiving the basic idea of the computer. He instead developed into one of the leaders of mid-Victorian industrial and agricultural expansion. He inherited his title and his great estates rather indirectly (through the death of a cousin), but he ran them very efficiently, introducing scientific agricultural methods. He founded the Royal Agricultural Society, and persuaded Cambridge to introduce the study of agricultural science. Likewise he became the second "Iron Duke" of the century, but he earned this title by the development of iron-ore mining on his Lancashire estates and building Bessemer plants to make steel. He built railways in Ireland and Lancashire; he founded the industrial port of Barrow in the north and the seaside resort of Eastbourne in the south. Speaking as President of the Iron and Steel Institute in 1869, he emphasised the importance of scientific discovery to the practical industrialist, probably thinking of the application of spectrum analysis to improve the operation of his Bessemer steel works.

It would be ludicrous to suggest that such a wealthy and powerful man supported the building of a laboratory in the hope that his business would benefit directly from its discoveries. His position socially and politically made him the ideal man to initiate a major change of direction in an important sector of British life. That he chose to do so shows that the social significance of the British aristocracy of the mid-nineteenth century was very different from the archetypal view of them as either deer-stalkers or grinders-of-the-faces-of-the-poor, though many of them undoubtedly did both. But perhaps the strangest fact about William Cavendish was that he refused to allow his sons to be educated in the way he himself had been; he felt that his education had deprived him of the power to exert the political influence which his rank and wealth bound him to exert.

On the strength of the Duke's gift the building of the Cavendish Laboratory was started in 1871 and the first Professor of Experimental Physics in the University of Cambridge was appointed. It was the greatest further good fortune that both Sir William Thomson of Glasgow (Lord Kelvin) and the great German physicist, Helmholtz, did not wish to leave their posts and homes. So the university got its third choice, James Clerk Maxwell, a minor Scottish laird, a man of poor health, who happened to be a genius.

Clerk Maxwell was not just a scientific genius whose great theory linked the phenomena of electricity and magnetism and showed they were the same thing and that light, too, was just one of the forms of electromagnetic radiation. He turned out, perhaps unexpectedly, to be a genius at founding a "school" in the academic sense, and the Cavendish Laboratory was the heart of that school.

Maxwell's inaugural lecture, delivered on October 25th, 1871 –

just a couple of months after Ernest Rutherford was born on the other side of the world – is an extraordinary intellectual feat. Looking back on it, we can see that he not only laid down the intellectual standards that the new laboratory would have to live up to, but gave some pre-vision of the great developments of physics which would occupy his successors for a century to come; we can find premonitions of the quantum theory, of the study of the nature of the atom which was to be Rutherford's contribution; and he was even able to foresee that the physical background of evolution would one day be studied, which surely was a foretaste of the Cavendish's greatest post-Rutherford triumph, the discovery of the structure of DNA.

Experiment, he pointed out at the start, was of two types. First the illustrative, used by the teacher to give the pupil a clearer conception of the phenomena or forces he was studying – and we too easily forget in the twentieth century that to most Victorians the word experiment meant what we should now term a lecture demonstration. Secondly, there are what we should now call experiments – Maxwell called them "experiments of research" – and he pointed out that, once this line of teaching had started, pen, ink, and paper would not be enough, and more room would be needed than a single desk and wider areas would be wanted than those of the traditional blackboard.

In research experiments "the ultimate object is to measure something which we have already seen – to obtain a numerical estimate of some magnitude," he continued – a phrase which consciously or not was to influence Rutherford. However, since this feature of measurement was so pronounced, "the opinion seems to have got abroad that in a few years all the great physical constants will have been approximately estimated, and that the only occupation then left to men of science will be to carry on these measurements to another place of decimals." This was an extremely accurate prophecy of what would be the main problem of physics twenty years later – in the 1890s – when Rutherford entered the field; J. J. Thomson and many others have recorded how at the beginning of that decade many were discouraged by the feeling that there was nothing important left to discover in physics.

Maxwell, in this founding lecture, explained that the new laboratory was to be far more than a place of measurement. It had a social duty to attend to the public's concern with science and to see that the scientific ideas communicated to the outside world were sound; it had an academic duty to see that the philosophy of science, the ideas of science, were discussed at a high intellectual level; it had a duty to science in other parts of the world and must build up intellectual contacts and cooperative experiments with other laboratories in other countries; and it must keep in touch with practical, technological work not only for the benefit of society, but also because such contact would improve the scientists in the laboratory and their work at a purely scientific level. (We can see all these trains of thought and action in Rutherford's own life, with the possible exception of the philosophy of science at the academic level – he simply enjoyed being a scientist.)

But Maxwell himself felt very profoundly that everything had by no means been discovered. It is a fascinating example of prevision that in 1871 he could suggest that studies at the atomic and molecular level would lead to a quite different understanding of nature. He explained that there was no solution yet to the age-old struggle between the theory of the plenum and the opposed theory of atoms and void as the basic explanation of the world around us. He announced that in his first course of lectures he would give some of the evidence for the existence of molecules which "as presented to the imagination are very different from anything with which experience has hitherto made us acquainted". He was particularly concerned with the philosophical implications of the fact that molecules of one sort are exactly the same all over the universe. By what theory of evolution could anyone account for this identity of properties, he asked – posing the questions which we now summarise as "the origin and nature of matter".

Could it be that our

scientific speculations have really penetrated beneath the visible appearance of things which seem to be subject to generation and corruption, and reached the entrance of that world of order and perfection which continues this day as it was created, perfect in number and measure and weight? We may be mistaken. No one has yet seen or handled an individual molecule, and our molecular hypothesis may, in its turn, be supplanted by some new theory of the constitution of matter, but the idea of the existence of unnumbered individual things, all alike and all unchangeable, is one which cannot enter the human mind and remain without fruit. But what if these molecules, indestructible as they are, turn out to be not substances themselves, but mere affections of some other substances?

A small query which has kept his pupils and successors at the Cavendish Laboratory busy for a hundred years and more.

It is believed to be an example of Maxwell's sense of humour that he delivered this lecture virtually unannounced and unpublicised to a very small audience in the teaching room of Professor Liveing's chemistry class. When he later gave the first lecture of his full teaching course many of the great figures of the university and the scientific world turned up in error at this lecture instead and had to sit through an elementary discourse on temperature and how to measure it.

Maxwell was in many ways untypical of the "great scientist". He took the post at the Cavendish largely because his personal finances were at a low ebb. He had done his great scientific work, on the kinetic theory of gases and the nature of electricity and magnetism, before he came to Cambridge, while he was professor at Aberdeen and King's College, London, though he published his "Treatise on Electricity and Magnetism" after he got to Cambridge. His work at the Cavendish was almost entirely devoted to resurrecting the world of Henry Cavendish, the eighteenth-century relative of the laboratory's benefactor. He repeated Cavendish's experiments and did others to elucidate what the earlier scientist had done; he published Cavendish's papers and became the first scientist of the first rank to devote himself to the serious study of another man's life and work.

The first four years of Maxwell's tenure as Professor of Experimental Physics were spent essentially in building the new laboratory, which was formally handed over to the university by the Chancellor, the Duke of Devonshire, the donor, all rolled into one man, on June 16th, 1874. In equipping the lab Maxwell kept in close touch with the Duke, though he found that the original estimate for apparatus could not possibly meet the needs. He bought the best he could obtain, and stopped buying only when he felt he could ask the Duke for no more. In the meanwhile he had to give his own lectures, moving, in his own words "like a cuckoo" from one borrowed lecture room to another. Not surprisingly there were few students; in 1874 only seventeen undergraduates took the Natural Sciences Tripos and very few of these did physics. The mathematical approach still reigned supreme in the place where it counted – the examination paper – and things became even more difficult in 1878 when the university moved physics into the Second, more difficult, part of the Tripos, so that very few, except those determined to do experimental work, found it worth their while to come to the laboratory.

Maxwell got a few people started, however, of whom the most notable was R. T. Glazebrook, and he encouraged those few who were interested in doing research by leaving them to select whatever problem interested them and exercising only the gentle supervision of making frequent rounds among the research workers accompanied by his dog Tobi. The first student to come to the Cavendish to do research was W. M. Hicks, who later became

77

Professor of Physics at Sheffield, and perhaps the most interesting and exciting was Arthur Schuster, later Sir Arthur Schuster, Professor of Physics at Manchester, and a central figure in Rutherford's life. Maxwell ran things on a loose rein, but in an intellectual atmosphere, until he died eight years after his appointment.

It will be remembered that, technically, the professorship had only been set up for the length of tenure of the first holder. When Maxwell died there was some fairly neat footwork involved in getting Lord Rayleigh to accept the office which did not exist and getting the professorship reestablished when he had agreed to take it. Plainly, however, it was generally agreed that the experiment had succeeded and these things can be managed when everyone wants the same result.

The small size of the scientific community in the 1870s can be judged from the fact that Maxwell was attended in his last illness by Sir George Paget, who was to become the father-in-law of J. J. Thomson. Maxwell's successor was John William Strutt, who had inherited the title of third Lord Rayleigh only two years before, and whose son, in turn, was to be one of the lights of the scientific world in the generation to which Rutherford belonged.

Rayleigh's family was altogether of a shorter history than either Maxwell's or the Cavendish clan. They were essentially Essex landowners who had made a large fortune but who had continued to draw their income primarily from agriculture. With the opening of the American western wheatlands, however, English agriculture suffered a severe depression in the mid-1870s. Rayleigh turned his farms to providing milk for the growing market of expanding London, but in the interval his fortunes, like those of Maxwell before him, were somewhat straitened. He, too, accepted the Professorship of Experimental Physics because the income would help, although his mother did not like the idea of his becoming a professor at all.

Rayleigh himself had had a brilliant undergraduate career at Cambridge, becoming Senior Wrangler (i.e. first in the entire list of the Mathematical Tripos results) and gaining the First Smith's prize. He then set about making himself a scientist at a time when the university provided no facilities whatsoever for this process. He became a Fellow of Trinity and extended many of Maxwell's earlier results, and helped to correct and improve the famous "Treatise on Electricity and Magnetism". He was only thirty-seven when he became Professor, and he made it clear that he only intended to stay in the post for five years or so.

What he gave to the Cavendish Laboratory was good organisation and sound financing, both developments being typical of his private life as shown in his reorganisation and redirection of the family agricultural business. His biography records the final stages of development of dairy farming on his Terling estate.

At the end of his life, in 1919, there were about 800 cows and sixty milkers. In order to be independent of middlemen, a shop with the legend "Lord Rayleigh's Dairies" was established in Great Russell Street in 1887 and this was followed by others in different parts of London. At the end of Rayleigh's life there were eight in all. Rayleigh's name perhaps gained more notoriety with the general public from these shops, and from the milkcarts seen in the London streets with his name upon them, than from his scientific activities.

To describe the importance of Rayleigh's work, it is best to turn back to the famous inaugural lecture of his predecessor, Maxwell. After the passage on measurement which has already been quoted (see page 75) he had remarked, "But the history of science shows that, even during that phase of her progress in which she devotes herself to improving the accuracy of the numerical measurement of quantities with which she has long been familiar, she is preparing the materials for the subjugation of new regions, which would have remained unknown if she had been contented with the rough methods of her early pioneers."

Rayleigh, in broad terms, set his Cavendish Laboratory workers to the more accurate determination of those physical units of measurement which are the basis both of experimental and technological progress. In particular they concentrated on the improved measurement of electrical units such as the ohm, the unit of resistance. This was of considerable significance at the time, for electrical engineering was beginning its impact on the world of industry and commerce in those years, and the importance of accurate measurement was made clear in developments such as the spread of the electric telegraph system and the provision of the first transatlantic cable.

Of course this sort of study required extra apparatus and instrumentation in the laboratory. One of Rayleigh's first acts was to set up a new "Apparatus Fund" and he opened this with a personal donation of £500; the ever-generous Duke of Devonshire responded with as much again; and soon £2,500 was raised. This enabled the Professor to get over his earlier problems the which had led him to complain in his first annual report that the laboratory had contained "no steam engine or other prime mover; nor among the acoustical apparatus is there any musical instrument!"

Yet surprisingly it was Rayleigh who started the Cavendish tradition that comparatively rough and simple apparatus, made preferably by the scientist himself, was the correct approach. This tradition strongly influenced Rutherford. Rayleigh set up his own private laboratory in his large house at Terling, and those who saw it record that it was by no means a shrine to "the brazen image which the instrument maker has set up"; instead Rayleigh worked where "sealing wax, string, rough, unplaned woodwork, and glass tubes joined together by bulbous and unsightly joints, met the eye in every direction". Only J. J. Thomson, his most distinguished pupil, seems to have seen further into the mess: "In the apparatus the part which really affected the accuracy of the results was sure to be all right and carefully made." It is also J. J. Thomson who analyses his master's strength as "the power of putting his finger on the really vital point of a question. He was therefore able to simplify the solution by taking a case where, though this point had not been affected, everything else which only increased the mathematical difficulties had been stripped off, and the mathematical difficulties reduced so much that a solution was possible." He follows this up by showing how Rayleigh followed Maxwell's and Kelvin's view of the importance of models by quoting one of his phrases: "There can be no doubt, I think, of the value of such illustrations both as helping the mind to a more vivid conception of what takes place, and to a rough quantitative result which is often of more value from a physical point of view than the most elaborate mathematical analyses." The word "model" here is used in the physicists' sense of a simple mental construct, not necessarily a physical "toy".

These points are important in Rutherford's development for they all appear in his own "style" of thinking and working – brought perhaps to a higher peak by his own extraordinary power of mind, but not springing up in him for the first time as so many seem to think; he was to show that he was firmly in the line of this "English" tradition which became more obviously opposed to the "European" approach as the physicists of the Continent produced the quantum theory and the theory of relativity in the first decades of the twentieth century.

There were two further quite specific instances in which Rayleigh plainly influenced Rutherford. The first of these occurred during his tenure of the Cavendish Professorship, when he induced his brother-in-law Mr A. J. Balfour to take a personal interest in the work of the laboratory. This came about because in 1871 Rayleigh had married Evelyn Balfour, and, through her, enlisted the aid of her sister, Eleanor, as a regular collaborator and note-taker in his experiments. Arthur Balfour, who remained a visited them in Cambridge. On several occasions he actually helped Rayleigh with his experimental work alongside Eleanor. J. J. records, "I remember seeing Mr Arthur Balfour set down to read a galvanometer."

The significance of this family involvement lies, of course, in Arthur Balfour's political achievements. It was not just that he was to hold many of the highest offices in the British government in the first three decades of the twentieth century, but also because of his intellectual stature, which made him the almost Olympian philosopher of the Conservative Party. It was through Balfour that scientists in general, and Cambridge scientists in particular, became involved in the work of government – firstly through war research in 1914–18, and later, during the time of Rutherford's ascendancy, on a much broader scale. Furthermore, as a Member of Parliament, Balfour sat for one of the Manchester constituencies at the time when Rutherford held the Chair of Physics in Manchester, so that they had many mutual acquaintances and a number of business contacts. Throughout Rutherford's adult career Balfour was always in the background, even in the mid-thirties when he had more or less retired from politics but was Chancellor of the University of Cambridge.

The second way in which Rayleigh influenced Rutherford was his discovery of the rare gas, argon, in 1892. There is some controversy still as to whether Sir William Ramsay should be awarded half the credit for this discovery, but the essential experiment was Rayleigh's repetition, in his private laboratory at Terling, of research originally carried out by Henry Cavendish, which showed a consistent difference, after the most careful measurement, between the density of nitrogen extracted from the atmosphere and the density of nitrogen produced from chemical compounds. This led to the conclusion that atmospheric nitrogen contained another gas which was isolated and called argon. Later Ramsay also isolated the other rare gases, neon, krypton, xenon and helium. The work of Rayleigh and Ramsay was not only a confirmation of Maxwell's belief that careful measurement opened up new scientific vistas, but it left Rutherford with the possibility always in front of him that there might be other rare and unidentified gases to be found (particularly isotopes of helium and hydrogen).

In order to provide a balanced view of Rayleigh's leadership of the Cavendish it is however necessary to stress that he started the laboratory workshop for the provision of new instruments and apparatus; that the work on determination of physical units and standards led to the setting up of the National Physical Laboratory, under Glazebrook, to carry on this type of work permanently and in all fields of measurement: Rayleigh strongly supported this development throughout his life and it has of course become a necessity for all developed countries to follow this example. Work on the study of the atmosphere and meteorology was also started in Rayleigh's time under William Napier Shaw. This was to be a continuing Cavendish line of research, leading up to some of the most exciting developments of our own time; Shaw himself became the first Director of the National Meteorological Office.

Rayleigh, however, could soon afford to look on a professorship with financial, if not aristocratic, disdain, as his switch to dairy farming took his family estates back into profitability; and at the end of his promised five years he duly resigned. His successor was the most brilliant of the young men working in the Cavendish, J. J. Thomson, although the choice was not in fact from a wide field.

Everybody called J. J. Thomson "J.J.", even his own son, and his arrival at the Cavendish signalled many changes, including a fresh informality. Firstly he was far from aristocratic – his father was a poor bookseller from Manchester, whose early death had so reduced the family circumstances that the fee for indenturing the young J.J. as an engineering apprentice could not be afforded. He therefore went to Owens College – the embryo Manchester University – and thus began his own distinguished academic career as well as the connection between the Cavendish and the North of England.

From Manchester J.J. came to Cambridge and to Trinity College; went into that strange obstacle-race, the Mathematical Tripos, and came out Second Wrangler. He then succeeded at the first attempt in the examination for a Fellowship at Trinity with a dissertation on the nature of energy – he then held that all energy was kinetic, the energy of motion. He had only started a little research in 1880, just four years before his election to the professorship. It is hardly surprising therefore that there was considerable consternation at his selection for such a major post. One senior university wag complained that things had come to a sorry pass when professorships were given to "a mere boy". Although Lord Kelvin had been approached for a third time to take the Cambridge post and for a third time had refused to leave his private interests as well as his professorship in Glasgow, there were several other possible candidates – notably Arthur Schuster who had done research at the Cavendish and had become Professor of Physics at Manchester. Most embittered were Glazebrook, the holder of the "number two" position in the laboratory, and William Napier

Shaw, the other demonstrator. Glazebrook could not even bring himself to acknowledge Thomson for some time, but eventually he wrote to J.J. "Forgive me if I have been wrong in not writing before to wish you happiness and success as Professor. The news of your election was too great a surprise to permit me to do so. I had looked on you as a mathematician not as an experimental physicist and could not at first bring myself to regard you in that light."

His judgment, though swayed by self-interest, was in essentials correct, for J.J. was no experimenter. J.J.'s first important paper had been essentially mathematical, drawing from Maxwell's theory the then extraordinary conclusion that a single electric charge moving in an electric field increases its mass as its speed increases, and this extra mass lies in the space just around the charge. When the electron was discovered (by J.J. himself some fifteen years later) this result became the foundation of the widely-held theory of "the electromagnetic nature of mass", and of course, it predates Einstein by many years.[1]

J.J.'s other important work before his election as Professor was an investigation of "the action of two vortex rings on each other". The importance of this apparently esoteric piece of mathematics was that Lord Kelvin had suggested (wrote J.J.), "that matter might be made up of vortex rings in a perfect fluid, a theory more fundamental and definite than any that had been advanced before. There was a spartan simplicity about it. The material of the universe was an incompressible and perfect fluid and all the properties of matter were due to the motion of this fluid."

In fact J.J.'s powers as an experimenter were of a very unusual nature. On the one hand he was clumsy – "J.J. was very awkward with his fingers and I found it necessary not to encourage him to handle the instruments," according to Newall, one of his personal assistants in his younger days. Nonetheless he had an uncanny understanding of experimental apparatus even if he could not handle it – as F. W. Aston who worked closely with J.J. for many years recalled:

> When hitches occurred and the exasperating vagaries of an apparatus had reduced the man who had designed, built and worked on it to

[1] One of the difficulties of writing this book on Rutherford is to avoid the appearance of "knocking" Einstein. Because his great Theory of Relativity embraces so many phenomena that seem incredible to the layman, such as the equivalence of mass and energy, formulated as the famous $E = mc^2$, and because so many modern scientific statements start from the firm basis of Einstein's work, it is not realised how many of the conclusions we associate with Einstein were in fact discovered by other scientists before him.

baffled despair, along would shuffle this remarkable being who, after cogitating in a characteristic attitude over his funny old desk in the corner and jotting down a few figures and formulae in his tidy hand-writing on the back of somebody's Fellowship Thesis or an old envelope or even the laboratory cheque book, would produce a luminous suggestion like a rabbit out of a hat, not only revealing the cause of the trouble, but also the means of the cure. This intuitive ability to comprehend the inner working of intricate apparatus without the trouble of handling it, appeared to me then, and still appears to me now, as something verging on the miraculous, the hallmark of a great genius.

Aston's work with J.J. resulted in the development of the mass-spectrometer which is still nowadays the most powerful machine in a scientific laboratory for analysing the constituents in an unknown sample.

J. J. Thomson continued the Cavendish Laboratory in the Maxwell tradition. Maxwell's great electromagnetic theory was accepted almost unquestioned in Cambridge, largely because of its mathematical elegance and power. It was little known and less regarded on the Continent of Europe where the main strength of physics was still to be found. Hertz's experiments demonstrating the existence of radio-waves, which had been clearly predicted by Maxwell's theory, had a great effect in bringing Maxwell's theory, and therefore the current trends of English physics, to continental notice. Hertz's success was in 1887 and by 1888 J.J. was demonstrating the radio-waves in his lectures and many experiments with them were performed in the laboratory. But the main drive of the Cavendish under J.J. was in the study of the conduction of electricity through gases, a subject which turned out to be very rewarding but extremely complex – and indeed still not completely worked out to this day.

In the decade from 1885 to 1895, that is in J.J.'s first ten years as Professor, the numbers of students in the laboratory increased dramatically, and a major extension had to be built at a cost of £4,000 to which J.J. was able to contribute £2,000 which he had saved from the fees paid for teaching and research. The first steps were also taken towards building up the international aspect of the Cavendish, as J.J. began to gather young men from many countries about him. His book, *Recent Researches in Electricity and Magnetism*, which can be regarded as an extension of Maxwell's work, did much to bring his work international renown. The year 1895 changed everything. It was, in J.J's own words, "one of the most important years in the history of the Cavendish"; in the words of J. G. Crowther, "it brought a revolutionary increase in strength";

and Rutherford himself included it in "the most interesting and important period in the history of research". There were two reasons for the importance of 1895: firstly it saw the arrival of the new research students at the Cavendish; secondly it saw the discovery of X-rays which marked the end of the period when even physicists felt that "everything had been discovered", and brought about the opening of modern physics, leading to the quantum theory, relativity and nuclear physics.

This admission of research students by the university accepted the existence of such mechanisms as the 1851 Exhibition Scholarships which brought "Empire" students to England; it also accepted, though covertly, that continental, and particularly German, universities, were attracting students who could call themselves "Doctor" when awarded their PhD after a course of advanced study or research. In Crowther's view it also meant that Cambridge accepted the change of role from being the provider of British science and scientists which it had held since Newton's day, to the wider role of being the leading provider of science and scientists for the international and growing British Empire.

Both contemporaries and historians agree that the new regulations did bring a great influx of scholars, scientists and strength, particularly to the Cavendish Laboratory. It is reasonable to view this laboratory as something more than just a university physics laboratory, as it did undoubtedly play a significant and leading role in British academic and scientific life at that time as well as subsequently. How much we may count it pure coincidence that Ernest Rutherford was the first to apply and to be received as a research student is difficult to calculate. Certainly he did not know about the scheme when he left New Zealand, and certainly J.J. Thomson was particularly pleased to receive so prompt an application from a youngster living as far away as New Zealand – to him it seemed to justify his support of the scheme that the New Zealander and an Irishman applied within an hour of each other at the earliest possible moment.

There were disadvantages, as well as advantages, for the laboratory in this access of strength. While there now started an "unrivalled magnitude of scientific output", the large number of advanced students added to the "teeming population" of the laboratory; there was now, it was said, "more physics being done per square centimetre at the Cavendish than in any other lab in the world". This put enormous pressure on the still inadequate amount of instrumentation and equipment: "Competition among the research students and between them and the advanced classes for instruments, retort-stands and rubber tubing waxed very keen, and it was occasionally conducted by means of raids, which forced one victim to describe himself as pursuing his investigations with his apparatus in one hand and a drawn sword in the other." This slightly bitter description of the state of affairs comes from L. R. Wilberforce, who was then a demonstrator at the Cavendish and who later became Professor of Physics at Liverpool. He was more interested in teaching than research, and saw the research students as examples of "teaching by research" in opposition to the teaching of advanced classes by "illustrative experiments", in which the students learnt by carrying out experiments which had been organised beforehand by the teaching staff to demonstrate the various principles of physics. Wilberforce felt that starting research too early could narrow a man's outlook and that there were arguments for both kinds of learning.

J.J. himself had the research students directly under him, and a strange sight he always was to meet. "His rather straggling aspect (his razor seemed always a little blunt), his shuffling walk, the little touches of his native Lancashire accent" gave him something of the appearance of "a grocer's errand boy" according to an obituary in *Country Life* in 1940. The anonymous author admitted however that he was the first man to become Master of Trinity who was not a clergyman, and he might have added that he was the first Professor of Physics of the university to come from the lower middle classes of the North of England.

But this was the world of Cambridge, the Cavendish and physics, into which Ernest Rutherford came in that remarkable year of 1895.

4

Science in Cambridge

Ernest Rutherford's detector of electromagnetic (radio) waves which he brought with him from New Zealand, plainly fitted perfectly into the Cavendish tradition and scheme of things.

A magnetised needle is partially demagnetised by the passage of electrical oscillations through a solenoid surrounding it. The decrease of magnetisation is always greater if the first oscillation of the discharge is in the right direction to reverse the magnetism. This property was used as a quantitative method of finding the damping of the oscillations in the Leyden jar discharge and of measuring the resistance of conductors for high frequency currents. Using large Hertzian vibrators the electrical waves emitted were observed by means of the magnetic detector for a distance of about half a mile. These experiments were made before Marconi began his well-known investigation on signalling by electric waves. This effect of electric oscillations of altering the magnetism of iron is the basis of magnetic "detectors" developed by Marconi and others which have proved one of the most sensitive and reliable of receivers in radiotelegraphy.

This he wrote in 1910, and J.J. Thomson frequently claimed that Rutherford had held the world record for distance of transmission and reception of radio-waves during his first few months' work at the Cavendish.

Rutherford's letters home to his fiancée and his mother tell how it was actually achieved. Just before Christmas he told his mother, "My own research work progressing satisfactorily but is pretty slow of course." On January 15th, 1896, after complaining that he only saw the sun once in his Christmas break in Edinburgh, he explains to his fiancée:

I started work at once in the lab on my electric waves and made my first experiment on long-distance transmission of signals without wires. I set up the vibrator in the Prof.'s room at the Cavendish and my detector 100 yards away in Prof. Ewing's lab. and got quite a large effect through the distance, traversing three thick walls by the way. The Prof. wishes me to continue at the work and see how far I can detect the waves, so I have been working lately to try and find the best conditions of sensitiveness of my detectors.

At first sight it may seem that this was taking a long time to achieve rather little, especially as Rutherford had arrived in Cambridge with his detector already in existence. This impression grows even stronger when one looks at the detector itself, reverently preserved in the museum of the Cavendish Laboratory: the greater part of the device is a large, roughish block of wood on which the very small glass tube (in which he put the magnetised needles) and the ancient brass connectors seem mere excrescences. But the idea that the detector could be used as a detector of radio-signals in our modern sense was slow in coming. Plainly Rutherford, and J.J. too, saw it first as a detector of a scientific phenomenon predicted by Clerk Maxwell. The first public demonstration was at a meeting of the Cavendish Physical Society, a body founded by J.J. himself on the lines which most modern laboratories would call "colloquia" – occasions when visiting scientists or the members of the laboratory's own staff would present their latest work, helping everybody to keep up to date. Rutherford demonstrated his detector at the beginning of December 1895 in a presentation entitled, "A method of measuring waves along wires and determination of their period".

But in that letter of January 15th, 1896, the young man confided, "My next experiment will be, I think, from the tower of the Cavendish to St John's tower nearly half a mile away." We do not know whether this was performed successfully, for the next episode is one of disaster, beautifully described in writing home on February 21st, 1896: "I am going to try my experiments tomorrow night on the Common. I hope they will turn out successfully. Townsend and McClelland are going to assist me. I am to use Townsend's room for my detector and go across the Common about half a mile with my vibrator and accessories. I will leave this letter open till tomorrow so that I may add particulars . . ."

Then comes a passage which all experimenters and inventors will feel strike a chord in their hearts.

My experiment came off tonight and was rather unfortunate. I am writing this about 1 o'clock just after my return from the expedition. I fixed up the receiver in T's room about 6.30 in the evening and had arranged for two men to come at 7.30 to take the apparatus on to the Common in a small go-cart. I waited till 8, and they did not arrive, and then I went and hunted round for the firm who had got the men for me, hunted up a clerk, who found that the pumpkinhead, to whom I had given the order, had put time down at 7 instead of 7.30. The men had come, waited and gone, and I did not know of it. However we hunted up two men from the street, got a hand-cart and got started about 9.15 instead of 7.30. We got the things ready and set them up at about 500 yards at which distance I anticipated a good large effect, but we could not get a sign. We tried different distances up to about 100 yards and not a sign. I knew then the internal economy of my detector had gone wrong at the critical moment for I had tried 100 yards with the same detector before and had got a large effect. This detector I had arranged in a specially sensitive method, so when it had gone wrong somewhere I had nothing to replace it. However I had another rough detector handy and on trying that we got a large effect up to 350 yards. We did not try any further as it was getting very late, and I had to take the apparatus back to the lab before 11.30. So we hurried back and got clear about 11.45. I was very tired as I had been running about from 6 to 12 and feel very much like bed now.

It is a pity my detector went wrong at the critical time. I had tried it in the afternoon and it was all right but one of the wires had evidently got broken somewhere inside. I may try the experiment again under more favourable conditions and am certain in my own mind I could get an effect up to a mile easily. My friends were very disappointed at first, but they rejoiced when they saw it was evidently due to a break in the detector. They both worked like Britons, and I could not have done without them.

But only a week later – on February 29th, 1896 – he was able to report triumph, if only very briefly:

When I left off last time I had just been out on the Common trying to detect waves at long distance. The next day I tried and got an effect from the lab to Townsend's diggings, a distance of over half a mile through solid stone houses all the way. The Professor is extremely interested in the results and I am at present very useful when he is writing to various scientific pots as he can mention what his students are doing at the lab.

This letter of February 29th, though rather brief, is of considerable significance in describing the development of Rutherford's life at Cambridge. After recording his success with the transmission and detection of wireless-waves over what was then the world's record long distance, the letter concludes, "I have sent a couple of the new photographs home to Taranaki – one of a frog, one of a hand, both taken at the lab. I am sorry I can't send you some, but these things are rather expensive and I have to be careful of the bawbees these days." These photographs were, of course, X-ray photographs, almost certainly among the first to be taken in England, so rapidly had J.J. and the Cavendish reacted to Röntgen's discovery of this new phenomenon. There will be more to be said about the discovery of X-rays, but the importance here is that they provided the main reason for Rutherford's decision to give up his radio work some two months after his first mention of their discovery.

At this stage of Rutherford's life, the six months from November 1895 to March 1896, there is so much happening that a biographer wishes he could present his material in three dimensions instead of a flat sequence dictated by print. We have Rutherford triumphing with his radio work, yet giving it up at the moment of scientific achievement and allowing others to develop it. Yet at the same time the radio-waves were his party-piece and it was in talking and writing of this work and demonstrating it that he attracted attention and first made his name. His first paper to the Royal Society, a sure sign of moving into "the big time" in scientific terms, was delivered in June 1896. It was about radio work, although he had already given it up. His first appearance months later on the larger stage of the British Association and its public meetings was devoted to the same subject. He was still interesting his contemporaries from other scientific disciplines and from even wider university circles by demonstrating radio-waves more than a year later. At the same time we have his own education in the social niceties of British upper-class life proceeding impressively as he came to the attention of people influential even beyond the boundaries of university scientific life. Yet, perhaps most important of all, it was at just this stage of his life that his immediate professional colleagues and his Professor began to realise that here was a man of extraordinary powers.

J.J. recorded that after only a few weeks of seeing Rutherford at work, "I became convinced that he was a student of quite exceptional ability". He added that even at this early period of his career "he possessed exceptional driving power and ability as an organiser", and J.J. quotes his ability to enlist the help of Townsend and McClelland and others as helpers as evidence of this power. Among his contemporaries, too, these powers were recognised. Andrew Balfour, who was later to work for the Egyptian government, wrote of "a rabbit here from the Antipodes who burrows mighty deep", while Langevin, the Frenchman who had a room next to Rutherford's at the Cavendish during these years, when asked if he had quickly become a friend, replied, "One can hardly speak of being friendly with a force of nature."

We have two good descriptions of Rutherford at work with his radio-detector. One, given by Eve, we can safely surmise came from Rutherford himself:

You have to think of the transmitter consisting of two large metal plates side by side and in the same plane with two short metal rods protruding from them and ending in polished brass knobs about half an inch or so apart. These were "excited" so that sparks passed between the knobs no matter whether by an induction coil or by an electrostatic machine, so long as there was a fairly steady shower of sparks, each of which caused electricity to swing to and fro between the plates perhaps a few million times a second . . . Half a mile away with houses in between, was Rutherford with his receiver made of two metal rods in line, each two feet long; joining their near ends was a fine wire going round a small coil wound about a minute bundle of fine magnetised steel wires. When a signal came the needles lost their magnetism and this was easily shown by the deflection of a mirror with a little magnet behind it.

The records of the Cambridge Natural Science Club show that on July 25th Rutherford invited the members into the rooms he was occupying in Trinity College during the Long Vacation and gave them a talk on "A New Method of Detecting Electric Oscillations"; this talk was "illustrated by showing some apparatus used in his experiments", and so presumably did not include a demonstration of the transmission and reception of wireless-waves. It was nearly a year later, in May 1897, that he entertained the same club in his lodgings in Park Street and finally showed them how he could receive signals from the Cavendish half a mile away. Sir Henry Dale remembers the occasion clearly:

As my memory pictures the scene, he had the receiving apparatus on the table of his living room, and he had contrived that the effect of the waves on his electromagnetic detector should somehow start the ringing of an electric bell. He explained to us that he had arranged with somebody in the Cavendish Laboratory to start the Hertz oscillator there at exactly 9 p.m. by the laboratory clock by which he had carefully set his watch; and we all sat round the table waiting for the great moment like a racing boat's crew counting the seconds to the starting gun. Exactly at 9 o'clock the bell began to ring and we greeted the demonstration with cheers; and then Rutherford gave us a short talk on the nature of the transmissions we had been witnessing. I wonder whether any of us had any sort of premonition of what might be the future of this wireless communication at a distance of which we had been given so early a demonstration.

Rutherford himself does not seem to have had at first any particularly visionary premonitions about the future value of his work and the same could be said about his attitude to nuclear science forty years later. His notebooks do not contain much of interest on the

91

subject – just solid and fairly unimaginative remarks on the various dimensions of what we should now call the aerials he was using and the methods of shielding from internal interference.

> Monday Mch 16. First Monday of Vac. Vibrator in advanced room; receiver in my own room. Tried the effect of various needles with different magnetising solenoids but with no very satisfactory results. Tried a specimen of chemically precipitated Fe_3O_4 but with no effect at all. Varied size of vibrator and distance between plates . . . Left subject of vibrator and turned to experiment on screening effects of wires . . .

In another, rather earlier, entry he records how he found it unnecessary to use parabolic reflectors and then, "It was expected that if a band of metal was substituted for a wire a larger effect would result, but contrary to anticipation no increase was observed." He also found that "not only the vibrator gives out waves but also the connections close to it".

His first paper to the Royal Society – his third scientific publication – was being written in the early months of 1896. It is titled, "A Magnetic Detector of Electrical Waves and some of its applications", and it is clear from this paper that Rutherford looked on his detector simply as a scientific instrument. It has four main uses, he writes: for detecting electromagnetic radiations in free space (he adds that he has done this over the range of half a mile); for detecting and measuring waves along wires; for investigating the damping of oscillations by the resistance of iron wires and the absorption of energy by conductors; and for the determination of the period of Leyden jar discharges. Indeed the paper, though rewritten in tauter style, is really little more than a thorough development of his New Zealand work. Incidentally in this paper he acknowledges the pioneering work of the American, Henry, performed fifty years before, perhaps a good example of the start of that most endearing of Rutherford's characteristics, his total willingness to give other scientists, even the most junior of his colleagues, their full measure of credit.

The section of the paper dealing with what are termed "Long Distance Experiments" tells, in those sober terms and with that infuriating continuous use of the passive which scientists believe gives objectivity to their work, of the experiments across Jesus Common so much better described in his letters. There are three interesting points, however. First he gives us a glimpse of the sort of apparatus he was using – zinc plates six feet by three feet in size for his vibrator, though at the other extreme of size his detector consisted of twenty tiny bits of steel wire, each only 0.007 inches in diameter and 0.393 inches (one centimetre) long, each separately varnished in shellac and the whole bundle wound in a coil of fine wire at eighty turns to 0.393 inches, and all kept in a glass tube. Here his skill at making his own experimental devices is clearly seen.

In writing about his results there is one point that is repeated five times in less than half a page. Clearly what impressed Rutherford about the electromagnetic waves was their ability to pass through thick walls. He writes that he started at a range of forty yards, "the waves passing through several thick walls between the vibrator and the receiver". In the experiments from the Park Parade house across Jesus Common or from the Cavendish to Park Parade, "the waves before they reached the receiver must have passed through several brick and stone walls, and many large blocks of buildings intervene between the vibrator and the receiver." Surprisingly, a wave "seemed to suffer very little loss of intensity in passing through ordinary brick walls". In the lab itself "the effect of six solid walls and other obstacles between the vibrator and the receiver did not diminish the effect appreciably. When the vibrator was working in the upper part of the laboratory a large effect could be obtained all over the building, notwithstanding the floors and walls intervening."

Rutherford's many scientific biographers and memorialists seem to have failed to realise the significance of these repeated remarks. They were written at precisely the time when X-rays were exciting the world of science, X-rays with their power of going through solid matter. Rutherford is insisting that his radio-waves also go through solid matter. Because we are nowadays so accustomed to the idea of penetrating radiation, because we realise, even if subconsciously, that the air around us is full of radio and television signals, full of radiation from rocks and space, it is difficult to imagine how stunning the idea must have been that "solid matter" could be penetrated, that there were invisible, intangible radiations and waves all round. But in the minds of the physicists, and in particular in the mind of Rutherford, the ground was slowly being prepared for the possibility that solid matter was not solid in the way our senses tell us it is.

Another significant point in this paper is the apparently simple statement in describing the mechanics of the experimental system that it is necessary that "the detector should be remagnetised and placed in position again after each observation". In other words, the detector could only signal the fact that electromagnetic waves were being received and measure their rough strength. There was no question at this stage of sending communications or a string of differing signals of differing strengths. The importance of this lies in the fact that it has always previously been stated that Rutherford gave up all his radio work at the start of the Easter Term, 1896, when he was invited by J. J. Thomson to join him in his personal research on the electrification of gases by X-rays. It is quite clear that this is not the case, and that Rutherford must have continued with at least some radio work long after April 1896.

It is in his letter home of December 8th, 1895, that Rutherford announces with justifiable pride that J.J. had suggested that he should publish his radio work to the Royal Society – and that meant, of course, that J.J., the Professor, would "communicate" it, or endorse it, in that most august set of scientific publications, *The Philosophical Transactions of the Royal Society*. "As only the best papers, or at any rate the papers of eminent men, are chiefly found there, I have nothing to complain of. As a matter of fact very few papers are recommended by Thomson for the Royal Society. I must apologise for the amount of ego that fills this letter, but human nature will out, you know . . ." Ernest Rutherford thus displayed his laurels to his fiancée, with that strange combination of full appreciation of his own worth and the plaudits which he felt he rightly commanded, mixed with an innate modesty and shrewdness which always prevented him from giving offence or arousing envy.

The paper was delivered to the Royal Society in June 1896, and from his letters we know he was working on radio until at least late March 1896. Yet later Rutherford was to make it quite clear that he had invented, independently of Marconi and others, a method of remagnetising the steel needles of his detector (by using a continuous flexible steel band) so that the device could receive a continuous stream of signals. This work can only have been done after March 1896. Furthermore there are letters from a Cavendish colleague, Erskine Murray, which have apparently never been published, but which will be referred to later, making it clear that Rutherford continued to interest himself in radio matters for some considerable time after he had turned his principal scientific interest elsewhere.

Finally this paper contains, at least in its final edited version (in CPR), an extremely interesting admission by Sir James Chadwick, who was responsible for the monumental publication of all Rutherford's scientific papers. It is a footnote to a table of values given by Rutherford as "examples of a few of the experiments on the demagnetisation of iron wires", and the final item is a "long soft hollow iron cylinder, ¼ millim thick and diameter 1.8 millims". Of this Chadwick has to write: "Rutherford was a somewhat indifferent proof-reader. There are many instances in his papers of inconsistencies in nomenclature, in abbreviations and some of arithmetical mistakes as well as obvious misprints . . . *Correction*: There can be little doubt that the dimensions of the cylinder referred to above should be 'centim', not 'millim'." And indeed only ten pages further on in these *Collected Papers* (and in a number of other cases) Chadwick has had to correct numerical values that were printed in the originals.

Failings of this sort could not check Rutherford's progress, nor did they bring up on him any sort of "professional" disapproval. So why did he give up his radio-wave detector work, at least as his main interest, in April 1896?

In January of that year he had written to New Zealand in high hopes that his work would bring him not only scientific success, but also financial benefit.

> I have every reason to hope that I may be able to signal miles without connections before I have finished. The reason I am so keen on the subject is because of the practical importance. If I could get an appreciable effect at ten miles, I would probably be able to make a considerable amount of money out of it, for it would be one of great service to connect lighthouses and lightships to the shore so that signals could be sent at any time. It is only in an embryonic state at present, but if my next week's experiments come out as well as I anticipate, I see a chance of making cash rapidly in future. I cannot say I am exactly optimistic over the matter, but I have considerable hopes of being able to push it a good long distance.

We know that, despite tribulations, his work showed every sign of success over the next two months. Nevertheless one of Rutherford's great failings can be seen clearly in his letter – his lack of imagination in translating the results of his work from the laboratory to the outside world of technology, profit and commerce, his diffidence in exposing himself and his work to physical contact with the needs and pleasures of the teeming world. Marconi, who was already working on radio but was considerably behind Rutherford technically at this time, had a far wider, more exciting and earthier vision of what could possibly be achieved.

Certainly Professor J. J. Thomson made every reasonable effort to help to convert the detector into a practical machine. He called in the advice of Lord Kelvin, the Glasgow Professor of Physics who had considerable experience in converting the results of his academic research into commercial profitability. Kelvin, as Rutherford reported in his letters, had been reasonably impressed with the detector, but had advised that £100,000 should be the limit for capital expenditure. That this turned out to be a ludicrous under-estimate should not be allowed to conceal the fact that nobody in the Cavendish, or indeed in Cambridge University, could lay hands on anything like this amount of money. The problem of finding capital to develop university research ideas into engineered and saleable products is one that still bedevils British science to this day.

The episode of calling in Kelvin's advice on the commercial possibilities of Rutherford's detector has been recorded by the fourth Lord Rayleigh who wrote a biography of J.J. It has been generally accepted as sufficient explanation for Rutherford giving up his radio work. But some other work was taking place in the Cavendish which must have had an effect on the decision, for this parallel work seemed to show that the detection of electromagnetic waves led into a scientific blind alley. The description of this negative experiment is given by Rutherford himself, although he did not participate in the work, in his chapter in *The History of the Cavendish Laboratory*. Along with his own radio work it is grouped at the end of the chapter with several other miscellaneous researches and is obviously regarded as somewhat outside the mainstream of Cavendish progress. He remarks that in 1897 W. C. Craig Henderson and J. Henry, at J. J. Thomson's suggestion, tried to find out whether the ether is moved by electric waves. (The ether, or aether as it was often spelt, was the medium through which electromagnetic waves were assumed to be propagated; it was a purely imaginative concept, brought into existence because scientists knew that electromagnetic waves existed yet were unable to conceive of waves that were not waves "of" or "in" something.) Whether Rutherford himself doubted the existence of the ether as early as 1895 we do not know; certainly writing in 1910 he says, "According to theory, when a wave passes through the ether, a mechanical force acts on the ether in the direction of motion of the wave. If the waves are undamped, the force is periodic and, on the average, zero; if damped, this mechanical force should have a finite value and it was of interest to try whether the ether was set in motion by this force." He explains that Henderson and Henry used an interferometer method but "No appreciable shift of the in- terference fringes was observed, indicating that, if the ether moves at all, its velocity was very small."

Remembering that Rutherford did not in fact give up his interest in his electromagnetic wave detector in 1896, it is clear from this brief account that J.J. had not given up all interest in the work either. This experiment in 1897 however showed that the electro- magnetic waves were not going to provide a tool for investigating the mysterious ether in which they were supposed to move. By 1897, too, Marconi was well launched into his practical development of the waves, and the subject was dropped, for the moment, from the Cavendish programme. Nevertheless the topic made repeated appearances, in the lives of J.J. and Rutherford, in their war work, in the programme of Cavendish physics in the 1920s and 1930s, and it is the basis of one of the mainstreams of Cambridge research at present under the title of "radio-astronomy".

Interlinked with this research activity, however, J.J. was steadily "promoting" his brilliant young New Zealander. We have seen how the Professor brought him to the notice of fellow physicists by asking him to address the Cavendish Physical Society – and Rutherford wrote home about this, "I am the first member of the Cavendish who has given an original paper before it, so I may consider the honour is greater than I can bear . . ." and supported his first paper to the Royal Society. There was also a wider public to be addressed, and J.J. saw to it that his new "find" was given a chance to shine at the "Science Conversazione" in March 1896 at the laboratory. Rutherford described it to Mary Newton, back in New Zealand, in a letter dated March 15th, 1896.

> I told you it was to be a very big affair and no pains or expense was spared to make the thing a success. From the entrance to the Free School Lane to the lab, about sixty yards, was covered in an awning and lighted with glow lamps and carpet laid down. Inside the lab itself carpet was laid down all over the staircases and everything was prettily lighted up. The new room where the guests were received had the usual display of biological, geological and physiological apparatus beside the physical experiments, but it did not look anything like as well as the big Canterbury College Hall on a similar occasion. The demonstrators presented Mrs J.J. with a bouquet for the evening which she held in her hand while receiving her guests. At 8.30 people began to arrive in great numbers and soon the place was pretty well filled up. There were between seven and eight hundred present, all more or less the very select part of Cambridge society. I was one of the stewards and wore a pretty favour of blue and pink. I had my new dress suit for the occasion, wherein I felt more at home, not to say at ease, than in my old ones. I think it was about time I went in for a new suit, on account of reaching the limits of elasticity. Mrs J.J. looked very well and was dressed very swagger and made a very fine hostess. J.J. himself wandered round looking very happy and grinning at everything and everybody in his own inimitable way. We first of all wandered round and had a look at the show generally, and I then proceeded to show my experiment to anyone who so desired. I had my vibrator in one room and my own special receiver in a room forty yards away with five solid walls in between, and showed quite a large effect at that distance. Some people

were very interested and reckoned it was <u>the</u> thing worth seeing in the Lab. I had quite a large number of distinguished people who came to see it and were very keen over it, including Sir George Stokes, Sir Robert Ball, Professor Vernon Boys and a good few other pots whom I did not know by name. Sir R. Ball was going round expounding the merits of the same in great style and some of the ladies who came round were tremendously interested or professed to be. Townsend assisted me and we both took turns at explaining and working the vibrator. My part went off very well, so I have nothing to complain about. In the morning before the conversazione Mrs J.J. brought her mother, Lady Paget, to see the experiment . . . Mrs J.J. seems to have got an absurd idea about my experiments and is nearly as good an advertiser as Sir R. Ball. Mrs J.J. told anyone she got a chance that they must go and see the wonderful experiment of Mr Rutherford and it was the only new thing being shown that night. It is lucky I am of a modest disposition, but these things don't affect me in the way you might expect. I really think Mrs J.J. regards me with considerable favour for several reasons. She always introduces me as Mr . . . who has come all the way from New Zealand. I think she appreciates the compliment to her husband by my coming straight from New Zealand to the Cavendish.

Then in June 1896, J.J. asked Rutherford to give an experimental lecture to the British Association for the Advancement of Science at its annual meeting which was held that year in Liverpool. And only a fortnight later, in the same week that he heard that the Royal Society had accepted his paper on the electromagnetic detector, Rutherford was invited to demonstrate the device at a conversazione at the Royal Institution in London on June 19th. Although this invitation was probably also a result of J.J.'s initiative, his pupil was by then working with him on X-rays and the invitation was turned down "as it is rather much trouble and hard to fix up in a hurry" according to Rutherford's explanation to his fiancée. However he admitted in this same letter (June 18th, 1896) that "My blushing honours are lying thick upon me."

The September meeting of the British Association was Rutherford's first appearance before a really large gathering of internationally famous scientists, for the "B.A." in those days was a vivid and active "shop-window" for science, at which the very latest discoveries were announced, essentially the same vibrant organisation which had seen the Huxley-Wilberforce debate on evolution thirty-five years previously. In fact the subject that dominated the physics section was the newly-discovered X-ray on which Rutherford was by then working, and for which he was publicly credited in J.J.'s opening lecture. His own contribution was not scheduled until the second week when he gave his lecture on his magnetic detector of waves. There can be little doubt of the truth of the story, told by Eve, of how Rutherford's apparatus refused to work during his lecture. The young man simply remarked to his large and distinguished audience "Something has gone wrong. If you would all like to go for a stroll and a smoke for five minutes it will be working on your return." The audience, doubtless, welcomed the break; they were even more impressed on their return when the apparatus did, in fact, work and they were able to see the proof. Not surprisingly a leading physicist remarked, "That young man will go a long way."

But at that same meeting of the British Association it was announced that Marconi had just that summer achieved a range of one whole mile with "wireless" transmission and reception, working on Salisbury Plain and mounting his equipment on top of wooden poles. It seems reasonably certain that Rutherford was not perturbed by this news – his main interest had indeed switched to the X-ray work with J.J. On the other hand, it was at this meeting that he started making friends with physicists and scientists from other countries – a process which he kept up throughout his life, so that in the days of his international fame he had a vast network of friends and correspondents who kept him remarkably well informed of all that went on in the fields of science surrounding his own. There is for instance a letter extant from Professor V. Bjerknes, a Norwegian physicist who later became one of the founders of meteorological services. Bjerknes writes in November 1896: "I see you are still anxious to persuade me not to judge English Science altogether by the B.A. meeting. Your fear is quite superfluous. I judge it from the work of Faraday, Joule, Maxwell, Kelvin; and even if I did judge it by the B.A. meeting only it would not be as unfavourable as you seem to fear."

J.J.'s help to his research student went even further than "promoting" his work to various publics – which, after all, was in the interests of his own Laboratory and Department and was a justification of the still-suspect introduction of colonials and foreigners into Cambridge. He also did everything possible to help the newcomer to integrate himself into the life of the university and to find the money to do so.

J.J. had urged his new research student at their very first meeting to become a member of a college so that he could take advantage of the regulations to gain a degree. The disadvantage of this proposal was that "the expenses will be rather high, I am afraid, unless they can be cut down considerably". But in his next letter home Rutherford writes (October 20th, 1895):

I have at last crossed the Rubicon and am now a regular undergraduate, or rather graduate of the university. I have been waiting for the last fortnight to see what allowances they would make for research students, as this is the first term the regulations have come into force. J. J. Thomson looked after our interests and obtained a substantial reduction in fees, and, as he strongly advised me to join a college, I at last decided to do so. He wished me to join Trinity, which is his own college, and also the best as well as the dearest in the university. It will not however make much difference to me which college I join, and Thomson seemed to think I would probably be able to command more influence from the people of Trinity than from any other college.

And how right J.J. was proved to be over the years.

Rutherford was joined by J. S. Townsend, his fellow research student in this venture, and together they went to see the tutor, Mr Ball, who told them what had to be done. The same letter continues:

So this morning, which is the general matriculation day as it is called, I went and got a cap and gown before 8.30 . . . At 8.30 all the freshmen of Trinity, about 200 in number, assembled in the Hall, had our names read out and then trooped off to the Senate House where we had to sign our name in the book kept for the purpose. By 10 o'clock I was formally initiated and then had to pay £5 matriculation fee and £15 caution money to be returned at the conclusion of the university course . . . The expense of joining a college will not be much more than if I were a non-collegiate student for I shall stay in the lodgings where I am but can turn up to dinner when I like and only pay when I do so. The great advantage of joining a college is of course the number of men you come to know and the social life which as a non-coll. you would miss entirely. Thomson very strongly advised me to join and reckoned the extra expense would not be thrown away in the time to come.

By April 1896, however, the expense was beginning to tell:

My £150 won't keep me going by any means and I do things on as economical a basis as possible. Everything is about as dear as it can be, as they pile it up on all sides for the students. I hope however that I will be able to make a little extra cash to keep me going for some time yet. In order to do that I told J.J. I was outrunning my allowance and asked if I could get any "pups" as people who get coached are called. He volunteered to do what he could, so at present I have one pupil for three hours a week and may get one or two more.

Coaching in Cambridge was more profitable than it had been in Christchurch he pointed out, bringing in about £9 a term for one "pup" at three hours a week, "so if I could get half a dozen men for three terms I could nearly make a living for one if not two".

In August of the same year J.J. was telling Rutherford that if ever the 1851 Exhibition Scholarship was renewed for a third year it would be so in his case, adding that there was also the possibility of a Fellowship at Trinity to be considered after his B.A. was attained. Then in December, after J.J. had returned from a trip to America, Rutherford reported:

> He has just been doing me another good turn. He generally examines the Sandhurst military candidates in physics, but not being able to go on Saturday he asked me to take his place. I am rather interested in seeing the chaps, etc. The fee will of course be two or three guineas for the day. J.J. has also dropped me a hint in regard to the Fellowship exams, to get up some general philosophy as they are examined in that in the Fellowship exams.

Only a few days later the Professor asked him to examine fifty candidates for the Artillery School at Woolwich, and for the two exams paid him six guineas. On Christmas day 1896 the letter to New Zealand says:

> J.J. seems anxious to help me in every way he can. He had been sent some books from *Nature* to review, but asked me if I would do it. I of course agreed and in consequence got a letter from *Nature* this morning asking me to review a couple of books. It is my first attempt at that sort of thing, and I hope I shall manage to do it decently. The books are scientific ones dealing with X-ray business on which I reckon I am an authority.

This letter also contains one of the most famous examples of Rutherford showing his sensual diffidence.

> At the dinner I mentioned some of the dresses were very décolleté. I must say I don't admire it at all. Mrs X., wife of a professor, wore a "Creation" I daresay she would call it, which I thought very ugly, bare arms right up to the shoulder and the rest to match. I wouldn't like any wife of mine to appear so, and I am sure you wouldn't either . . . Mrs J.J. is generally about half and half and looked very well indeed.

It is difficult not to feel that J.J.'s influence must have helped a year later when Rutherford's 1851 Exhibition Scholarship was renewed for a third year and he was awarded the Coutts Trotter Scholarship, founded by a famous mathematician at Trinity College, and worth £250, and "nearly enough to get married on". The happy letter announcing this triumph is dated December 12th, 1897:

> On Friday J.J. casually remarked to me that the Council met that day and would probably fix up about the Coutts Trotter. So I trotted round but said nothing about it . . . About 6 a Trinity College clerk came round and presented me with two letters, one from my Tutor congratulating me on getting the scholarship and the other asking me to Trinity Commemoration dinner which took place that evening. I handed my Tutor's note to J.J. and after a minute he came round and, after congratulating me, calmly remarked that he had known about it all day but couldn't tell me.

It would, however, be quite wrong to view all Rutherford's progress and success as being due to J.J.'s pushing or Mrs J.J.'s advertising. The young man was quite capable of impressing people himself, through his own work mostly, but also showing a keen and ambitious appreciation of those who were likely to be helpful to him. It was his friend J. S. Townsend who introduced him to Sir Robert Ball, a Fellow of King's College and Professor of Astronomy. As a result, Rutherford wrote the most outstanding letter of his life, the only time he seems ever to have been so affected by what had happened to him as to provide us with a real glimpse of the man who was writing. Dated January 25th, 1896 it reads:

> Sir. R. (Ball) developed a tremendous interest in my experiments on the detection of electric waves for long distances and must needs make an appointment to come down to the lab and see the effect and the apparatus in general. He turned up one morning and I showed him how easily I could detect a wave through half a dozen walls and rooms and he was very much interested. He is especially keen on it as he thinks that experiments of the nature I am doing will solve the difficulty of lighthouses in time of fog, when the light does not penetrate any distance. His idea is to fix up a vibrator in a lighthouse and as a fog does not stop an electric wave a suitable detector on board a vessel should tell her when she is within, say, a couple of miles of a lighthouse. Of course the arrangement would be very useful for signalling at sea at night between vessels and informing each other of their close proximity in times of fog. Sir Robert wanted to know about the whole matter and ended by asking Townsend and myself to dinner at King's College tonight . . . We turned up in cap and gown at 7 p.m. and first of all met Sir Robert and all the Fellows in the Combination Room as it is called. I was introduced to those whom he considered most interesting from my point of view, and he spoke of me in such flattering terms that I felt inclined to disappear out of sight. However I recovered my normal

modesty and walked into Hall with Sir Robert marshalling me in front as an honoured guest. The 150 students of King's and the Fellows all dine in the large hall together. The Fellows, i.e. all the professors and lecturers, etc. of the college dine at a separate table at the top of the room. All the students stood up as we entered and naturally they all wanted to know what in the deuce a youngster like me was doing among the Fellows. To add to the shock which I had already suffered I was placed in the position of honour at the table, on the right of the Provost, as he is called. Grace was said and dinner was served. Conversation was kept up pretty well but the Provost (Augustus Austen Leigh) himself had an impediment in his speech and was not particularly brilliant. I was able, however, to keep up a conversation with some interesting men on the other side of the table. I really felt a great deal like an ass in a lion's skin after the way I was treated, for I did not see what I had done to be one of those whom "the King delighteth to honour" but Sir Robert Ball's remarks had evidently fallen on good ground, for I was looked upon as a scientific expert. Dinner lasted just an hour at the Fellow's table and then we all retired to the Combination Room, and, after a little desultory talk, all sat down at a long table and chestnuts and cakes were placed on the table as well as wine and cigars. Conversation was very interesting and all the men were of course more or less famous in various branches and their conversation was worth listening to. Here again I was seated alongside Sir George Humphry, a great medical man, about seventy-five, although he looks much younger, and we had an interesting chat on various topics. I was sandwiched between Sir G. Humphry and Sir R. Ball and tried to look as dignified as the circumstances permitted and really felt at ease notwithstanding my usual shyness or rather self-consciousness.

Seated opposite me was a Mr Browning – a lecturer at King's – and he made himself very agreeable. He seemed an extremely well informed man – in appearance he was a good deal like the typical John Bull one so often sees in *Punch*. He evidently knew such men as Goschen and Chamberlain [eminent British politicians and Cabinet Ministers] personally very well, so I thought he must be rather a prominent character. In the course of conversation he asked me to lunch tomorrow where, of course, I am going. I believe he has the most beautiful set of rooms in King's and is a very clever man as well as a good conversationalist. Excuse the way I put the above, it reminds me a good deal of Boyle – you remember "He was the Father of Chemistry and Brother to the Earl of Cork".

About ten o'clock we rose to leave the Combination Room and a Professor Oldham – Professor of Geography – asked us to his rooms where we stayed half an hour telling stories . . . I felt such an impostor masquerading before the learning of King's, but it is a good thing my modesty does not allow me to take all the above as my right. It really has been a very eventful evening to me both as regards the people I met and the general impression I have received.

103

The Mr Browning here mentioned, was of course Oscar Browning, one of Cambridge's most famous, if not most admirable, characters around the turn of the century – a man of whom many tales are told, none better than than recorded by J.J. who described how Browning complained that he did not know what to do in his beautiful rooms because his collection of books was growing so fast, and was "shot down" by Joseph Pryor, a Trinity tutor and mathematician, who replied that he might try reading them. Nor was Rutherford taken in for long by the poseur, though at first he was plainly slightly dazzled. In the same letter he described his lunch at Browning's with three others, like himself all very young:

> One young fellow Goldsmid by name, and a Jew, had just come into a fortune of £30,000 a year. A Howard of Howard Castle was also present and seemed a very intelligent type. Conversation was very shoppy and I was glad to get away. Browning is quite a character here. He is a bit snobby. From what I have heard he professes to know all the people worth knowing in Europe. It is a common yarn about him that he said "The German Emperor was about the pleasantest emperor he had met", which I should judge quite characteristic of him. His rooms were very fine and he has a very good library of which he is very proud.

It is worth noting one particular feature of this letter back home to New Zealand, and that is the remark about his own sentences reminding him of the statement about Robert Boyle. The letter is a purely personal one back to his fiancée and sweetheart; it is one of a series in which there are no literary "quotes" or allusions, and which mainly detail his small triumphs and minor adversities. Yet up comes this literary/historical remark which he seems to expect Mary Newton to remember or associate with some occasion. Plainly this young man would be able to hold his own in the civilised, perhaps over-sophisticated conversation of a Cambridge college High Table or Combination Room.

There were, indeed, significant signs that Rutherford was making a major impression on Cambridge society. Shortly after his dinner at King's, Sir George Humphry called round personally at the lab to invite Rutherford and Townsend to dinner; Sir Robert Ball came round again to look at the detector; the Science Conversazione was a big success; there was even an invitation to breakfast with John M'Taggart Ellis M'Taggart (just one person), who was a Fellow of Trinity, a Hegelian philosopher (according to Rutherford), a man who proved that time does not exist (according to modern Cambridge myth), a man who avoided compulsory games at his public school by lying down on the field and refusing to move (according to J. J. Thomson). Rutherford was not impressed here either – ". . . he gave me a very poor breakfast, worse luck. His philosophy doesn't count for much when brought face to face with kidneys, a thing I abhor . . ." Nor did M'Taggart's philosophy impress Rutherford the more when twenty years later they found themselves on opposite sides when Cambridge debated the advisability of offering new degrees to research men.

Among those who were more nearly his contemporaries Rutherford also overcame the initial hostility to the research students. In May 1896 he reported, "I have just had what I suppose is considered rather an honour conferred on me, i.e. I have been elected a member of the Cambridge University Natural Science Club which has a membership of about a dozen and is fairly select at that. They meet every Saturday night in one of the members' rooms and get someone to read a scientific paper and then an hour or two is given to social converse." This was the club at which he demonstrated his detector, as described by Sir Henry Dale.

There was also the Ray Club, even more exclusive, founded in 1837, consisting also of a dozen members but equally divided between senior and junior members of the university. This club met for dinner in the rooms of one of the senior members at least a couple of times a year. Rutherford seems to have started attending these dinners in March 1897 and to have continued for much of the rest of his life. The objective was to introduce younger men to their most distinguished seniors, but it was probably in this company that he first made friends with Grafton Eliot-Smith, who is known professionally as a brilliant pioneer of neuroanatomy, the study of the architecture of the brain, and whose broader "fame" is chiefly due to his unwise support of "Piltdown Man", but who became one of Rutherford's earliest allies and dearest friends.

Clearly they were originally joined as allies for Eliot-Smith was the first of the new research students on the biological side, working under Donald McAllister who had been one of the protagonists of the new system. When the time came to memorialise Eliot-Smith, an Australian, Rutherford himself remembered that they "saw much of each other outside laboratory hours". "We had many interests in common for we had both been brought up in the country, were very much of the same age, and both had to rely on ourselves to make our own careers. In addition we were the only advanced students from our respective universities and we were conscious that it devolved on us to make good and to show ourselves worthy of the opportunities given to us," he wrote. And of their joint membership of the Ray Club he recorded, "It was a great privilege to attend these meetings which offered an opportunity, not only of scientific discussions, but of meeting socially many of the most distinguished scientific men in Cambridge." Obviously he felt that it was in such club meetings that the prejudice against the advanced students was largely dissipated, and goes on, "In many cases, too, the research student from a university in our Dominions had not had the opportunity to obtain such a complete training as the Cambridge man and there were obvious lacunae in his knowledge to be filled up." However in a few years the "brilliant work of Eliot-Smith and others . . . led to a general recognition of the great value of the new scheme and whatever criticism there had been died away. On looking back over the perspective of forty years one cannot rate too highly the wisdom of the pioneers who recognised that the university must open its doors more widely to qualified students of outside universities if it was to play its part as a great centre of research and learning." Nevertheless most of his friends and close comrades were "Colonials" or men from other universities such as Townsend the Irishman. His letters contain mentions of R. S. Maclaurin, brother of the man who originally beat him for the 1851 Exhibition Scholarship, and another New Zealander, the engineer W. S. La Trobe.

For his first Christmas holiday in Britain he went up to Edinburgh where we know he had several New Zealand friends studying medicine. For his Easter break in 1896 he went to Lowestoft in a party of "two Afrikaners, one Armenian and two New Zealanders". The two from South Africa were both students of law, de Waal and de Villiers – "both very much in favour of the Dutch element in the Cape and of course very hard against Jameson in stirring up the Dutch against the English". This was presumably a reference to the Jameson Raid which preceded the outbreak of open hostilities between the Boers and Britain. Maclaurin was the other New Zealander and the Armenian remains unnamed.

It was on this holiday that Rutherford seems first to have come across British prudery and hypocrisy over the innocent business of a morning dip. He described the difficulty of obtaining a swim even in April:

The first morning we went in, in front of our digging, dressed *en règle*. A policeman came next morning saying he had been specially sent by the superintendent to tell us to go further down as one of the landladies in the neighbourhood had objected. After two mornings a bit further down the beach, another policeman gave us the hint to move still further down, as another modest female had raised objections. So we finally had to walk about one-third of a mile to get a dip in the tide. It was pretty cold of course but one felt in good trim after it. The alarming modesty of the British female is most remarkable – especially the spinster, but I must record to the credit of those who were staying there that a party of four girls used to do the Esplanade at the same time as we took our dips and generally managed to pass us returning, but we were of course in eminently respectable attire by that time.

*   *   *

Returning from this brisk, young man's, Easter break in 1896, Rutherford made the decision which was to reshape his whole life. He relegated the work on his magnetic detector to second place and joined his Professor in work on the new X-rays. It is difficult nowadays to realise just how revolutionary, how astounding, the concept of the X-ray was at the time of its discovery. It was the single concept that shattered the view that there was really nothing left to be discovered in physics. It was the discovery that started physics on the dramatic progress which made it the premier science of the first half of our century. Looking back, we can see that even on this most material level it was important to Rutherford, for, at the time of the discovery, chemists were at the top of the scientific pile as the men of most usefulness, working on the most exciting science, and X-rays changed this. Even the pages of *Punch* and popular journals carried jokes about the new rays "revealing all", and quacks advertised lead-lined clothes to protect female modesty against rays which could penetrate any number of Victorian petticoats.

Rutherford himself wrote fifteen years after the event, "It is difficult to realise today the extraordinary interest excited in the lay and scientific mind alike by the discovery of the penetrating X-rays by Röntgen in 1895. Almost every physical laboratory in the world started some work on them." J. J. Thomson was certainly one of those who leapt into the new field without waiting for any of the courtesies. "I had a copy of his apparatus made and set up at the laboratory and the first thing I did with it was to see what effect the passage of these rays through a gas would produce on its electrical properties."

Even if that was the first thing that J.J. did with the X-rays, it was not what interested the public and the rest of the laboratory most. It was photographs of the insides of things that interested everyone – even Rutherford – and he reported to New Zealand on January 25th, 1896:

The Professor has been very busy lately over the new method of photography invented by Professor Röntgen and gives a paper on it on Monday to which I will of course go. I have seen all the photographs that have been got so far. One of a frog is very good. It outlines the general figure and shows all the bones inside very distinctly. The

Professor of course is trying to find out the real cause and nature of the waves and the great object is to find the theory of the matter before anyone else, for nearly every Professor in Europe is now on the warpath.

The discovery of X-rays did not come entirely out of the blue. It was, in many senses, a logical step in the investigation of the strange behaviour of discharge tubes – Crookes' tubes as they were often called – in which electric currents were passed through gases at low pressures inside partially evacuated glass vessels. Strange lights and colourings, often very beautiful and striking, appeared when this was done, and there were also areas where no light was generated. The whole subject was an interesting, though apparently not a very significant, corner of that almost closed world of physics where everything worthwhile seemed to have been discovered.

But this subject was one of the main features of J.J.'s programme at the Cavendish – the conduction of electricity through gases. The subject was in many ways a provocative one, for gases had always been considered to be insulators; they would not allow electricity to pass through them unless very high voltages were applied, as in a lightning flash. The subject had only been opened up by the discovery by Heinrich Geissler in 1855 of the mercury pump which enabled scientists to reduce the pressure in glass vessels, and it was then realised that gases would pass electricity if they were kept at reduced pressure. The development of a better pump – the Toepler pump – speeded up work in the field, though these pumps were incredibly slow and tedious to operate by modern standards. Inside these tubes were two metal terminals, a "positive" called the anode and a "negative" called the cathode. They were connected to the outside world by fine wires passing through the glass, and the electric potential was applied by connecting up these wires to a power source. It eventually became clear that electricity was carried somehow in straight lines from the cathode to the anode, and these lines, which could only faintly be seen, came to be called "cathode rays". There was a great difference between English and continental physicists about the nature of these rays – in England they were considered to consist of very small particles (perhaps molecules of gas) which carried charges of electricity, whereas the continentals regarded them as ethereal waves, electromagnetic types of wave.

The background to this difference of interpretation probably arose because Faraday's work on passing electricity through liquids, in which the liquids broke up into their chemical components, had implied that there was some kind of basic unit of electrical charge. When he split water up into its components of hydrogen and oxygen he showed that a certain quantity of electricity would always liberate a precise amount of hydrogen gas at one of the electrical terminals, implying that each hydrogen atom should be associated with a certain basic unit of electricity. Furthermore Maxwell's theoretical work, which was little known on the European mainland, had introduced the concept that there might be an "atom" of electricity, a minimum basic unit of electric charge which could not be further subdivided into smaller units.

It was Philipp Lenard, the German physicist at Heidelberg, who first got "cathode rays" to emerge from the discharge tubes in an attempt to study their properties. Hertz had spotted that the rays would pass through very thin films of material inside the tubes, just as his "radio"-waves would pass through matter. Lenard made windows of very thin metal in the glass of the discharge tube and got the rays to pass through to the outside world, and he showed that the ability of the rays to penetrate matter depended, quantitatively, on the mass of the matter present. "The idea of a type of radiation capable of penetrating matter opaque to ordinary light was quite new to science while the remarkable law of the absorption of the radiation by matter at once attracted attention," wrote Rutherford in 1910 in his own summary of the history of this development. We have seen how keen he was to emphasise this matter-penetrating ability of his own electromagnetic waves in 1895 and 1896 in his papers on the radio work.

Wilhelm Conrad Röntgen, another German physicist, working at Wurzburg, was following up Lenard's work and had covered his discharge tube with thick black paper to cut out unwanted effects and was working in a darkened room when, on November 8th, 1895, he noticed that a screen of fluorescent material (barium platinocyanide) glowed brightly every time the discharge was passed through the "blacked-out" tube. Even when he moved the screen several feet away from the tubes the effect continued. He immediately realised that some sort of ray or radiation was passing right through the black papers, and we now know that the X-rays were caused by the cathode rays striking the metal of the "window". In a short spell of intense work Röntgen established the existence of his "X-rays", that they passed through wood, metal and human flesh and that they acted on a photographic plate. He also showed that they could discharge electrified bodies by their effect on air, that is they made air into a conductor of electricity so that an electrically charged object would discharge its electricity through the normally insulating air around it. But it was the photographic aspect that naturally most impressed Röntgen, and he announced his discovery with a photograph of the bones of his hand.

(Incidentally a physicist at Oxford, one Frederick Smith, had previously observed that photographic plates in a box near his Crookes' tube tended to become fogged when he used the tube, and ordered his laboratory assistant to keep the plates somewhere else, because the fogging of his plates was certainly not the object of his research and he failed to note the interesting connection.)

Röntgen revealed his discovery to the Wurzburg Academy early in December 1895, but it was not until January 4th, 1896 that the public announcement was first made and the early X-ray photographs first shown to a large scientific audience, when he spoke before the German Physical Society in Berlin. Copies of his paper reached England in the same month and a translation appeared in *Nature* on January 23rd. But Sir Henry Dale recalled that reports of the extraordinary rays had reached Cambridge before then – they had appeared in the popular newspapers which had picked them up from German newspaper articles. "The first reaction to the news in Cambridge, as I remember it, was one of amusement and incredulity, but it soon became known that J.J. had taken the report seriously," he wrote in his own unfinished and unpublished book on Rutherford's life. Since Rutherford's first mention of X-ray photographs came in his letter of January 26th, J.J. must have moved very quickly indeed to get his copy of Röntgen's apparatus made.

Today we still think of X-rays almost entirely in terms of their ability to penetrate matter and to provide us with pictures of the inside of the human body. Indeed the Cavendish X-ray tubes were also used for medical emergencies, but the main interest in the laboratory was in the power of X-rays to make a gas at normal pressure into a conductor of electricity. What was a major development in scientific knowledge, the revelation of something entirely new in nature, was immediately turned into a tool by the Cavendish. J.J. was soon able to confirm Röntgen's statement that X-rays turned a gas into a conductor, and furthermore he obtained strong evidence that both positive and negative "ions" were formed to enable this to happen, and this in turn strengthened the belief that cathode rays were streams of negatively charged particles.

It was to pursue this line of research that "J. J. Thomson was joined by Rutherford who had just completed his experiments upon the magnetic detector of electrical waves," to use the laconic phrase employed by Rutherford himself. They planned "a systematic attack" (a favourite Rutherford description) on the nature of the process in the gas that made it act as a conductor under the influence of the rays.

Rutherford was well aware that his future prospects lay very largely in Thomson's hands – he had written to Mary Newton several months before (November 28th, 1895), ". . . if one gets a man like J.J. to back one up one is pretty safe to get any position for which he will put himself out to aid you".

Of all those who have written about Rutherford only Henry Dale knew him personally at this time of his life and Dale wrote of his giving up the radio work "without any thought of a direct profit for himself or any immediate concern indeed for the bearing of his researches on material benefit to mankind". Dale remembered that Rutherford was perfectly willing to leave the exploitation of scientific discoveries to others who had natural aptitudes for such work – "his own were for free exploration into the unknown". And Dale adds, "On the other hand I never observed in him any sign of regret or resentment such as some fundamental investigators have been prone to exhibit at the thought that others might have made profits from the use of any of his own discoveries." Dale and Rutherford apparently discussed such matters when they were both senior officials of the Royal Society in the 1920s and 1930s and even then Rutherford speculated quite happily on what he might have become if he had stuck to the radio work and its development.

Another expression of the same thoughts comes from J. G. Crowther, who knew Rutherford in his later years:

> Instead of fostering this first child of his mind [the radio detector], with the possibility of wealth and an early marriage, he changed to a very different and purely academic field, not following his own ideas but assisting in the development of those of another. The incident reveals Rutherford's fundamental humility and social discipline. Few men of proved original power choose to subordinate themselves to the aims and ideas of others when they have already found a fine line of their own. It reveals also his profound admiration for J.J. and the ideal of pure, fundamental research.

We may hold that both Dale's and Crowther's interpretations present a somewhat ideal version of the true Rutherford; both interpretations hold something of the truth, and there was also something of a young man seizing his best opportunity. There must have been a mixture of motives, as is true for most human decisions. To Rutherford himself it did not seem to be a great decision at all – it is only with hindsight that we see it as a watershed in his life. He simply reported to New Zealand on April 24th, 1896, among many other things:

My scientific work at present is progressing slowly. I am working with the Professor this term on Röntgen rays. I am a little full up of my old subject and am glad of a change. I expect it will be a good thing for me to work with the Professor for a time. I have done one research to show I can work by myself. The Professor has an assistant, Everett, a very smart manipulator and supposed to be the best glass blower in England. He fixes everything up and helps generally, but of course doesn't understand the theory very much. I generally turn up at 10 and the Professor comes in from 12 to 1.30. I then start at 2.30 and continue till 6.30, the Professor coming in at 4 or 5. It is rather interesting work so I don't mind the long hours.

These working methods, however, produced five major papers by Rutherford in the next two years, five papers which established him on the international scene as a major scientist. The first, and only the first, appeared jointly under the names of J. J. Thomson and Rutherford; the remaining four bore his name only. This first paper was probably the most important since it, almost by itself, established the basis of a new branch of science. The last of the five launched Rutherford into a world of radioactivity and nuclear science which he was to make his own special preserve.

To understand the crucially important first of these five papers, for which it is plain that Rutherford did most of the experimenting and measuring while J.J. supplied the theory, the discussion and perhaps most of the mathematics, it is essential to go back in time and in scientific theory just a little.

In the 1880s the Swede Arrhenius (Svante August) had proposed the theory of ions to describe electrolysis (typically the breaking up of water into oxygen and hydrogen under an electric current, but more generally the passage of electricity through a liquid). Ions are atoms which have become changed by the removal or addition of electrons; under the influence of an electric current the ions proceed to the anode or cathode according to the nature of their charge, thus forming a flowing current of electricity. It was a useful concept; time has shown that it is an accurate one, too, and "ionisation" is a common concept in modern science. At the end of the last century the Cavendish scientists, with their leaning towards "corpuscular" theories of electricity, applied the concept of ions to the conduction of electricity through gases, visualising the molecules of the gas acquiring positive or negative charges under the influence of the electric potential in the discharge tubes and flowing towards anode or cathode. In particular they thought the particles of the cathode rays should be negatively charged ions flowing towards the positive anode. They admitted, however, that if a positive and a negative ion came into contact, they would neutralise each other's charge and become normal neutral molecules again in a process of "recombination".

The Cavendish scientists were, however, having trouble with ions in electrified gases before X-rays and Rutherford came on to the scene. So large were the electric forces they had to apply in their discharge tubes to "ionise" the gas that the simple theory seemed to produce anomalous results. In Rutherford's own description, the problem was that all the negative ions did not turn up at the positive anode, to which in theory they should have been attracted, and correspondingly the positive ions did not all arrive at the negative cathode. But some progress had been made – in Germany Hertz had been unable to bend the beam of cathode rays by electric or magnetic forces, which made him believe that the rays were waves of some type; however in 1895 Perrin, in France, demonstrated that if he collected the cathode rays in a metal cylinder they deposited a negative charge, implying that they were negative particles. In Cambridge they had established that when gas conducts electricity it does not behave like a metal conductor (it does not obey Ohm's Law) – in fact the amount of current carried by the gas never increases beyond a certain amount however great the electric potential applied to the gas, and this maximum possible amount of current they called the saturation current. As soon as his X-ray tubes were made, J.J. showed that the rays did indeed make gas into a conductor of electricity at quite normal pressures and he got strong evidence that this was because the X-rays acted upon the gas to form positive and negative ions which could then carry a current.

When he "was joined by Rutherford" there existed also "several suggestive observations that threw a good deal of light" on the process – a very typical Rutherford phrase – and the most important of these was that the gas retained its conductivity for quite a measurable time after the X-rays had been cut off. This was just the sort of oddity that Rutherford was so good at picking out and exploiting both theoretically and experimentally. He arranged his apparatus so that the gas that had been exposed to X-rays was blown along a simple metal tube. He immediately showed that the gas retained whatever property the X-rays had given it for several feet along the tube. But if the gas was subjected to a strong electric field in the tube it lost this property, and likewise it lost it if it was forced through a plug of cotton wool or if it was bubbled through water.

So "it was thus clear that the rays produced some kind of structure in the gas which could be removed or destroyed. The view that the rays ionised the gas . . . was soon seen to offer a satisfactory explanation of the experimental facts". Furthermore, this theory was also "seen to give" a simple and rational explanation of the saturation current problem – namely that the very strong electric fields swept the ions so rapidly towards the electrodes that only a limited number of ions were available at any time to carry the current and therefore there was a maximum value of the possible current. Rutherford rather neatly turned this problem into an advantage; with J.J. he showed that passing an electric current through a gas "structured" by X-rays destroyed the structure. This of course implied that the structure was electrical and that X-rays had "ionised" the gas.

Their first paper on the subject, published in November 1896, seems strangely tentative by modern standards: "We tried whether the conductivity of the gas would be destroyed by heating . . ." (it was not); "A very suggestive result is the effect of passing a current of electricity through the gas . . . the analogy between a dilute solution of an electrolyte and gas exposed to the Röntgen rays holds through a wide range of phenomena . . ." (the gas is ionised). On the other hand, it examines the behaviour of seven different gases, all of which are measured carefully and the existence of the saturation current phenomenon is shown in them all with its variation from gas to gas. There is considerable mathematical work on such subjects as the rate of disappearance, or recombination, of the assumed charged particles after the rays have been switched off, and there is evidence of a great deal of hard work, though with inconclusive results, in such statements as, "We have made a large number of experiments with a view to seeing whether there is any polarisation when a current of electricity passes through the gas."

One of the favourite stories told by scientists about Rutherford concerned his indignation when someone suggested that ions might not exist. He insisted loudly that they did exist; they were "jolly little beggars, so real that I can almost see them". It is an informative little story. It tells of how intensely the man flung his whole person into his research. It also shows the way his mind worked, the basis of his many successes; it shows how he thought so continuously and so "physically" about the subject in which he was working that in a sense he lived in a world in which he was himself a part of the processes he was observing and measuring. It was this sinking of himself into the world of atomic particles that provided him with the "intuition" for which he was later so famous. We shall see this perception of the physical reality of the minute world he was to investigate again and again throughout his life and work. His discoveries were indeed "the invention of a fragment of a possible world", in Medawar's words, and his experiments were measurements to prove that this imagined world really existed and was self-consistent and consistent with observation.

It had indeed been a summer of hard work. Rutherford himself had obviously prepared for it mentally in his usual way by giving the problem some very deep thought before he even began the experimental work. This is proved by the first note in his books on the subject, which is far from tentative: "If two plates be taken, one connected to earth and the other charged to a definite potential, under the action of X-rays, the potential rapidly falls, the rate of leak varying with the distance between the plates. The peculiarity is that the rate of leak increases with the distance between the plates although the electromotive intensity is much less . . ." Once again we have the "spotting" of an oddity which seems to have sparked his mind; in this case the observation led to nothing in particular and was irrelevant to the main course of the research, though it was the type of work he would return to in his next papers.

It seems likely that the work reported in this paper, massive though it was in terms of measurements made and readings reported, had been completed by early September, for at the meeting of the British Association, which was held in Liverpool in September, Thomson gave his presidential address to the physics section on the 17th and dealt with his work with Rutherford on X-rays. Five further papers on X-rays were given the following day and two more on Saturday, September 19th, showing clearly that Röntgen's discovery dominated the thoughts and work of physicists in 1896. (Rutherford's paper on his wireless detector was given the following week.) In his presidential address Thomson reported:

> Mr Rutherford and I have lately found that conductivity is destroyed if a current of electricity is sent through the Röntgenised gas . . . When a current is passing through a gas exposed to the rays the current destroys and the rays produce the structure which gives conductivity to the gas; when things have reached a steady state the rate of destruction by the current must equal the rate of production by the rays.

This statement is quite as definite as anything on the same subject in the printed paper and the implication must be that the great bulk of the work, published in November, had been completed by the beginning of September.

It was this paper that established Rutherford's reputation in the world of science. He says himself, ten years after the work was done, "The simple mathematical theory of the conduction of electricity through gases has formed the basis of all subsequent work on this subject and has been found to account in a satisfactory manner not only for gases made conducting by X-rays but also by a variety of other ionising agencies." J.J. recorded, "Rutherford devised very ingenious methods for measuring various fundamental quantities connected with this subject, and obtained very valuable results which helped to make the subject 'metrical' whereas before it had only been descriptive." Here J.J., who was not, unfortunately, famous for giving others credit where it was their due, very accurately points to Rutherford's real ability, the ability to make measurements of reality where others merely speculated; and it is generally recognised by most philosophies of science that the move from the merely descriptive observation to the measurement or "metrical" stage of investigation marks the beginning of a true science. Eve was probably justified when he said, ". . . Rutherford was fully conscious that Prof. J. J. Thomson and he, working together, had made a discovery of great importance. The fact is that between Easter and November 1896 they had laid the sure foundations of a new subject."

J.J. went on from here to the greatest discovery of his life, the discovery of the electron. Several of the most distinguished of his colleagues in the Cavendish, notably Rutherford's friend and contemporary, Townsend, and McClelland and Zeleny moved into the study of ionised gases, and Rutherford continued the direct line of work.

Rutherford's unique contribution to the founding of the new subject was undoubtedly his "very ingenious methods for measuring"; it was this development that turned the subject into a "metrical one". Basically this was the use of the electrometer – in these first experiments the published paper says it was a quadrant electrometer. This was an instrument which was being steadily developed all through the first half of Rutherford's scientific life, and can nowadays be seen in many forms in scientific museums. (The electrometer was still part of school physics in my student days though it seemed to play no part in the education of my children.) But all these instruments, and also the electroscope which Rutherford sometimes used, were the same in principle – two surfaces or arms, physically connected but electrically insulated, were charged up with electricity of the same sign (positive or negative) so that they repelled each other; normally the air would act as an insulator so that the charge would leak away only slowly, but as it leaked the repulsion between the two surfaces would reduce and they would return to some original position close to each other. The various types of electrometer arranged this system in various different ways but the rate of leakage of the charge could be measured by the clock and some movement of the charged surfaces. When Rutherford introduced ionised air into the electrometer or connected one side of the electrometer to part of his apparatus which was leaking electric charge through some container of ionised air or gas, and compared the rate of discharge of his electrometer with the natural rate of leak, he could get a very good estimate of the amount of current being carried by the gas at any time. This may not seem to us, nearly a hundred years later, in the age of electronics, a very sophisticated or accurate system of measurement, but it certainly provided experimental results accurate to within three or five per cent and that was almost always good enough for Rutherford. It has to be emphasised that throughout his career he was always, by choice or otherwise, pioneering in entirely new fields. This order of accuracy was enough for him to describe, almost always without error, the outstanding features of whatever territory he entered – pioneer farmers do not need exceedingly accurate maps.

Of course he became more and more skilled at using these electrometers as time went on, and newer and better models of the machine were being invented (the Dolazalek electrometer of the early years of the twentieth century was more accurate, more reliable and easier to work, for instance). But essentially this method of measuring the ionisation of a gas and the current that flowed through it was the chief instrument with which Rutherford unravelled the mystery of radioactivity and established the main lines of the families of radioactive elements and their relationships with each other. It was not until ten years later that he would develop the second great scientific tool of his career, the particle-probe which is now one of the chief instruments of science.

This 1896 paper, "On the Passage of Electricity Through Gases Exposed to Röntgen Rays", also reveals another technique which was to become one of Rutherford's favourites, though modern scientists might regard it with suspicion. Among the dozen tables giving his reading of the leakage through various gases, in terms of the electromotive force applied and the leakage current observed, there is a scattering of asterisks, apparently at random, but with the note "the observations marked with the asterisks were used to calculate the constants in equation (4)". These are examples of what Rutherford called "good" results. He had this feeling for good results, results that he relied on, probably because they were taken when his apparatus was working well, perhaps because he could get the same reading several times. His notebooks are scattered with remarks to this effect in bold writing – "Good results" or just "Good": and sometimes, of course, they carry other messages, "poor" or "inconsistent" or even "useless as the tube kept breaking" or similar expressions of exasperation.

Exasperation may well have played an even larger part in research in the nineteenth century than it does today. Towards the end of his life, J.J. wrote, "Another instrument which was exasperating to work with was the old quadrant electrometer. This not infrequently refused to hold its charge and neither prayers nor imprecations would induce it to do so." And Lord Rayleigh also recorded the problems of these machines: "J.J. himself used what was called the Elliott pattern, after the name of its inventor. I do not know who designed it, but . . . I suspect that it was primarily the Devil." The X-ray tubes could be even more temperamental, made as they were in the Cavendish workshops by J.J.'s personal assistant, Everett.

> Running these primitive tubes was not quite the simple matter it might appear to be. To carry the discharge, the gas in the tube must be at a low pressure, but not too low, for if there are too few molecules of gas present there are too few carriers and the discharge ceases to pass. The pressure may be initially just right, but under the influence of the discharge the gas gradually becomes absorbed by the walls of the tube, until finally there is not enough of it and the tube ceases to work. Heating the walls of the tube releases some of the gas and so puts things in order again, but all this means that constant expert attention was required for a satisfactory production of rays.

The man who wrote these words, E. N. da C. Andrade, worked with Rutherford in Manchester before the First World War, and obviously wrote it from personal experience. Even in the flat prose of science a little of the feeling about early X-ray tubes creeps into the Thomson and Rutherford paper: "As these measurements require the intensity of the radiation to be maintained constant during each series of observations, a condition which it is very difficult to fulfil, we think the agreement between theory and observation is as close as could be expected." These problems with apparatus also help to explain Rutherford's personal decisions as to which set of readings to accept as the best; he was perfectly capable of taking the statistical mean of sets of readings, and in later life often did so, but in the early stages only the experimenter himself working at the bench could have the feel for when everything was going right.

Incidentally, the blowing of the air along the tube in these experiments was achieved by using household bellows, which were then a feature of most English homes where they were used to keep the coal fires burning.

Throughout the winter of 1896–97 Rutherford continued to work in the development of this subject that he had opened up. Now he was working alone, and the papers from this period are in his own name only. J.J., however, continued at first to take an interest and added a note to the first of these independent papers pointing out that the results implied that X-rays must be a form of electromagnetic radiation similar to light, though this problem of the nature of X-rays was to remain a controversial one for many years. By 1897 however, J.J. was hot in pursuit of the electron and it must be assumed that he took less interest in other people's work as he reached the peak of his own career. There is good reason to believe that Rutherford, in one of those spells of intense effort which were his greatest strength, did all the work which formed the basis of three papers during this winter and the spring of 1897. The grounds for this belief, which is contrary to the statements of the standard biography of Eve, are that in June 1897 when Rutherford applied to the Commissioners of the 1851 Exhibition Scholarship for an extension of his award to a third year, he mentioned in his formal report of the work he had done that he had begun to examine the radiations from uranium. This was the subject of his final research in Cambridge and the work was published in a paper very much longer than the three which preceded it. Furthermore Becquerel's discovery of uranium radiation was made and announced in 1896 and it would have been very un-Rutherfordian to allow a year to elapse before he followed up something which seemed interesting. We can therefore assume that all Rutherford's remaining work on the ionisation of gases was for all practical purposes completed by the end of 1896.

The three papers which he published reporting this work appeared in the *Philosophical Magazine* in April and November 1897 and in the *Proceedings of the Cambridge Philosophical Society* of 1898. Starting with apparatus very similar to that he had used with J.J., Rutherford showed in the first paper that those gases which were better conductors of electricity when ionised were also the ones that absorbed X-rays more efficiently; a result which clearly supported the ionisation theory. Indeed, one of the first sentences of the paper says that the experiments are "to investigate more fully the way in which electrified gases can be obtained by means of the Röntgen rays and also to examine the properties of the charged gas". However, the second half of the paper is more occupied with the velocity of the ions after they have been created and Rutherford demonstrated that "the velocity of the conducting particles for air is about one cm per second for a potential gradient of one volt. This is of the same order as the rough determination made in the previous paper by Professor Thomson and myself." And the next paper proceeds from this. It is titled "The Velocity and Rate of Recombination of the Ions of Gases exposed to Röntgen Radiation". Note that ions are now so well accepted that they can appear in the official title. A major variation of the experimental method is introduced half way through this paper when the electrified gases are no longer blown along a tube but fed between two flat plates. This led him into an interesting situation:

> The velocity of a molecule of hydrogen through hydrogen and carrying an atomic charge is thus 340 centimetres per second, while the experimentally determined value is only 10.4 cm per sec. The disagreement of theory and experiment seems to point to the conclusion that either the charge is less than the charge carried by an ion in ordinary electrolytes or that the carrier is larger than the molecule. We have not sufficient experimental evidence to decide between the two suppositions . . .

He gets out of the dilemma by deciding (and producing evidence for the supposition) that charged ions collected aggregates of neutral molecules about themselves and thus the ion was slowed down. Modern theories of drop formation and the functioning of the Wilson cloud chamber and modern bubble chambers rely on this explanation that Rutherford provided for a difficult situation. But the essential thing is that he placed complete reliance on the experimental evidence as he was to do for the rest of his life. If theory and experiment disagreed then the actual measurements, the known facts, must hold and the theory must be adapted.

This paper contains the final proof that Rutherford had started his work on radioactivity by the summer of 1897, for though its date of publication is November 1897, the date given at the end is "Cavendish Laboratory, July 19th, 1897". Yet the last paragraph reads:

> Further experiments which are not yet completed have been made to find the velocity of ions of a gas conducting under the influence of the radiation given out by uranium and its salts. It can be shown that the velocity of the ions in the conducting gas is the same as when the gas is acted upon by Röntgen radiation so that the carrier in the two cases is identical. Further results however must be reserved for a future paper.

The third, and least important, of the ionisation papers written by Rutherford independently of J.J. does little more than show that a gas ionised by ultraviolet light falling on it behaves in much the same way. The knowledge that ultraviolet light has some effect of this sort was not new, and Rutherford deals rather cavalierly with the several German investigators who had worked on it: "Most of these papers have dealt with the general character of the discharge, but the subject of the nature of the conduction and of the carrier that discharges the electrification has not been specially attacked." He goes on to suggest that some results obtained by the fierce and autocratic Lenard at Heidelberg "are capable of other interpretations". It is plain that this work was conducted in parallel with the early work on uranium which he had started in the summer of 1897. The results he obtained are not very exciting and are not stated very firmly. The fact is that by 1898 Rutherford was no longer chiefly interested in ionisation and electrification of gases. He had looked at uranium originally as no more than a means of ionisation but his interest was captured by uranium itself, or at least the radiations that came from it.

In the last years of his life, when he was old and famous, Rutherford was asked by a German correspondent what had been the most important moment in his life, and he chose this moment – the time when he took up the study of uranium and thus of radioactivity. It is worth adding that the study of the ionisation of gases is still not complete – it has turned out to be a most complex and difficult subject. Townsend, Rutherford's contemporary, who became Professor of Physics at Oxford at a very young age, became bogged down in it. But it is generally agreed that one of the hallmarks of distinguished science and great scientists is to ask of nature questions that can be answered.

While he was making his scientific reputation and finding the source of his life's work, Rutherford was also living his life. In November 1896, just as his first great paper with J.J. was being published, the young Rutherford confided his great secret to Mrs J.J.

> . . . Yesterday after my paper to the Physical Society which passed off all right before quite a distinguished audience, I asked Mrs J.J. to allow me to see her home, as it was quite dark and J.J. was busy over some experiments. On the way home I broached the subject of our engagement, and she was very pleased to be told about it, and said some very nice things, and asked all sorts of questions about you, and says she is very anxious to see you when you come. I know she will do anything she can for you, so we will trust in the future that you may visit her when you arrive. She remarked that when I was giving my paper, not being

learned enough to follow it completely, she had been wondering what was to be my fate in that respect. Truly the ways of women are not for men to understand. I think Mrs J.J. looks on me as a very nice boy and after I have taken her into my confidence will take an affectionate interest in my welfare.

Mrs J.J. was present at the meeting of the Cavendish Physical Society, which had been founded by her husband several years previously, because it was her job to dispense the tea at these afternoon meetings. Rutherford's attitude to "the ways of women" may seem marginally offensive nowadays, but it was the standard attitude of the British upper middle class of the time, and can be seen reflected in any of the popular writers of the era from Conan Doyle to John Buchan. It was to remain Rutherford's style for the remainder of his life, but it was almost certainly no more than a style or set of manners, for his attitude to women in his laboratories was egalitarian and well in advance of his time.

The reflection on his manners is quite important because it was during this period that he was being educated in the polite uses of the society in which he was to live. Although he remained brash and noisy all his life, he adapted quite remarkably well to English university life; social inadequacy or colonial rough edges never presented him with problems. Indeed in the same letter home he records a visit that he and two other research students paid to Mrs J.J. – a formal visit to welcome her back from America. "As there were a large number of ladies present we stayed over an hour assisting to entertain, for which Mrs J.J. was very grateful. I am now getting quite an expert at afternoon calls and don't feel at all shy usually."

And still in the same letter we hear of social progress in his own professional world: "On Tuesday night I gave an inaugural paper to a Physical Science Club just formed among us researchers. It was held in my rooms and I supplied coffee, biscuits, baccy and cigarettes – the usual thing in these cases." The slightly defensive note here is part of the curious relationship Rutherford had with his fiancée who came from a strongly teetotal, puritan background. Only a few months earlier, in August 1896, the young man had written:

. . . A good long time ago I gave you a promise I would not smoke, and I have kept it like a Briton but I am now seriously considering whether I ought not, for my own sake, to take to tobacco in a mild degree. You know what a restless individual I am, and I believe I am getting worse. When I come home from researching I can't keep quiet for a minute and generally get in a rather nervous state from pure fidgetting. If I took to smoking occasionally, it would keep me anchored a bit and generally make me keep quieter. I don't think you need be the least bit alarmed with regard to yourself. For I don't think I will ever become a confirmed smoker, but seriously I believe it would be a very good thing for me in many ways. Every scientific man ought to smoke as he has to have the patience of a dozen Jobs in research work.

Eve printed this letter in his life of Rutherford and Lady Rutherford, as she was then, must have insisted on the inclusion of the footnote which reads: "Lady Rutherford explains the apparently arbitrary exaction of a promise not to smoke as follows: 'Ernest suffered from a persistent irritation of the throat causing a slight cough which worried me in the light of the fact that his eldest sister was delicate and one always had the fear of tuberculosis.'"

He did of course become a confirmed smoker, and his pipe and the dry tobacco with which he stuffed it and his continual search for matches became part of the Rutherford legend among his fellow scientists. It is difficult to explain why it is so delightful to catch out one's "hero" and his partner in such egregious arguments – perhaps it is related to my own anti-social pipe. Another letter of this period, written to his mother in May 1897 after a visit to the Crystal Palace in London, reveals a certain conservatism.

The chief point of interest to me was the horseless carriage, two of which were practising on the ground in front. One was capable of seating two persons and the other five. The engines were placed in the rear part of the vehicle and did not occupy much space, and the oil which is used for the motive power was kept in a cylinder at the back. They travelled at about 12 miles an hour but made rather a noise and rattle. There was a collection of these horseless carriages in the Crystal Palace itself, some evidently very old and correspondingly clumsy and heavy. I was not very much impressed with the machines as vehicles, but I expect they will come into very general use shortly.

The visit to Crystal Palace may well have been in company with his fiancée, for Mary Newton visited England in the spring of 1897. The Thomsons put her up for May week in Cambridge and she was able to see her partner take the first B.A. research degree awarded by the university. The two later visited Killarney in Ireland with a party of friends before she returned to New Zealand for another long spell of separation.

The most important feature of 1897, however, was Rutherford starting work on uranium. This importance is best expressed by Alfred Romer.

Modern nuclear physics had its beginning in radioactivity. Yet radioactivity made an interesting science in its own right . . . above all it was a science of surprises, and to those who followed it an eye for the unexpected was worth quite as much as any foresight in planning. It began in 1896 because Henri Becquerel hoped to produce X-rays by purely optical means, and it reached a high point in 1904, when Ernest Rutherford established the possibility of the spontaneous transmutation of elements. If its end was not implied in its beginning, if half its investigations veered off unexpectedly, progress in this field was nevertheless rational. However odd the phenomena and however contradictory the results there were always intelligence and imagination to bring them into order.

We know, because so many scientists have celebrated it, that the possession of an eye for the unexpected was precisely Rutherford's greatest gift, allied with that combination of imagination and intelligence that brought order into situations of apparent contradictoriness; what we can never tell is whether Rutherford developed these characteristics because he stepped into this particular branch of science, or whether this branch of science developed in the way it did because Rutherford stepped into it. The latter is quite as likely an explanation as the former.

What is certain is that this science also sprang from Röntgen's discovery of X-rays and the excitement the discovery caused. In this particular case the excitement was aroused in Paris. As early as January 20th, 1896, the Académie des Sciences was shown a picture of the bones of a human hand by two doctors and to the same meeting Henri Poincaré, a mathematician and physicist, brought his own copy of Röntgen's first paper on the subject. Present at the meeting was Henri Becquerel, one of France's most distinguished scientists; he was Professor of Physics at the Museum of Natural History in Paris and a member of the Académie, and in both positions he followed his father and grandfather. Amongst other things this very assured position meant that anything that Becquerel cared to present to the Académie would be in print in the world's official scientific literature within ten days. But at that meeting of the Académie at which X-rays were reported, Becquerel seems to have drawn an association between the production of the rays and the fact that the glass of the discharge tube fluoresced. He set out to find if he could produce X-rays by purely optical means using the then mysterious techniques of fluorescence and phosphorescence.

The story of his discovery of the radiation from uranium is well known; he was trying to see whether various salts known to possess fluorescent properties would affect photographic plates if they were briefly exposed to sunlight and then kept in the dark in lightproof packets that contained both the exposed salt and the photographic plate. Due to the lack of sunlight in Paris in February he postponed an experiment with a double salt of uranium and potassium, but on returning to his packets he found that the uranium salt had affected the plate although it had been kept in darkness in its packet in a drawer. In a series of four short papers published by the Académie between March and May 1896, Becquerel proceeded from "On the radiation emitted in Phosphorescence" to "On the invisible radiations emitted by Phosphorescent Substances" and ended with "Emission of new radiations by metallic Uranium". He showed among other things that the new radiations had the effect of causing electrified bodies to discharge, which Rutherford must have immediately spotted as meaning that the radiations ionised air. Becquerel however was not clear about the nature of his radiations and his final words were that "the emission produced by the uranium . . . is the first example of a metal exhibiting a phenomenon of the type of an invisible phosphorescence".

The fact that Rutherford had entered this field by the middle of 1897 is again emphasised by his citations of Becquerel's papers – he quotes seven of them, five dated 1896 and two in 1897. The result of his work was not however published until 1899, by which time he had moved to Canada. But it is a very long paper, taking up forty-six pages of his *Collected Works*, and the change of the weight of his interest during the course of it is revealed on the first page; we know that he came to the study because of his interest in the ionisation of gases, but he declares that the object of the paper "is to investigate in more detail the nature of uranium radiation and the electrical conduction produced".

From our point of view in the late twentieth century the most significant feature of the paper is the discovery and naming of alpha- and beta-rays as two distinct types of radiation, the alpha-radiation very easily absorbed and the beta-radiation much more penetrating. The paper however begins with a clear statement of the advantages of Rutherford's electrical method of measurement: "The photographic method is very slow and tedious and admits of only the roughest measurements . . . On the other hand the method of testing the electrical discharge caused by the radiation is much more rapid than the photographic method and also admits of fairly accurate quantitative determination." But it is not made clear whether he realised himself that he had thus turned the ionisation of gases into a method of measuring the properties of uranium radiation rather than being itself the chief object of study.

The fourth section explains why he was led to his remarkable conclusions about the nature of uranium radiation: "Röntgen and others have observed that X-rays are in general of a complex nature, including rays of wide differences in their power of penetrating solid bodies." But while we now know that this difference depends on the differing wavelengths of essentially similar X-rays, Rutherford showed that the radiation from uranium, though complex, consists of two entirely different types – and he achieved this with a convincing simplicity. He simply covered his uranium with thin foils of aluminium, gradually increasing the number of foils. For the first three layers of foil the radiation escaping from the uranium decreased progressively in such a way as to suggest an ordinary law of absorption – i.e. that the thicker the layer of aluminium the less radiation penetrated to ionise the air. More thicknesses of aluminium, however, had little further effect in reducing the radiation at first, but eventually the intensity of radiation began to diminish again as even more foils were added. "These experiments show that the uranium radiation is complex and that there are present at least two distinct types of radiation – one that is very readily absorbed, which will be termed for convenience the alpha-radiation, and the other of more penetrative character which will be termed the beta-radiation."

We now know after eighty more years of investigation that these two different forms of radiation are caused by two different forces which are among the most fundamental features of the physical world. It is an extraordinary tribute to Rutherford's imagination, and his sheer power of measuring things previously unmeasured, that he was able to distinguish between them in his very first investigation of the subject.

This paper contains yet another discovery which is never credited to Rutherford in official scientific histories. It is quite plain that he also observed gamma-radiation for the first time. He establishes that "All compounds of uranium examined gave out the two types of radiation and the penetrating power of the radiation for both the alpha- and beta-radiations is the same for all the compounds." But then comes a sudden excited gulp of breath: "While the experiments on the complex nature of uranium radiation were in progress the discovery that thorium and its salts also emitted a radiation (similar to uranium) was announced. [Thorium radiation was discovered by G. C. Schmidt in May 1898.] A few experiments were made to compare the types of radiation," he writes.

These experiments show that thorium radiation differs in

penetrative power from uranium, but more important, "With a thick layer of thorium nitrate it was found that the radiation was not homogeneous but rays of a more penetrative kind were present." These are clearly gamma-radiation, but "On account of the inconstancy of thorium nitrate as a source of radiation, no accurate experiments have been made on this point."

Many scientific reference books credit the Frenchman, Paul Villard, with the discovery of gamma-rays in 1900. But the evidence in favour of Rutherford's right to the credit is strong. If we take the published statement referred to in the previous paragraph and then look at Rutherford's first major book, *Radioactivity*, written in 1904 we find, "On the discovery of very penetrating rays from uranium and thorium, as well as in radium, the term 'gamma' was applied to them by the writer. The word 'ray' has been retained in this work although it is now settled that the alpha- and beta-rays consist of particles projected with great velocity." This is in a footnote to page 91, and on the main body of the page, where he first mentions the gamma-rays, he gives no attribution to Villard and giving proper attributions to other people's work – "citations" is the modern word – was a matter in which Rutherford was peculiarly careful. Later in the same book, on page 141, we find "Villard, using the photographic method, first drew attention to the fact that radium gave out these very penetrating rays . . . This result was confirmed by Becquerel." On the following page Rutherford claims clearly that he himself was the first to discover that uranium and thorium gave out gamma-radiation, and since he had started work on these two elements two years before Villard reported his findings on radium it is clear that priority should go to Rutherford. There is no trace in the records of anyone disputing his statements or claiming that he had not given Villard credit for priority, yet in his Nobel lecture of 1908 he says, "When a very penetrating type of radiation from radium was discovered by Villard, the term gamma-rays was applied to them."

The strongest evidence of all comes, however, from two of Rutherford's collaborators. A. S. Eve, his life-long friend and his official biographer, was set to work on gamma-rays by Rutherford at McGill as early as 1904, and in writing Rutherford's obituary for the Royal Society (with Sir James Chadwick), after describing the discovery of the alpha- and beta-radiation Eve declares, "He noted in the case of thorium a yet more penetrating type, but the name gamma-rays was given later by Villard." Finally Edward N. Da C. Andrade, whose work with Rutherford in the Manchester Laboratory in 1914 finally settled the nature of gamma-rays as similar to X-rays, although he gets the date wrong, says firmly, "They were

127

first so named by Rutherford . . . after being called simply 'very penetrating rays'."[1]

In his official scientific publications Rutherford in a short letter to *Nature* in 1902 credits Villard again only with first observing the rays from radium by the photographic method and goes on to describe some recent work of his own with thorium and uranium, but he only uses the term "gamma-rays" for the first time in a note attached to Eve's published results in 1904. It seems that Rutherford was sure enough of his own position simply not to bother with priority claims, even when his friends wished him to assert his rights. Certainly in later life he made it a positive policy never to indulge in public controversy on scientific claims, and we will come across at least one other claim to a major discovery which he could have made yet which he deliberately chose to leave unpressed.

The very short section of this letter dealing with thorium also contains a sentence in which much future work and important discoveries were implicit: "It was found that thorium nitrate when first exposed to the air on a platinum plate was not a steady source of radiation, and for a time the rate of leak varied very capriciously." This capriciousness was to be another of those "anomalies" which Rutherford was to exploit so ruthlessly, but it is interesting to realise, again in contradiction to many accounts of his life, that the work on thorium and the discovery of this strange behaviour of thorium was made as early as 1898, before he had left Cambridge.

There is a very strangely placed letter, quoted by Eve and attributed to October 1896:

> I am working very hard in the Lab and have got on what seems to me a very promising line – very original needless to say. I have some very big ideas which I hope to try and these, if successful, would be the making of me. Don't be surprised if you see a cable some morning that yours truly has discovered half a dozen new elements, for such is the direction my work is taking. The possibility is considerable but the probability is rather remote.

It seems likely that this is a wrong dating by Eve, possibly even a slip of the pen by Rutherford himself, and yet the letter contains references to J.J.'s return, presumably from America, which are consistent with Eve's date. But Becquerel's basic discovery of radioactivity had only been announced six months before and there is no trace of Rutherford working on uranium so early. Could he really have imagined the entire theory of radioactive disintegration and the families of radioactive elements so early and simply

[1] *Rutherford and the Nature of the Atom*, p. 69.

NEW ZEALAND BOYHOOD

Right. Ernest Rutherford, aged twenty-one, the brilliant student of Canterbury College just starting his first research.

Below: Serious Scottish immigrants in the New Zealand woodlands. James and Martha Rutherford, Ernest's parents.

Above: The pioneering life in which Rutherford was brought up. This is the main street of Havelock in the 1880s.

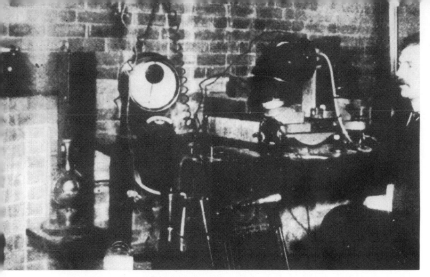

Left: The young Professor at work with his radioactivity apparatus in the MacDonald Physics Building at McGill University. Note the obvious white "cuffs".

Below: Reproduced from Rutherford's notebook – his first serious thoughts on the "structure of the atom", written in Manchester in 1911.

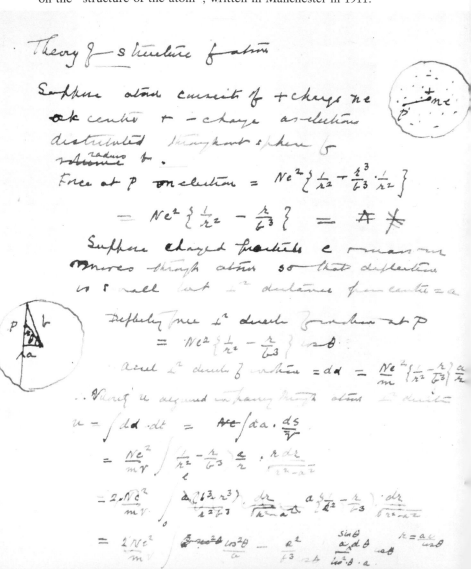

spent the next eight years proving what he had already "seen" in nature?

His notebooks give no indication of any of these ideas, although they do provide one illuminating sidelight on the workings of his mind. Among a great deal of rough working on the mathematics of ionisation – enough mathematics to disprove the commonly quoted sneer that Rutherford was no mathematician – and what seem to be lecture notes taken while attending some of J.J.'s lectures there are his reasons for rejecting right from the start Becquerel's idea that the radiations from uranium were connected with phosphorescence. These are really very rough notes of a mind turning over a problem. "If effect is due to phosphorescence effect should be determinate (?) but irrespective of length of time taken," and more conclusively, "Effect independent of phosphorescence – Nitrate/uranium in solution is still active though not phosphorescent." It is also worth noting that he had so little uranium available for his experiments that he was unable to carry out work on thick films of uranium salts which would have told him how much of the radiation was self-absorbed.

This work on uranium, culminating in his demonstration that uranium radiation was something entirely new and consisted of at least two different components, was completed before he left Cambridge – for although it was not published until 1899, by which time he was in Canada, the paper is clearly marked as having been completed on September 1st, 1898.

Considering that while Rutherford had been doing this basic work on radioactivity, J.J. had been discovering the electron, the younger man's summary of the years 1895 to 1898 was justified:

One of the most important and interesting [periods] in the history of research in the Cavendish Lab. It is remarkable not only for the number of new discoveries of the first importance but also for the inception of those newer ideas of the connection between electricity and matter which have so greatly influenced the trend of research . . . It was essentially a period of pioneer advance into a new and fertile territory when new ground was broken day by day and when discovery after discovery followed in quick succession.

5

Radioactivity

At the age of twenty-seven, in 1898, Ernest Rutherford became a Professor – the Second MacDonald Professor of Physics at the University of McGill in Montreal, Canada. His salary was to be £500 a year. But since his first-class fare across the Atlantic, sharing a cabin with another young Cambridge man who was on his way to become a Professor of Zoology in Canada, was only £12, this was a step to reasonable affluence.

This spectacular promotion, this three-thousand mile move to yet another, different society, caused no interruption in his research. He had found his field, the radiations from uranium and thorium, the subject which we now call radioactivity, and he rushed into it, scarcely pausing for a week or two over such matters as professorships. In 1898 and 1899 there was really only one thing that interested him – his research. He was involved in one of his most profound, almost manic bursts of work, and, though he did not know it at the time, he was about to revolutionise physics and chemistry, about to cause "one of the biggest revolutions in scientific thought", according to the chemist Lord Fleck.

One of the chief reasons for his wanting the post at McGill was that the research facilities in the university's new MacDonald Physics Building (always referred to locally as MPB) were possibly the best in the world. And one of the chief reasons for accepting the post when it was offered to him was that there was no very great emphasis on the teaching duties. Before he left the Cavendish Laboratory Rutherford used the good offices of the secretary to order fresh supplies of uranium and thorium from Germany which were to be sent to Montreal.

His first letter to New Zealand firmly put off his proposed wedding for eighteen months, mostly on financial grounds. His first lecture was given on October 3rd and the following day he wrote again to his fiancée, "I am in such a rush of work that you will have to be content with a short letter from me today. I am now in full swing of work and have not a moment to spare as I am getting classes in order, lecturing, etc." He was also immediately involved in what might be termed practical research, forensic research, which he was invited to undertake, for a fee, by Professor Bovey, Dean of McGill's Faculty of Law. This concerned a court case in which certain residents of Montreal were trying to sue the local tramway company for damage to their houses caused by vibrations from the company's main power house.

It is only by deduction from his laboratory notebooks and from a few references in his correspondence with J.J. that we know that Rutherford at the same time was conducting his own personal research in the MacDonald Physics Building and that this work was an intense and detailed study of the energy mechanisms involved in the processes of ionising gases by "Röntgen and Becquerel Rays". The results of this work did not appear in public until 1901, when he published three papers on the subject. It was all a direct continuation of his Cambridge work, as the opening sentences of the main paper – a massive thirty-five page contribution to *The Philosophical Transactions of the Royal Society* – make clear:

> The primary object of the investigations described in this paper was the determination of the amount of energy required to produce a gaseous ion when Röntgen rays pass through a gas, and to deduce from it the energy of the radiation emitted per second by uranium, thorium and other radioactive substances. In order to determine the ionic energy . . . it has been necessary to make a special investigation to measure accurately the heating effect of X-rays when the rays are absorbed in metals and also the absorption of the rays in gases.

That this was Rutherford's basic work in his first months at McGill is proved by the subsidiary dating at the head of the paper which says quite clearly that it was received on June 15th, 1900 – and by that time Rutherford was finally in New Zealand to marry Mary Newton. The paper is important for several reasons. It is Rutherford's first jointly-authored paper with himself as senior partner; his assistant was R. K. McClung, one of McGill's very small number of demonstrators in physics, who became, as did nearly all his research partners, a life-long friend and correspondent. But the method of attack on the problem shows Rutherford's astuteness, his first real demonstration of worldly wisdom, for it capitalised on the known strength of McGill physics which was in the accurate measurement of heat and heating effects. This had been the specialised field of H. L. Callendar, Rutherford's immediate predecessor in the post of Junior Professor, and the physics building contained a considerable amount of specialised heat-measuring equipment and an active school of researchers following Callendar's line of work.

Several of Rutherford's biographers have implied that it was only when he worked with H. T. Barnes on the heating effects of radium in 1904 that he tapped this source of expertise, but it is clear from a careful reading of this paper, and the true dating of the work involved, as well as from photographs of apparatus preserved at McGill that the methods and equipment and the expertise of MacDonald Physics Building personnel were immediately put to work in pursuit of radioactivity. The paper itself is not always an exemplar of the Rutherford genius. He could not solve the practical problem of finding out how much of the radiation from uranium was absorbed by the outer layers of the uranium material itself and never allowed to escape. He concentrated on the alpha-rays that were emitted and showed that, although they were so very easily absorbed, and could be cut off by covering his radioactive sample with nothing more than sheets of thin paper, nevertheless they had by far the greater power of ionising any gas, and therefore carried by far the largest share of the radiated energy. The much more penetrating beta-rays were very poor ionisers and he ignored them in his final calculations. He was obviously, however, in a state of considerable confusion about the beta-rays and their nature (see CPR, p. 292), and this in itself betrays the early date of this work.

Nevertheless the work done here, overshadowed as it must be by the simultaneous start of the great series of investigations which led to the discovery of the laws and nature of radioactive disintegration, was absolutely essential as the quantitative basis of so much that followed. Firstly the work provided the basic experimental justification for the methods of measurement that Rutherford was to use in his later work – it proved that the electrical effects caused by ionisation of gases under radiation could be used as a demonstration of real phenomena whose comparative sizes could be measured, and this was to be vitally important when Rutherford was insisting that materials existed which could be detected and identified only by the precise nature of the radioactivity they showed. Secondly he established, however roughly, the major quantitative parameters of the work he was going to do. He showed what order of magnitude the energies and velocities of the radioactive phenomena must have. Perhaps this was most important to his own mental picture of the world he was examining – it was the foundation of that grasp of the quantities involved that later appeared as almost miraculous "intuition" and prediction of the results that would in future be obtained.

The final paragraphs of this paper are truly remarkable, especially when we remember that it was written in early 1900. Here Rutherford faced up to the extraordinary, dogma-shattering implications of the earliest discoveries by all who were involved in radioactivity. He first points out that Becquerel's original findings about the constancy of uranium's radioactivity, whatever the scientists might do to it, implies that radioactivity depends on the "uranium molecule alone and not what it is combined with". He then asks whether the radiations come from the surface of the uranium only. At first sight it appears to be so, since the radioactivity does not increase with thicker layers of uranium, but it is quite likely that the reality of the situation is that the uranium, or its compound, is absorbing the radiation that comes from deep within the sample and that the energy is dissipated as heat within the material (this point was shortly proved to be correct). He goes on, "If the radioactive power is possessed by the whole volume it follows from the above supposition that the mineral pitchblende must have been radiating energy since its formation as a mineral. If we suppose the radiation has been going on at its present rate in the course of 10,000,000 years each gramme of uranium has radiated at least 300 calories." This figure was actually far too low, for Rutherford, as we have seen, ignored all but the alpha-radiation in his own calculations based on the ionisation of gases. Nevertheless, here was a bold, almost literally an earth-shaking, concept. But it was followed by an even more astounding statement: "It is difficult to suppose that such a quantity of energy can be derived from regrouping of the atoms or molecular recombinations on the ordinary chemical theory."

This is the first glimpse, the first implication, of what we all know now as "atomic energy".

He immediately qualifies this by pointing out that if the present trend of the Curies' work on radium is upheld, and it proves in fact to be a hundred thousand times more active than uranium, and if experiments prove that radium, too, emits radioactivity at a steady rate, then "the difficulty is still further increased. . . . In order to account for such a rapid emission of energy it would be necessary to suppose that the radioactive substance in some way acts as a transformer of energy. Such a supposition does not seem probable and leads us into many difficulties." So he hesitates to draw the conclusion that we, with hindsight, can see so easily: that the energy must come from inside the atom. Yet he discounts the only possible alternative, that radioactive atoms somehow collect energy from somewhere unknown and re-emit it. This explanation was, not unnaturally, the one favoured by Becquerel, and to some extent supported by the Curies, since it was analogous to the well-known phenomenon of phosphorescence. Marie Curie claims the credit for pointing out that radioactivity must be a property of the atom itself, but here Rutherford clearly shows that by the slightest consideration of the energy-regime involved one must at least think about the possibility that the atom itself contains energy. But he still dares not take the final step – indeed the evidence would not allow it at the time. The best he can offer is what we now must regard as some rather muddled thinking based on J.J.'s discovery of the electron (of which much more a little later in this chapter).

Nevertheless Rutherford's views in 1900 were bold, in the right direction and well ahead of all others:

On the view . . . advanced by J. J. Thomson, that an atom is not simple, but composed of a large number of positively and negatively charged electrons, the possible energy to be derived from the closer aggregation or regrouping of the components of a molecule is very much greater than on the atomic theory, as ordinarily understood. The energy required to completely dissociate a molecule into its component electrons would be many thousand times greater than the energy required to dissociate a molecule into its atoms. The energy that might be derived from a greater concentration or closeness of aggregation of the components of such a complex molecule would possibly be sufficient in the case of uranium to supply the energy for the emission of radiation for long periods of time. The sudden movements of electrons would set their charges in oscillation and give rise to a series of electromagnetic pulses corresponding to X-rays. The remarkable property of some of the radioactive substances in naturally emitting a kind of cathode ray shows that the present views of molecular actions require alteration or extension in order to explain such phenomena. The energy that might possibly be derived from regrouping of the constituents of the atom would not, however, suffice to keep up a constant emission of energy from a strong radioactive substance like radium for many years . . .

We do not write scientific papers in such strong language nowadays. Here is this young physics professor of barely a year's standing demanding that everyone, and chemists in particular, reconsider the laws of molecular action, and even suggesting that that might not be enough. Nowadays we would call it "speculative", the most profoundly pejorative adjective in the scientist's vocabulary. But Rutherford's unerring sense of the orders of magnitude involved let him safely and correctly point out that the latest figures on radium would not allow either physics or chemistry to "get off the hook" and that revolution, the breaking of the paradigm, was at hand.

This major paper was, then, the backbone of Rutherford's work in 1899. We know that in the autumn of 1898, on his arrival at McGill, he had to give his first series of lectures to student classes – no mean experience for anyone – and he also had the commercial vibration work to do for Dean Bovey. We know that the paper must have been completed by April 1900, for in that month he set off for New Zealand to get married, and the paper was received by the Royal Society in June 1900. It would have been a good year's work by itself, but it was overshadowed by much more exciting discoveries made at the same time.

These discoveries all resulted from a strange anomaly and a pair of spare hands. The anomaly was the "capricious variation" in the radiation from thorium oxide which Rutherford had noted in his last few weeks at Cambridge. The spare pair of hands belonged to R. B. Owens, who held a Fellowship from Columbia University and who had been appointed Professor of Electrical Engineering at McGill at the same time as Rutherford had been given his post. Owens had an arrangement to go and spend some time at Cambridge to work in the Cavendish under J.J., and it seems reasonable to accept the notion that he asked to work with Rutherford on radioactivity as a brief preparation for the kind of work he would encounter when he went to England. The research was, as they described it in their very brief joint paper, "an extension of the investigation [of] the radiation emitted by thorium and its compounds".

Owens confirmed that thorium gave out radioactivity of two sorts, as Rutherford had suggested in his Cambridge work, using exactly the same methods of masking and absorbing the radiations with thin metal foils. But the measured intensity of the radiation and its ability to penetrate paper and metal covers was quantitatively different from uranium radiation. He also confirmed that it was thorium itself that was responsible for the activity since different compounds – the oxide, the nitrate and the sulphate of thorium – all showed activity. But while the oxide showed the greatest intensity of radiation it again "varied in a most capricious manner", and this "was the more peculiar" since the nitrate and the sulphate were constant and since Becquerel had found constancy in his uranium compounds. So "the inconstancy of the radiation of thorium oxide was examined in detail as it was thought it might possibly give some clue as to the cause and origin of the radiation emitted by these substances". This pursuit of the oddity, the anomaly, is pure Rutherford, and unfortunately nowhere does he give us a hint of what mental process persuaded him that here was the start of the route into the mysteries of radioactivity. It may have been no more than mere puzzlement, for the paper goes on:

> It was found that if the substance was enclosed in a lead box with a door the rate of leak was much slower with the door open than closed. The addition of a slight draught of air caused by opening or shutting the door of the room diminished the rate of leak still more. Under similar conditions the rate of leak due to the sulphate and nitrate of thorium and the uranium compounds is not appreciably affected. The sensitiveness of thorium oxide to slight currents of air is very remarkable and made it difficult to work with. With the air quite still the substance in a few minutes regained its normal activity.

(The rate of leak here equates exactly with "the amount of radiation being produced by the sample", for they were measuring the ionisation of a gas by measuring the amount of current it allowed to pass between two plates and this was given by the leak of electric charge from the charged quadrants of an electrometer which appeared as movement of the electrometer needle.)

The two men then demonstrated the leak was, in fact, caused by the movement of the air between the plates. Then they showed it was nothing to do with the nature of the air itself, by changing the amount of water vapour present in the air and even by substituting coal gas for air and finding that the effect remained the same – a draught reduced the radiation but when the gas was still the full effect returned steadily and gradually. They then passed the same air to and fro through the apparatus and still the effect was the same: "This seems to show that it is not the presence of any substance in the air existing in small quantity that produces the effect. A large number of experiments of various kinds have been tried, but so far no clue has been obtained as to why this action should be so manifest in thorium oxide." The only suggestion they offer is that there must be "some change in the condition of the radioactive substance at or near its surface".

At this point Owens departed for Cambridge, leaving the problem to Rutherford. Their joint work was published locally in the *Transactions of the Royal Society of Canada* and dated May 26th, 1899, which means virtually the end of the academic year, since the vacations started at the end of April. Rutherford solved the problem in his summer holidays, and in so doing shook the foundations of chemistry as it was then known.

It is quite clear that Rutherford must have decided in his own mind that the "change in the condition of the radioactive substance" was a giving-off, an emission, of a new radioactive substance, and he had decided this by the start of July 1899. He called the substance that was given off the "emanation"; he wrote about it to J.J. in the mid-summer because, although Rutherford's letter is lost (and evidence about the state of J.J.'s desk plays a part in this story some thirteen years further on), we still have J.J.'s reply dated July 23rd, 1899, in which the word "emanation" is used for the first time in discussing Rutherford's results. This discussion is otherwise unimportant and the significance of the letter in the history of science is that it contains our first information about some of J.J.'s own results which will be alluded to later.

A few dates emphasise the point. Rutherford's paper announcing the discovery of emanation was completed on September 13th, 1899, according to the official version which was published in the *Philosophical Magazine* in January 1900. His work with Owens had been completed by about April of 1899 and contains no hint of anything being given off. So we see that he had decided that emanation was the answer to the puzzle sometime in May or June 1899. The rest of the summer was taken up with proving it. And this illustrates very clearly the nature of Rutherford's scientific method. A problem, often an anomalous result, is seized upon; it is worried by that rather slow, but very powerful, mind until an answer emerges; and then in a furious burst of laboratory work the man whom all acknowledge to be the greatest experimental scientist of the century eliminates every other possible explanation and rams his "intuition" down the throat of every other scientist by irrefutable evidence. There is usually a section of fairly straightforward mathematics that provides a simple law or theory to which the process conforms, at least to a first order of accuracy, and there are clear measurements which quantify at least the order of magnitude of the effects that are being described.

At this stage of his career, what is more, Rutherford presented his results in more or less this order, too. The second sentence of the paper announces the existence of the emanation:

In addition to this ordinary radiation (from thorium) I have found that thorium compounds continuously emit radioactive particles of some kind, which retain their radioactive powers for several minutes. This "emanation" as it will be called for shortness, has the power of ionising the gas in its neighbourhood and of passing through thin layers of metals, and, with great ease, through considerable thicknesses of paper.

137

He then runs through the evidence gathered by Owens, including the mystery of the change of activity when the door of the laboratory causes a draught. And he continues, "The phenomena exhibited by thorium compounds receive a complete explanation if we suppose that, in addition to the ordinary radiation, a large number of radioactive particles are given out from the mass of the active substance." A few lines later:

> The explanation of the action of slight currents of air is clear on the "emanation" theory. Since the radioactive particles are not affected by an electrical field, extremely minute motions of air, if continuous, remove many of the radioactive centres from between the plates. It will be shown shortly that the emanation continues to ionise the gas in its neighbourhood for several minutes, so that the removal of the particles from between the plates diminishes the rate of discharge between the plates.

He does not dare to say that the emanation is itself a gas, for he has no proof, but he points out that it behaves like a gas in its characteristics of being able to diffuse slowly through thin metal foils and rapidly through paper. But he proves his point about the reality of the particles by a new version of the trick he had used in his Cambridge experiments on the ionisation of gas by X-rays – he blows the air away from the thorium, up a long tube and into a charged collecting chamber which contains the insulated electrode connected to an electrometer. As soon as the emanation starts arriving the electrometer needle starts moving, and not before, showing that the particles of the emanation are ionising the air around them.

But here he got a surprise, when he found a state of affairs that no one had yet encountered when dealing with radioactivity. The radioactivity of the emanation gradually decreased. Being Rutherford, he immediately measured the rate at which it decreased, using a slight variation on his experiment. The radioactivity decreased "in a geometrical progression with time". In this particular case the radioactivity reduced to half its original value in one minute. It then decreased by half of that half value in the next minute, so that there was only a quarter of the original value after two minutes and only an eighth after three minutes.

(For those unfamiliar with this type of thinking it can be explained that this type of decrease, or even increase, is typical of many physical processes where the rate of change at any moment is proportional to the amount of the substance that is changing and is present at that moment. Think of a large vertical glass tube filled with water which has a small horizontal tube at the bottom through which the water can escape – it is rather similar to bathwater draining out through the plughole. At first the water will flow out very fast, forced out by the weight or pressure of the water high above the exit in the vertical tube. As the water level falls, however, the pressure will decrease and the water will come out of the bottom more slowly. Or if half your bathwater drains out in five minutes then after ten minutes there will be only a quarter of the original amount of water left in the bath, and in fifteen minutes there will be only one eighth remaining, and so on. Incidentally, baths don't behave exactly in this way – for theoretically there is still some water left in the tub after an infinite amount of time. Such rates of increase or decrease are termed "exponential" because the mathematical equations describing the process involve the "exponential function" which is always written as the symbol "e".)

This discovery of the loss of radioactivity by the emanation left Rutherford with a further problem – why did the radioactivity normally shown by thorium compounds, when there was no draught, appear at a steady rate? He showed that if the emanation at any one moment was blown away and then the air blast was stopped, more emanation would be produced by the original sample, and furthermore that this production of emanation balanced the loss of radioactivity by the emanation. Exactly half of the fresh production of emanation occurred in the first one minute (later refined for both measurements to fifty-four seconds). In other words, the normal steady state of thorium radiation was in fact an equilibrium, with an exact balance between the loss of radioactivity by the emanation and the production of fresh emanation.

Rutherford rather left this extraordinary situation "in the air" in this first paper. He tried some physical experiments aimed at showing that the emanation was a gas, but could not get any results. He was left with two possibilities – either it was particles of radioactive dust emitted by the thorium compounds, though he had some evidence against this, or it was a "vapour" of thorium. But this matter was left unresolved in a rush of enthusiasm for an entirely new discovery which was announced in the final paragraph:

Experiments, which are still in progress, show that the emanation possesses a very remarkable property. I have found that the positive ion produced in a gas by the emanation possesses the power of producing radioactivity in all substances on which it falls. This power of giving forth a radiation lasts for several days. The radiation is of a more penetrating character than that given out by thorium or uranium. The emanation from thorium compounds thus has properties which the thorium itself does not possess.

His last sentence after this outburst is almost like that of a writer of thriller serials. "A more complete account of the results obtained is reserved for a later communication."

Much of the work behind this paper can be found in a very good sequence of preserved Rutherford notebooks – in fact laboratory notebooks rather than Rutherford's own, for they contain notes and readings by other hands. The Canadian series begins with the date of March 1899 and the first entries are almost certainly by Owens. But soon there is work which is recognisably Rutherford's on "the effect of moisture on thorium oxide". The result of the work is "Effect apparently quite independent of state of moisture of air, but depends only on the rate of flow (?) of air. The effect of a disturbance of the air soon reaches a maximum and is apparently very little altered with increase of the disturbance."

It is possible to see the elimination process at work, too. A little later there is a series of jottings: "Independent of CO_2 in air; independent of dust; independent of moisture; independent of direction of motion of air." These are notes for the mind rather than words that will appear in an eventual paper. The work continues quite regardless of the departure of Owens. Another notebook starts just three days after the publication of the first Canadian paper, written jointly with Owens; the date is May 29th, 1899. It is quite obvious that Rutherford is working like a fanatic. "Tuesday evening" is marked against one set of readings, "Thursday evening" against another. About this time he wrote to his fiancée that he was working every evening often until eleven or midnight. On page 67 of this notebook the word "emanation" first appears several pages before the next date, which is July 7th, 1899. There is here a loose sheet stuck into the notebook with some mathematics on it, including versions of the "exponential equation" which later appears in the published paper.

Another of the rare dates appears – August 3rd (1899) and a few pages later we have calculations that obviously refer to the decay times of the activity deposited by the emanation. Readings are recorded for three days, Tuesday, Wednesday and Thursday, and there is the jotting, "Therefore radiation falls to one quarter in 22 hours, therefore falls to half in 11 hours." This makes it clear that what came to be called "induced" or "excited" radioactivity was discovered after August 3rd, but was included in that thunderous last paragraph of a paper dated September 13th, 1899. The next notebook opens in June 1900.

We are therefore without much further insight into the workings for the main paper on Rutherford's newly-discovered type of radioactivity unless we are willing to redate a notebook clearly marked "December 10th, 1900" and put it a year earlier because it contains work on "excited radioactivity". But the paper itself, dated November 22nd, 1899, was published in the February 1900 issue of the *Philosophical Magazine* and entitled "Radioactivity produced in Substances by the Action of Thorium Compounds".

Once again we can rely on Rutherford's magnificently straightforward presentation of his case to find out what it is all about. The paper begins:

> Thorium compounds under certain conditions possess the property of producing temporary radioactivity in all solid substances in their neighbourhood. The substance made radioactive behaves . . . as if it were covered with a layer of radioactive substance like uranium or thorium. Unlike the radiations from uranium or thorium, which are given out uniformly over long periods of time, the intensity of the excited radiation is not constant, but gradually diminishes. The intensity falls to half its value about eleven hours after the removal of the substance from the neighbourhood of the thorium. The radiation given out is more penetrating in character than the similar radiations emitted by uranium and thorium . . . and radium and polonium.

Again the discovery was the result of a typical Rutherford following-up of something unexpected: "Attention was first drawn to this phenomenon of what may be called 'excited radioactivity' by the apparent failure of good insulators, like ebonite and paraffin, to continue to insulate in the presence of thorium compounds."

Having shown that the phenomenon occurs, Rutherford then links it firmly to the presence of his emanation, rather than to the original thorium samples themselves. And he did this by another variation of his old favourite of blowing along a pipe. He blew air across the top of a thorium oxide sample and along a tube which had flat plates spaced out in a line along the inside. Each of these plates was connected to a separate electrometer. As the radioactivity from the emanation disappears, each successive plate gets a small deposit of excited radioactivity, which can be shown to remain radioactive long after the emanation has passed away, proving that "There is a very close connection between this emanation and excited radioactivity – in fact the emanation is in some way the direct cause of the latter." And of course it proves also that the excited radioactivity is not the immediate product of the original thorium sample.

There is then the usual long series of experiments to show what

141

cannot be the cause of the phenomena, leaving only three possible explanations: either there is a kind of phosphorescence effect from the original thorium; or the ions produced in the air by the active emanation make the surroundings radioactive; or there are actual radioactive particles being deposited that were emitted by the original thorium. He sums up the evidence for the first two possibilities, and finds them unlikely though he cannot conclude definitely against them. He prefers the third possibility, and proposes further experimental tests. The paper ends, however, in a bitter footnote. "As this paper was passing through the press," the November edition of the French scientific publication *Comptes Rendues* carried reports by the Curies and Becquerel that they too had found "excited radioactivity" caused by radium and polonium. Rutherford points out that the Curies' radioactive bodies are between ten and fifty thousand times as active as his own uranium sources, and the phenomenon seems to be similar to that produced by his thorium. And then he criticises his rivals – "but there are not sufficient data on which to base any comparison". There is no mention of the effects of the electric field, there is no mention of whether there is any emanation from radium and polonium to produce the activity. He ends dismissively, "Curie concludes that the results obtained are due to a kind of phosphorescence excited by the radioaction; while in the case of thorium the author has shown that such a theory is inadmissable." But dismissive or not, scornful of the lack of reliable measurement in the French work as he was, Rutherford learnt the hard lesson of the sheer distance of Canada from the centres of scientific activity in Europe where researchers could get their results printed, and claim priority of discovery, within a few days of submitting their work.

Rutherford published nothing more in 1900. Any work he did in the first few months of the year went into the paper we have discussed on the energy involved in radioactivity. His main activity in 1900 was the vastly important one of getting married and, as it turned out, rapidly fathering his only child. He was away from McGill and his laboratories for nearly six months of the year. Thus his work came to a halt.

In Europe, after the original discoveries of the radioactivity of uranium and thorium, Marie Curie, a Polish girl who had moved to Paris in her determination to become a scientist, spotted that pitchblende, the ore from which uranium was usually extracted, was more radioactive than uranium itself. She concluded that there must be other radioactive materials in the ore and was eventually able to extract first polonium and then radium. She was joined in this work by her husband, Pierre Curie, who had already made his reputation by discovering the phenomenon of "piezoelectricity" (the ability of certain crystals to produce an electric current when physically stressed; a property which later was to be turned to practical use by Rutherford himself) and by distinguished work on the effects of temperature on magnetic materials. Their struggle in humiliating conditions, without proper support, and against the weight of establishment opposition was a dramatic story in itself. Soon their colleague, Debierne, discovered yet another radioactive material which he named actinium. Rutherford's discovery of the different types of radiation was followed up largely in continental Europe and it was shown that his beta-rays were effected by magnetic fields – the rays were "deviable". This led to the discovery that beta-rays were streams of negatively-charged electric particles, similar to the cathode rays, though moving more slowly. No one, however, could deviate the alpha-rays from their path, and these radiations were generally considered to be some form of weak X-ray excited in the radioactive material by the emission of beta particles. But results were inconsistent, which to us can be hardly surprising in view of the weakness and impurity of the early radioactive source materials, and the insensitivity of the radiation-detection devices which were often unable to sense radiations which we know now to be of vital importance.

The whole picture became an incomprehensible puzzle, with different radioactive materials producing different types and intensities of radiation. And while uranium, radium and thorium produced steady radiation, polonium, Rutherford's thorium emanation, the radium emanation discovered by Dorn in 1900, and the excited radioactivity all decreased with time. Even more puzzling, the excited radioactivities decreased at different rates again, some more slowly, some more quickly than the emanations. Then, again in 1900, the British amateur chemist, Sir William Crookes, separated a substance from uranium – he called it "uranium-X" – which appeared to carry with it all the radiating power of the uranium, leaving the pure uranium inactive. Perhaps all radioactivity was the result of some impurity which acted with the parent element to produce radiation. At the same time Marie Curie found it difficult to convince her fellow scientists that radium was indeed a new and separate element.

Parallel to all this interest, and contemporaneous with it, came the long series of discoveries of definite new elements, the rare gases, by Sir William Ramsay, the leading British chemist of the time. This work was highly significant and influential in Rutherford's life. His many scientific memorialists have ignored the clear evidence in his notebooks that he was ever conscious of the possibility of discovering a new light gas – indeed at one time he thought he had found one, when he had actually succeeded in being the first man to split the atom. Knowledge of the new gases and their inertness was essential to his next major discoveries, and the rapid developments in the science and techniques of spectroscopy were to be of vital importance both in the confirmation of his views on radioactivity and in the first stages of the nuclear science he was to found some ten to fifteen years later. (When a chemical element is heated or excited so strongly that its atoms emit light, this light can be seen by using a spectroscope to consist of a series of very narrow bands of differently coloured light – it is rather as if a rainbow was split up into its different colours with a clear gap between each coloured band. The number of coloured bands and their precise wavelength is absolutely specific to each of the chemical elements, and the presence or absence of that element can be unequivocally demonstrated by examining the spectrum of any sample, even if that sample be a star millions of light years away.)

By far the most important scientific discovery of the last five years of the nineteenth century – as Rutherford himself would emphasise in lecture after lecture – was the discovery of the electron by his patron and teacher, J. J. Thomson. The work was done in the Cavendish Laboratory in 1897, when Rutherford was still pursuing the ionisation of gases by X-rays with Thomson, but was not directly involved in the crucial measurements.

The argument between British and continental scientists as to whether the cathode rays in a discharge tube were particles or waves raged fiercely and Thomson's objective was to settle this argument. If the stream of cathode rays could be diverted or bent by electrical or magnetic forces, that was strong evidence that the rays were particulate in nature, and Lenard in Germany showed that they could be diverted magnetically. Perrin in France, showed that the cathode rays carried a negative charge, which was further evidence for the particulate or "corpuscular" theory. J. J. Thomson, using the Cavendish knowledge about ionisation, concluded that the previous failures to deviate the rays by electrical fields (notably Hertz's negative result) were due to the pressure of the gas in the discharge tube. He therefore set about producing a higher vacuum in his tubes, which was a task that took many days. When he had done this he was able to show that an electric field did, in fact, deviate the cathode rays and in such a way as to show that the rays carried a negative charge. A remarkably neat piece of experimentation measuring the heat caused by the impact of cathode rays enabled him to calculate their energy and hence their speed. Once this measurement was established he could find the crucial figure, e/m, the ratio of the electric charge of the particles, "e", to their mass, "m", by comparing their deflection by electrical forces to their deflection by magnetic forces. Even his first rough experiments came up with an astounding result – they gave e/m a value of about ten million, and this was the case whatever gas was in the discharge tube.

The smallest value anyone had ever found for this ratio e/m previously had been a value of about ten thousand for the atom of hydrogen in the process of electrolysis, the splitting of water into oxygen and hydrogen. But all physicists believed that the amount of electricity carried by a hydrogen atom in electrolysis was the smallest amount of electricity that could exist, and many considered it the "atom" of electricity. If this was the case then the mass of J.J.'s "corpuscles" must be ten thousand divided by ten million – they must each be only one-thousandth as heavy as the atom of hydrogen, the smallest known body. J.J. announced his results at the regular Friday Evening Discourse of the Royal Institution in London on April 29th, 1897. His "corpuscles" are now called electrons and nothing has shaken the validity of his work.

Of course the discovery of the electron has led to all the electronics and computerisation of our own times; it also allowed science to explain and improve all the devices of the electrical supply industry which were just at that time being developed as practical machines to revolutionise the domestic life of industrialised societies. But its intellectual impact in the scientific world was its most profound effect. The methods of physical science had shown that the atom of chemical science is not the basic unit of which all matter is made; the atom was divisible; all atoms, of whatever element, contain identical bodies, electrons, which are all of the same mass and carry the same electrical charge; and these electrons are only one thousandth the size of the smallest atom. It must be emphasised that chemistry in the late nineteenth century was the premier science, and that the atomic theory, first enunciated by John Dalton to nine members of the Manchester Literary and Philosophical Society in October 1803, although it had recently been challenged by Ostwald and his school, reigned supreme. This theory stated that the basic constituents of matter were the atoms, that each chemical element consisted of enormous numbers of identical atoms, but that atoms of one element were different from atoms of another element in various ways such as weight and chemical activity. Atoms were the smallest units which could be found. Most scientists thought of atoms, in terms first suggested by Sir Isaac Newton, as hard spheres rather like billiard balls.

So the paradigms of science were collapsing in 1900. The atom was divisible and from this point Thomson and many others, such as Larmor and Lorentz, went on to develop the theory that all mass was electromagnetic in origin. Radioactivity, though apparently less important, was the first new property of matter to have been discovered since Newton established gravity two hundred years before.

There was also the phenomenon of public interest in the new scientific discoveries. X-rays, with their immediate medical usefulness and their fascinating ability to reveal that which is hidden, had given the first taste of this social phenomenon, for these were precisely the years in which the "mass-media" were first appearing, the very years in which mass-circulation newspapers and popular magazines were satisfying the needs of mass-literacy. Marie Curie, the slight but steely figure of the pioneer woman scientist, was beginning her rise (quite unwanted) to international fame. Radium was the word which carried the connotation of radioactivity in the popular view. What most scientists fail to mention is that radium glows in the dark, in practice providing luminous dials at the cost of death and ill-health to early radium workers, but to the imaginative showman and publicist a wonderful new "gimmick" which was rapidly exploited. In San Francisco showgirls appeared on a completely darkened stage yet were visible by the light of their radium-painted costumes. Cocktails that glowed in the dark were served to New York gamblers who played "radium roulette' with luminous chips. Medical uses from the grotesque, such as whitening the skin of negroes, to the serious and valuable such as the treatment of cancers were suggested, and caught the public imagination as all proposed cures seem to do. And the energy calculations typical of the popular press, such as the power of a gram of radium to drive a fifty-horsepower vehicle around the world at thirty miles an hour, caused scientists including Rutherford's friends and correspondents to chaff, but showed that the imagination of the science-fiction writer is probably a better guide to the future than the pompous carefulness of the great men of the day.

It was into this atmosphere of intellectual ferment that Rutherford returned to McGill for the start of the winter session in autumn 1900, bringing with him his new wife, already pregnant with the daughter who was to be their only child. But he was not happy. He had already established himself and the MacDonald Physics Laboratory as one of the very small number of world centres for radioactivity research, but he felt cut off, too far from the focus of intellectual activity, hampered by the slowness of a publication system which had to cross and recross three thousand miles of ocean. He plainly felt what he wrote to his mother a year later in one of his best known letters: "I have to keep going as there are always people on my track. I have to publish my present work as rapidly as possible in order to keep in the race. The best sprinters in this road of investigation are Becquerel and the Curies in Paris who have done a great deal of very important work in the subject of radioactive bodies during the last few years."

So in March 1901, with the birth of his first child due any day, he wrote to J.J. about the possibility of getting the post of Professor of Physics at Edinburgh University, just vacated by P. G. Tait. On March 26th he wrote again:

> I think you know fairly well my position here. The laboratory is everything that can be desired and I am not overburdened with lecturing. I am very comfortably situated and have practically a free hand in the laboratory. On the other hand the salary is not very great and does not go very far in Montreal, where I think the cost of keeping a household is considerably greater than in England.
>
> After the years in the Cavendish I feel myself rather out of things scientific, and greatly miss the opportunities of meeting men interested in physics. Outside the small circle of the laboratory it is seldom I meet anyone to hear what is being done elsewhere. I think that this feeling of isolation is the great drawback to colonial appointments, for unless one is prepared to stagnate, one feels badly the want of scientific inter-course.

J.J.'s reply about the prospects of getting the Edinburgh Chair was not optimistic – there was a very large field of candidates and "The election, too, is made by a body of local men who do not know anything of physics so that local influence will be especially power-ful in this case." (Rutherford took Thomson's advice to apply for the job as a gesture, but the electors followed J.J.'s predictions, and he did not get the post.) But the basic problem was acknow-ledged: "I quite appreciate the isolation of scientific workers in the colonies; it was that made Threlfall give up Sydney," J.J. wrote. The problem was to persist in Rutherford's mind virtually through-out his life, for new universities were regularly founded in Canada, Australia, New Zealand, South Africa and India, many of his pupils and correspondents obtaining posts on these far-flung cam-puses, and the same refrain of isolation from scientific contact is

147

repeated in letter after letter when he as Professor at the Cavendish had to comfort his protégés.

On the other hand, he would have done well to pay more attention to a letter he had received the previous year from his old friend, J. S. Townsend in Cambridge, who said, "I wish we had the same opportunities here of doing work outside the Cavendish as I must confess that eternally working at ions is beginning to be tiresome." By a delightful irony, however, the very letter to J.J. in which Rutherford complains of his isolation contains the first hint of the happenings which were to hold him in Montreal for six more years. He had written, "Your corpuscular theory seems to take the field in physics at present . . . We are having a great discussion on the subject tomorrow in our local Physical Society when we hope to demolish the chemists." The chemists were not so easily demolished for their champion in the debate was the young Frederick Soddy, newly arrived from Oxford. The debate that began between the two men the next day went on for two weeks, and probably neither of them accepted the other's point of view for years, yet for eighteen months they joined in one of the most successful and significant scientific collaborations ever known.

Frederick Soddy was a brilliant young chemist, at this time twenty-three years old and six years Rutherford's junior. But it was not only his chemical expertise which complemented Rutherford's physics; he also brought to their partnership the cultural background of an English upper-middle-class education at public school and Oxford. Soddy was the clever member of the pair where Rutherford provided the power; Soddy was much the quicker of mind and much the better writer, lacking Rutherford's penetration but better at seizing the wider philosophical implications of their joint work.

He certainly proved a remarkable, even a bitter, debater when the two men first crossed swords at that Physical Society meeting on March 28th, 1901. The subject was "The existence of bodies smaller than an atom" and the society, of which Rutherford was chairman, had issued invitations fairly widely. The hand-written minutes comment, "Invitations having been previously sent to several other members of the University, the Society was favoured by quite a large attendance." Soddy called his paper "Chemical evidence of the indivisibility of the atom" and lashed out at the physicists.

Recent advances connected with the discovery of the cathode, Röntgen and Becquerel types of radiation have led physicists to the belief that in these phenomena they are dealing with particles of matter smaller by a thousand times than the absolute mass of the atom. So certain are they of the interpretation of their experimental results and of the inability of the present hypotheses to explain them that not a few of them have definitely abandoned the accepted notion of the structure of matter and have boldly attacked the atomic theory which as everyone knows has been the foundation of chemistry as a science from the time of Dalton to the present day. Physical theorising such as this, beyond affording unbounded satisfaction to its promoters, has so far passed unnoticed by the main body of the chemists.

Soddy acknowledged sarcastically that now the physicists had called him to a "pow-wow" to present the views of the "conservative" chemists "in the malicious anticipation, I have some reason to suspect, that I shall thereby deliver myself and my case over, bound into the hands of the adversary or shall at least afford an easy target for the modern artillery at the disposal of the ionists". Far from presenting a target, he went on to the attack pointing out that the physicists had landed themselves in a dilemma: either their radiant matter had some relationship with gravitational matter in which case the physicist had to bridge the gulf between energy and matter, or they could admit that the gulf was not bridged and then "the connection between the carrier and the atom is one of imagination and analogy rather than of fact". He went on to point out that chemists had been quite clear that they had no idea what was the absolute mass of the hydrogen atom – "that, and similar speculations has been almost the sole province of the physicists". Indeed he accused the physicists of making "a long series of unwarranted assumptions" in arriving at a value for the mass of the hydrogen atom, "a value repudiated by chemists as the result of fantasy overshadowing the critical faculty". And even more in the same vein, "As an estimate of the immeasurably small, as an incursion into the fascinating realm of the imagination, the value has its interests to chemists and physicists, but as a basis on which to found arguments profoundly affecting our conception of matter it is to my mind ludicrously inadequate."

Admittedly there were chemists who questioned atomic theory: "Thus Professor J. J. Thomson and Professor Rutherford – to whom we all owe so much in introducing us to and inspiring us with this fascinating subject – have been known to give expression to opinions on chemistry in general and the atomic theory in particular which call for strong protest." But then Soddy cooled things off – he would "leave these somewhat polemical, not to say recriminatory topics which are by way of retaliation for a certain undue levity about things chemical on the part of the Physical Society of late", and suggested some ways in which the old and new views might be reconciled.

But soon he was back to the attack – he slaughtered Prout, who had put forward a theory that the hydrogen atom was the basic unit of matter and that all the other heavier elements have atoms built up of combinations of hydrogen atoms: "If it requires one genius to formulate one hypothesis that is sound, it requires ten to nail one that is unsound in its coffin," adding, "Prout was a man of apparently no scientific attainments or activity whatever." Then he demolished Sir Norman Lockyer: "Astronomers often appear to lose themselves in the contemplation of the celestial, and such mundane considerations as logic sometimes escape them."

This was all excellent debating stuff – the sort of thing which the sixth form debating society at public school and the Oxford Union teaches a young man. And there were enough good scientific points there, too, to make Rutherford think. But, in that same English tradition, Soddy was careful not to wound his man personally and he ended with a tribute which was also a strange prophecy:

I have purposely refrained from entering into a consideration of the new work which has given rise to the question before us. Professor Rutherford has taught me what little I know of it, and I, of course, cannot presume to criticise it, *a priori*. My object has been throughout to show that matter is possessed of very positive attributes which lead us to consider the atom as a material entity. If, as appears the case, radiant matter has lost almost all its distinguishing properties and appears hardly differentiated from electrical energy, I think the onus of proving it is really a form of matter rests on the new school, and until it is shown that it is affected by gravity or otherwise possesses features distinct from the ether, there will be no necessity for chemists to modify the atomic theory. Possibly Professor Rutherford may be able to convince us that matter as known to him is really the same matter as known to us, or possibly he may admit that the world in which he deals is a new world demanding a chemistry and physics of its own, and in either case I feel sure chemists will retain a belief in, and a reverence for, atoms as concrete and permanent entities, if not immutable, certainly not yet transmuted.

It is known that Rutherford was quite taken aback by the fury of Soddy's attack, but he read his own paper, a version of which appeared in 1902 in the *Transactions of the Royal Society of Canada*, and which appears in his *Collected Papers*, ironically in the middle of the great series of papers he was soon to write with Soddy. Here we must give Soddy the better of the argument, not because he was right, quite the opposite, but because we have already presented the views and experimental evidence for "The Existence of Bodies Smaller Than the Atom". But then, to quote the minutes of the McGill Physical Society, "The hour being rather late it was unanimously decided to adjourn the discussion until the following Thursday." And on the following Thursday the minutes record "a renewal of the interest evinced in the subject under discussion was shown by the larger attendance of members and friends". Professor John Cox, the senior physics Professor, was the chief speaker, and he reviewed the physicists' evidence again. Rutherford and Soddy both spoke again in the general discussion and the minutes concluded: "The steady increase in the interest of other members of the University, as shown by the large and general attendance of the two closed meetings indicated that the influence and usefulness of the society had been very much extended." As any television producer could have told them, there is nothing like a good row to pull in an audience – but that was before the days of show-biz.

There is no doubt that it was some time after this meeting that Rutherford asked Soddy to join him in a physico-chemical examination of the thorium emanation he had discovered. Rutherford could track the emanation by its radioactivity, since it left its traces in the deposition of excited activity, but what was it? Rutherford had first asked the Professor of Chemistry for his cooperation, but he had politely (and very foolishly) rejected the invitation. Soddy, who had come to Canada in the vain, but adventurous, attempt to obtain the Professorship of Chemistry at Toronto, and had decided to stay on in a demonstratorship at McGill, was next to be asked and he jumped at the chance; he had already been attending Rutherford's advanced course lectures on radioactivity. Soddy wrote in his "Personal Records for the Royal Society":

> He invited me to join him in a physical and chemical study of the thorium emanation, a radioactive vapour or gas emitted by the thorium compounds . . . So far as I remember the whole subject of radioactivity was practically new to me at the time, but, being afforded a demonstration of the thorium emanation, I at once joined Rutherford in the attempt to determine its chemical character, and we quickly discovered that it had none, but was in this respect an argon gas.

Elsewhere Soddy makes it quite clear that it was *not* Crookes' discovery of uranium-X that started the two men on their enterprise, though they named a new substance thorium-X by analogy. He also makes it clear that he did the chemical work in the Chemistry Building of McGill while Rutherford did all the physical measurements of the radioactivities of the specimens Soddy produced in the Physics Building. But Soddy does not do himself full justice in this account, for he was in fact predisposed to take advantage, ahead of most chemists of his time, of combining physical and chemical methods of investigation. In his essay which won him the Evans Prize at Oxford in 1895 he had written of a "vast field of research on the borderland, so to speak, between chemistry and physics, [which] is almost virgin soil holding out a bountiful harvest to those who, not content with treading the well-beaten paths of science, are enterprising enough to attack the problems and patient enough to overcome the difficulties which pioneers of scientific research always have to encounter".

Rutherford, in that same spring when he was debating with Soddy, had started a further line of research which the two together would find most fruitful. Working with Barnes and his young pupil Harriet Brooks, he had been measuring the rate of production of emanation from various thorium compounds as the physical conditions were varied. In particular it seemed as if the production of emanation could be completely stopped if the thorium compound was heated to a very high temperature. It seemed to Rutherford as if this "de-emanating" must be a chemical process, and since it was a complete puzzle, it also demanded the help of a skilled chemist.

Rutherford and Soddy did not begin working together until October 1901. Soddy had spent the long summer vacation in England and had attended the meeting of the British Association in Glasgow. Rutherford had been for his first family holiday with his wife and small baby daughter. The notebooks show clearly that October 12th, 1901 was the day that the great Rutherford–Soddy partnership started work. The work is described as "Comparison of emanations from thorium oxide and thorium oxide heated", and, in truth, their work was not only an examination of emanation but also an examination of the process of "de-emanating". The first results were confusing. It turned out they could not completely de-emanate their thorium compounds; some sixteen per cent of the activity was always left behind and Rutherford concluded that the power of emanating was a specific property of thorium. Certainly neither man could separate anything else that gave emanation from the thorium.

Then Soddy turned to the emanation itself. He put it in the presence of the most powerful reagents known to chemistry, such as platinum black, zinc, magnesium and lead chromate. He even heated his reagents to white heat but the emanation remained totally unaffected. Then he did it all again in an atmosphere of carbon dioxide instead of air; again nothing happened, thus proving that emanation was not just a special state of ordinary air excited by the radioactivity of the thorium. It seemed then that emanation must be a totally inert gas which would not react with anything or combine with anything, just like the recently discovered argon types of gas. Rutherford and Soddy became convinced of this – they have since been proved right and the emanation is now called thoron and is classed with radon and actinon, the emanations of radium and actinium, in the family of inert gases.

But at the time the implications of the discovery were, literally, frightening. Thorium was an element, normally a solid. Emanation was a gas, an argon type of gas, probably a new element. One element was turning into another. This was heresy. This was transmutation, the alchemist's dream. Soddy recalled how he was "standing there transfixed as though stunned by the colossal import of the thing" and how he blurted out, "Rutherford, this is transmutation: the thorium is disintegrating and transmuting itself into an argon gas." This was what was frightening; professional reputations were at stake.

Soddy went back to checking the source of the emanation. Could there be something other than thorium at work? Crookes' discovery of uranium-X, the apparently active impurity in uranium came to mind. The de-emanating situation was confused; results seemed to depend on which thorium compound was used and whether it was in the solid state or in liquid solution (the actual difficulty we now know was the occlusion, or trapping, of the gaseous emanation in some compounds more than in others). Soddy tried his methods of separating thorium from its compounds yet again, but the results were still confusing. And then they tested the "left-overs", the precipitates from which all the thorium had been removed. And they found that there was emanating power in these, though there was no thorium. Further, these thorium-free precipitates had the ordinary radioactivity of thorium. So there had been an "impurity" in the original thorium all along, it seemed, and they had discovered a thorium-X similar to uranium-X. However there was no suggestion as to what sort of material thorium-X might turn out to be, and there was no mention of transmutation in the first Rutherford–Soddy paper which was sent off just before Christmas 1901 and was printed in the *Transactions of the Chemical Society* in 1902.

Soddy had spent the last few days before the Christmas break trying to get all the thorium-X away from the original thorium, but however many chemical precipitations he tried he could never get the activity of the thorium to go below twenty-five per cent of his original value. Then when they came back in January there was another shock – the deposits of thorium from which Soddy had been trying to strip off the active thorium-X were if anything even more active than when they had left them, while the deposits containing the active thorium-X had almost completely lost their activity. The facts were getting even more uncomfortable, even more like transmutation. It looked very much as if thorium naturally produced its own radioactivity and produced something (thorium-X), which was an entirely different chemical and was radioactive in its own right. This thorium-X then seemed to produce the gas emanation, which was radioactive, and the emanation itself produced yet another chemical, probably a solid, which it deposited as "excited activity" and which was also radioactive. And all these different chemicals lost their activity with time, except perhaps the original thorium.

It must have been about this time that rumours of transmutation began to spread in McGill. It so happened that Soddy had given some lectures on the history of chemistry before ever he teamed up with Rutherford, and in these lectures he had dismissed alchemy as totally unworthy of the chemist's attention and not to be treated as part of the history of his science. But in a semi-popular lecture (illustrated with slides) called "Alchemy and Chemistry", which is not precisely dated but belongs to this slightly later period, he defends alchemy as a true origin of chemistry; as a study, wrong, but not to be despised. He attacks those historians of chemistry, especially French ones, who claimed that Lavoisier was the founder of chemistry, and he reports favourably on the physicist's approach to the matter, regarding it all as fundamentally the same in being subject to the same laws of motion and gravitation and heat and electricity. The physicist therefore "works nearer the origin of things" while the chemistry of the elements "is a mere collection of facts and analogies, each of which is a law unto itself". His conclusion is that "The constitution of matter is the province of chemistry, and little indeed can be known of this constitution until transmutation is accomplished. This is today, as it has always been, the real goal of the chemist before this is a science that will satisfy the mind." And in his Notes for the lecture he has added, "Recent physical work, done nowhere in the world more than in Montreal, enables us to detect changes in matter heretofore only appreciable after geological epochs of time. Possibly it may lead to new and positive evidence of the nature of the elements."

Certainly Soddy spoke to the McGill Physical Society in February 1902 and described the thorium emanation as "having apparently the properties of a radioactive chemically inert gas of the argon family". Whether it was this that brought transmutation out into the open is unknown, but there was much disquiet about it. Professor A. Norman Shaw recorded (in the *McGill News* of 1937):

There were several occasions when colleagues in other departments gravely expressed the fear that the radical ideas about the spontaneous transmutation of matter might bring discredit on McGill University! At one long-remembered meeting of the McGill Physical Society he [Rutherford] was criticised in this way and advised to delay publication and proceed more cautiously – this was said seriously to the man who has probably allowed fewer errors to creep into his writings and found it less necessary to modify what was once announced than any contemporary writer. At the time he was distinctly annoyed and his warm reply was not entirely adequate, for in his younger days he sometimes lost his powerful command of ready argument when faced with unreasonable or uninformed criticism.

Immediately John Cox quietly rose to his support and gave a clear review of the new ideas . . . [and] ended rhetorically with a stirring prediction that the development of radioactivity would bring a renown to McGill University by which in future it would be widely known abroad. He ventured also to predict that some day Rutherford's experimental work would be rated as the greatest since that of Faraday.

Cox, of course, proved quite right.

Rutherford at this time of his life gave quite a different impression from that of the powerful but bucolic figure we now remember. He could be subject to black despairs and monumental rages, to enormous enthusiasms and to brilliant public exhibitions. A classics professor at McGill, John McNaughton, a man wont to describe scientists as "plumbers" and "destroyers of art", described an encounter for the *McGill University News* in 1904.

We paid our visit to the Physical Society. Fortune favoured us beyond our deserts. We found that we had stumbled in upon one of Dr Rutherford's brilliant demonstrations of radium. It was indeed an eye-opener. The lecturer himself seemed like a large piece of the expensive and marvellous substance he was describing. Radioactive is the one sufficient term to characterise the total impression made upon us by his personality. Emanations of light and energy, swift and penetrating, cathode rays strong enough to pierce a brick wall, or the head of a Professor of Literature appeared to sparkle and coruscate from him all over in sheaves. Here was the rarest and most refreshing spectacle – the pure ardour of the chase, a man quite possessed by a noble work and altogether happy in it.

In slightly different mood Soddy recalls the occasion he came into the lab just as Rutherford had gripped one of his alpha-ray measuring devices – without remembering to switch off the electricity. He leapt about, howling and swearing, in what Soddy believed were Maori oaths, before dashing the delicate piece of apparatus, which he had made with his own hands, to the floor and to pieces. On another occasion, actually the last moment of contact before Soddy went off to England, he found Rutherford excitedly pacing up and down the path outside the lab, with a big smile on his face and a roar of, "Soddy, the darned thing's going up" – referring to a specimen which was recovering its activity against all expectations.

The strength of conviction about transmutation, which the two men based on their experiments with thorium, was increased at the very end of 1901 by results from Europe. Becquerel found that old specimens of uranium which he had treated to remove uranium-X had recovered their activity. Crookes told Rutherford of this in a letter dated December 18th, 1901, and Crookes immediately set about re-examining his own old uranium specimens. It was also Crookes who obtained for the Canadians a better supply of pure thorium nitrate from Knofler, the German chemical firm, and this arrived in March 1902. With this Soddy set out to prove chemically that thorium-X could only come from thorium. This was necessary to eliminate the possibility that radioactive products could come from some other impurity, normally quite inseparable from the thorium and which could have radioactivity induced in it by the radioactivity of the thorium and which could only then be separated from the original mass. First he proved conclusively that thorium-X was different chemically from thorium and separated from all other impurities in the original sample. Then he repeated the separation several times at fixed intervals of time, and each time the amount of thorium-X was the same. Finally he changed the molecular condition of the thorium by changing the compounds chemically and by changing the form in which the compounds appeared by heating. But still there was no change in the production of thorium-X. The two men were forced to conclude that the production of thorium-X was the result of a "sub-atomic chemical change".

This was the stage of the argument at which first-class chemical evidence was absolutely vital. It was the portion of the partnership where Soddy was crucial. But, of course, each time I write "the amount of thorium-X was unchanged" the reference is to Rutherford's electrical measurements of radioactivity. Soddy clearly understood, and was fascinated by, the fact that electrical measurements allowed them to say conclusively that a certain substance was present when the amount of that substance was far too small for any balance to weigh. Because the electrical measurements followed the new property of matter, radioactivity, they were "capable of exact quantitative determination . . . to follow chemical changes occurring in matter".

So now they were faced with this problem:

> The problem was to imagine where the thorium-X came from. It might come from nowhere or it might come from the thorium. Both guesses were absurd, but the first was also incredible. If thorium-X was a substance, if it was made of matter, then the only matter it could possibly come from was the thorium. Since the two were chemically different they must be made of different kinds of atoms. To put it bluntly, thorium was an element, thorium-X another, and the atoms of thorium must be steadily transmuting themselves into atoms of thorium-X. No other conclusion was possible.[1]

Actually this sort of conclusion had been predicted some five years before by two Germans, Elster and Geitel, who offered no evidence except that obtained by Becquerel and Marie Curie in their very first work. They had written:

> The source of energy must be sought in the atoms of these particular elements. This suggests that a radioactive element consists of atoms similar to the molecules of an unstable chemical compound, capable of undergoing a transformation to a stable state with the release of energy. If this be the case then the active substance must gradually become transformed into an inactive one with a corresponding modification of its properties as a chemical element.

(Julius Elster and Hans Geitel were among the great pioneers of radioactive research whose names have tended to be lost behind the glamour of Marie Curie and the success of Rutherford. At the time when Rutherford and Soddy were achieving their remarkable success the two Germans showed, by exposing a naked, negatively-charged wire, that the air was radioactive, which we now know demonstrates the existence of the emanation gases in the atmosphere in tiny quantities. But it caused a great furore at the time and scientists all over the world looked for radioactivity in nature and found it in water supplies and the Niagara Falls and in the air of every country. J. J. Thomson himself showed that there was radioactivity in the water supply of Cambridge.)

[1] Romer, *The Discovery of Radioactivity and Transmutation*.

But a little mild speculation about a newly discovered pheno-menon is very different from a formal scientific publication claim-ing that one has proved that transmutation can occur. Again the help of Sir William Crookes was sought, for the paper was being sent for publication in the *Transactions of the Chemical Society*. So on April 29th, 1902 Rutherford wrote:

> All these processes are independent of chemical and physical con-ditions and we are driven to the conclusion that the whole process is sub-atomic. Although of course it is not advisable to put the case too bluntly to a chemical society, I believe that in the radioactive elements we have a process of disintegration or transmutation steadily going on which is the source of the energy dissipated in radioactivity.

He then emphasises that the evidence is all *experimental* – which word suggests in ordinary English a certain tentativeness, but in scientific meaning implies absolute definiteness. Both Rutherford and Soddy asked for Crookes' influence in pressing the publication of their paper "if difficulties arise over 'atomic' views".

Thus their paper does not use the dread word "transmutation" – instead it speaks again of "sub-atomic chemical change". But the meaning is quite clear, for under the section heading "General Theoretical Considerations" they write, almost as if placating the wrath of scientific gods, "The idea of the chemical atom in certain cases spontaneously breaking up with evolution of energy is not of itself contrary to anything that is known of the properties of atoms, for the causes that bring about the disruption are not among those that are yet under our control, whereas the universally accepted idea of the stability of the chemical atom is based solely on the knowledge we possess of the forces at our disposal." This was no special pleading. Rutherford's work, although it helped to shatter the dogma of the indivisible atom, made no difference at all to the atomic theory as used by chemists, where the atom remains indivisible in all chemical reactions. The supreme irony of the historical situation is that Rutherford's work led him to provide, *en passant*, far clearer evidence of the actual existence of atoms than any chemist had ever been able to produce: Rutherford showed that atoms exist by splitting them.

This paper by Rutherford and Soddy which thrust the reality of transmutation on the world ("The Radioactivity of Thorium Com-pounds, II") covered more than just the chemical proof that thorium-X was produced by thorium. It of course calls in the support of uranium and uranium-X, but only briefly. The other main feature of the paper is the consideration of the rate of decay of radioactivity of thorium-X and the rate of recovery of radioactivity of the original thorium sample – which meant the rate of production of more thorium-X. This was expressed in a graph:

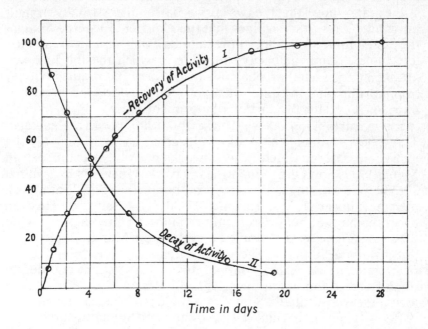

Time in days

The measurements on which this graph was based were taken, not for the sake of producing the graph at all, but to demonstrate the presence of thorium-X. However the measurements produced a major advance in the theory of radioactivity. They showed, just as Rutherford had earlier shown in the case of the emanation, that the rate of change of activity followed a geometrical progression with time. And they showed that all these radioactive changes followed a mathematical formula which is expressed as

$$\frac{I_t}{I_0} = e^{-\lambda t}$$

or some simple variation on the theme. The symbols mean that the rate of change of radioactivity follows an "exponential" law, and, what is more significant to most people nowadays, the equation introduces the idea of the "half-life" of a radioactive substance, that is the time it takes for the substance to lose half of its original activity.

Still at this stage Rutherford and Soddy considered that the energy of the radioactive atom produced in a transmutation, the energy that it gave out as radioactivity, was given to it by the transmutation – "the process of disintegration or transmutation [is] the source of the energy dissipated in radioactivity". Another look at the graph shows that the two curves both end as virtually straight horizontal lines – this signifies that the radioactivity of the original thorium sample, which they now claimed would consist of radioactive thorium and radioactive thorium-X which the thorium was producing, consisted of a balance made up of the radioactivity of true thorium plus an equilibrium between the decaying activity of thorium-X and the production of more thorium-X.

It is worth looking again at Rutherford's notebooks at this stage. They show (notably the books indexed as CLM 10 and 12) the incredible rate at which he was working. Almost every day in March 1902 shows great quantities of work. On Thursday, March 27th he was working until 10.00 p.m. but further results are given next morning at 11.30 and again at 11.30 on the Saturday. He even takes readings of beta-rays from uranium on Christmas day. One set of readings covers a sample every day from March 27th to April 7th including Sundays. There is excitement when "new radium from Paris arrives", and disappointment with "result bad due to small trace of radium in lower vessel", "certain, no effect when new vessel substituted". Sometimes there is a little confusion – the "effect of time on thorium-X" is measured on Sunday, March 29th, say the notes, but "the decay of uranium-X" is measured on Saturday, March 29th. There is some of Soddy's work here, too, because there is a note stuck inside with the brief legend "Please leave. F.S." And there are also the rough and finished versions of the graphs of radioactive decline and increase.

These notebooks also make it clear that at the end of March and into April 1902, the interest was shifting towards an examination of the radiations themselves and away from the chemical nature of the substances. But the combination of chemistry and physics was still vitally important. At this point Soddy returned to the chemical separation of uranium-X and thorium-X from their parent bodies and it was discovered that all the beta-radiation came away with the X-products but that the parents, uranium and thorium, emitted alpha-rays, and apparently alpha-rays only. Up until now it had been generally believed that alpha-rays were caused by the electrons of the beta-rays striking the parent material in their escape, just as X-rays were caused by the electrons of cathode rays striking material in the discharge tubes. This analogy was strengthened by the fact that neither X-rays nor alpha-rays could be "deviated" by electrical or magnetic fields. But if all the beta-radiation came

160

THE MANCHESTER SCHOOL

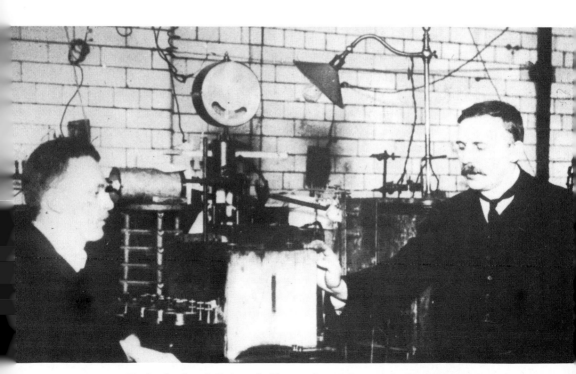

Above: Rutherford and Geiger (left) counting alpha-rays in Manchester.

Above: The Physics Board of the DSIR visits the RN Mining School at Portsmouth in 1921. In the centre row F. E. Smith is second from the left next to Rutherford, J. J. Thomson and W. H. Bragg. Tizard is on the extreme right.

Above: Rutherford in the centre of a group of friends in a garden at Broadway in the Cotswolds. The garden belonged to Madame de Navarro.

away with the X-products what could be causing the alpha-rays to be produced by the parent material? And what could the alpha-ray be?

At this point Rutherford produced his greatest experimental skill. He designed, made and operated, all in the course of a few weeks, an entirely new device. In essence it was simple, as all his experiments were. He made a series of narrow, parallel, rectangular channels of metal – like a metal box without two of its opposite sides split into narrow slices by metal partitions. Down these parallel channels he fired streams of alpha-particles from his strongest alpha-particle source, which was the French radium. At the opposite end to his source he blocked off half the exit from each channel with tiny strips of metal – the whole of the little box construction was only about five centimetres high. Then he fed electric current on to the metal of the box so that there was a strong electric field exerting its force across each channel. If the alpha-particles could be deviated by an electric field some would be caught behind the strips blocking half the exits and the total flow of alphas through the channels would be reduced. He had terrible problems getting a strong enough electric field yet one that would not spark across from one channel wall to the next, but he succeeded in cutting the flow of alpha-particles and thus proved that they could be deviated. Next he put his box between the poles of a powerful magnet. Here again he had troubles and had to borrow a stronger electromagnet from the electrical engineering department. But again he succeeded in deviating the alphas and trapping some of them behind the blocking-strips.

Then he applied the same sort of mathematics as J.J. had done with the cathode rays and he came up with some surprising results. The alpha-rays were streams of definite material particles and they were the size of whole atoms – certainly as big as hydrogen atoms and carrying either the same charge as the hydrogen atom or even twice as much. And finally, what surprised Rutherford most, although it was not relevant to the immediate investigation and was not pursued until later was that the alpha-particles were positively charged. These results confirmed a suspicion that had been steadily growing in the minds of both Rutherford and Soddy and provided concrete evidence for the final form of their "Theory of Radioactive Disintegration". They abandoned the idea that radioactivity was the dissipation of energy obtained by the daughter-product as it transmuted from the parent. Because the graphs of the change of radioactivity were so typical of those processes in which the rate of change is proportional to the amount of the changing substance that is present, they came to realise that the process of emitting radioactive particles is the same thing as the process of one element transmuting into another. Radioactivity and transmutation are the same process. The alpha-ray experiments confirmed this in a way that could easily be grasped – for if an atom of radium which is a heavy atom spits out something the size of a light atom, it is only reasonable to expect that the remainder of the original radium atom will have become a different atom of some other chemical element which is rather lighter.

In his Nobel lecture Rutherford recalled, "It had been considered probable that the radiation from any particular substance accompanied the breaking up of atoms. The proof that the alpha particle was an ejected atom of matter at once strengthened this conclusion and at the same time gave a more concrete and definite representation of the processes occurring in radioactive matter." Rutherford always wanted this concrete picture of what was happening even at a level where no pictures could be taken. He confided to J.J. as he sent him the results, "This result is of great importance in its application to radioactive bodies as it gives a mental picture of what is taking place in the succession of changes that is going on."

J.J. played an important part in their story at this stage, providing support and comfort in what was professionally a dangerous adventure. He was securing early publication of the full statement of the great new theory with all the supporting evidence. Rutherford and Soddy in the spring and early summer of 1902 completely rewrote the two papers they had sent to Sir William Crookes and which were going into publication in the *Journal of the Chemical Society*. They now presented the whole matter in the logical order of the final theory of disintegration and the additional matter – the full statement of the theory – they added as almost an appendix. Thomson got these published in the *Philosophical Magazine*, then the world's prime vehicle for announcing scientific discoveries and results. They appeared as two papers, "The Cause and Nature of Radioactivity" in the September and November issues of the magazine.

The long summer holiday had intervened and both men had taken a break but when they returned Rutherford and Soddy provided themselves with a beautifully conclusive piece of evidence to face the controversy they knew would come. Sir William MacDonald, the rich and eccentric benefactor who had built and equipped the MacDonald Physics Building in which so much of the work had been done, was prevailed upon to present the university with a liquid-air machine. This was at the time the very latest thing in scientific equipment, a machine which cooled air to such a low

162

temperature that the air became liquid. Using the supply of liquid air as a coolant Rutherford and Soddy were able to condense the emanations of thorium and radium into liquids, thus proving even to the most sceptical that these emanations, the undoubted products of solid matter in the shape of thorium and radium, were themselves real matter of a quite different chemical composition to the originating bodies; they were true gases at normal temperatures and like all other gases they could be liquefied at low temperatures. The typical and specific radiation of the emanations could be obtained from the drops of liquid in the glass tubes just as they could from the gaseous phase, and though these radiations were present in the air fed into the apparatus they were not present in the gas that came out beyond the cooling.

The machine was presented in October 1902, and Soddy reported the essence of the results to the McGill Physical Society in February 1903. It was formally published in the *Philosophical Magazine* in May of that same year. By then Soddy was working in London, at Sir William Ramsay's laboratory, and the proof of the material nature of emanations (which the Curies had doubted) was an armour against criticism, an armour they had briefly warned the world about in a short letter to the Chemical Society at the end of 1902. Three other papers with Rutherford and Soddy as joint authors, tidying up their work or giving additional results, were also published in 1903.

In later years the two men drifted apart and there was at times some acerbity between them, though never a serious row. Soddy's scientific spark went out after he had discovered the existence of isotopes in 1913, whereas Rutherford's interest never flagged. Yet it had been a happy and hectic partnership which Soddy remembered gratefully:

My relations with the then youthful Ernest Rutherford were uniformly cordial and inspiring, and I came fully under the influence of his magnetic, energetic and forceful personality, which at a later date was to cast its spell over the whole scientific world. My recollection of him was of an indefatigable investigator, guided by an unerring instinct for the relevant and important, and as an unequalled experimentalist seeing, amid all the difficulties, the simplest lines of attack . . . By the time our cooperation ended, radioactivity, which had already become a considerable jigsaw puzzle, had been put together, and my chief impression of those days remains of an intense mental exaltation as the pieces came together and they were fitted by the single theory of atomic disintegration into a convincing whole.

A convincing whole it was. It is still held as the basic description of radioactive matter today. Rutherford always held that the value of a theory or hypothesis was to be found in the amount of good work it inspired. On that count the disintegration theory is certainly a good one, too.

Both men always refused to be drawn into any controversy or even discussion over questions as to who did what or who thought of it first. Their correspondence shows that Soddy did much of the writing of the final papers but equally shows Soddy accepting and welcoming crucial phrases and explanations provided by Rutherford. But Rutherford, in writing a reference for Soddy who was applying for a post in Glasgow, insisted that it had been a partnership of equals from which any credit should be equally shared. And it may be added that they continued to disagree about the electrical theory of mass which had first brought them into contact with each other.

The theory of disintegration is universally accepted today. Soddy added to its strength immediately by helping Sir William Ramsay to show that radioactive change produced helium. In principle they put radium into a tube and showed that helium, which had not been there before, appeared in an evacuated glass container around the radium. This is actually caused by the alpha-particles escaping from the radium and alpha-particles are helium atoms. The experiment, although it confirmed that transmutation happened, sent many people off on the wrong track since it was not then known that alpha-particles were helium atoms. Soddy has a graphic description of the experiment in Ramsay's Mortimer Street laboratory:

> My impression of the actual first experiment is of being entirely occupied with the manipulation that was ultimately to admit the infinitesimal bubble of gas into the micro-spectrum tube, with Sir William observing with a small pocket spectroscope. As I worked Sir William's lab silently filled . . . Word had got round that the crucial experiment that was to settle this mad prediction was about to be made and the room filled up with students. Then came Ramsay's exclamation "It's helium" followed by that of the others who had brought pocket spectroscopes, which were passed round from hand to hand. I was forgotten and I suppose I was the last person in the room to see those significant lines of helium.

Soddy then examined the other rare gases that Ramsay had discovered for signs of radioactivity, but found nothing.

Rutherford's main line of research in the immediately following years was to try to sort out the complicated chains of elements that were produced by the radioactive decay series.

But there is one important theoretical matter to be mentioned. The graphs and the mathematical expressions of radioactive processes carry some implications. They say that radioactive decay is a matter of individual atoms changing. Any one atom of thorium or uranium may have a life-time infinitely long or infinitely short. But half of them will change to something else in the time given by the "half-life".

No one can say when or whether any individual atom will change. This is clearly different from the views held in the early days by many scientists, including the Curies, that radioactive energy is issued by all the atoms all the time, even if slowly. It is still unknown why any individual atom should radiate and change itself at any given moment. It was soon appreciated that the law of disintegration was, therefore, statistical – an expression of random occurrences – and this had considerable effect on the thinking of theoretical physicists; and it must be remembered that this was the period when Planck first introduced his quantum theory.

So to quote Soddy for one last time, in extracts from an article he wrote for the *McGill University Magazine* in December 1902, just a few weeks before he left, giving "An Account of the Researches of Professor Rutherford and his Co-workers".

> The researches have led to a new and more or less fundamental view being put forward of the nature of matter . . . The key to the history of its development at McGill is summed up in one word, measurement. Not perhaps of extreme accuracy, for pioneer work rarely requires it, but sure measurement, within a known small percentage of error, of effects often almost inconceivably minute . . . so the consideration of a curious set of phenomena in a small corner of science has led to a view, which, correct or not, has already outgrown the facts it was put forward to explain.

Almost the last joint words of the partnership, a sentence in the last paragraph of their last paper reads: "All these considerations point to the conclusion that the energy latent in the atom must be enormous compared with that rendered free in ordinary chemical change."

6

Life in North America

In following the torrential rush of Rutherford's research, all detail of his personal life, his appointment to McGill, his marriage and the birth of his child, his reaction to North American society and his effect upon it, has been swept aside. At least as far as his domestic life is concerned this strangely reflects the record, for there is no one left alive who remembers him from his Canadian days, and the flow of letters to Mary Newton naturally ceased when he married her. Only his letters to his mother might have told us more, and they have disappeared except for those few that were printed, against Rutherford's wishes, in the local New Zealand paper, *The Taranaki Herald*, in 1936.

But fortunately we do know a great deal about his appointment to McGill and about the institutional facets of his first professorship. McGill University in Montreal was one of the senior universities of the North American continent, having been founded in 1821. It received an enormous increase in its resources at the beginning of the 1890s when William MacDonald started taking an interest in it; he was an immensely wealthy man with a fortune made from the wholesale tobacco trade, despite the fact that he was a non-smoker who would have no one smoking about him. The list of his benefactions to the university is long, but in particular he agreed to finance the construction of an Engineering Building when all Canada could see the importance of providing that expanding country with a supply of well-trained engineers. In order to provide a basic course in Applied Physics for this great new engineering department, a physics building was also necessary, and this physics department also later proved useful in supplying basic scientific courses for the medical school. MacDonald, despite his own austere life, was exceptionally generous to the university. The appendage to the Engineering Building, the Mac-Donald Physics Building was the largest and finest physics laboratory in the world when it was completed in 1893. And, junior though the role of physics might have been, it was equipped to the highest possible standards. MacDonald paid its two professors and hired three demonstrators, a mechanic and an assistant. "Let us

166

have everything of the best," he said, giving £6,000 for equipment when he had been asked for only £5,000. And the money continued to pour out – there was £25,000 worth of apparatus in the MPB before the end of the century. These figures mean more than their mere money value. They demonstrate that the meaning of "apparatus" was changing and that Maxwell's distinction between the two types of experiment – the demonstration to students as opposed to the enquiry into nature – was coming to have a very practical meaning. By the time Rutherford reached the MPB it was dealing with about 300 students a year, but very few of these were looking for degrees in physics (indeed probably less than ten each year aimed at Honours degrees in physics), and somewhere around 200 were students of the applied sciences faculty, that is to say mainly engineers. A large proportion of the apparatus was therefore research apparatus, and it has been noted how readily MacDonald gave a liquid-air machine in 1903.

The first MacDonald Professor of Physics was John Cox, a man with strong Cambridge connections, a former Fellow of Trinity College, a quiet, reliable man, who gave Rutherford not only the moral support he needed in his scientific adventure, but also the time and freedom his research required. Cox designed the MPB and supervised its building, but it was not for himself that he installed the magnificent apparatus; he preferred teaching and administration, the dull jobs which the research man scorns. When the building was ready to start work in 1893 Cox went to Cambridge and recruited one of the best men from the Cavendish Laboratory on J. J. Thomson's advice. This was Hugh L. Callendar who was then involved in the task of developing the platinum resistance thermometer into the reliable technical tool it remains to this day. Callendar was just the sort of man Canada wanted, scientifically distinguished enough to be made a Fellow of the Royal Society a year after he arrived at McGill as second MacDonald Professor of Physics, the first of J.J.'s pupils to receive this honour although he had started with a degree in classics. He was an athlete and a skilled teacher, yet combined these talents with a modest and pleasant personality. He not only built up a powerful research programme of heat measurement, but he went outside the university and did research into highly practical problems among the steam engines and internal combustion engines of Montreal's industry. MacDonald, in particular, approved of this practical activity, and there was real concern both inside and outside McGill when Callendar left in 1898 for a Chair at University College, London, where the expenditure on apparatus was less than a quarter of McGill's generous financing.

Cox went back to the Cavendish Laboratory and J. J. Thomson to find a replacement, and he was joined there by Principal Peterson; so the intention must have been to reach a firm decision on a new professor of physics. The two men interviewed candidates in Cambridge in July 1898, but long before that Rutherford had mounted a serious campaign to get the job.

Among Rutherford's earlier biographers only one seems to have realised that the young man carved his way into McGill – nearly all the others see the picture as McGill picking the brightest young man from J.J.'s stable. Rutherford's letters to his fiancée tell a different story. He became aware of the vacancy at the earliest moment, and he was writing about it by April 22nd, 1898, just as McGill's academic year came to its end. But "I think it doubtful whether J.J. will want me to go in. There will probably be big competition for it all over England as the average man does not mind going to Canada, though he would bar Australia . . . The salary is not large, only £500, but still I would not mind taking it, but I think it extremely doubtful that I will compete for it. I will not go in unless J.J. advises me to." He thinks his friend, Townsend, will try, as he is "anxious to get something to do". And finally, "Personally, next to New Zealand, I would rather like Canada as I believe things are very jolly over there."

Three weeks later the tone is still similar but there comes a more sinister note: "I don't think I shall go in for this Montreal Chair. J.J. does not appear to wish me to, and as my chance would depend entirely on his backing it would not be much use my trying . . . It is possible J.J. may want me to stay on here. It will be advisable to see who are going in rather than to go in anyhow on one's individual merits which in these days do not count much unless one can obtain strong backing."

But by May 18th there is a very considerable change of tone and an obvious tugging at J.J.'s apron strings: "I have made up my mind to go in for McGill chiefly as a business matter because it will probably do me much good even if I don't get it. I settled to go in last Saturday and have been very busy about it since." We then see that the young Rutherford was very far from a callow and un-worldly researcher from the far Antipodes. He showed the political awareness that was to make him such an influential man in his later years. He went to see Sir Robert Ball who is "going to work a friend at Montreal". He went to see Dr MacAllister "who knows all the pots at Montreal and may say a good word for me". He collected testimonials which he soon had printed and bound in a booklet form. Above all, "There is one good thing about my going in, it may make J.J. act in regard to getting me something to do in Cambridge. I will probably go in for a Fellowship this year if I don't get the appointment, but my chances are very problematical."

Whether J.J. exerted himself over Rutherford's Fellowship is not known but he certainly wrote him a glowing testimonial: "I have never had a student with more enthusiasm or ability for original research than Mr Rutherford, and I am sure that if elected he would establish a distinguished school of physics at Montreal . . . I should consider any institution fortunate that secured the services of Mr Rutherford." In his own letter of application the young candidate had proudly mentioned his seven published papers and his current work on radioactivity. J.J. gave these papers his highest commendation: "These papers are distinguished by the importance of the results obtained, by the ingenuity displayed in the design of the apparatus, and by the grasp of the physical principles shown in the interpretation of the results." Sir Robert Ball had been "invariably struck by his devotion to science and his great skill", while R. T. Glazebrook, the senior demonstrator at the Cavendish wrote, "All this work is of a very high order of merit. Mr Rutherford shows great power of grasping the point which needs investigation and of arranging his experiments to elucidate it." E. T. Griffiths, who became Principal of the University College of Wales, added that he had "that valuable quality which I can best indicate by the phrase 'scientific keenness'."

The point has been made many times that Cambridge should have made more effort to keep this paragon. James Rutherford seems to have inspired the *Taranaki Herald* to make the charge when he published his brother's letters there in the late 1930s, but it is unlikely that James himself thought of this academic point – it is far more likely to have come from one of Rutherford's many pupils and friends who then held posts in his native land. The charge was repeated by Lord Snow and by C. L. Boltz in the 1950s, and they were both men who knew the scientific scene well. Snow wrote, "Cambridge however did not treat him particularly well: it was obvious that he was an experimental physicist of the highest powers, but, instead of being offered a Fellowship and showered with resources he was allowed to go to Canada."[1] This roused Sir Henry Dale to devote much effort in his draft biography to countering the charge, "to contradict a suggestion which appears to have gained some currency already and which might become an accepted tradition. This represents Rutherford as having been driven, by lack of proper encouragement to stay in Cambridge, to accept a position in Montreal of which the status and stipend were

[1] *Go* (Travel Magazine), June/July 1951.

by implication unworthy of the promise which he had already shown at that stage." Dale pointed out that a Cambridge professor of those times only got £600 a year compared to Montreal's £500. He asserted that Rutherford would easily have got a Fellowship the following year in the same way that Townsend did. And in any case Cambridge had no resources to shower on any young man – which was certainly true of the university of the time but not so true of its constituent, and often well-endowed, colleges. Dale, however, seems to contradict himself when he makes the strange statement that "the magnificence of the Montreal equipment was not a determinant or even an important factor in the decision".

Rutherford himself completely supports the charge against Cambridge, though on rather different and much more bitter terms. On June 2nd he wrote to New Zealand that he felt his chance of the McGill appointment had gone "up the spout", and his chances for a Fellowship were little better:

> As far as I can see my chances for a Fellowship are very slight. All the dons practically and naturally dislike very much the idea of one of us getting a Fellowship and no matter how good a man is, he will be chucked out. There is a good deal of friction over this research business which was intensified by my getting the Coutts Trotter [Scholarship]. I know perfectly well that if I had gone through the regular Cambridge course, and done a third of the work I have done, I would have got a Fellowship bang off . . . one has to face the situation squarely and not look always on the rosy side.

He was obviously in one of his black moods, and it is worth emphasising what his closest friends knew, that he was what they called "highly strung".

Nevertheless on July 10th, 1898, he got a cool note from J.J. "Cox and Peterson are in Cambridge and wish to personally interview the candidates for the Montreal Chair. Cox, and I daresay Peterson, will be at the lab tomorrow (Monday) morning and would like to see you if you could manage to be there. I find they attach much more importance to original work than Callendar." Rutherford knew perfectly well that Cox and Peterson were there, for later that week he wrote, "I was in Hall, on Sunday night and saw J.J. pointing me out to some man and was told it was the Montreal delegate." But J.J.'s note shows that Callendar had been doing some prospecting on McGill's behalf and had perhaps not been impressed by Rutherford, for the same letter describes the interview with Cox and Peterson and concludes, "I don't suppose I'll be appointed as I believe they want the man they appoint to be a lecturer with experience as there will be a good deal of that type of work to do . . . My chances are almost entirely due to research and that is only a partial consideration with them. Besides I think they consider I am rather youthful."

However by July 30th he had heard from various sources such as his sponsor, Griffiths, at Sidney Sussex College, and from a visiting American, Professor Ames of Johns Hopkins, that he had got the post. Cox had also called him to an unofficial meeting, and while unable to tell anything definitely, had reassured Rutherford that the bulk of the research work would be left in his hands. By this stage Rutherford was provisionally discussing marriage plans and research plans in alternate, excited sentences.

McGill is a very important place to be at, for Callendar was an FRS . . . and quite a pot in the scientific world, so I will be expected to do great things. I am expected to do a lot of original work and to form a research school in order to knock the shine out of the Yankees . . . Living, I should imagine is much the same as in New Zealand, but we should of course have to keep up a certain amount of style, and we ought to do it very comfortably on £400 and put by the rest.

And on August 3rd, "Rejoice with me, my dear girl, for matrimony is looming in the distance. I got word on Monday from Dr Peterson to say I was appointed to Montreal". His feelings were mixed. His friends were pleased and teased him by calling him Professor when he hadn't "a boot to throw at their heads". He was sorry in many ways to leave Cambridge, but "I think it would probably be much better for me to leave Cambridge as, on account of the prejudice at this place, it would be difficult to get anything to do, and if I did, it would be a rather unpleasant position". At no time in his life did Rutherford ever show any signs of having a persecution mania and we must take his word as evidence that there was, indeed, a considerable amount of jealousy and ill-feeling against the successful newcomer.

There was little left to do in England. Within a month he had shaken its prejudiced, but admired, dust from his feet. In the meantime he had reassured both his parents and the Newtons that he would be far better off accepting the £500 job at McGill rather than the upcoming possibility of a £700 job as Professor at Wellington, the nearest university to his home in the North Island of New Zealand. "My chances of advancement are much better in McGill than if I go out to New Zealand. There is also a certain amount of satisfaction in having a swell lab under one's control, and probably in New Zealand my chance of research work would be very small." It was a point he was to bear in mind all his life, and when he was a man of pull and importance he spent a great deal of time and influence trying to ensure that professors at Wellington and other colonial universities got time and support and even supplies of radioactive material in order to do modern research.

So with a fellow Cambridge adventurer, A. E. MacBride, sharing his cabin Rutherford sailed in the 5,000-ton passenger ship *Yorkshire* and had a thoroughly bad, stormy voyage, made slower still by fog along the first stretch of the St Lawrence River. Professor John Cox with his wife and son were also aboard the ship.

Rutherford actually landed in Canada on September 21st, 1898, but that was only for a short look at Quebec, since there was a further day's sail up the St Lawrence before he arrived in Montreal, where he was received with the greatest hospitality and taken to the home of Dean Bovey, head of the Applied Science Faculty. He was introduced to the laboratory benefactor, Mr MacDonald, at the first possible moment and quickly found lodgings on McGill College Avenue. But then his troubles started, both professionally and personally, and for Mary Newton, back in New Zealand, her hero's first letter from Canada must have come as a considerable shock.

Now I have to consider the weighty question of when I am to come out for you. I have been collecting data since I have been here and have come to the conclusion that I will have to stay here a year before I go out to fetch you. I start about zero as regards cash, and it would hardly be right for me to fix up till I have settled my bill at home. They say living is a third dearer in Canada than in England. Rents are very high in the town and a small pillbox house costs £100 a year. Supposing I stay here now for eighteen months and come out in April or May, I will probably have saved £400 and will then have enough to start house-keeping after paying off my educational expenses. If I went out there for you this year (as of course I should very much like to do, needless to say) I would hardly have enough cash after six months to go out and bring you back and we would start with short funds at once, so I really think, my dear, we will really have to postpone our partnership for eighteen months from now. It seems a very long time when the opportunity is so to speak at hand, but I think it will be much better in the end. I will have had time to get settled and things in order and so be able to devote more time to you when you do arrive. Moreover, I am afraid I could not get away early this year, as they don't like the young professors running off at once . . . I am sure you will like Montreal. A professor here occupies a very prominent position in society, far more so than in New Zealand, and one is not so buried as in New Zealand.

This is a difficult letter to interpret today, as the mores of society have changed so much. Plainly the cost of living in Montreal had come as a shock to the young man – earlier in the same letter he had complained about the difficulty of renting any apartment with a good sitting room. In our time we are accustomed to thinking that the cost of living is higher in a metropolitan area, and cheaper in the country, that a young man may take a job in some compara- tively remote place in order to save and return to life at the centre with some financial backing. But before the First World War, before the USA had become the greatest industrial centre of the world, the transport costs of capital equipment and even of tech- nical expertise from Europe put a considerable margin on top of Canadian costs – the economic balance was still weighted in favour of Imperial Europe. The salary of £500 offered by Mr MacDonald plainly looked different in Montreal than it did in Cambridge, and Mr MacDonald, who himself lived on £250 a year, considered that his professors should not need any more. Furthermore Rutherford here refers, for the only time in his letters, to the money he had borrowed in New Zealand to finance his break out into the intellectual world of one of Britain's older universities. It seems likely that this borrowing may eventually have amounted to the £400 he mentions and that most of the money came from his elder brother, George, now established as a successful farmer in New Zealand. It almost certainly did not come from his parents, for there is no mention of loans or borrowing in the letters to his mother, and it seems unlikely that any came from Mary Newton's family. However Rutherford clearly decided that the honourable course was to repay his "grubstake" before taking on any further commitments.

This is not a romantic letter, even by the standards of an age when a young man did not marry unless he could support his wife. It illuminates the true nature, then, of the relationship between Rutherford and Mary Newton, which was for both of them the relationship of two shy people clinging to the one person they really knew. Rutherford at this stage of his life was a good looking, well-set-up young man, with enormous faith in his own powers and the knowledge creeping upon him that he was destined to be highly successful. He must have known that if he had waited another couple of years he could have had the pick of the girls of English middle-class intellectual society. But he chose to stick to a girl who was virtually the provincial sweetheart of his youth. There is not a trace of sexual adventure in the record of the remainder of his life, and those women who remember him personally insist that, though always polite to them, he never showed any trace of interest in them or in any other woman. We must agree with C. P. Snow that there was a deep streak of sexual shyness or diffidence in the man throughout his life.

Again we have to rely on the memories of those who knew Mary Newton when she was Lady Rutherford, and again we have a diagnosis of a deep shyness given to account for her abrupt, sometimes haughty, and often difficult, manner. She would have been happier living out her life in her native, provincial New Zealand society, and she was never really comfortable in the mannered, affluent circles of Cambridge in the 1920s and 1930s. As a widow she finally chose, after living some years in Cambridge, to return to New Zealand.

None of this is to say that Rutherford's marriage was not successful – probably it means just the contrary, for in times when divorce was rare and difficult, and when sexual athleticism was not considered a desideratum, the partnership was stable and a protection for the inadequacies they both, presumably, felt.

There were professional problems, too, which rapidly took the brightest shine off the prospects offered by the New World – the first was his need to learn to lecture, the second was his need to face comparison with his predecessor, Callendar.

There has always been considerable dispute about Rutherford's ability as a lecturer. Some men have recorded memories of his brilliance, clarity and enthusiasm. Others have written of him as dull, appearing to be bored, confusing, and inaccurate. Many of the "Rutherford stories" concern his lectures and the way in which he would often get lost in the mathematics on the blackboard and then carve his way out of the difficulties either by a cavalier change of a minus into a plus, or by turning to the class and telling them to work it out for themselves. Yet there are any number of distinguished physicists who admit to having had the direction of their lives and work changed by the sheer enthusiasm and evangelism of Rutherford's lectures. When all the evidence is assembled it becomes clear that the dividing line between those who found his lecturing brilliant and those who found it dreadful lies not so much in the lectures themselves as in the attitudes of the auditors. Those who were, or were destined to become, distinguished scientists or physicists of Rutherford's own school or cast of mind found the lectures marvellous; those, especially undergraduates, for whom physics was simply a course to be taken as part of a syllabus, whose prime interests were elsewhere, found them bad. Undergraduates in his Cambridge days complained that he did not even cover the subjects advertised and necessary for the examinations, yet the chorus of praise for his great public speeches at scientific gatherings was loud and apparently unanimous.

Professor Norman Shaw, who heard him lecturing at the start of his McGill career, sums the matter up:

> When the assurance is recalled with which Rutherford discussed the major problems of physics in learned societies, and the ability with which he handled men, it is surprising to remember that as a young man he was a nervous lecturer, particularly when dealing with elementary topics for undergraduates. His lectures on electricity and magnetism to large classes of second-year engineering and arts students at McGill were at first above the heads of the students and revealed a feeling of despair in regard to the previous mathematical and scientific training of his class. Those of his students who were interested in physics, however, caught something of the fire with which he inspired all his advanced collaborators.

The difficulty was not just his own – McGill physics students had organised a round-robin petition when Callendar first started lecturing, demanding that he bring his presentations down to the level they could understand. And over the years the admirable John Cox took over more and more of the lecturing as Rutherford's research results earned him the justification of more and more time spent in the laboratory.

Sir Henry Tizard, who will figure largely in this story when Rutherford becomes one of the leading figures in the new world of government-science, gives the most penetrating account of Rutherford as communicator.

> He took great pains in making his written papers clear and concise; they are models of what scientific papers should be. But although his writings were uniformly good and clear, Rutherford was never a good speaker except on his own subject. On formal occasions, or when he had to speak extempore, he would hesitate, fumble for words or repeat himself. When he had to make an official speech with notes perhaps supplied by others, those who knew him well would fidget in their seats, waiting for the time when he would put the notes aside and say to himself, almost audibly, "now let's tell them about something interesting", and become himself again. But when he lectured about his own work he was superb and unique . . . I can almost see him now, standing in this famous lecture room, massive and commanding, demonstrating to an enthralled audience the transmutation of matter by high-speed protons and deuterons. A train of valves amplified the effect of the transformation of a single atom to such an extent that it could operate a counter within sight and sound of the audience. He showed first the transmutation by protons. The counter ticked slowly as the process went on. "Now," said Rutherford, "if you will allow me we will

bombard the same target with deuterons and I think you will observe a greatly accelerated rate of transformation." The assistants made the necessary adjustments and the current was turned on again. The counter ticked if anything rather more slowly than before and the audience began to titter. The assistants came forward to see what was wrong. "No," roared Rutherford defiantly, "leave it alone". The words were hardly out of his mouth when the ticker immediately went off with a rush and the audience dissolved into laughter and cheers. I was told afterwards that some of the valves used were sensitive to sound.

Nearly all this exposition is a clear statement of the failures and successes of Rutherford in communicating. The last sentence must not be ignored, for it is a scientific joke, not easily understood by anyone who is not accustomed to scientific phrasing – the implication is that it was Rutherford's shout of protest that made the counter tick and satisfied the audience and was not really a successful demonstration of deuteron transformation of matter. Tizard here was delivering a Rutherford Memorial Lecture to an audience of scientists and he made exactly the same assumption as Rutherford himself – he spoke in language only scientists would understand.

Tizard's point is amply demonstrated in the only recording of a Rutherford lecture that is extant. It was recorded in secret at Göttingen University when Rutherford was being given an honorary degree in 1932. Rutherford's British admirers only discovered its existence several years later and then they persuaded the record company "His Master's Voice" to make a professional pressing of the speech. It is evidence of the extraordinary width of the circle of Rutherford's admirers and friends that this transaction was made possible by Sir Ernest Shoenberg of the EMI Company, the man whose inventions made television a practical public service, and whose son, now Professor David Shoenberg of the Cavendish Laboratory, was a pupil of Rutherford's. This heavy and expensive "album" of records achieved a satisfactorily large sale and copies of the records are still in existence, in some cases preserved by the families and descendants of the original purchasers. The recording shows exactly what Tizard describes – a slow and hesitant opening delivered very much in a public-oration style, full of historical references to the connection between Göttingen and the English royal family. But by the end Rutherford is talking freely, talking personally, about sub-atomic particles and gamma-rays. Even now the listener is entranced because the speaker gives the feeling that he is opening his mind, letting you into the secret of his thoughts.

All trace of the public oration has gone; he is speaking as an individual to individuals. The New Zealand accent seems to have disappeared; the voice is standard middle-class English with an occasional tone, especially in the vowel sounds, that might be mistaken for a West-Country intonation.

It was at the end of this lecture that Rutherford made his most famous statement about himself. Without overtly criticising the theorists who thronged the physics department at Göttingen, he demanded simplicity. "I am a simple man," he said.

Rutherford must have been warned about the difficulties students were having in understanding his McGill lectures, for there was never any official complaint or recorded student unrest. He must therefore have adapted his style to those who were not going to make physics their career, and he continued to give lectures to general elementary classes all his life – plainly it was regarded by him and others as one of the Professor's duties. At McGill he introduced, on his own initiative, courses of advanced lectures, notably the course Soddy took on radioactivity, and a postgraduate course on electrical waves and oscillations – a return to his old wireless interests. Eve says that Rutherford overcame his initial problem by dogged perseverance at learning the craft of lecturing. But Cox took over more and more of the teaching work as F. P. Walton, Dean of the Law Faculty, remembered: "Cox was an accomplished and versatile man and a good teacher but, as I was always told, had no flair for research. When he realised what a young eagle they had got, he took over a lot of the drudgery which embitters the lives of professors and gave Rutherford much more leisure than a junior professor usually gets."

Rutherford's second major problem lay in the enormous reputation that his predecessor, Callendar, had built up.

Callendar essentially inherited the Rayleigh tradition of the Cavendish, the belief that meticulous measurement of physical quantities was the next step towards developing the mechanisms and devices that were useful to mankind. This was the attitude of the 1880s when Callendar started research, and when it was felt that there was not a great deal more to be discovered in physics. The same decade had been, for Rutherford, the period of his basic education, the time when attitudes are formed. And among sciences less depressed than physics, in biology, anthropology and geology, for instance, this was the period when, with the Darwinian view of evolution established, the publicists and formers of opinion were proclaiming most loudly that science could and would demonstrate truth. The idea was not itself new, having been propounded most notably by Francis Bacon two hundred and fifty years earlier. But in the latter half of the nineteenth century it was the most shocking and most exciting feature of the argument between science and religion, the claim that truth was revealed not only, or not at all, by religion and the Bible, but instead by science. In the world of the physical sciences the situation had been somewhat different; here men who were primarily engineers, much such as Joule and Carnot, had uncovered truths about heat and energy by working primarily with machines and measuring the quantities involved. Rutherford was plainly influenced by this intellectual climate of his student days. He was never interested at all in religious matters, and he adopted, apparently without ever arguing the matter, the position of the "pure" scientist, the man whose goal is the uncovering of scientific truth.

And so his attitude to his predecessor at McGill changed from the letter written at Cambridge when he had just been appointed. "I think my appointment is a very much discussed matter at Cambridge as Callendar was considered a very great man . . . Your acute mind will at once gauge my importance if they place me in his shoes when the beard of manhood is faint upon my cheeks," to "I am getting rather tired of people telling me how great a man Callendar was, but I always have the sense to agree. As a matter of fact, I don't quite class myself in the same order as Callendar, who was more an engineering type than a physicist, and who took more pride in making a piece of apparatus than in discovering a new scientific truth." That was written within a year of his arrival at McGill. It is interesting to note that this was Rutherford's attitude to machinery throughout his life – he did not despise it, he was fully aware of its importance in the laboratory, but it was not what interested him in his work. An American scholar describes this as "an expression of snobbery", whereas I suspect that most British scientists would regard it as a perfectly reasonable, and perhaps rather admirable, statement of personal preference.

It appears too that some people at McGill were a little worried by the choice of Rutherford to succeed Callendar, for it is well-recorded that Principal Peterson openly expressed the University's regret at losing Callendar, only to be told rather sharply by J.J.: "I don't know why you should; you've got a better man in his place." It also seems certain from Callendar's correspondence that he had met Rutherford, probably during the summer of 1898 between his departure and Rutherford's accession, and the older man had not taken to his successor.

Callendar had left two important things at McGill. The first was a small but strong research group of three demonstrators and three graduate students working in the field of calorimetry (heat measurement). The group was led by Howard T́. Barnes. Taking the most easily demonstrable measurement, the number of scientific papers published, this group was more active than Rutherford and his early recruits throughout the first four years of Rutherford's period at McGill – though no one could claim that their work was more significant than the fewer papers on radioactivity. We have here a classical recipe for a great and long-lasting academic conflict, which would have put Barnes, the man-in-place superseded by the youngster from outside, in continuous opposition to Rutherford. And Barnes was no mean scientist – his work, begun in 1895, on the mode of formation of ice in rivers in relation to water temperature and the development of frazil-ice which blocked water intakes and shipping routes, was a major contribution to scientific knowledge and to the practical solution of keeping Canada running throughout its bitter winters. Later he went on to develop a method of measuring air temperatures so accurately that a ship could detect the approach of an iceberg in fog or darkness – an invention that seemed of great interest coming at the time of the sinking of the *Titanic*.

It is the best evidence that Rutherford was truly a man of fine, generous and straightforward character, with an extraordinary ability to win the affection and regard of others, that out of this potential disaster area he first formed a scientific collaboration with Barnes in which they did important work on the heat production of radium; and he went on to form a life-long friendship with his potential rival. Of course Barnes himself must take half the credit for this outcome – perhaps even more than half the credit, for he also took on a larger share of the teaching load in the MPB and reorganised the practical classes. Yet in their subsequent years of correspondence Barnes, though never subservient, is always the junior figure, calling on Rutherford for aid to get government support and finance for his practical experiments on the river and at sea.

The second thing that Callendar left at McGill was "a piece of apparatus" in which he took great pride, a patented resistance box which was the centre-piece of the calorimetry work, measuring the specific heat of water by a complicated equilibrium of compensated resistance coils. It had been specially made, at some expense on the MPB account, to Callendar's instructions. When Callendar left McGill he made an arrangement with the patient John Cox that Barnes should continue to use the box in order that the main flow of the work might go on – but McGill would have another identical box made so that the original could be sent on to Callendar in London. When it did not come Callendar, quite wrongly, felt that

Rutherford was "partly to blame" and wrote angrily to him to this effect. It seems likely that he felt Rutherford had captured the powerful support which Cox could offer and that without this support Barnes, a mere demonstrator, could not command the funds to get the duplicate box made. Cox however had continued as always to give Barnes every possible support and when Callendar's letter arrived, the box was already being constructed and the original was about to be sent off. Callendar and Rutherford never became friends or colleagues.

Rutherford was faced with the task of building up his own research school in radioactivity from zero. His first aide, we have seen, was R. B. Owens, who started the research on thorium emanation in the few months before he left for Cambridge, but who was in any case from the Faculty of Applied Science. Soddy, of course, was a chemist, and in no way part of Rutherford's laboratory. So the team had to be built from those very few students who were taking the Honours Physics course. In due course Rutherford recruited four of them, R. K. McClung, Harriet Brooks, S. J. Allan and A. G. Grier. McClung was the first member, joining in Rutherford's early work on ionisation, then winning an 1851 Exhibition Scholarship himself and going to the Cavendish for a year; he later became Professor of Physics at the University of Manitoba. Harriet Brooks, the first example of Rutherford's willingness to place women on an equal footing with men, at least in the laboratory, worked with him on trying to estimate the atomic weight of radium emanation by measuring the diffusion characteristics of the gas. When Rutherford insisted on naming her as co-author of the paper giving the results, she protested that he was too "generous" in giving so much credit to someone who had been no more than a "humble assistant". Harriet Brooks was then at Bryn Mawr College, and on the strength of her paper with Rutherford she was awarded a scholarship to Cambridge. Rutherford personally made arrangements with J.J. for her admission, reception and accommodation. After Cambridge she returned to Canada where, fairly soon, she decided to marry Pitcher, one of the demonstrators at MPB, an older man who worked in the Callendar group. Rutherford expressed some disapproval of this marriage in cautious terms to his friends, and it looks as though he was possibly correct on personal grounds and certainly correct on scientific grounds, for Mrs Harriet Pitcher moved into the Montreal social scene and eventually became famous for her gardening after leading a not very happy life.

The third of the group of undergraduate collaborators, Samuel J. Allan, was a most regular correspondent for ten years. Rutherford had obviously taken great trouble to help with the correcting of his first paper which was to be submitted to the *Philosophical Magazine* and Allan was contrite: "I was very much chagrined over the return of my paper . . . I never was very good at English composition . . . I am greatly obliged for your kindness and foresight in this matter and will take your criticisms to heart." In 1903 Allan took a job as demonstrator at Johns Hopkins University in Baltimore, where the head of the Physics Department, Professor Ames, wished to start a programme of radioactivity research and offered to purchase instruments and radium. Ames, reported Allan, was primarily interested in the mathematical side of physics but was "no good on the experimental side". Furthermore the majority opinion at Johns Hopkins "did not agree very well with some of your latest papers". However Ames "seems very interested in radioactivity but somewhat hazy in his notions of it. They are fully alive to the importance of it, however."

Arthur G. Grier is the least known of Rutherford's young collaborators, though ironically it was his work that was most important. He was studying "the analogy of radiations", concerning the similarity or otherwise of the alpha, beta and "penetrating" radiations from the different radioactive substances, uranium, thorium, and so on. There are hints of the disintegration theory ideas in the Rutherford–Grier paper and the results of the work were also useful in showing that the various radiations, far from being "analogous", were more likely to be specific for each radiating element. Yet Grier disappears from the records after becoming a demonstrator in the physical and electrical engineering laboratories of McGill.

Rutherford's letters to his old master, J.J., tell much about the conditions in the MPB at that time. "I like the life here very well. No one bothers me in my department and I have my own sweet way in most things. Cox . . . does what he can to cut down my lecture work so I have a somewhat enviable time of it. We have rather too small an endowment to buy much in the way of new things but I can rub along very comfortably for a few years," he wrote in January 1900. And in that letter on the same subject of equipment, there was another passage which makes one wonder about our present attitude to radiation and its safety. Rutherford complains that he is having some problems with his source of X-rays which gives out radiation far stronger and more penetrating than he really wants: "I can see the shadow of the hand in open daylight thirty feet off." In other words, the X-rays were throwing pictures of his own hand on to a wall thirty feet away, yet there is no record of Rutherford ever suffering from burns or any other radiation damage.

In these years Rutherford not only had his four students doing experiments, he also had all his own radioactivity work at hand, and the continuous measurement of decline or increase of activity from the many samples Soddy had prepared. These would be in different rooms on all the five floors of the MPB. The chief memory of one of the early visitors, a memory Mr R. S. Willows retained forty-five years later when he spoke at the Physical Society in London in 1948, was of the "frantic rush" round the entire building and up and down the stairs as Rutherford careered around recording all the readings from the electroscopes and electrometers.

At the same time we can see the enormous network of friends and correspondents that was to be such a feature of Rutherford's later life, beginning to build up. While from Cambridge W. La Trobe writes in the style of almost boyhood friendship to "Dear Old Rutherford" and C. C. Farr writes from New Zealand about the domestic and political activities of the group centred round Mrs Newton (now Rutherford's mother-in-law), the first of his Montreal pupils, R. K. McClung writes from Cambridge about a recent visit to Paris where J.J. had given him letters of introduction to Langevin and Perrin, and more significantly, a visit to the laboratories at Manchester: "It is a very fine lab . . . However so far I don't think I have seen yet a lab that can beat the MPB."

Meanwhile John Zeleny, who had been one of the advanced students at the Cavendish with Rutherford, and who both trusted his scientific judgment and admired his work, had been appointed Professor at the University of Minnesota at Minneapolis. As early as 1900 he had written, "I am about ready to believe that most anything is possible. The facts you present are certainly strange and more light on the nature of these things will be of still greater value." Much of Zeleny's time was taken up supervising the construction of the new physics lab and in 1902 he had a budget of 10,000 dollars to equip it. He sought Rutherford's advice in great detail about the number and types of instrument he should buy and he remarked in July of that same year that Rutherford had "scientific London at your feet".

Rutherford had been offered at least two professorships in America, at Yale and at Columbia when that university attempted a major coup: it sought to get J.J. as its senior physics professor and Rutherford as second professor. There was therefore a most illuminating letter from McGill to J.J., where Rutherford first says that he is going down to Washington for a meeting of the Physical Society – "It is my way of keeping in touch with what is going on down South." Americans, he declares, are very keen on meetings and societies, and there will be forty-six papers in the physics section of the AAAS (American Association for the Advancement of Science) but "The papers rather lack individuality and are in many cases a rehash of old work." Only then does he come to the subject of Columbia's offer to them both, which had "somewhat surprised" him. Rutherford does not feel he would "better himself either financially or from a research point of view" especially as the cost of living in New York is "ridiculously great". He has therefore "refused to consider the proposition" and he does not imagine it will draw J.J. from Cambridge either. But obviously he has, in fact, given the matter some thought and made some enquiries: "Besides I think that the present status of Columbia in physical sciences is miserable and a very drastic weeding is required of the junior staff to make the place worth having. I speak from experience as I have sized them up individually and collectively."

Columbia was quick off the mark in offering Rutherford a Chair in 1902 for it was in the years 1902 to 1904, with the publication of disintegration theory, that Rutherford first became internationally known to a wider public, and it was also the time when scientific honours started to be showered upon him. Writing and giving lectures in response to the many invitations he received began to take up a considerable portion of his day – even to the extent that friends warned him against letting such functions occupy too much of his time and energy. And there can indeed be found a falling off in the amount of work he published in the years 1905 and 1906, which represent the laboratory work of 1903 and 1904. Not only did the number of papers fall from twelve or thirteen a year in 1902–3 to nine or ten a year, but few of these 1905–6 papers are major ones and a number are little more than short notes.

More important, as a result of his growing fame and his increasing scientific reputation, the first foreign students began to apply to work with him. First of all was a woman, Fanny Cook Gates, who had been at Bryn Mawr, Göttingen and Zurich, and who came to McGill – perhaps attracted by the success of Harriet Brooks in a world where women scientists still found life a hard uphill push. She spent the year 1902–3 with Rutherford while on leave from Goucher College, Baltimore. She was followed by H. E. Bronson, as soon as he had obtained his PhD at Yale in 1904. Tadeusz Godlewski was the first of the Europeans, and he eventually became Professor of Physics at the University of Lemberg in what was then his native Poland. Greatest of all Rutherford's McGill collaborators was Otto Hahn, who became the world's leading radio-chemist, a Nobel Prize winner, a man whose experiments first showed the natural fission of uranium, the crucial piece of work which opened the door to the atomic age in 1939. With Hahn came Max Levin, also from Germany, later to be followed by his brother-in-law, Gustav Rumelin. There was also Arthur S. Eve, technically a Canadian, a former Cambridge Wrangler who was teaching in secondary schools, but was attracted to research by Rutherford's name although by then aged forty-two.

The work these scientists performed as members of Rutherford's first international "school" was not of outstanding interest, rather of solid worth in building up the main structure of radioactive knowledge. Fanny Gates showed that the loss of "excited activity" by heating was due to the deposited radioactive product being driven off – volatilised – rather than an ordinary physical process affecting the basic atomic property of radioactivity. Hahn had already discovered radio-thorium, a new radioactive element between thorium and thorium-X, a discovery made almost by accident when he was working in Sir William Ramsay's laboratory in London, and while at McGill he discovered radio-actinium. Godlewski had missed this during his research into the decay products of actinium. Eve's work was mostly on gamma-rays and he managed to get a long way towards showing that they are, in fact, electromagnetic radiations similar to X-rays.

Eve tells his own story of his failure to make a sensitive small-capacity gold-leaf electroscope when requested to do so by Rutherford. Rutherford asked "Lester Cooke used to make them, why can't you?" – and told Eve to get the help of the laboratory mechanic. The mechanic made an instrument that was beautiful to behold but difficult to work with – once charged up, its leaf would collapse into uselessness within twelve hours. Eve continues:

> This puzzled me. One night I could not sleep and got up in my diggings and made an electroscope of a tobacco tin, an amber mouthpiece of a tobacco pipe and some Dutch-metal foil; charged it with sealing wax and went to sleep. The leaf of this home-made freak electroscope remained open and charged for three days and solved the problem . . . all the material inside the [laboratory] building was contaminated and coated with active deposit including the slow period transformations of radium. Rutherford said, "Good Boy" though I was eight or nine years his senior in age.

Contamination had not before appeared as a problem – from now on Rutherford knew he had to take action to keep contamination of the laboratory under control for the rest of his working life.

Hahn remained Rutherford's admirer throughout his life, despite the interruption of the First World War, and he continued to give public thanks for the opportunity he had been given to work with him. "Everyone who has had the good fortune to have worked directly under or with him has experienced with joy the magical influence of his magnetic personality and has felt something of it flashing over to himself, as it were," he wrote in *Naturwissenschaften* in celebration of Rutherford's sixtieth birthday; and "The writer of these lines considers it the greatest fortune of his life that it was granted him for nearly a year to work under Rutherford at Montreal when a young beginner and to tread humbly in his footsteps," were his words on Rutherford's death.

But Otto Hahn, certainly the greatest of Rutherford's "foreign" pupils, also had less formal memories of this formative year. He remembered the dark cellar full of nooks and crannies where much of the radioactive lab equipment was kept, particularly a Toepler air pump which worked slowly and not always satisfactorily – "But Rutherford did not let himself be discouraged by initial failures . . . This enthusiasm, which nothing discouraged, infected all the fellow workers.' There was also the occasion when a photographer wanted a picture of the young Professor in 1906; various pictures were taken but somehow the photographer felt that his subject was not appearing "grand" enough – perhaps it was because he was not showing enough of the white cuffs below his coat sleeves that were then considered the sign of a gentleman at work. Rutherford had no such attachments, but he borrowed Hahn's and it was in Hahn's cuffs that he finally appeared to the wider world, for the photo achieved considerable circulation.

Of course Rutherford's enthusiasm for his own subject could be rather irritating to others whose interests lay elsewhere, as Hahn remembered:

> Although at this time scarcely thirty-four years of age, Rutherford would impress the stamp of his powerful personality on any meetings whether scientific or social in character. There were meetings for the discussion of physical and chemical problems. It happened not infrequently during the discussions on some problem of organic chemistry, the beloved alpha-particles were smuggled in, and quite unexpectedly, though welcome to all, the latest problems of radioactivity were the centre of the discussion.

Nevertheless the facet of Rutherford that most impressed Hahn, as it was to impress so many others in the years to come, was the personal attention and kindness that Rutherford extended to young newcomers – "which aroused especially in us Germans, astonishment and amazement, as we had expected that so renowned a professor would be conscious of his own dignity, and be rather unapproachable".

Just how different Rutherford's methods and manners were

from those pertaining in a German university is well illustrated by Andrade. He described the Heidelberg University of 1911, when Lenard was Professor of Physics, a man with a mighty reputation as the discoverer of the release of electrons by light, the first man to show that particles could pass right through atoms:

> Twice a week at fixed times Lenard went round the research laboratory followed ceremonially by his Assistant Professor . . . In the course of such a tour of inspection . . . every research worker was visited and his work discussed, with everybody standing, briefly or at greater length according to the progress and the problems it presented. Very remote from this formality were Rutherford's daily rambles round the laboratory, in the course of which if there was a difficult or interesting point to discuss, he would sit on a stool for some time throwing out incisive comments and suggestions. He talked to the research worker as to a friend or collaborator . . . The physics colloquium was also a contrast . . . At Heidelberg . . . Lenard took the chair, very much like the master with his class. He had a habit, if any aspect of his work was being treated by the speaker, of interrupting with "And who did that first?" The speaker would reply with a slight bow, "Herr Geheimrat, you did that first," to which Lenard answered "Yes, I did that first." . . . Lenard much resented that the first discovery that the electricity released by light consisted of electrons was often claimed for J. J. Thomson, whereas he actually had prior publication. At the Manchester colloquium, which met on Friday afternoons, Rutherford was, as in all his relations with the research workers, the boisterous, enthusiastic, inspiring friend, undoubtedly the leader but in close community with the led, stimulating rather than commanding, "gingering up" to use a favourite expression of his, his team . . . He was always full of fire and infectious enthusiasm when describing work into which he had put his heart, and always generous in his acknowledgment of the work of others.

Rutherford's behaviour at McGill must have been very close to his later manners at Manchester, but at McGill Rutherford himself was subject to occasional visitations. On one occasion, Eve records, the young professor rushed into his room saying breathlessly that he was to open his windows, put away his pipe and hide any trace of his tobacco. Eve complied immediately, but not unnaturally asked what all this was about. Rutherford responded, "Hurry up, MacDonald is coming round the laboratory." Sir William MacDonald, as has been mentioned, considered smoking a filthy habit although he made his fortune out of selling tobacco.

There are two important aspects of Rutherford's impact on McGill as an institution that must be considered. It has been argued, very cogently, by Dr Lewis Pyenson, that there was an attempt to transmit physics, as it was known in Europe, to the Western Hemisphere around the year 1900. He takes as examples the very considerable investment in money and manpower by the Germans into Argentina, and by the English into McGill, culminating in Rutherford's spell in Montreal, but he argues that this transmission failed, or at least was incomplete. Dr Pyenson holds that although Rutherford gained international fame for McGill, the very brilliance of his results and their universal appeal to scientists of all disciplines weakened and virtually destroyed the foundation of a true "school" of physics at the university. "Though it enchanted scientists abroad, however, the laboratory suffered an underlying malaise, one due in part to the spectacular success of Rutherford's experiments," he opines and points out that physics at McGill was problem-orientated and that the broad education offered by German universities was lacking.

What is more the problems on which McGill scientists worked were divergent – on the one hand was the Callendar–Barnes stream of careful, accurate, heat measurements:

> It represented work well suited to the MacDonald Physics Building, and it was an orientation guaranteed to appeal to large numbers of future engineers, chemists, geologists and doctors. In joining this line of research demonstrators and lecturers would find an affinity between the teaching required by the Director of the Laboratory, John Cox, and their own graduate research. At the same time they could anticipate a career in setting up an analogous, if less ostentatious programme at another institution. Reasonable men and women, most advanced physics students at McGill, followed this trajectory. Many were deflected however by the second line of research at McGill, that of radioactivity. As a convinced scientific imperialist Rutherford drew bright younger colleagues and graduate students from their more conventional thesis research to assist in his experiments.

And Dr Pyenson adds, "For many teachers and students physics at McGill was a problematic science: it lacked distinctive characteristics . . . In part because of its universal appeal Rutherford's research offered only a vague identity for physics," and he points out that many of Rutherford's collaborators – Boltwood, Hahn, Soddy – were chemists and Rutherford's own Nobel Prize was to be given for chemistry. It is certainly true that the physics department at McGill suffered a long and disastrous decline after Rutherford's departure. There were never enough advanced students attracted to Honours courses or further degrees in physics even during his time in Montreal. There were barely twenty MScs in the ten years after he had left. It is also absolutely correct that he did not leave behind him a flourishing group of workers in radio-activity.

As far as McGill itself is concerned more of the responsibility for the decline of its physics department can be laid at the university's door than at Rutherford's. Very shortly before he left, terrible fires destroyed the medical and engineering buildings. Nor did McGill settle on a satisfactory institutional system for higher and research degrees until long after Rutherford had left. During his entire tenure there had been fierce arguments about the nature of the degrees to be awarded – the original B.A., requiring a knowledge of Greek and Latin was supplemented by BSc, MSc, and even DSc degrees and the PhD was also started as part of a graduate school, but only one PhD was taken in physics between 1906 and 1926, and that was by R. W. Boyle, one of Rutherford's recruits. (Rutherford was to face similar arguments and campus politics at both Manchester and Cambridge when he moved to these universities.) Furthermore, McGill's financial situation was insecure; as Dr Pyenson points out the university was largely financed by private donations from businessmen of British stock, in the predominantly French-Canadian society of Montreal where secular education as a whole was viewed with considerable distrust. Sir William MacDonald himself was the archetype of these benefactors and later in his life (he died in 1917) he concentrated most of his benefactions on the School of Agriculture. Partly because of the shortage of funds, the faithful John Cox took an early retirement in 1909, in far from happy mood. Finally, at a time when Rutherford's pupils and inheritors were establishing themselves as professors and lecturers at McGill and other Canadian universities there came the world war, disrupting not only the flow of possible research students, but actually bringing many of his friends across to Britain to join him in performing distinguished work in scientific research into antisubmarine warfare – Eve and Boyle were outstanding examples.

Even when the war was over McGill did not recover preeminence in physics – this passed to Toronto where John C. McLennan took over as Professor. McLennan was himself Canadian-born, but what is more relevant he had become a friend and ally of Rutherford even before the war and when he moved to Toronto he took with him not only Rutherford's moral support but a great deal of practical help and considerable funds from the Royal Navy, which Rutherford and W. J. Pope, the Cambridge Chemistry Professor had organised, for a plant to obtain helium from the Canadian natural gas wells.

But by looking at McGill alone a false perspective is given of

Rutherford's impact on physics in North America. The wider view covering all Canada and much of the USA must be taken. Rutherford pupils from McGill, Manchester and Cambridge eventually occupied influential posts at Queen's University, Ontario, at Saskatoon and Alberta, at Dalhousie in Nova Scotia and in the Canadian National Research Council laboratories. Nor can it be coincidence that at Chalk River the Canadians have achieved a nuclear power centre to match any in the world and that Canadian expertise and designs in nuclear reactors and nuclear power stations give that country virtually the only "independent" nuclear engineering tradition outside the USA, Russia and Britain.

However it is in the USA that Rutherford's enthusiasm seems to have struck most sparks. Yale, where his friends Henry A. Bumstead and Bertram B. Boltwood worked, was the chief centre of his influence. Boltwood in particular, who was soon to become one of the world's leading radiochemists, continued Rutherfordian radioactivity research there for twenty years and also worked with Rutherford in Manchester. Rutherford's friends and pupils succeeded to Chairs at Princeton and Columbia, and while it is impossible to prove a direct historical line of descent from Rutherford's visits to California it was there that the first major non-European contributions to nuclear physics – as radioactivity had by then become – were made in the 1930s with Anderson's discovery of the positron and Lawrence's development of the cyclotron.

The early years of the present century were a time of laboratory building, in which McGill, with the MPB led the field. But much of Boltwood and Bumstead's efforts around 1910 were taken up with Yale's new laboratory and we have seen Zeleny's preoccupation at Minnesota with building and Columbia's ambitious plans, which can be taken as examples. In Europe at this time the only really large laboratory for physics was being built up at Leyden in Holland by Kamerlingh Onnes, the prototype of the "Big Machine" physics lab which is now so familiar. J.J. had a truly international research team at the Cavendish, but this was the only one of its kind in the world for physics, and in physical terms the laboratory was still small and poor, its annual bill for equipment and supplies only a few hundred pounds. Probably J.J. put his finger most accurately on the basic weakness of Canadian physics as an institution when he wrote in 1912:

It is a weak point in the universities overseas that so much value is attached to buildings. The buildings and laboratories make one's mouth water, but with all that there is a great dearth of any means of enabling poor students to continue their work and education after they leave university. I am quite sure that if that could be remedied, that if you could only constrain your millionaires to spend their money on men instead of on buildings, the results would be very much better.

"If I were founding a University, I would found first a smoking room; then when I had a little more money on hand I would found a dormitory; then, after that or probably with it, a decent reading room and a library. After that, if I still had more money that I couldn't use, I would hire a Professor and get some text books." This quotation comes from one of McGill's most famous sons, a faculty colleague of Rutherford's. The author was Stephen Leacock, a scientist, by title "political scientist", but such an iconoclast that he made his name as a humorist rather than as a university worthy. And the ghost of Rutherford might well ask what a man should do in order to win the respect of his academic descendants. For while Dr Pyenson has concentrated on the failure to found a "school of physics" comparable to those in Europe, Professor Lawrence Badash gives the quotation from Leacock in a paper in which he "accuses" Rutherford of being one of the founders of "Big Science" and of planting the seeds of this development while he was at McGill.

Badash admits that Rutherford, throughout his own scientific life, remained a "string and sealing wax" experimenter, indeed "few have done so much with so little in the way of elaborate apparatus". And he admits that during his McGill period the MPB was so lavishly equipped that Rutherford could avoid "the seamier side of Big Science", the money grubbing and grant-securing politics that figure too largely in the life of today's successful professor. But he continues:

> Another characteristic of Big Science is the ability to convince your colleagues and the public that your speciality is of great importance. This has the obvious benefit of encouraging a flow of funds and graduate students, and support for them, as well as bestowing less-apparent institutional leverage in academic infighting and in the distribution of prizes, medals, honorary lectureships, and memberships in prestigious societies.

Rutherford's achievement at McGill according to this interpretation was to pull physics out from under the shadow of engineering, to which it was so plainly subordinated in all the original planning of the MacDonald buildings. Badash continues:

Rutherford in particular raised the respect toward research to that accorded to a natural resource. He did this in two general ways: by actively proselytising – being a missionary – and by happily, but without conceit, accepting the honours that came his way. He was clear in his own mind that physics was the fundamental science; chemistry was "stinks" and all other branches of natural philosophy were merely stamp-collecting. Within physics, radioactivity was of primary importance because it was a basic phenomenon of Nature, to be compared with electricity, magnetism, heat and gravitation, and because it promised to unlock that puzzle confronting all physicists, an understanding of the nature of matter.

Rutherford's many lectures at US universities and his frequent talks to the American Physical Society, the long list of scientific papers published by himself and his collaborators, his occasional popular articles, including even one in *Harper's*, and his frequent missionary visits back to Britain, all are seen by Badash as significant, perhaps especially as signifying the transference of science into the hands of middle-class people and away from the blue-blooded amateurs of earlier years.

The argument is summed up by quoting Eve's famous remark charging Rutherford with always riding the crest of a wave, and Rutherford's even more famous reply, "Well, I made the wave, didn't I?" Badash concludes by suggesting "that Rutherford not only made and rode the wave of modern physics, but of Big Science as well. In both cases there were other important figures, to be sure, but Rutherford's own career seems to hold the seeds of this transition, a change that effectively began in Montreal."

In my view Badash cuts out the heart of his own argument in a passage where he quotes Alvin M. Weinberg, Director of the Oak Ridge National Laboratory, who pointed out that Big Science all too often spends money rather than thought, using everything that money can buy in a shotgun assault on a problem, hoping that something will hit the target. Consequently Big Science is all too often wasteful science, "thriving on publicity and often emphasising the spectacular". Rutherford always told people to "Go home and think" instead of spending late hours in the laboratory. And so Badash writes:

But if the research in his laboratory covered a spectrum of topics and if he was sometimes willing to try any "damn fool experiment" on the off-chance that it might work, the shotgun approach was not his style. He was far more inclined to plan his laboratory's attack on the unknown . . . And if he was aided by publicity that emphasised the spectacular, its origin could not be attributed to him. Such publicity moreover was not always beneficial, for the rising medical demand for radium resulted in steep price increases and limited availability for the scientists. In fact Rutherford was a sober, conscientious leader of his field, and leadership is a vital feature in our discussion of Big Science. Rutherford was instrumental not only in setting the direction of radioactivity research, he strongly influenced the style of scientific activity.

7

Last Years in Canada

Rutherford's influence was international; he would and did cooperate with and correspond with any scientist of any nationality – he genuinely believed in the search for scientific truth anywhere, and he would later send all Kapitsa's equipment to Russia so that the work could go on. He had no interest in big machines, only in the results they could provide. In his Cambridge days he actually refused to act on offers of financial support and he delayed far too long in starting the building of a cyclotron although pressed to begin the work by Chadwick and Cockcroft. But he did believe in the development of science, both pure and applied, as a national resource. If Rutherford did anything to bring about the development of Big Science it was not in his reluctance to build big machines but in his enthusiasm for government support for science.

In this respect his missionary journeys and proselytising back to Britain were more important than his parallel work in the USA, for with his rise to international fame on account of his scientific results, he began to move into influential circles in the British scientific "establishment". At the same time, he needed to secure the support of this establishment in the acceptance of his disintegration theory, which, we must remember was not only revolutionary but an attack on the accepted paradigms, at least of chemistry. It seems clear that Rutherford must have been angling for election to the Royal Society, the greatest honour that British science can bestow, as early as 1901, or even 1900. For in February 1901 J.J. wrote to him, "With regard to the RS, Williams is up again and will I hope get in. I think it would hardly be politic to run another candidate prominently connected with the Cavendish Lab as there would be very little chance of them both getting in, while one alone would be almost certain, so I should think the best policy would be to run one at a time." But a year later Rutherford had obviously obtained J.J.'s support and his nomination for the great honour – and still he was not accepted. On May 2nd, 1902, a chagrined J.J. wrote:

I am exceedingly surprised and vexed that you were not elected to the Roy. Soc. I put you up and everyone I spoke with regarded your election as certain. I think myself the result is a great scandal but the council of the Society is a very bad one this year; there is no one on it who knows anything about the modern developments of physics . . . They just seem to have taken the longest standing candidates and Larmor always supports a Belfast man.

This last comment on the mutual support of Ulstermen may have been a little unfair, for Sir Joseph Larmor was to become one of Rutherford's most powerful allies and supporters, especially when the question arose of a successor to J.J. at the Cavendish.

However, it was in 1902 that Rutherford' and Soddy's results really began to flow into the literature of science. J.J. seems to have accepted their revolution with little difficulty. In May 1902 he acknowledged receipt of the latest papers on radioactivity, saying, "It seems to me that your explanation clears up a great deal of obscurity." 1903 was the peak year in this matter – the year in which the full disintegration theory was published. It was also the year in which Rutherford was elected Fellow of the Royal Society; and at the same time it was the year in which he first returned to England to "defend his thesis" in public. J.J. was a supporter of the revolution from the start and the experimental evidence from Canada seems to have carried conviction to most men immediately – men such as Rayleigh, Larmor and Sir Oliver Lodge for instance. Within a month of the great Rutherford and Soddy paper of May 1903, Lodge devoted half of the Romanes lecture at Oxford to the subject. He left himself plenty of escape routes but he appreciated immediately the importance of the theory: "Plainly if an elementary form of matter is found to be throwing off another substance . . . here is a fact, if fact it be, of prodigious importance . . . Assuming the truth of this strange string of laboratory facts, we appear to be face to face with a phenomenon quite new in the history of the world." A week later Pierre Curie, lecturing at the Royal Institution in London, was much less interested in the subject and said little about it except to throw some doubt on the wisdom of accepting radioactive evidence only for the existence of new forms of matter.

Despite Sir William Crookes' willingness to publish the Rutherford Soddy papers in the journal of the Chemical Society, the chemists, led by H. E. Armstrong, provided what little opposition there was to the disintegration theory. They were helped among the physicists only by Lord Kelvin, the seventy-nine-year-old doyen of the profession, and by the reluctance of the Curies and Becquerel to accept the fact that the energy evolved by radioactive substances must all arise from a purely internal source. Soddy was by now in England working with Sir William Ramsay to provide supportive evidence for the theory, and to England Rutherford came, too, towards the end of the summer of 1903, to open a discussion at the Southport meeting of the British Association.

This was one of the very few occasions in his life when Rutherford seems to have been truly nervous about his position and his reputation. Kelvin had gone so far as to prepare a short "broadsheet" outlining his opposition, and Rutherford wrote to a number of friends asking for their presence and support. This lobbying turned out to be unnecessary, however. Larmor had welcomed the news that he was coming to England by writing to him, "You may be the lion of the season for the newspapers have become radioactive. I see that you are again monopolising the *Phil. Mag.*" Sir Oliver Lodge spoke strongly to support Rutherford's opening statement at the meeting on "the emanations from radium". And Lord Kelvin did not appear, pleading ill-health. The only outspoken opposition came from H. E. Armstrong, who supported Kelvin's written case that radium must receive its energy "by absorption of ethereal waves" and said he was "astonished at the feats of imagination to which I have listened". The chemists, he declared, "certainly had no evidence of atomic disintegration on earth". But the day was Rutherford's and he returned to Canada with his position secured.

In the late months of 1903, Rutherford indulged in some controversy in the letters column of *The Times* newspaper, engaging a Professor Meldola who had suggested that radium was no more than a compound of helium formed in prehistoric ages and slowly decomposing. It was the only time in his life that he joined in public controversy, and for ever afterwards he preferred to settle arguments in private, or even to avoid arguments altogether and wait for experimental facts to settle the dispute by themselves.

The acknowledgment of his triumph and his position came in 1904. First he was invited to give the famous Bakerian Lecture to the Royal Society. He was later presented with that Society's Rumford Medal. For these two occasions he made two transatlantic trips. It was also in this year that he wrote and published, for the Cambridge University Press, his first book, *Radioactivity*.

To write a book, especially a complete review of a whole subject, carried much more prestige in those days than it does now, when there is so much more scientific publication. It was to be acknowledged as the authority on the subject. The book is dedicated to J. J. Thomson – "A tribute of my respect and admiration." Rutherford's position is made quite clear in the Preface:

The phenomena exhibited by the radioactive bodies are extremely complicated and some form of theory is essential to connect in an intelligible manner the mass of experimental facts that have now been accumulated. I have found the theory that the atoms of the radioactive bodies are undergoing spontaneous disintegration extremely serviceable, not only in correlating the known phenomena, but also in suggesting new lines of research. The interpretation of results has to a large extent been based on the disintegration theory, and the logical deductions to be drawn from the application of the theory to radioactive phenomena have also been considered.

The strictest modern philosopher of science could hardly find fault with this – the theory is a working hypothesis with the great advantage that it really works. He set about calming the fears of chemists and other opponents right from the start, beginning the first chapter with:

The close of the old and the beginning of the new century have been marked by a very rapid increase of our knowledge of that most important but little known subject, the connection between electricity and matter. No subject has been more fruitful in surprises to the investigator, both from the remarkable nature of the phenomena exhibited and from the laws controlling them. The more the subject has been examined, the more complex does the constitution of matter appear, which can give rise to the remarkable effects observed. While the experimental results have led to the view that the constitution of the atom itself is very complex, at the same time they have confirmed the old theory of the discontinuous or atomic structure of matter. The study of the radioactive substances and of the discharge of electricity through gases has supplied very strong experimental evidence in support of the fundamental ideas of the atomic theory. It has also indicated that the atom itself is not the smallest unit of matter but is a complicated structure made up of a number of small bodies.

There is much of the essential Rutherford in this first paragraph alone: the delighted enthusiast getting pleasure from "fruitful surprises"; the man who goes right to the heart of the matter, showing that the fundamental views of the nature of matter are at stake; the essentially kindly man, reassuring those who might feel that the whole of their professional world was being struck down; the absolute belief in experimental evidence overruling any theory; and the clear, bold writer.

The full statement of the disintegration theory comes on page 2 and is followed by the clear declaration that this involves "a veritable transformation of matter". Then he takes energy in his stride.

On this view the continuous emission of energy from the active bodies is derived from the internal energy inherent in the atom, and does not in any way contradict the law of the conservation of energy. At the same time, however, it indicates that an enormous store of latent energy is resident in the radioactive atoms themselves. This store of energy has previously not been observed on account of the impossibility of breaking up in to simpler forms the atoms of the elements by the action of the chemical or physical forces at our command.

In our nuclear age we may find this a pregnant paragraph. Rutherford himself was to be the very first man to find a way of breaking up atoms, but until the very last years of his life, he would see no way of doing this on a large enough scale to make it possible to tap this enormous latent energy for man's use or self-destruction.

The book is virtually a complete review of the current state of knowledge of radioactivity. One particular item of interest is that it contains the first steps towards building up our modern picture of the radioactive "families" of elements – the forms of matter produced by continuous radioactive changes from the parent elements of uranium, thorium and actinium. The decipherment of these "genealogical trees" in their final form would not be completed for many years. At various times Rutherford would exhibit some especial interest in these problems, but on the whole his attitude towards the study of "radiochemistry" was one of fatherly goodwill towards the workers such as Boltwood, Hahn, Soddy and the French collaborators of the Curies who pursued the matter. It would require the enunciation of the Radioactive Displacement Law and the discovery of the idea of isotopes (Soddy's second major achievement) and ten years' more work before the matter was virtually concluded.

In the following year, 1905, Rutherford, with considerable labour, produced a second, and much larger, edition of *Radioactivity*, and it was this second edition which was normally regarded as the "standard textbook".

In England Kelvin remained unconvinced for a few more years, and he argued mightily in places such as Strutt's country home – Terling, the seat of the Rayleigh family – and in the letters column of *The Times* in 1906 (on the specific subject of the transformation of forms of matter) but the weight of experimental evidence bore him down and likewise overwhelmed the few remaining chemists who continued to be sarcastic about the "suicidal tendencies" of radioactive atoms. Disintegration theory had won the battle with great ease and remarkably little loss of blood. Rutherford was now ensconced in the ranks of acknowledged "great scientists" in

America, Britain and even in Europe where at just this time Planck's Radiation Laws, with their strange "quanta" of energy were beginning to make an impact on theoretical thinking about the nature of matter and energy, and where the young Einstein was beginning to publish revolutionary papers.

Indeed by the end of 1904 Rutherford was suffering more from his friends than from any enemies. Soddy had, quite naturally, been lecturing about radioactivity and a series of these lectures was being reprinted in *The Electrician* as they were given. It was part of the agreement that they would appear as a book when the series was complete. Rutherford became very irritated about this. Although Soddy made full acknowledgment of Rutherford's role in the partnership – "this task could hardly have been possible but for the training I have received in the subject from Professor Rutherford" – Rutherford wrote some very strong letters, objecting that Soddy's effort was bound to affect the sales of his forthcoming first book, quite apart from the propriety of Soddy writing first about it at all. Whatever Rutherford may have said in his calmer moods about a truly equal partnership, and whatever modern scholars may argue, there can be no doubt that in 1904 he felt very much the senior partner with almost proprietorial rights over the subject. Soddy agreed not to publish the lectures in book form until three months after Rutherford's book appeared.

But J.J. had to be called in to smooth things over, and he wrote on February 18th, 1904, that Rutherford must not worry about either the lectures or the rival book. "I don't think you have any reason to fear that people will be misled because no one familiar with your papers can fail to see that Soddy's lectures were taken straight out of them." J.J. added that he had recently heard Soddy lecture in Cambridge and had been "not very favourably impressed", finding his delivery "tired" and with "no enthusiasm". (Rutherford's book did in fact come off very well. He was able to tell his mother that it had received "very good reviews", and many of his personal friends wrote to him in most flattering terms, even predicting that it would be a "classic".)

This letter of J.J.'s, however, introduced another subject that was to be a constant irritant over the coming years – the behaviour of Sir William Ramsay. J.J. wrote, "I have been utterly disgusted with Ramsay's behaviour and have said so on many occasions. I always, when speaking on the subject, point out how you had foreseen and indeed planned the experiments, and that if it had not been for you they would never have been made. It is a great pity his morals are so inferior to his manipulation."

Ramsay was an immediate convert to the disintegration theory –

indeed he was too enthusiastic. We have seen that, with Soddy, he provided enormous support for the disintegration theory by showing that helium was evolved from radium emanation. It would be tedious to recapitulate the many scientific disagreements between Rutherford and Ramsay over the next ten years; it will suffice to say that Ramsay, or men in his laboratory, claimed to have found transformations of many elements into quite different elements under the influence of radioactivity or even of X-rays. These claims almost invariably arose from slack experimental methods such as allowing air to enter the equipment or failing to realise that there is always a certain amount of helium in the metal parts of a discharge tube. Rutherford and his true allies such as J.J. had to spend considerable amounts of time and effort disproving Ramsay's claims with scientific evidence and countering the effects of his many excursions into the more popular media. Ramsay had brought to a high pitch of efficiency his methods for securing scientific priority for his discoveries of rare gas elements by sending in notes and preliminary announcements to the Royal Society publications. He tried to use this device to secure his priority for a claim to be the first to obtain the spectrum of radium emanation, after Rutherford had described his own work on these lines in a private conversation. But he failed to account for the hydrogen present, since Rutherford had not mentioned his method of eliminating this particular problem and so his "preliminary announcement" was discredited when Rutherford published his own more careful and definitive results.

The anti-Ramsay theme is fairly continuous in Rutherford's private correspondence – there are no less than twenty critical passages in the Rutherford Boltwood letters for instance, including accusations against third parties of "Ramsamania" and pugnacious descriptions by Rutherford of how he and others, including Sir James Dewar, "danced on" Ramsay during a public argument at the 1908 British Association meeting in Dublin. Sometimes this led to misjudgments, as when he and Boltwood described Hahn's radio-thorium discovery in Ramsay's laboratory as "a mixture of thorium-X and stupidity", a judgment they were compelled to withdraw when Hahn came to McGill and convinced them with hard evidence.

But it may be that Rutherford learnt from Ramsay's lordly disdain. In the five years after the publication of the disintegration theory, we have seen that he did once indulge in a newspaper controversy, and that he did get into public disagreement with Ramsay at the British Association meeting in 1908, an occasion on which Ramsay spoke "in an airy manner" against the deduction of atomic weights by the alpha-particle method, described his own results showing that radium emanation changed into neon, and then left. Battle was resumed on a later day when Rutherford openly said that the neon Ramsay had found "was due to the air let into his apparatus". A report in a Dublin paper, carefully cut by Rutherford and sent to Boltwood ended: "Sir William Ramsay was called upon, somewhat maliciously I thought, to reply to each attack as it was delivered. He parried them with an easy air of superiority, expressing his hope that 'out of so much heat, light may come'. It was a drawn battle."[1] Rutherford certainly did not consider that it was a drawn battle, but he may well have concluded that it was a useless battle. In his other controversy with Ramsay, in that same year, over the demonstration of the spectrum of radium emanation, acknowledging that Ramsay had shown a photograph of his spectrum to the Royal Society on June 25th, 1908, some ten days before his own preliminary notice was sent off to *Nature*, Rutherford merely comments that "it will be of interest to compare the two spectra". It seems that he had decided that an "easy air of superiority" was the best public attitude when he knew that the experimental facts would convince his scientific peers that he had the better case. Certainly he never again entered into public controversy.

Consideration of the Rutherford–Ramsay relationship has taken the story far ahead of the McGill period, but there is another long-running thread in Rutherford's life that also requires treatment outside the strictly sequential. This is his first love, radio transmission. It continued to occupy a place in his thoughts and even to take up some of his time all through the first three years in Montreal. It had particular advantages as a popular subject for lectures and he got exciting experiments which could easily be shown to a large audience. Thus in August 1902 he wrote to his mother about a popular lecture he had given: "I had the largest audience they have ever raised at McGill. People were stored everywhere, including some who were looking through a radiator in the roof. The experiments went off very well." He could also mount a lecture on this subject at short notice when another speaker fell out at the last moment. These lectures probably led to his restarting some experiments in practical radio and Rutherford has the strange distinction of being the first man ever to transmit radio signals to and from a moving train. This work was done (we do not know for what reason) on the Toronto Grand Trunk line and Rutherford was able to write to his mother: "They came off

[1] L. Badash, *Rutherford and Boltwood; Letters on Radioactivity*, Yale, New Haven, 1969, p. 188–189.

very well with the train travelling at 60 miles an hour. We got signals over eight miles from the station." That was in October 1902, and presumably the work was dropped under the pressure of the final rush of work on the disintegration theory.

Rutherford's correspondence with scientific colleagues reveals however that he had not only kept his interest in the work, but had even considered taking out patents on his original detector, and had made some improvements to it. There are letters, for instance from J. A. Pollock of the University of Sydney in Australia in both 1901 and 1902, discussing his use of the magnetic detector in experiments on the conduction of waves along wires, where it had been found "most excellent", and Pollock describes his plans for further work with it. A series of letters from J. Erskine Murray, writing at first from Bedford and later from University College, Nottingham (now the University of Nottingham), tell a more exciting story. The first letter is dated June 12th, 1900, and after referring to their acquaintanceship at Cambridge says: "My latest invention has been a receiver for wireless telegraphic signals based on your electromagnetic detector. In its very rough state it works very well, though it is not so rapid or sensitive as Marconi's latest. I would say it is as good as his when he brought it over to this country." Erskine Murray describes how he has just been sacked by the Marconi company as an economy measure, so he can hardly be considered an unbiased witness when he describes the management of the company as "commercial men" and "rather a shady lot", but he suggests, "If you care to take the thing up we might patent it in the States, Canada, etc. I can hardly use my name as I have so recently left the company and like Marconi personally." He goes on to explain that Marconis are already having a number of patent fights in the USA with Dolbear of Tufts College, in Massachusetts, who is "backed by the Whitney syndicate", and so in a postscript he advises Rutherford again to patent the device at once "as it is much simpler than Marconi's and, if worked out properly, may be quite as efficient".

In a second letter in October 1900, Erskine Murray writes in such a tone that Rutherford has plainly given consideration to patenting his detector (though Rutherford's letter is not extant), and explains that he had done his first experiments while actually on the Marconi company staff. However there are no drawings, the equipment was broken up at the time of Erskine Murray's dismissal and no other experiments can have been done because all the other experimental staff have been dismissed. Furthermore Marconi's existing patents will not cover the device because they specify a "coherer" as the essential receiver. The essence of

201

Erskine Murray's idea was to remagnetise the Rutherford steel needles by "thick wire coils projecting over the outer end of the detectors". Erskine Murray wrote again in November and December 1900, but must have received no further encouragement from Rutherford for he then announced that he was going to put in a provisional patent himself. Eighteen months later, in July 1902, he admits he never did take out the patent as he had neither the money nor the opportunity to experiment at the right time.

This last letter arrived precisely when Rutherford's interest in the matter had been re-aroused. Marconi had lectured on "space telegraphy" at the Royal Institution on June 20th, 1902, and he had described a magnetic receiver which he had successfully employed to receive messages between Poldhu in Cornwall and Poole in Dorset. Rutherford sent off a stately letter to *The Electrician* which was published on July 25th, 1902. In it he recites and gives references for his early work on radio-wave detection and adds that he had decided that his detector "was inferior in delicacy to the coherer". Rutherford says that his magnetic detector is well-known among physicists and is constantly used in many laboratories in various parts of the world. But "In the press of other scientific work I have not devoted much further attention to the experimental side of the subject." (A coherer detects radio-waves by their effects on fine particles of magnetised iron in a container; it will be remembered that the disadvantage of Rutherford's receiver was that it needed to be remagnetised after it had demonstrated the arrival of each signal.) The vital sentence of the letter follows: "I have however used in the laboratory (for more than a year) a device very similar to that employed by Marconi in his latest form of receiver – viz, an endless moving band of steel passing through the solenoid in which the electrical oscillations are set up." This steel band performed the remagnetisation necessary to make Rutherford's detector of single signals into a receiver of continuous radio messages.

But there is no claim against Marconi, and the writer hopes he will have further successes. It is a simple claim for scientific priority and quite remarkably unworldly. Yet this was the ground on which Rutherford stood fully ten years later when there was yet another Marconi patents case pending in the courts, this time in England. Rutherford's evidence as an expert witness was sought on the basis of this letter to *The Electrician*. But a letter to the solicitors for the British Radio Telegraph and Telephone company made it quite clear, on November 5th, 1910, that Professor Rutherford was most emphatic in his desire not to be associated with either side in the dispute; that he did not want to be subpoenaed; that he would most strongly resent any attempt at compulsion in securing his evidence on what was a straight matter of historical fact. Professor Rutherford had done his work on the moving steel band for remagnetisation a year before he had written his public letter, but when he saw Marconi's paper of 1902 he had "thought it only right to draw the attention of scientific men to his previous work and his experiments with moving steel wires". The original phrase used here, amended in Rutherford's own hand was "let the world know he had openly used in his laboratory a detector of the type described by Marconi in his paper".

Professor Rutherford, the solicitors are told, holds a position "of complete detachment and disinterestedness . . . he adheres to his letter to *The Electrician* . . . He has never exploited any of his scientific work for commercial purposes or with a view to patenting . . . He fully and freely publishes results of all his scientific work and therefore his research possesses greater value." "This work is not done behind closed doors but with a view to adding to the world's knowledge," it was added.

Plainly the well-established Ernest Rutherford, Professor of Physics at Manchester University, had conveniently forgotten the hopes of commercial exploitation briefly held by the young immigrant from New Zealand some fifteen years earlier. But he cannot be condemned as hypocritical. The attitude expressed here is precisely the attitude held by the vast majority of British académics throughout the first forty years of this century. It is an attitude which is documented as being the position held by the Medical Research Council when the question of patenting penicillin was raised in 1940, an attitude which prevailed until the British Pharmaceutical Industry found itself forced to pay licence-fees for the deep-fermentation process developed by American industry for the manufacture of penicillin. It can be argued that it is still the prevailing attitude, and that the setting up of the National Research Development Corporation in the immediate post-war years, precisely to take over the patent rights and development financing of inventions made by academics, is an attempt to preserve the moral position of the publicly funded research worker while retaining the commercial benefits of his work for the public funds.

Certainly Rutherford never took out a patent in his life and was remarkably uninterested in obtaining personal property. To return to his McGill period, it was only in the last couple of years in Montreal that he acquired a plot of land on the West Mountain outside Montreal. It was described as having "a fine outlook towards the Lake of the Two Mountains" and Rutherford had plans for a house on the site. Rutherford left Montreal before any building started, and got Eve to manage such affairs as the mortgage, from which in later years he received a small income of less than fifty dollars a year. So, apart from a country cottage in Wiltshire which he acquired in the very last years of his life, Rutherford never even owned his own home.

In his seven years in Montreal there seem to have been only five things which affected him enough to bring forth a personal comment in his correspondence, apart from matters scientific or official. There is not a single word anywhere about political or social matters: the Boer War receives passing mention, but only in terms of pride that New Zealand has been well forward in sending contingents, and in qualified approval of the Boer "stand" at the St Louis International Exposition (where he also approved of the magnificent electric lighting). His daughter's birth was treated in the "chaffing" tone again of a man who was not truly interested in such "women's matters". He wrote to his mother:

> The baby, to Mary's delight, is a she and is apparently provided with the usual number of limbs . . . It is suggested I should call her "Ione" after my respect for ions in gases. She has good lungs, but I believe uses them comparatively sparingly compared with most babies. The baby is of course a marvel of intelligence and we think there never was such a fine baby before. I hope Pater is well – enjoying life – and meditating on the probable number of his grandchildren.

In August 1902, however, when he was in the middle of proving disintegration, writing his book, and renewing his interest in wireless, he did take some relaxation in the form of visits to the theatre, and Mrs Patrick Campbell in *The Second Mrs Tanqueray* drew from him one of his very few literary comments: "Very well done, and was the best thing we have had in Montreal for some time."

Rutherford was always impressed by the grandeur of nature. One of his first letters from Canada to his mother, written on February 19th, 1899, describes the St Lawrence in winter:

> Three of us walked over the frozen St Lawrence . . . and a long walk it is. The river is two miles wide at Montreal and is frozen solid the whole way across. It is rather interesting to walk and see sleds and sleighs going where three months before great ships were lying at the wharves. The river does not freeze flat like one would naturally expect, but like one's back lumber yard. Great blocks and hills of ice are scattered over it where the pressure of the river has piled up ice blocks on top of one another.

Later he was also impressed by the Grand Canyon (August 1906).

He was however more interested in a new friend, Jacques Loeb, the biologist at Stanford University, and he exhibited the accuracy of his scientific "nose" in spotting the importance of Loeb's work, which remained largely unrecognised until our own day. Rutherford reports delightedly on a day spent in Loeb's laboratory where he was allowed to do his own experiments in parthenogenesis of sea urchins: "I first tried the normal method with the spermatozoon and saw the egg develop in the course of a few hours. I then developed the beggar from the egg till he swam by adding the appropriate chemicals without calling in the aid of the male." For once the physicist enthused about another science: "They are really wonderful experiments and he appears to me to be on the right track for great discoveries. Loeb is a short dark man, German Jew by origin, but not semitic looking, about forty-five and getting grey. I like him very much and find we have a lot in common. He is very modest but a terrible enthusiast, worse than I am."

Rutherford helped with the washing up as the Loebs had no maid, and was fascinated by his host's description of his escape from the terror of the " 'Frisco quake", carrying his sick wife on a wagon, being held up by hoodlums and "calming them with a rifle". The same letter (July 6th, 1906) carries an unusual complaint: "I am worried with reporters but trust I shall escape without much damage," and indeed his summer lectures at Berkeley were well received and drew satisfactory audiences.

Rutherford was remarkably free from scientific prejudice, keeping his mind flexible and unencumbered by presuppositions. The disintegration theory had a logical corollary that Rutherford was the first to see: if radium, thorium, actinium and uranium had been steadily disintegrating and producing daughter elements in families ever since the elements had been "born", then the proportions of elements in the rocks of the earth would give a measurement of the age of the rocks. Radioactivity was a steadily ticking clock that should be capable of telling the lapse of time since the earth was formed. There were two ways in which the clock could be "read" with the knowledge available in the first decade of the century. Either the amount of helium present in radioactive minerals could be compared with the amount of parent radioactive element, and a calculation could be done based on the belief (Rutherford's belief at any rate) that alpha-particles were helium atoms, and that each helium atom represented the decay of one alpha-producing radioactive atom since the beginning of the earth. The discovery that most minerals containing radioactive materials also contained unusually large amounts of helium supported both the disintegration theory and the idea that such a clock would give correct results. The second method was based on the heating effects of radium, and involved a consideration of the earth as a radiating body in space. It was this second method that attracted the earliest attention.

Lord Kelvin had worked on this problem long before radioactivity was discovered. He calculated that the earth must be between twenty and forty million years old by simply working out its heat loss radiated out to space. These figures worried most geologists who wished to date rocks to at least 100 million years of age. If however radioactive decay of the original atoms of radium, uranium, and so on was taken as an extra source of heat to the planet then the earth must be several hundred million years old to have reached its present state and temperature.

Rutherford announced this result in a lecture at the Royal Institution in London in May 1904; it was a rather hurried arrangement, made to fit in with his visit to England to give the Bakerian Lecture at the Royal Society. The popular press, which had become deeply interested in matters of radioactivity, welcomed his statement with at least one headline which read "Doomsday Postponed", for the less rapid rate of cooling of an earth with an additional radioactive heat source made our planet longer-lived into the future as well as older in past age. But during the lecture Rutherford had avoided another doom.

> I came into the room, which was half dark and presently spotted Lord Kelvin in the audience and realised that I was in for trouble at the last part of my speech dealing with the age of the earth, where my views conflicted with his. To my relief Kelvin fell fast asleep but as I came to the important point, I saw the old bird sit up, open an eye and cock a baleful glance at me. Then a sudden inspiration came and I said Lord Kelvin had limited the age of the earth, <u>provided</u> no new source was <u>discovered</u>. That prophetic utterance referred to what we are now considering tonight, radium! Behold! the old boy beamed upon me.

During this short visit to England Rutherford was invited to Lord Rayleigh's home at Terling for a full English Edwardian "weekend". Lord and Lady Kelvin were among the many other guests and Rutherford reported to his wife: "Lord Kelvin has talked radium most of the day and I admire his confidence in talking about a subject of which he has taken the trouble to learn so little . . . . He won't listen to my views on radium, but Strutt gives him a year to change his mind. In fact they placed a bet to that effect . . .." The argument was basically whether the energy emitted by radium was derived solely from the atom's internal energy or whether the energy was somehow collected and redistributed by the radioactive atoms from some "ethereal" source. It has to be said to Kelvin's eternal credit that he publicly abandoned his theory at the British Association meeting later that year (1904) and, what is more, paid up his five-shilling bet.

The other method of "radioactive dating", the method of counting the proportion of elements in rock samples, could not be brought to any degree of accuracy until much more was known about the full range of radioactive families of elements. Nevertheless Rutherford himself started some work on it, and encouraged even more work by friends and colleagues such as Boltwood. The first preliminary ages of minerals and rocks were drawn up and they showed even greater ages for the rocks and therefore for the earth than the simple overall heat equations. Eve records that Rutherford once stopped Professor Adams, Professor of Geology, on the McGill campus and showed him a piece of pitchblende he was carrying and asked him how old he believed it to be. The geologist replied that best estimates gave an age of probably a 100 million years. Rutherford declared, "I know this piece of pitchblende is 700 million years old."

Very little of the work done in this field in Rutherford's time is of value nowadays. What is important to us is that Rutherford proposed and started the development of "radioactive clocks" to date our earth and what we find in it, and this method, or its modern variants, carbon-dating, potassium-argon dating, uranium-lead dating, have revolutionised archaeology and palaeontology. In fact Rutherford wrote one paper on the subject, the last of his Canadian period. It is titled "Some Cosmical Aspects of Radioactivity". It is remarkable for the breadth of interest it shows in scientific problems outside his own field. But it is even more remarkable in its last paragraph which shows that Rutherford was capable of speculating boldly far beyond the range of his experimental results. Here he considers the mode of heat production by the sun, and he suggests that in the intense conditions of the sun's interior all matter might undergo radioactive changes and break up into simpler forms with the emission of "a great quantity of heat". This sort of speculation was again good material for more popular publications and Rutherford earned himself 350 dollars for an article in *Harper's* monthly magazine on these matters. Rutherford was however saved from becoming a populariser (if a scientist needs to be saved from such a fate) by the sheer pressure of invitations that poured in upon him in his last years in Canada to give lectures and write articles.

It was in this period that an old colleague of the Cavendish days wrote to him skittishly about "your playful suggestion that, could a proper detonator be found, it was just conceivable that a wave of atomic disintegration might be started through matter, which would indeed make this old world vanish in smoke". We may not regard this suggestion as so playful, but it emphasises that Rutherford could range very widely in his thoughts. What he published was firmly tied to laboratory results, which was its great strength when attacked; but his mind was always operating out towards the frontiers, as far as he could see, and that was a great deal further than most other men.

The science that Rutherford worked on in his last years in Canada – 1905, 1906, 1907 – was almost entirely concentrated on the alpha-particle, his alpha-particle, as he plainly considered it. The problem with which he struggled was the nature of the alpha-particle. Rutherford believed the alpha-particle was a helium atom, indeed he "knew" it was a helium atom as he wrote to Boltwood privately, but he could not prove this unambiguously and publicly, and he was not to achieve this goal until he was well settled into the next period of his life, the Manchester period. He returned to the type of experiment in the deflection of alpha-particles by electric and magnetic fields that he had used in clinching the disintegration theory. He tried to get better values for e/m, the ratio between the electrical charge and the mass of the alpha-particle. His results left three alternatives. The alpha could either be a molecule of hydrogen, that is two atoms of hydrogen linked together; or it could be one half of a helium atom carrying a single unit of charge; or it could be a helium molecule (one helium atom) carrying twice the charge of the hydrogen atom. This was the position he had reached when he gave the Silliman Lectures at Yale in 1905. (The Silliman Lectures were typical of the "Honorary lectures" said by Badash to be typical of the successful Big Science man, a series of annual lectures paid by a legacy, and delivered each year by the most distinguished scientist the university could find. J.J. had been a Silliman lecturer very shortly before Rutherford. Rutherford's lectures were published as his third book, *Radioactive Transformation*, Constable, London, 1906.)

In these lectures, after presenting the three alternatives (pp. 183–186), Rutherford gave the evidence for preferring the helium atom alternative, admitted that final proof had not yet been obtained, and used the concept openly for the remainder of the lectures and the arguments therein.

Apart from Rutherford's own work, there were two other "inputs" of great importance to his thinking in 1905 when he gave these lectures. They both came in letters he had received the previous year. Quite out of the blue there came a long epistle which must have seemed to arrive out of his past. It came from W. H. Bragg, Professor of Physics at Adelaide in South Australia, the man whom Rutherford had visited briefly during a stop-over from the boat that was taking him from New Zealand to London and Cambridge in 1895. Bragg was therefore much the older man, yet he ended his letter, "I have read of you so often since you went through Adelaide many years ago that I seem to have never lost touch with you entirely. But let me take this opportunity of congratulating you on your magnificent work." Bragg was basically a mathematician and a clever one, shooting from the obscurity of a not-very-prosperous English rural background to his professorship at Adelaide at the age of twenty-three. A true "gentleman", modest to the core, he did not even consider the possibility of doing research until he was forty.

His letter of August 31st, 1904, which started yet another life-long friendship and collaboration for Rutherford, gave the results of Bragg's first research. What Bragg had done was to build a small ionisation chamber which could detect alpha-particles and which could be moved towards or away from a radioactive source. With this device he had shown that the range of flight of alpha-particles differed according to the source of the particles. The importance of this discovery was two-fold: firstly it showed that different radioactive elements emitted alpha-rays with different velocities, or energies; secondly the range of an alpha-particle, when carefully measured, could be used to decide which sort of atom had emitted it. Both these aspects of Bragg's discovery were to be exploited for many years to come, but Bragg himself had started from a third implication – if the alpha-particles were atoms, flying in a straight line through hundreds of thousands of air molecules, then the alphas must frequently actually pass through the atoms in the air and there must be times when two different atoms actually occupied the same space. It was this thought that set Bragg off to start research, for he had included the concept, based on some research by Marie Curie, in a lecture he gave when he was President of the Physics section of the Australian Association for the Advancement of Science; he had read of the matter when "trying to think of something to say". At first the matter remained speculation for Bragg could do no research as he had no radium.

However a rich benefactor purchased a small quantity of radium for Bragg's use and immediately "I found that helium atoms of four different ranges . . . were shot out from the radium preparation, which must belong to the four different active substances that Rutherford has shown to exist". (In 1903 Rutherford had shown

that the radium family tree went Radium-Radium emanation-
Radium A-Radium B-Radium C- on to other elements notably
Radium-F which turns out to be Marie Curie's "polonium". Four of
the first five members gave out alpha-particles when they decayed
into the next member of the family. Radium-B was thought to
decay with a "rayless" change, but this was only a measure of the
inadequacy of the detection methods of the time.)

Bragg continued:

> Then I got a hint from Professor Soddy, who was passing through
> Adelaide, that I should dissolve the preparation in water, which would
> wash away three of the active substances, but leave radium itself, the
> parent of them all. So I did, but, horror of horrors, as I brought my
> measuring apparatus up towards the radium in the way I had learned to
> do, there was no radiation at all . . . when I was well within the old
> range. However with a very downcast spirit I pushed the apparatus
> closer still, and closer: and suddenly a tremendous effect flashed out.
> The radium itself sent out the particles of the shortest of the four
> ranges, not the longest as I had thought . . .

And so Bragg was another convert to Rutherford's growing army
of enthusiasts for radioactivity research.

Bragg wrote these reminiscences in 1926, so he could safely write
that alpha-particles were, indeed, helium atoms. In 1904 and 1905
it was by no means certain. Indeed Soddy, back in England, had
taken on the task of writing an annual report for the Chemical
Society on the year's progress in radioactivity research, and in his
1906 report he made the position quite clear – that there was no
certain answer yet. Rutherford, Hahn and Bragg had by their work
shown that all alpha-particles were identical except in the speed
with which they were all initially ejected, and Rutherford had
shown that e/m was the same for all alphas and was almost exactly
half of the same measurement for hydrogen in hydrolysis. Soddy
reduced the possibilities to two – alphas were either helium atoms
with a double charge or half-helium atoms with a single charge. Or
just possibly the whole thing was wrong: "There is very little actual
evidence at the present time on which to base a conclusion." (The
view finally accepted, and currently held, is that the alpha-particle
is the nucleus of a helium atom, a helium atom stripped of the two
negatively charged electrons it normally has, and therefore showing a positive charge of two units.)

In this annual report Soddy includes one half-sentence which
reveals what Rutherford thought was the essence of the matter,
whatever the alpha finally turned out to be (and Rutherford
"knew" it was helium). "It is pointed out (by Rutherford) that,

whatever its nature, the alpha-particle must be a fundamental constituent of the atoms of all the radioactive elements."

The kernel of Rutherford's Silliman Lectures, the main object of his thought for the next thirty years – "The greatest problem that at present confronts the physicist" – was the structure of the atom, the nature of matter. English physicists had already started the game of "atom building" – in Rutherfordian terms that meant building concrete, though mental, pictures of what an atom might be like and seeing if the results fitted with experiments. The second of the 1904 letters mentioned above seems to have started this process. It was from J.J. and he wrote that he had been working on the structure of the atom "regarding the atom as built up of a number of corpuscles in equilibrium or steady motion under their mutual repulsions and a control attraction". A more sophisticated version of this model had the corpuscles (electrons) in precise orbits, and J.J. pointed out that orbiting electrons could account for many chemical facts – a profound foresight of the truth for which J.J. is rarely given the credit which is instead diverted to Niels Bohr. However there was one weakness to the model – under classical mechanics these orbiting electrons would steadily radiate energy and slow down until they fell below some critical velocity when the whole structure would become unstable and the electrons, in the old theory, would crash down into some sort of explosion. This mechanical problem was the one that Bohr did eventually solve, but J.J. in 1904 tried to turn the problem to advantage, by wondering if this was the basic mechanism of radioactive disintegration.

It was about this time, although the item is undated, that there appeared in Rutherford's laboratory notebooks a careful typewritten list of French radioactivity papers, annotated in his handwriting. Most of his notes are on the subject of the energy involved in radioactive processes, but he also notes that Jean Perrin had published a paper in the *Revue Scientifique* of 1901 in which it had been suggested that each atom was a planetary system. J. J. Thomson and Kelvin went the other way and eventually produced what became known as the "plum-pudding" model of the atom – which consisted of a "sphere of action" of positive electricity in which were embedded, like fruit in a pudding, negatively charged electrons, perhaps to the number of a thousand or more, since the weight of one electron was only one-thousandth of the mass of a hydrogen atom.

Rutherford highly approved of this method of thinking. He said in his Silliman Lectures:

211

Such representations of the nature of the atom and of the processes occurring therein are, in the present state of our knowledge, somewhat speculative and imperfect, but they are, nevertheless, of the greatest assistance to the investigator in providing him with a working hypothesis of the structure of the atom. The behaviour of such model atoms can be compared with that of actual atoms of matter under investigation, and in this way it is possible to form gradually a clearer and more definite idea of the constitution of the atom.[1]

He emphasised, for the benefit of chemists and other scientists, that this work was in no way an attack on standard atomic theory – on the contrary it was an extension into an entirely new field:

Modern physical and chemical theories are all based on the assumption that matter is discontinuous and is made up of a number of discrete atoms. In each element the atoms are supposed to be all of the same mass and of the same constitution, but the atoms of the different elements show well-marked differences in physical and chemical behaviour. It has been quite incorrectly assumed by some that the study of radioactive phenomena has tended to cast doubt on atomic theories. Far from this being the case, such a study has materially strengthened, if, indeed, it has not given proof of, the atomic structure of matter.

And again a page later:

While the study of radioactivity has emphasised the ideas of the atomic constitution of matter, it has at the same time indicated that the atom is not an indivisible unit but a complex system of minute particles. In the case of radioactive elements, some of the atoms become unstable and break up with explosive violence expelling in the process a portion of their mass. Such views are rather an extension than a contradiction of the usual chemical theory, which supposes that the chemical atom is the smallest combining unit of matter in ordinary chemical change. The atom may be the smallest combining unit and still be a complex system which cannot be broken up by any physical or chemical agencies under our control. In fact the great emission of energy in radioactive changes shows clearly the reason why chemistry has failed to break up the atom. The forces which combine together the component parts of the atom are so great that an enormous concentration of energy would be required to break up the atom by the action of external agencies.

He goes on to explain the history of the discoveries that led to the belief that an atom is a complex structure, and he deals fully with the Thomson–Kelvin "plum-pudding" model. He explains the difficulty involved in the theoretical loss of energy by orbiting

[1] *Radioactive Transformations*, Constable, London, 1906, p. 256.

electrons and admits "This necessary loss of energy from an accelerated electron has been one of the greatest difficulties met with in endeavouring to deduce the constitution of a stable atom" and he too plays with the idea that the spontaneous break-up of some atoms could be connected with the loss of electron energy: "It is thus not so much a matter of surprise that some atoms spontaneously break up as that the atoms are such stable arrangements as they appear to be." And he goes on to speculate that perhaps all atoms really are breaking up, but so slowly that no one has yet detected it. He foresees a "great attack" on the problem of atomic structure to which a "promising beginning has been made".

There is no trace here of the mathematical-theoretical approach to physics which was favoured by the German tradition and which has been so strongly represented in modern American thought. Model-building controlled by experiment is Rutherford's ideal, the English tradition, and especially the Cavendish tradition, a tradition which has remained at the core of British scientific thinking right up to our own time – witness a much later Cavendish achievement, the double-helix model of DNA.

The writing in *Radioactive Transformation* is, in my view, Rutherford's best. It is remarkable for its clarity, for it can be understood by the complete non-scientist. The argument is well phrased, and, above all, well organised, driving continuously and straight to the point. A truly powerful mind is clearly at work, bringing forth detailed evidence to make a point, but ranging both wide and far ahead. Yet clear distinction is drawn between the known and the unknown, the whole is anchored to the firm rock that is Rutherford's vision of the atomic world he was investigating. It is a quite extraordinary reversal of the usual order of things that a man who could write so vividly and feelingly about his scientific mental processes, should write so wearisomely in his private correspondence.

It was Rutherford himself who went on to lead the "great attack" on the constitution of the atom. He directed it from Manchester, setting his collaborators and students to work on many crucial points of leverage. But he also exercised a sort of unofficial coordination role through his massive correspondence. He continued to encourage Boltwood and seek his help. He exerted a quiet guidance over Bumstead, who at one stage believed he had split the atom with X-rays. He kept in touch with Hahn and the growing school of radiochemistry on the European continent; he established friendly relations with the Curies and the French workers, notably through his old friend, Langevin. The work at McGill, as we have seen, declined but there were a number of exciting developments at the Cavendish with which Rutherford kept closely in touch. And finally he made contact with the German physicists of the theoretical school – Planck and Einstein and the Dutchman Lorentz. And, of course, it was a Dane, Niels Bohr, the greatest of all Rutherford's pupils, probably the only one of those pupils to equal the master in stature, who finally brought together the experimental model makers with the mathematical theoreticians.

But just before this new era was opened, and just before he left Canada, Rutherford himself stumbled on the clue. It all stemmed from W. H. Bragg's observations on the different ranges of alpha-particles and the same man's later finding that the alpha-particles ceased to be able to ionise a gas before they had lost all their energy and before they had reached the end of their flight. Rutherford was busy at his alpha-particle deflection experiments in his attempts to determine e/m in the winter of 1905 and was finding the electro-static part of his work "devilish difficult" when Becquerel pub-lished a paper on alpha-rays. Rutherford wrote to Bragg on November 4th, 1905:

> You will probably have seen by this time that Becquerel has run amok and has laid out yourself as well as myself to his own satisfaction. He says the "rays from radium are homogeneous (ye gods!) and are all equally deflected and do not suffer decrease of velocity after passing through matter but rather increase"!!! I have never seen a more damnable confession of intellectual impotence in one who considers himself a leader of science . . . I consider Becquerel is an ass of the first water and I trust he will be snowed under. How any man in his senses could declare the rays are homogeneous after your work passes my comprehension.

(It should be said that Rutherford had, in fact, had McClung repeat Bragg's work and found it completely confirmed. He went on to promise a public reply to Becquerel "from which he should never rise again. He will be atomised or rather electronised".)

Rutherford had no explanation, he admitted, of why the alpha-particles should suddenly stop ionising before they stopped mov-ing. But he guessed that it was a fundamental relationship and he went on, "I believe that the alpha-particle at that speed is taken into the atomic system" and he thought this was a possible method of building up new atoms. In the immediate sense this was not correct, yet years later something very like this would become enormously important. What was significant at that time was that Rutherford was made to think in this direction and made to work on this subject to deal with Becquerel. He started to use photographic methods to record the arrival and deviation of his beams of alpha-particles, using tiny photographic "plates" only one centimetre square. He started to measure the loss of velocity caused by the alphas passing through thin mica plates and in one series of experiments he sought to compare on one and the same plate the arrival of particles slowed down and the arrival of unhindered "control" particles. He therefore covered just half the channel through which the alphas were fired with his mica plate. He noticed that the two images caused by the alphas were different in another way – those that had passed through the mica gave a more fuzzy, blurred image. When he switched off the magnetic field that had been deflecting the alphas the difference in the two images was even more pronounced – indeed it was undeniable. Rutherford had been troubled by fuzzy images before and he had put it down to the presence of air in his apparatus "scattering" the alphas slightly. But this time a different conclusion was forced into his mind. The most intense magnetic field he could produce acting all along the flight path of several centimetres' length had deflected some of the particles about the same amount as the "scattering" caused by the particles passing through a sheet of mica about three-thousandths of a centimetre thick. Such a deflection, if he had to achieve it electrically, "would require over that distance an average transverse electric field of about 100 million volts per centimetre". So it followed that "such a result brings out clearly the fact that the atoms of matter must be the seat of very intense forces". That was published almost as an aside, and certainly as a support for the electronic theory of matter in a paper which was largely an extension of Bragg's work and which dealt with the slowing-down of alpha-particles as they passed through matter.

But Rutherford was always the pursuer of interesting and unusual sidelines – and this one led him eventually to the structure of the atom.

8

Starting at Manchester

The mode of Rutherford's translation from McGill to Manchester would make any present university administrator wonder whether he was necessary. The incumbent Professor of Physics at Manchester decided that he wanted Rutherford as his successor; he approached Rutherford personally and negotiated with him; he offered his own early retirement if, and only if, Rutherford could succeed him; when he had secured the university's formal agreement to his plan, he proceeded to give major financial aid to his successor by paying the salary of a new reader out of his own pocket. This cavalier benefactor was Sir Arthur Schuster, who was not only wealthy, but talented and cultured. He had learned his research techniques at the Cavendish at about the same time as J.J. He had acquired a reputation for providing brilliant ideas which he was not always successful in confirming, and for occasionally grasping the wrong end of the stick with what J.J. described as "Schusterisms".

Rutherford had, of course, met Schuster at scientific meetings and probably socially at the Cavendish but it is not known when they had the crucial first conversation on the subject of the Manchester Chair. However Sir Arthur wrote to Canada from his holiday home at Goathland in North Yorkshire on July 7th, 1906:

> When we last met I mentioned the possibility of my retiring from my Professorship in the near future and you did not seem disinclined to return to England if the Chair were offered to you. May I now ask you (in confidence) whether I am correct in this recollection or whether your own ideas have altered. My work is drifting away from the experimental side and I want more leisure for working at a book . . . I have not mentioned the matter to anyone in Manchester yet and I should like you to treat it as quite confidential. But any further steps would be made much easier if I could feel that you were ready to step in should a vacancy occur. I really do not know of anyone else to whom I would care to hand over the office.

Schuster himself would like to stay in touch with the laboratory and its work, doing a little lecturing, supervising some research work on the mathematical side, but he would hand over control of the laboratory, the directorship as well as the professorship, completely – "I do not believe in dual control." He went on to say that conditions of employment were not too bad and there was not too much lecturing involved; he admitted that at first he had to struggle to maintain order in the elementary lectures, especially when medical students were present. The salary would be £1,000 rising later to £1,200 – which would make the holder one of the best paid professors in Britain.

This letter must have arrived while Rutherford was still in California, but on September 26th, 1906 he replied:

> I was very glad to receive your kind letter in reference to the Chair of Physics at Manchester, as it came at a time when I was seriously considering my future plans. I have had to decide during the past year between the attractions of McGill and Yale University and finally decided to remain here. My chief reason for this was my hope to return ultimately to England to a position where I would not have to sacrifice laboratory facilities by doing so. The position at King's College seemed to me to invite the probability of the latter.

This seems to make clear that Schuster had certainly not offered Rutherford anything concrete in their previous conversation, but it also points up the known fact that Rutherford had considered applying for Callendar's vacated Chair at King's College in London and had been warned off by his friends because of the poor laboratory facilities there.

To Schuster he continued, "I very much appreciate your kind and cordial letter and am inclined to consider very favourably the suggestion of becoming a candidate for the position you propose. The fine laboratory you have built up is a great attraction to me as well as the opportunity of more scientific intercourse than occurs here." Apart from the major fact of his virtually accepting Schuster's proposal, this paragraph also carries all Rutherford's motives – his determination to break out from the scientific isolation of his transatlantic position, and the fact that Manchester, under Schuster's leadership, had built the finest physics laboratory in Britain, not up to the standards of the MPB, but vastly superior to the Cavendish or anywhere else at the time.

There were a number of fairly pointed questions about the amount of lecturing to be done – Rutherford tried to stipulate not more than five lectures a week – the amount of committee work and the provision of funds for laboratory needs. But Rutherford was obviously learning to be diplomatic too:

I need hardly tell you how much I appreciate the suggestion coming in the way it has. I fully recognise the spirit of self-abnegation displayed by you in the letter. Nothing could give me greater pleasure than to have you a member of the department to add your strength in the branch – mathematical physics – in which I should be only too delighted to have your assistance and advice and I am sure that, as far as I am personally concerned, you would never regret the arrangement.

Schuster's response was immediate, and very lengthy. On October 7th he wrote giving full descriptions about the amount of lecturing, the nature of the courses being run, the personnel of the department, the research work in hand and the side interests of the department, including an outstation working on atmospherics and meteorology. He wrote of the present financial state of the university and the independence of the head of the department in financial matters with the current running costs of the physics department standing at about £440 per annum. He was frank about the problems too, the "complication" arising from the fact that the electro-chemistry and electro-technology departments had to share the same building with physics, though this had the counter-vailing advantage that "strong currents" were available whenever physics wanted them. There had also been a recent battle between the municipality and the university over the Technical School but "with the help of a few stalwart friends, entirely on the Arts side, I succeeded in showing what I believe to be the right solution. The Technical School was taken into the university as a Faculty of Technology having their own degrees. This is still a burning issue." However, "The Council are fully alive to the fact that the reputation of the university and its power to attract students from outside depends entirely on the research work carried on in the laboratories, and they do what they can to encourage research by providing funds as far as their means allow and also sufficient assistance."

The letter ended by dealing with the moves that would have to be made. "My formal resignation will go to the Council this week, but it will be a few weeks before definite steps can be taken towards filling the vacancy . . . You will, in all probability, be communicated with officially if the Senate accepts my view of the procedure and does not advertise." Schuster could not be quite certain that he would carry the day, nor did he know whether the Council would waive the formality of an interview before making a definite appointment.

At exactly this time Rutherford was clearly still considering the possibility of going to King's College in London, for the advice he had sought from his friends was still coming in. J.J. had written on October 1st that he would tell all he knew "and try to be as unbiased as possible although my keen desire to have you back in England makes this difficult". Headlam, the Principal of King's had spoken personally to J.J. about the possibility of getting Rutherford, but "the laboratory is poor, almost destitute of modern apparatus, probably not £5 spent in the last three years". The students were "not a very high class lot, poor in intellect and rather rowdy in behaviour and discipline". J.J. admitted that it would mean a full year's interruption in Rutherford's research work to build up the equipment and standards of the lab and he could not bring himself to advise either way.

Rutherford's position now became exceptionally strong, for that same year, in addition to the London post, he was approached for the second time to take the Secretaryship of the Smithsonian Institute in Washington, and for the third time to take a chair at Yale which he refused because, as he explained to his mother, he did not want to become an American citizen. More than that, Montreal offered to increase his salary to a figure above that offered by Manchester. A number of letters passed between Rutherford and Schuster in October and November (letters which have not previously been reported). The younger man was quite frank, and from all the other evidence quite honest, in telling Schuster about these other offers, and with each one he gave his reasons for not wanting it and preferring Manchester. Yet nevertheless this enabled him to bargain very sharply.

He did not want to go to Manchester for a personal interview, nor did he want the university to advertise the job. On October 26th he told Schuster that he could not spare the time or the money to go over for an interview: "I am not altogether and [sic] untried man and have had a good deal of experience in looking after a laboratory and in teaching . . . I have lectured to large classes and have never had any difficulties in regard to discipline in the classroom." But he was worried about the amount of money available for equipment and supplies in the Manchester lab: "The sum you mention seems small compared with what is spent on the same lines in many American Universities . . . but quite prepared to take your view that such a sum is sufficient." He was anxious that the development of the laboratory along research lines (and he surely meant along radioactive research lines) might be held up for want of a hundred or so pounds a year. What about some special grant from the government or from some other fund?

On November 8th he told Schuster about the Smithsonian

approach, although he said he did not feel such a job was "exactly my metier". In the same letter he mentioned the approach from Yale and mentioned also "another American university", adding "it never rains but it pours". His view of all these offers was basically that "we are too near the scientific periphery here, for America as yet does not count very seriously".

On this very day the University of Manchester had sent off an official offer of the post. A telegram, dated November 8th, read "University offers Professorship. Conditions follow. Schuster." Three weeks later Rutherford was still bargaining, for he wrote to Schuster on November 27th, referring to an official letter received from the Vice-Chancellor of Manchester on November 20th. The argument now was about salary. Rutherford did not "wish to appear grasping" but "the locals are not altogether pleased" and had offered a salary increase. "My acceptance of the Manchester offer could not however be seriously influenced by mere questions of salary," said Rutherford, but he pointed out that Manchester's conditions could leave him permanently paid less than heads of other departments.

Schuster bore with all this, acting more as a broker than a man who was handing over an important Chair to a successor he had chosen himself. Rutherford remained grateful to him all his life, and one of his last letters, not long before his own death, was a letter of condolence on her husband's death to Lady Schuster in 1937, in which this gratitude is again expressed. But on December 10th, 1906 Schuster had to write to Canada: "You will receive by the same post a letter from the Vice-Chancellor which I hope you will find satisfactory. The conditions are now as favourable as I think they can be made . . . I hope you will now see your way to give a definite answer in the affirmative."

Very few people knew about these negotiations; everybody involved kept their cards very close to their chests. Rutherford did not tell his closest friend of the time, Boltwood, anything until Boltwood came for a short visit just after Christmas 1906. This visit, which was arranged at fairly short notice, was almost certainly Rutherford seeking his friend's advice, though there is no firm evidence of this. J.J. only heard about the Manchester possibility from Schuster. He wrote to Rutherford on December 18th, 1906, immediately after his return from Stockholm where he had been awarded the Nobel Prize: "I hope it is true. I think you would like the Manchester people and the climate is not nearly so bad as its reputation." Obviously he had not been told all the details because he warned: "The only spot I see on the horizon is the possibility of a dual management of the lab by Schuster and yourself proving a little inconvenient. I think it would conduce to smooth working in the future to have the respective parts to be played by Schuster and yourself defined as clearly as possible at the outset. It will be very delightful to have you back again." There is considerable irony in this warning for Rutherford and J.J. were to face exactly this problem between themselves in thirteen years' time when Rutherford came to take over the Cavendish. And one cannot help wondering whether something of this sort was not already in the mind of J.J., unworldly in appearance, but wily in politics. His letter goes on to describe the extensions that were then being built to the Cavendish with a grant of £5,000 from Lord Rayleigh's private purse, and the plans included certain rooms "so that if any time there are two Professors of Physics each can have a laboratory". (There is a Cambridge story that one day J.J. bought a pair of trousers on his way home to lunch, having been persuaded that his existing garments were finally too baggy and worn. Apparently he changed into the new trousers immediately, for his wife found the old pair in his bedroom that afternoon, and immediately telephoned the laboratory in alarm, convinced that her husband had gone back to work without any trousers at all. Yet his handling of the finances of the Cavendish at this time was very sure and he could obviously plan very far ahead.)

It was at moments like this that Rutherford revealed that deep-seated uncertainty, that diffidence that only C.P. Snow seems perceptive enough to have spotted. It seems he negotiated no more – there are no traces in the documents of any further arguments. He simply wavered – and he was to do exactly the same when the time came for him to decide whether to leave Manchester. Later he described it to Boltwood as his "time of worry". He took about two weeks to make up his mind, and plainly it was an agonisingly undecided Christmas for him. It was not until January 15th, 1907 that Schuster could write that he was "much relieved" by Rutherford's final decision to come to Manchester. Immediately that extraordinary man temporarily endowed a Readership in mathematical physics, and told Rutherford about it in the same letter: "I hope you will approve of this. Ultimately we shall need a Professor of Mathematical Physics." He went on to discuss the proper relationship between mathematics and experimental physics at some length, deploring their growing separation and adding, "Cambridge is a good deal responsible for this, the Natural Sciences and Mathematical Triposes having no, or very little, connection with each other." And this again was a point Rutherford had to take up when his own turn came to go to Cambridge.

McGill had not let him go without a struggle. Rutherford's formal resignation to Principal W. Peterson on January 1st, 1907, expresses regret at having to leave "the institution in which I have been eight years Professor and in which I have been granted such exceptional opportunities for carrying out my special scientific work. In deciding to go . . . the determining factor has been my feeling that it is necessary to be in closer contact with European science than is possible on this side of the Atlantic." He expresses particular thanks to John Cox – "it is difficult to overestimate the value of such unfailing consideration and support" – and to Sir William MacDonald, whose donations "alone rendered possible much of the research work". Peterson's personal reply on January 5th, 1907 said that the Board was reassured "that everything had been done that could have been done to make you content to stay. I made it clear in what I said that you had not been just allowed to 'slip through our fingers'."

The terms he finally received from Manchester, in a letter from the Vice-Chancellor, Alfred Hopkinson, dated December 11th, included a salary rising to £1,250 after five years and a promise that he need not give more than five lectures a week. The university declared itself "strongly in favour of the development of research work" and said there was hope of government assistance to promote research and the likelihood of special grants for apparatus. The Vice-Chancellor was "glad you consider the conditions offered are generally satisfactory and the work here is attractive to you".

His American friends bewailed his departure: Boltwood said, "I hope that even though an ocean separates us that you will not quickly forget your friends in America"; Zeleny wrote of "the great loss to American science. But you came and conquered and built up a great monument here for science and yourself, and why shouldn't you go nearer the centre of things?" Godlewski, who found himself "almost completely alone" in Poland with "All the people which is working on the science they are generally chemists, and they alike as the chemists in all the world don't understand radioactivity at all" had written, "I am very proud that I was the first who showed for the European people the way to Montreal, and I am so happy that the first was from Polish not from the horrible German nation."

At McGill the attitude seemed to spread downwards from Principal Peterson that Rutherford's departure was inevitable. Dean Henry T. Bovey conveyed the official views of the Faculty on January 26th. A long resolution had been passed congratulating him on "nine years crowded with epoch-making research" which had "extended the fame of McGill to all parts of the world and attracted to her laboratories distinguished men from the US and Europe". But their tribute to Rutherford, the university politician, was even more remarkable. "His energy, directness and independence of view . . . made his opinion on questions of university policy one of the first and most valuable to be obtained: witness the leading share he took in organising the courses for higher degrees." Furthermore his "sincerity and geniality", they declared, "enabled him to take a strong line without risk to the warm personal friendship of every member of the staff".

McGill, after all those plaudits, gave him an Honorary Doctorate of Laws in a final April Convocation before he left. This was his third honorary degree although he was not yet forty years old – both Pennsylvania and Wisconsin Universities in the USA had similarly honoured him – and he told his mother that he felt he was rather youthful for such compliments, since they were normally reserved for "septuagenarians". That the feeling of love for the man was real is confirmed by a letter from H. T. Barnes, the man who could well have been a bitter rival as the inheritor of Callendar's mantle. It was the first letter Rutherford was to receive from Canada when he arrived in England and it was written by Barnes just after he, and a large crowd of others, had seen the Rutherfords off from the railway station on May 17th. It reads: "I am writing you this feeling exceedingly lonely. The time of parting came at last as it always does . . . and now I return to this building that is associated in every corner with you and your splendid work and with the most delightful memories of my relations with you."

Rutherford simply hurled himself into life and science in England. He arrived in Manchester on May 24th, 1907. Within three weeks he had set up his first "emanation electroscope" in his Manchester laboratory, and continued the series of experiments he had been doing in Montreal. He swept his new colleagues into his enthusiasm for radioactive research; before the end of June he had persuaded J. E. Petavel (later Sir Joseph Petavel, Director of the National Physical Laboratory) the senior of the men left by Schuster, to join in an experiment with his "bomb". Petavel's specialty was the study of high pressures and their effects; these pressures were achieved by causing small explosions inside a reinforced steel sphere or bomb. Rutherford put radium inside the bomb but was able to measure its presence by the highly penetrating gamma-rays it emitted. There was no change in the radioactivity when the radium was subjected to the 1200 atmospheres of pressure produced in the bomb. It was a small experiment, of no great significance, but it provided material for a short paper which Rutherford could give at the British Association meeting at Leicester later that summer, and it reinforced the general position that no known force could affect the behaviour of radioactive substances.

With the same gusto Rutherford threw himself into the scientific society and talk he so much craved. Barely three weeks after his arrival in England he was visiting the Cavendish Laboratory where he gave a lecture (June 16th) and in early July he was visiting London, attending Royal Society meetings, talking to Schuster, listening to Ramsay's accounts of his latest and most surprising results. And, of course, he was being introduced to his new colleagues in Manchester.

The Rutherford family very soon found a house that suited them – a comfortable and fairly large villa at 17, Wilmslow Road, Withington. It was just a couple of miles from his laboratory, a few minutes' tramride, in an area which was then fairly well-to-do, if not smart, but which is now dominated by the tower of Manchester University's Owens Park student residences. Like all Rutherford's homes it was graceless within, a fact he did not notice, but its decor was at least bright, for Bohr recalled the white-painted sitting-room, and American visitors in particular noted that it was warm by English standards. Certainly Rutherford reported to his mother long before the First World War that he was having a small central-heating system put in, which was for those times a most advanced improvement.

This was the start of the happiest period of Rutherford's life, just as the twelve years in Manchester covered the very peak of his scientific career. He loved Manchester, its people, his laboratory and his work. He made a series of discoveries in these twelve years unmatched by any other scientist at any other time. "This is a pretty active place and, but for its climate, has a number of advantages – a good set of colleagues, a hospital and kindly people and no side anywhere," he wrote to Boltwood before he had been six months in his new position.

Manchester in the first decade of this century was indeed a lively and active place, still the city that had coined the phrase "What Manchester thinks today, England thinks tomorrow." And if the city had lost a little of its political power since the days of Cobden and Bright to the rising star of Birmingham under the Chamberlains, it had lost none of its cultural supremacy. Manchester had the Hallé Orchestra, and Miss Horniman's famous repertory theatre. It also had a particularly strong tradition of music hall in the era when music hall was in its prime. And of course it had the *Manchester Guardian*, the greatest Liberal newspaper of those days.

Furthermore the university had at that time a particularly brilliant staff. "They were a remarkable group of men, and made up, I believe, as distinguished a faculty as was then to be found anywhere in any English or European school," is the evaluation of Chaim Weizmann, at that time a lecturer in the chemistry department at Manchester, though he was already committed to spending all his spare time building up the Zionist cause and campaigning in particular for the foundation of the Jewish University which now bears his name.

The university boasted Tait and Tout the historians, Samuel Alexander, the philosopher with his pupil, Roth. Eliot-Smith, Rutherford's old Cambridge friend, came to the anatomy department and Perkin was working in the chemistry laboratory. Rutherford dined regularly with the historians and philosophers, while his predecessor, Sir Arthur Schuster, introduced him to the powerful and cultured society of Manchester Jewry, whose leaders included Nathan Laski, father of Harold Laski.

No one could call Rutherford a "cultured" man in the normal sense of having musical or artistic or literary tastes that were well-developed. Yet he mixed continuously with those who were cultured; he enjoyed their society, and plainly they, enjoyed his company, too. Some polish he had undoubtedly picked up at Cambridge, but it seems it was the extraordinary warmth of his character and his unbounded enthusiasm that won him so many friends among men of such different tastes. Weizmann himself became a life-long friend and describes Rutherford as

Youthful, energetic, boisterous, he suggested anything but the scientist. He talked readily and vigorously on any subject under the sun, often without knowing anything about it. Going down to the refectory for lunch, I would hear the loud, friendly voice rolling up the corridor. He was quite devoid of any political knowledge or feelings, being entirely taken up with his epoch-making scientific work. He was a kindly person but he did not suffer fools gladly. Also he was rather contemptuous of persons who spoke a few languages. "You can express yourself well in one language and that should be English," he used to say. Any worker who came to him and did not prove to be a first-class man was out in short order . . . With all this, Rutherford was modest, simple and enormously good-natured.

Rutherford supported Weizmann's Hebrew University campaign, and did so far more willingly than Albert Einstein, who was an even closer friend of Weizmann – which led the Jewish leader to a most interesting comparison of the two men:

> I have retained the distinct impression that Rutherford was not terribly impressed by Einstein's work, while Einstein on the other hand always spoke to me of Rutherford in the highest terms calling him a second Newton. As scientists the two men were strongly contrasting types – Einstein all calculation, Rutherford all experiment. The personal contrast was not less remarkable: Einstein looks like an etherealised body, Rutherford looked like a big, healthy, boisterous New Zealander – which is exactly what he was. But there is no doubt that as an experimenter Rutherford was a genius, one of the greatest. He worked by intuition and whatever he touched turned to gold. He seemed to have a sixth sense in his tackling of experimental problems. Einstein achieved all his results by sheer calculation. Rutherford was considered the greatest chemist of his day. He obliterated the line of demarcation between chemistry and physics and discovered the transmutation of the elements, turning chemistry back to alchemy. But he knew no chemistry in our accepted sense of that science and method. Nor was he a great mathematician, in which he again stood in contrast to Einstein.

Weizmann's statement about Rutherford's complete lack of interest in politics is fully borne out by Rutherford's correspondence. There is not, I think, a party-political reference in the whole of that vast mass of letters, and despite his clubbableness and gregariousness he numbered very few politicians among his friends. Nevertheless it was in this Manchester period that Rutherford first came up against the world inhabited by politicians, the world of government. And it seems most likely that it was in Manchester that Rutherford formed his only alliance with a politician, though the man in question was no less a personage than A. J. Balfour. Balfour, we have noted, was connected to the Cavendish through the marriage of his sister to Lord Rayleigh and the politician had actually helped to carry out experiments in the laboratory. But it was not only on the strength of this small experience that Balfour spoke in public on matters scientific, for he was a man of very considerable intellectual stature, and he published books of his own cool, sceptic philosophy and political views.

Balfour had, in fact, delivered the presidential address at the meeting of the British Association in 1904, which Rutherford had attended during the long summer break from McGill. After this meeting Rutherford wrote to Balfour congratulating him on his address and "admiring the scope of his scientific knowledge". Whether Rutherford was entirely disinterested may be doubted –

presumably he was as aware as any man of the advantages of having a well-placed political ally. At any rate the reply he received from Balfour on September 13th, 1904 was most satisfactory, including: "Your own name will for all time be associated with the growth in our conceptions of the physical universe, and it is with the utmost satisfaction that I learn of your approval of my attempt to deal from the outside with the problems to whose solution you have so greatly contributed." Balfour's biographer adds, "Whether at this time Balfour grasped the destructive and military implications of Rutherford's work on nuclear physics it is not possible to say, but it is very possible that he had heard and pondered Rutherford's 'joke' to the effect that 'some fool in a laboratory might blow up the Universe unawares'."

Just before Rutherford's arrival in Manchester, Balfour moved into the politics of that city. He tried to win one of the Manchester parliamentary seats in the election of 1906 and failed, but later he sat for Manchester East, and was well known in Manchester society. In particular Balfour was popular in Weizmann's circle, and indeed had first met Weizmann during one of his political campaigns. Thus he and Rutherford must have had social contacts as well as formal knowledge of each other's public activities. It is significant that, in reporting home to his mother on his first weeks' activities in England, Rutherford mentioned particularly the trip he made to London in the latter part of June which included a visit to a very smart Royal Society soirée. Here he met not only Rayleigh, who was President of the Royal Society, but, as he is most careful to mention, "Balfour was present as well as most of the scientists and their wives." He was forming the most valuable political connection of his life.

One of the very few people still living who knew Rutherford in his domestic setting at Manchester was then a little girl whose father, a general practitioner and brother of the Vice-Chancellor of Manchester University, lived in the same road as the newly-appointed Professor of Physics. She was invited to a fancy-dress children's party at the Rutherfords' home and went, dressed as a violet, arriving a little late. She therefore missed the rules of the game of hide-and-seek that was being played, and trespassed into a room which had obviously been put "out of bounds". There she found a very large man working at a desk, who was plainly surprised to see his visitor. Nevertheless he turned out to be extremely kindly, and hid the child in the knee-hole of the desk while he continued writing, although she had at first been terrified by his greeting of "Who are you? A little girl?"

Later the Professor's wife called upon the doctor's wife, bringing her only daughter with her. Apparently Rutherford's daughter, Eileen, was at this time a very spoilt child, the sort of child who licked the icing off the fancy cakes and refused to eat the plain remains. At one stage of this not-very-promising relationship the doctor's young son actually bit the Professor's rather unpleasant daughter. It is sufficient commentary on the very small number of people in the leading circles of British scientific life in the first quarter of this century that the doctor's daughter, when the whole family had moved to Cambridge, married a young scientist Lawrence Bragg, the son of that W. H. Bragg who had already figured in Rutherford's story. The younger Bragg followed Rutherford as Professor of Physics at Manchester and, in turn, at Cambridge, so the little girl of Manchester days is now Lady Alice Bragg and was able to observe the Rutherford family for forty years.

It is part of the Bragg family tradition that Rutherford was an enormous power in the Senate of Manchester University – "He rode rough-shod over them by his sheer size" – and this belief would not only have come from his successor, Lawrence Bragg, but also from the information of Sir Arthur Hopkinson, the Vice-Chancellor of Manchester and Lady Bragg's uncle. The tradition is fully supported by Eve, the mildest of biographers. He records that the first faculty meeting that Rutherford attended was marred by a fracas. The chemistry department had annexed certain rooms in the laboratory building which had previously belonged to the physics department – Schuster had warned his successor of the uncomfortable arrangement by which electro-chemistry shared the physics building. Apparently there was much sympathy for the newcomer who had thus found his patrimony diminished but nevertheless the action of the chemists was officially condoned. "When the time came for Rutherford to speak he brought down his fist on the desk with a resounding bang and exclaimed 'By thunder' – and that was only the opening of a 'vigorous speech'. He finally pursued the Professor of Chemistry to his study protesting that he was a nightmare – 'like the fag end of a bad dream'." When in his more sober days the story was recalled to him, he was quite unrepentant and hailed it with delight. Thus began his long war with Harold Baily Dixon, the Professor of Chemistry, a war which still raged fifteen years later when W. Lawrence Bragg succeeded to the Chair.

Throughout his first year in his new appointment Rutherford had to fight very hard to keep Dixon at bay. In March 1908 Rutherford wrote to Schuster that Dixon had "sprung" on the Council a scheme for a large increase in the chemical laboratory, and he continued:

Dixon is very close to the V.-C. [Vice-Chancellor] – too close in my opinion. I expressed to the V.-C., when Dixon had worked a job for his own department, that I was a little tired of Dixon's sublime assumptions that all the research work in the University was done in the Chemical Lab. I thought it was about time he had a few straight remarks on that subject. Dixon advertises the research in his department on weekdays and Sundays in the most unthinking way. I told him, when he informed me of the great list of researches his department had published last year, that he had better prepare for a shock from the Physics at the end of the year.

Schuster, then travelling in India in his new post as Secretary of the Royal Society, encouraged his successor in this line – "If the Professor of Chemistry speaks again about the number of researches in his department, you might hint that researches should be weighed, as well as counted" – a fine academic counterthrust.

Battle continued. In May Rutherford described another encounter: "I have had some little trouble with friend [Dixon]. He saw Pring (one of the lecturers) and complained that his lectures were not up to standard . . . He had forgotten to see me first in his usual way, so next day he had to metaphorically crawl in my office. I gave him a little to chew over in private for some time to come." On another front the wrangle over the electro-chemistry sub-department also raged: "I have no doubt he has in view taking over the whole of physical chemistry and isolating the electro-chemistry completely." Slowly Rutherford gained the upper hand – he won a battle over an appointment for Petavel and he was victorious in the infighting over appointments to geology and engineering after some scrapping in sub-committees. It seems to have taken him about a year to establish himself safely in the political structure of the university, for by June and July 1908 his letters to Schuster take on a calmer tone and Dixon seems to have become no more than a steady irritant. At this time Rutherford reveals that he has been offered, in confidence, the post of Research Professor at Imperial College in London. He turned this down, largely because Callendar was director of the laboratory there and "I would sooner have Dixon in my department than the combination of Callendar and Watson".

Sir Arthur Schuster's protection and benefactions went much further than providing Rutherford with advice and political support. To his mother, Rutherford reported on grand dinners at the Schusters' at which he was introduced to the society of Manchester, both civic and academic. Schuster, as we have seen, had persuaded the university to provide a post of Reader in Mathematical Physics, for which Schuster himself paid the salary for the first few years. It was the availability of this position that enabled Rutherford later to bring Sir Charles Darwin and Niels Bohr into his team. But in his first letter of thanks to Schuster (written from McGill, January 27th, 1907) Rutherford had said:

> I quite agree with you about the divorce between experimental and mathematical physicists. The latter as a rule know a certain amount of theoretical physics, but as far as I have seen are almost entirely destitute either of experimental knowledge or instinct. Even the best of them have a tendency to treat physics as purely a matter of equations. I think this is shown by the poverty of the theoretical communications on the problems which face the experimenter today. I quite recognise that the experimenter is inclined to drop his mathematics also . . . As a matter of fact it is extremely difficult to keep up the latter when all your energies are absorbed in experimentation.

This passage, written at a time when he was at the height of his powers as an experimentalist, sums up Rutherford's position more clearly, and more honestly, than many of the better-known "tales" in which his boisterous, teasing manner is allowed to figure too largely. It also expresses his self-knowledge of his own mathematical strengths and weaknesses better than the glib dismissals of his maths which can also be found. From the start of his Manchester tenure, Rutherford always had a mathematical physicist, or even several, available. With the exception of Niels Bohr, however, none of these men were an original genius, though all of them were highly accomplished in their field. Rutherford, quite frankly, used them to explain and support the facts produced by experiment, and there is only one major advance in his work which was facilitated by theoretical considerations *before* experimental work – this was Gamow's prediction which led to Cockcroft and Walton building the accelerator that first artificially smashed atoms. The European approach on the other hand – exemplified by the work at Göttingen which has so much influenced modern American physics – was to take the minimum number of assumptions about the physical world, to operate on these assumptions with the known laws of physics, such as the Law of Conservation of Energy, and to deduce mathematically what the experimentalist should be able to observe. If, then, experimentalists did observe what was predicted, the original work is taken to be correct. Thus Einstein's special theory of relativity, with all its accompanying paradoxes, can be regarded as a deduction from the assumption that the laws of physics apply equally everywhere in the universe.

There was, however, a much more pressing and practical problem awaiting Rutherford as he arrived in Manchester. The laboratory, well equipped as it was (including the possession of its own liquid-air machine), had almost no supply of radioactive source material. In particular it had virtually no radium. Again Rutherford turned to Schuster, and he must have done this on his first visit to London in June and July immediately after his arrival in England. After making his first impact Rutherford went off for a long holiday in Cornwall, where he stayed in the village of Mullion. From there he wrote to Sir Arthur thanking him for writing to "the Vienna people" about a supply of radium, and remarking, "If I get it, it will be a great score for you and indirectly for the University." He had also persuaded Sir Arthur to obtain from the Royal Society the loan of some actinium residues which were in the Society's possession.

"The Vienna people" were the Radium Institute of Vienna. They were at that time an imperial organisation of the Austro-Hungarian Empire, and, as such, were able to control the output of the Joachimstal mines in what is now Czechoslovakia. From these mines came the pitchblende which had long been the chemists' chief source of uranium, and from which the Curies had extracted radium. Rutherford was doing his best to scrape together whatever material he could find in England. On July 5th he had written to Lord Rayleigh, as President of the Royal Society, asking for the actinium and polonium residues then held by the Society, and asking for a research grant of £75 from the "Radium Fund" for expenses at the same time. Rutherford said he had already tested the actinium residues with the permission of Larmor and checked that there was still activity in them, and now he asked for them to be handed over to him for temporary use. "I have given evidence which leads me to think that there exists in ordinary actinium residues a new element which is transformed slowly and directly into radium . . . Although such work may appear more suited to a chemist than a physicist, I may mention that the most important part of the investigation is physical." In fact, Rutherford was not particularly interested in the difficult business of tracing the radioactive families for himself, though he put some of his students on to such work. He kept very closely in touch with the work, however, through his long and intensive correspondence with Hahn and Boltwood, the two most prominent workers in the field, and it was Boltwood who eventually pinned down the elusive "parent" of radium, the element "ionium".

Manchester University was not, however, the only establishment wanting radium. From his holiday cottage in Cornwall,

Rutherford confided to Schuster: "From a remark Ramsay let fall I gather he is trying to tap the same source of supplies. He told me there was a very large number of medical applications. I did not let on about you. Ramsay, by the way, was very decent and presented me with all his actinium." The strange, ambivalent relationship between the great chemist and the pushing young physicist continued, with Sir William Ramsay apparently totally unaware of what the physicists were saying behind his back. But when the holidays were over and Rutherford's first term at Manchester began, the rivalry became more acute. The two men had met at the British Association meeting at Leicester, where there must have been some discussion about how to allocate any supplies of radium that did arrive. Certainly nothing had yet come from Austria, and Manchester University started its new academic year without radium. Rutherford was worried. As soon as term started he wrote directly to the Austrian Academy, the scientific body which was the parent of the Radium Institute, making a "formal application" for the loan of some of the radium preparations in their charge. He wanted half a gram of pure radium, but it was not necessary that this should be sent in the pure form of radium bromide; he was perfectly prepared to accept preparations of as little as twenty-five per cent purity as long as the radium itself was there in sufficient quantity. He stated his experimental aims – to examine the physical and chemical properties of the emanation, to examine the ionisation of gases exposed to intense radiation, and to try to discover what element was eventually left when all the radioactive transformations of radium and its daughters were completed.

More than a month passed before any news was received, but in the meanwhile Rutherford must have heard from Vienna that a highly embarrassing situation was about to arise – the Austrians were sending the radium, but it was being sent for the joint use of himself and Ramsay. It is well worth noting that this somewhat unperceptive response from Austria did not prevent the establishment of yet another life-long friendship between "the Vienna people" and Rutherford, a friendship that withstood the impact of the First World War, the dismemberment of the Austrian Empire, and a major scientific dispute between the Vienna Institute and the Cavendish, which was settled by Chadwick and Rutherford in their own favour but without public controversy and without forcing any loss of face on the losers.

Eventually the radium famine seemed about to end. On November 11th, 1907, Rutherford wrote to Ramsay: "I wrote to you a fortnight ago about your views in the arrangement between ourselves and for the use of the radium from the Austrian Academy, but I have not received a reply. Today I received the official communication from the Academy notifying me that Dr Brill is taking over to you the 350 milligrams of radium bromide and that it is loaned for our common use." He wanted to know whether anything was known of Brill's arrival and asked what arrangements Ramsay proposed, as "I presume that the Academy will expect to know the general arrangements between us, more especially as the results of the experiments are to be communicated through the Academy". (Incidentally, this agreement about publication was to result five years later in the first "proof" of the existence of the atomic nucleus appearing in German.)

To Ramsay he continued:

> I do not know whether you are aware of my state of poverty in regard to radium. The maximum quantity available for experiment is 7 milligrams: I am unable to do any experiment that involves the use of even 30 milligrams and am consequently hampered all round. I have a number of physical experiments I wish to carry out as soon as I get a workable quantity . . . I shall be very pleased to consider any proposition that is mutually convenient.

On that same date Ramsay wrote announcing the arrival of the radium. With Olympian disregard for other people's arrangements or requirements he said, "It would be an infinite pity to divide the radium. It is so infinitely more valuable as a whole. So please agree to my retaining it for some time. I will measure the emanation volume, try its action on salts, get its spectrum (and all the machinery for this is nearly ready) and after a year or a year and a half, hand it over to you." For his part Ramsay offered Rutherford a regular supply of the radium emanation provided not only by the Austrian radium but also from his own supplies of radium. Apart from the experiments he wanted to do himself, Ramsay continued, "I don't see how the solid can otherwise be used. It would be better to run no risk of wasting it till all experiments on emanation which suggest themselves have been made – or at least a fair number." And he ended magisterially, "Have no hesitation in demanding from me frequent assignments of the emanation plus mixed gases. It will give me very little trouble. I draw every four days, and if inconvenient I will tell you frankly . . . Let me hear what you think."

What Rutherford thought went into the post on the same day he received Ramsay's letter, November 13th:

I frankly state that I am somewhat disappointed over the radium affair. I was given to understand some months ago that I was to be loaned a considerable quantity of radium by the Austrian Academy; then the wish to please two people leads to a joint distribution; and now I find you wish to have practically the whole use of the loaned sample for a considerable time. I had made preparations for work in several directions which must of necessity be postponed until I get a supply of radium at my own hand. In this connection also, I recognise very clearly that there is no time like the present. However I am desirous of meeting your wishes as far as I can, for I quite recognise the importance of not dividing the specimen. My initial proposal appeared to me to be reasonable since by an occasional interchange of the emanation from a half quantity, each of us would be able to get the advantage of the whole.

You suggest that you would like the use of the 350 milligrams of radium bromide for a year or so. I think it very desirable in the interests of both of us to have a very definite agreement on this point. I am willing to forego my claim on the radium in the meantime if you would hand over the whole preparation to me not later than Jan 1st, 1909. This would allow you time to carry out a good many of the experiments you have in view with a large quantity of radium. The date of handing over must not depend on whether any particular work is finished or not, for I recognise there will be plenty of work to do on this subject for a decade or more.

In return for this concession Rutherford offered to limit his demand for emanation to eight days' collection per month and offered the same to Ramsay when the basic material should change hands. But, "I told you I am practically destitute of radium and this arrangement would help me in several directions. Naturally, however, I could not hope to do much serious work under such conditions."

Ramsay's reply, two further days later, was perhaps carrying superiority too far:

I am glad you are frank; it makes things so much easier . . . I would rather not promise January 1909, because I might be tempted to break my word; but I will take the date as provisional, and try to meet your wishes. Let me be as frank as you. I am used to handling this stuff in quantity; I have learned by long experience; and I am a little afraid that you or your assistant would find it difficult to get the apparatus as perfect and workable as I.

And just four days later (November 19th) Ramsay was saying that because he wanted to do certain experiments very carefully and because the apparatus was not quite ready yet, he could not guarantee the first supply of emanation to Rutherford even during the next week. Rutherford did indeed receive several supplies of emanation from Ramsay in the next month or two. One of the tubes even went astray and was lost for a while in the railway system; what nowadays would be treated as a major "scare story" was solved by an exchange of postcards. But Rutherford had obviously decided that working with Ramsay was a "no win" operation. He made up his mind to start again and obtain his own supply of radium, so he wrote to the Austrian Academy once more. Many years later Professor Stefan Meyer, of the Radium Institute, remembered:

> The first loan of about 300 milligrams of radium bromide (equivalent to about 175 milligrams of radium) was given to Ramsay who had to participate with Rutherford in the use of it. But there was always friction between the two, and Rutherford could not work in this way as unrestrainedly and profitably in his own way as he, and with him the whole scientific world, would have liked. So the Vienna Academy resolved to give him a preparation for his exclusive use.

This new supply of radium came to Rutherford in January 1908. In February of that year Ramsay wrote calmly, "Brill told me of your present and I congratulate you on it," and then proceeded to give some chemical advice on how the radium solid should be managed. In private Rutherford rejoiced rather more than perhaps he should. He wrote to Schuster on March 1st: "The Austrian Academy, I understand, were not quite pleased with the Ramsay distribution of the radium and have forwarded me for my own use about 500 milligrams of radium bromide . . . I have now more radium than Ramsay and so should virtue always be rewarded!" It was with this radium and its emanation that Rutherford carried out his major work in Manchester.

Yet he and Ramsay maintained outwardly friendly relations. Another letter to Schuster of about the same time reports:

> This evening Ramsay is taking me to the Chemical Society meeting, to sacrifice me, I judge, on the altar of Ramsay. He tells me privately he has made revolutionary discoveries, which he briefly outlined, but I don't know how much he will state when the meeting comes off. I presume I am to be on hand to do polite homage. I am looking forward to the meeting as it is the first time I have had an opportunity of listening to *ex cathedra* utterances of the said gentleman.

Similarly Rutherford kept everything publicly correct – in writing his official thanks to the Vienna Academy of Sciences for the radium which Ramsay had kept, he only remarked (December 3rd, 1907): "The radium is at present in the hands of Sir William Ramsay, with whom I have made a working arrangement for mutual use of the material."

The great discoveries that Ramsay claimed were his "observations" of the transformation of emanation into neon and the transformations of copper and sodium into other elements under intense radiation. Rutherford "knew in his bones" that these things were not so, but he could not possibly say so without proof. Letters from his scientific friends all raise the question of whether Ramsay was to be believed. He could only advise them all, from J.J. in Cambridge, to Hahn in Germany, to Boltwood in America, to Bragg in Australia, to remain sceptical. The problem was that it would take many months of careful work to prove Ramsay wrong and neither he, nor his friends, had the time or facilities to do anything in less than a couple of years. Meanwhile Ramsay's work had been given enormous publicity round the world, and it was indeed some years before even the majority of the scientific community became convinced by evidence that Ramsay and his pupils had made serious experimental errors. It was, for instance, as late as January 1909 before T. Brailsford Robertson would write from Pacific Grove, California, "We have been watching the gradual disappearance of Sir William Ramsay's stupendous discovery with considerable amusement."

One result of the increased availability of radium in England was the battle between Rutherford and Ramsay for priority over the claim to have demonstrated the spectrum of radium emanation. It has already been noted that Rutherford considered that Ramsay had indulged in very sharp practice in picking up a private conversation and rushing back to his laboratory to use the hints Rutherford had given. Ramsay similarly rushed his results into print with the Royal Society, and it was only some weeks later that Rutherford was able to publish his own results, gained with his colleague, Royds, in Manchester. However, their work showed clearly that Ramsay had done a far from thorough job. Rutherford was again comforted by the support of his scientific friends. J.J. wrote in July: "Ramsay's conduct about the spectrum of emanation seems to have been inspired by the Olympic Sports to beat his own record in matters of this sort." A few months later, when Rutherford was sure of his own results, and sure that Ramsay had got it wrong, but was unsure as to how to tackle this refutation of the master, J.J. was firm. On October 25th, 1908, he wrote: "I think that if the spectrum of the emanation given by Ramsay contains extraneous lines, it is most desirable in the interests of science that this should be pointed out otherwise people might be discovering radium emanation from sources quite innocent of it."

Ramsay was, however, perfectly correct in one thing – his estimate of the competition for scarce radium supplies by the medical profession.

9

Science International

It had been recognised from the days shortly after the Curies' first discovery of radium that radioactive energy could be used for the treatment of cancer. Broadly speaking, the first results of using radium to treat tumours almost always seemed amazing – tumours which would yield to no other treatment apparently disappeared, or at least were much reduced in size by comparatively few exposures to radioactivity. It was only later that the problems and the lack of permanent cure became clear.

It would not be fair or accurate to say that Rutherford took little interest in this work, but it must be admitted that it was always to him a sideline, something not in the mainstream of interest. He frequently involved himself in the arrangements necessary for using radium for medical work, and later he became highly influential in moves to provide really large radium sources for research and hospital treatment. And always he provided advice and encouragement. He first became involved shortly before leaving McGill when he was approached by a keen "amateur" scientist, Charles E. S. Phillips of Shooters Hill, in Kent. Phillips was a founder member of the Röntgen Society of London and was also a member of the Royal Institution. It may be said that he became the first medical physicist. In 1906 he visited Rutherford and attempted to interest him in a committee he was trying to get together to look into the possibility of setting a standard unit for radioactivity. They corresponded for several years until Phillips finally succeeded in getting a properly scientific research laboratory and treatment centre at the Free Cancer Hospital. Then he was able to write to Rutherford in May 1911, "The state of things as regards the medical attitude towards physics is deplorable . . . But it is an innovation on the hospital's part to have a mere physicist to advise them . . . When I tell you that the original scheme was to set up their emanation apparatus next to their best electroscope you'll see about how they stand." Phillips' work and advice bore fruit. A year later he was able to write again describing a very satisfactory set-up at the hospital which had cost £3,000 to build and equip, and he was very pleased with the design and facilities. He added, "Some splendid results have been achieved with radium and X-rays – results which never get in the papers but which none the less are of the greatest value."

All during his Manchester period Rutherford was also in correspondence with J. Joly of Trinity College, Dublin (then the chief University of Ireland). Joly was primarily interested in radioactivity and geology. He did much measuring of radioactivity and radium in seawater, studying samples sent from as far away as the Antarctic and, despite his writing of "our deplorable backwardness here", he pioneered a form of cancer treatment with radium in which the radioactive sample was enclosed in glass tubes which were then enclosed in platinum needles which could be inserted surgically into the outer, growing, edges of tumours. Since Rutherford was given this information by Joly, there was some tendency for him to become an informal information centre. A Dr Sinclair White, from Sheffield, for instance, wrote to Rutherford asking not only how to keep and handle their supply of radium, but also for details of Joly's method of using the substance on tumours. Rather similarly, Dr L. Wickham wrote from a clinic in Paris asking what thickness of lead was necessary to eliminate all beta-rays.

Just before the start of the world war Rutherford arranged for one of his laboratory technical staff, Walter C. Lantsberry, to go to work in a clinic in Philadelphia run by a Dr Kelly, which specialised in treating patients by the Joly method. Lantsberry reported back on the "astonishing results" recorded in the first four months of the treatment. He also told the tale of the disappearance of one of the tubes containing radium, and his ability to recover it from the city rubbish dump, to which it had been taken accidentally, by searching for it with his electrometer. He had remembered a story Rutherford had told at one of the famous laboratory tea-time meetings, when a similar radium-containing tube had been recovered from a British dustcart by Professor Wilberforce wielding his electrometer.

In a rather similar manner Rutherford became during this period a sort of informal employment and advice agency dealing with enquiries from every level. Both J.J. and Larmor (Sir Joseph Larmor, then Secretary of the Royal Society) wrote to him about the election of Crowther to the Royal Society. Thanking Rutherford for his helpful letter J.J. added, "I never struck such a difficult and unsatisfactory job as this election in my life." Larmor, writing in October 1908, was rather more blunt; he wanted to know whether Crowther was "an independent thinker or could the work have been done by J.J.'s laboratory assistant". It speaks volumes for the trust that the greatest of his contemporaries already had in Rutherford's judgment and fairness that they should ask him questions such as this when Crowther, who worked at the Cavendish, was already beginning to figure as a character in the debates between J.J. and the Manchester school about the nature of the atom.

In these years Rutherford was advising W. H. Bragg to try to get the McGill professorship he had vacated; was corresponding with Soddy over his former colleague's dissatisfaction with Glasgow University, and was successful in proposing Soddy's election to the Royal Society in 1909. From 1909 onwards he was in frequent correspondence with J.J. and others about nominations for Nobel Prizes. In 1909 he was corresponding with R. Kleeman, formerly of McGill, about the physics Chair at Wellington College in New Zealand, and three years later with the same man about various posts in the new University of Western Australia. The old subject of the varying merits of the research man as against the teacher in "colonial" universities crops up again, and in the following year there is a list of violent complaints from Kleeman about his treatment, and that of his friends, by the University of Western Australia. Indeed he writes, "A lecturer is likely to spoil his chances of advancement in his own and other universities by carrying out research – and that through exciting the jealousy and resentment of the fogey professors."

In 1910 Rutherford was officially consulted about the appointment of a new physics professor at the University of Queensland in Brisbane. And only two years afterwards he was engaged in a long correspondence with his former McGill pupil, R. W. Boyle, about his move to the professorship of physics at the University of Alberta. Boyle wrote of the dissatisfaction at McGill, particularly over the language problems in Montreal, and remarked that "many are going to Toronto, Canadawards". Rutherford went to some trouble to get Boyle a supply of radium for his work at Alberta and discussed various candidates for the lecturers Boyle would need there.

Again in 1910 Rutherford was invited to take the physics Chair at Columbia University in the USA. When he turned down the offer, Nicholas Murray Butler wrote to him, "So tell us who else we can ask." Columbia had plenty of good teachers but they wanted to build "a great school of research in experimental physics". Columbia, in fact, went on to offer the post to Harold A. Wilson, another of Rutherford's most regular correspondents. Wilson, another product of Trinity College in Cambridge, had taken the post at King's College in London which Rutherford had rejected in

1906, and he wrote extensively when he heard that Rutherford was leaving McGill: "King's College here is not a very promising institution and the lab is very small. Also there is so much teaching work that one cannot do much else. The worst thing about the place however is its strong theological atmosphere and the terrible set of crocks one has to get on with." Wilson did go to McGill, and stuck to the Canadian university despite Columbia's offer and despite the fact that he "was becoming a theorist. There are no research students at the moment, only demonstrators." However in 1912 Wilson asked advice about going to the Rice Institute in Houston, Texas. He did make this move to the USA after complaining that "Barnes has kept the research students and monopolised the workshops".

At the same time Rutherford was writing regularly to T. H. Laby in Melbourne in Australia and supported strongly the claims of his Manchester pupil, D. C. Florance, a New Zealander by birth, for a post in the physics department at his own Canterbury College in Christchurch. There were ever-growing relationships with Japanese physicists, too. These seem to have begun within a few weeks of Rutherford's arrival in England, and in a way which he found very flattering. In his first letter to his mother from Manchester he not only said, "The laboratory is very good, although not built so regardless of expense as the laboratory at Montreal" but reported how his predecessor, Schuster had introduced him to the visiting Japanese Minister of Education, Baron Kikuchi and added with obvious glee that the Japanese had later remarked to Schuster, "I suppose the Rutherford you introduced me to is the son of the famous Professor Rutherford" (June 30th, 1907). Certainly within the next few years Rutherford received visits, apparently short working visits, from a number of Japanese physicists. The subsequent correspondence shows very clearly how Rutherford spread the "doctrine" of radioactivity by his personal influence. In 1909 S. Kinoshita, from Tokyo Imperial University, spent a short time in Manchester and in 1910 wrote back reporting that he was now giving a course of lectures in the physics department on radioactivity, but "We have practically no active matter and no decent instruments for investigations of such kinds. Everything is inconvenient here to do anything, and I often wish I was able to go back again to Manchester." A couple of years later, after Rutherford had helped to get one of his papers published in the *Philosophical Magazine*, Kinoshita wrote again, "I wish I could go back again to your lab so that I shall be able to do some decent work."

In the meanwhile Professor Nagaoka had visited Manchester in the course of a long tour round most of the major physics laboratories of Europe, and, in sending his thanks for Rutherford's "great kindness and hospitality", wrote, "I have been struck with the simpleness of the apparatus you employ and the brilliant results you obtain. Everybody engaged with the investigation on radioactivity seems to be impressed with the same fact and expresses admiration of the splendid results which you obtain with extremely simple means." In response to this welcome, if well-deserved, flattery Rutherford replied with the firmest of all his statements about simplicity: "I do not know that the simplicity of my experiments was so unusual. As a matter of fact I have always been a strong believer in attacking scientific problems in the simplest possible way, for I think that a large amount of time is wasted in building up complicated apparatus when a little forethought might have saved much time and expense."

This little exchange carries, however, a much greater significance. Nagaoka's first letter is dated February 22nd, 1911. In the course of it he implies that he left Manchester about five months before, and he also mentions that in 1904 he had proposed a "Saturnian" model of the atom, an atom with a large heavy centre surrounded by rings of electrons. Rutherford first announced his own discovery of the general structure of the atom just two weeks after this letter was written. It will be shown that Rutherford had worked long and hard on his own discovery and it is not being suggested for one moment that Nagaoka's atomic suggestion was the source of Rutherford's ideas. However, the coincidence of timing is interesting for it suggests that Rutherford may have discussed the problem with the Japanese scientist during his stay in Manchester and may have been influenced by the trend of Nagaoka's thoughts. But the full story of Rutherford's scientific work at Manchester will come in the next chapter.

The feelings of the Japanese scientists when they took back the ideas of radioactivity to their homeland were strangely paralleled in Germany. Otto Hahn, writing to Rutherford in June 1907, announced that he had just succeeded in becoming a *privat dozent* (roughly, achieving a licence to teach at university level). He had passed his viva examination because "all the people were so terror stricken that they asked only some simple radioactive questions about the matter and did not want to hear anything else". Despite his success, however, Hahn still felt "lonely among all these chemists who don't really believe in radioactivity". As his long series of letters show, Hahn was rapidly carving out his place as the world's leading radio-chemist, with a series of new discoveries of radioactive daughter elements. He also showed a wisdom and humour which impressed Rutherford, for when the New

Zealander suggested "paradium" as the name for one of Hahn's newly discovered elements – meaning "parallel to radium" Hahn rejected the suggestion on the grounds that the name was too reminiscent of military activity and goose-stepping.

Filed among these letters is a sad, touching and brief correspondence with a Montreal doctor, Dr G. E. Armstrong, who appealed to Rutherford in 1909 for a supply of radium to treat his wife's inoperable stomach cancer. John Cox and Eve would apply whatever suggestions Rutherford might make – but "These things are more than awful when they come into our own family and philosophy gives but cold comfort. Try and suggest something. It seems to me radium offers the only hope." Rutherford did supply radium and advice, but the poor lady benefited but little and the correspondence soon came to an equally inevitable end.

From the USA to Manchester came Jacques Loeb's son and Alois F. Kovarik, just two among many visitors and short-term workers, but they both help to illustrate how broadly Rutherford's influence was spreading. After two years in England, Kovarik moved as Assistant Professor of Physics to the University of Minnesota in Minneapolis, and wrote back, "In the laboratory we had nothing, and so I had a good deal of work getting things together. Now we are fairly well equipped." But now he had to return to work: "Work, teaching I mean, so hard on my time that my research time is dwindling down." Jacques Loeb confided that he had heard and been "entertained" by reports of Ramsay's alleged transformation of copper into lithium, "which goes to prove that our reporters have invaded England". He had heard, too, of the professional problems of biologists where "a small clique of men, not too prominent scientifically, dominate supremely. I wish I might one day have the chance of stirring them up. It is remarkable that a country which is so unquestionably in the lead in physics, has so little use for experimental biology."

Rutherford also heard some inside views from laboratories nearer home. Louis L. V. King, a graduate from McGill, working at the Cavendish from 1906 onwards (and who later in the decade returned to McGill on the teaching staff) found that it was considered "bad form to depart from the beaten track of traditional mathematics". He also found that Cambridge was tending to fall away from the experimental approach – reading for the Tripos was largely covering what others had already done, with little chance of experimenting. "The result is that one gets more and more disposed to do armchair physics, a trait characteristic to the Cambridge school of mathematicians." He felt that Rutherford's arrival in Manchester might make this worse: "The attendance of research men will fall off in consequence of the attraction to Manchester. Several of the men are thinking about it."

King, a little later in 1908, brought Rutherford up to date with the atmosphere in Paris, describing his meetings with Madame Curie, Langevin, Perrin and others, and telling of the Curies' original "rickety old shed, with holes in the roof". He found very good attendances at the lectures of both Langevin and Poincaré, attendances which made Cambridge audiences seem very small "so that Cambridge will have to adopt very extensive reforms in its mathematics course and make things more attractive to advanced men if it intends to keep alive". A couple of months later he was seeking Rutherford's advice as to where to go next, since he was suffering from the "deadening effect" of reading for the Tripos and finding mathematical physics at Cambridge "very stagnant".

Another of King's letters (in 1909) drew one of Rutherford's most powerful and typical replies; King had complained of some ill-health due to overwork, and his mentor replied, "Never work so hard at a problem as to lose your sleep. If I find I am carrying any work to bed with me, I always make a point of easing up for it is a bad sign. I know how interesting it is to keep hammering at a thing when it begins to go – but it doesn't pay in the end." Plainly he had forgotten his midnight and weekend work at Montreal, but this was to become more and more his view. In Manchester he began to try to shut the laboratory at night, but was in some sense held at bay by Harry Moseley, the most brilliant of his collaborators, who would often choose to work right through the night. By the time he moved back to Cambridge, Rutherford had made it a rigid rule that the laboratory was shut at six o'clock. The evening and the night were for reading, digesting the day's results and above all for thinking. Many of his memorialists who knew him in his Cambridge days have retailed stories of his ferocity against those who did not want to "knock off" at the end of the day. More and more, as he grew older, he developed the practice of alternating furious bursts of work, when his energy and concentration astounded much younger men, with frequent and complete holidays.

This attitude was certainly not so clear at the start of his Manchester period. The janitor of the physics building was asked by a *Manchester Guardian* reporter about the new Professor's hours of work, and replied, "You can never tell when he'll leave his lab and go home" (March 21st, 1908). And at the end of his first year at Manchester Rutherford himself wrote to Bumstead and others that he had never worked so hard in his life before. He coupled this statement with the fact that he was now growing fat, or at least that his wife accused him of so doing. Certainly he was never a man for exercise or physical fitness. He also had to admit to his mother that he now had to wear glasses for prolonged spells of reading. It seems likely that as he approached the age of forty he was beginning to show those signs of middle-age which come to all but the most fortunate or the most energetic. But similarly there came the signs of affluence and influence. It was at this time that Rutherford was elected to the Athenaeum, perhaps the most famous of all the London clubs, which offered a home from home particularly for those whose position in society depended mostly on intellectual effectiveness – academics, clergymen, senior civil servants. This club, and the fellow members he met there, were to become most important to Rutherford's lifestyle and political effectiveness in the inter-war years.

In 1908 there came a new fame, the certain mark of success, the Nobel Prize. But it was the prize for chemistry. Rutherford himself led the chorus of jokes and good-humoured mutual congratulations with cracks about his "instantaneous transmutation" from a physicist into a chemist. There was a marvellous irony about it; that the man who so battled against chemists as a profession, the man who was chiefly opposed by chemists, the man who was leading his own science of physics to an intellectual and academic domination of chemistry, should be given the prize for chemistry. Yet the appositeness of the decision by the Nobel Prize Committee cannot be doubted, for the science that Rutherford had revolutionised, albeit with physical methods, was indeed chemistry. The earlier Nobel Laureates in physics had been Röntgen, Lorentz, Becquerel, Rayleigh, Lenard, J. J. Thomson and Michelson, and these were precisely the men whose work had laid the foundations upon which Rutherford built – indeed in Rayleigh and Thomson they were his direct predecessors and teachers at the Cavendish.

The announcement that he had received the greatest public recognition that can come to any scientist, and won it before he was forty, came in the usual way, by telegram at the end of November 1908. As usual, too, those with their ears close to the ground had guessed he might be one of the recipients after certain discreet enquiries had come from Sweden. Congratulations poured in from his friends and colleagues, but perhaps more importantly, from major scientific figures in other countries with whom he had not previously been in contact – men such as Svante Arrhenius, the Swedish chemist and Kammerlingh Onnes, the great Dutch physicist and pioneer of low-temperature research work. Typically Onnes offered the facilities of his large and powerful laboratory at Leyden University for any experiments Rutherford might wish to do on the effects of low temperatures on radioactivity.

Rutherford took his wife with him, of course, to Stockholm for the award early in December 1908. They both seem to have been impressed by the ceremonies, by the banquets in the palace, by the attentions of Swedish scientific dignitaries, by meeting the King and the Crown Prince and Princess. It must be remembered that Rutherford was not yet forty years old – indeed an eye-witness recorded that he looked "ridiculously young" – and that Mary Rutherford had only left New Zealand seven years before. The joke about his instant transformation from a physicist into a chemist stood him in good stead in the many speeches the Laureates are called upon to make, but Rutherford had the immense good sense to stick to what he was best at when it came to making his official Nobel address. Its title was "The Chemical Nature of the Alpha-Particles from Radioactive Substances", and it was, quite simply, a description of his latest work at Manchester, in which he had, at last, shown beyond doubt that alpha-particles were helium atoms, stripped of their electrons. In the course of this work he had become the first man to demonstrate, equally beyond doubt, the existence of individual atoms as material particles, thus succeeding in something that no professional chemist had ever performed, and establishing the essential truth of the atomic theory. His conclusion was that "the alpha particle is a projected atom of helium which has two unit charges of positive electricity . . . The alpha-particle is released at a high speed as a result of an intense atomic explosion and plunges through the molecules of matter in its path. Such conditions are exceptionally favourable to the release of loosely attached electrons from the atomic system. If the alpha-particles can lose two electrons in this way the double positive charge is explained."

The award of the Nobel Prize had two important effects for Rutherford, in addition to the obvious establishment of his undisputed place among the great scientific powers of his day. Firstly it brought him nearly £7,000, which was a very large sum of money for those days – something more than five years' salary. For the first time in his life he was reasonably wealthy – and it is one of the most interesting pieces of evidence about his true disregard for money, that the amount he left when he died was almost exactly £7,000. (In harsh exactness the Nobel Prize came to about £6,800 when it finally reached him.) The first thing Rutherford did with the money was to send fairly substantial cash presents to all his brothers and sisters, and to his parents, back in New Zealand. Virtually the only letters from his immediate family that have been preserved in his correspondence are the letters of thanks, simple, pleasant and plainly revelling in his triumph, that arrived early in 1909.

More practical is the letter from Rutherford to a Mr Becker (presumably a Manchester stockbroker) dated February 1st, 1909, ordering him to buy "£2,000 Queensland 3½ per cent stock, £1,000 New Zealand 3½ per cent stock, £1,800 Central Argentine Railway Consolidated Ordinary Stock and £1,000 Baltimore and Ohio 4 per cent Preferred stock". It seems then that Rutherford had decided to spend or give away £1,000 of his prize and invest the remainder. It was certainly about this time that he bought a car – he wrote to his mother early in 1910: "We have decided to order a motor car – a Wolseley Siddeley. It is very desirable to have some means of getting fresh air rapidly. We have chosen a four-seated car, 14–16 h.p., which suits our requirements for quiet travelling." After Easter he reported further: "We got our car on Good Friday and spent three days running around while I practised driving." There follows a typically dull piece of Rutherford writing detailing miles and route and stopping points and places where they had lunch. But after 500 miles he says:

> I have learnt to drive fairly well without a single incident, even of running over a chicken. A car is very easy to manage and far more under control than a horse. We average about 17 miles an hour over country and on a good road run along fairly freely at 25. We can go 35 or 40 if we want to, but I am not keen on high speeds with motor traps along the roads and a ten guinea fine if I am caught. These are the woes of motorists that I hope to avoid.

In his next letter he actually remarks, "I am feeling unusually well and fit for work due to motor exercise." His car was a great delight to him, providing him with the frequent holidays that he obviously found necessary, and he gradually became more and more adventurous with the pioneering machine, tackling first the Lake District, and eventually even Europe.

The second, and probably more important effect that winning the Nobel Prize had on Rutherford was that it introduced him personally to many of the great figures of continental European physics. There were, of course, the notabilities he met at the ceremonies in Stockholm; the Parisian Gabriel Lippmann was the physics prize-winner; Paul Ehrlich, the German pathologist, and Ilya Metchnikoff, the Russian immunologist, had won the biological prizes. But of more significance were the men he met in Germany and Holland on his return journey. Otto Hahn was now based in Berlin, and he organised the Rutherfords' visit to Berlin. This visit was largely social, and the celebrations were in the German style which we now regard as rather heavy. Nevertheless this was the Berlin of Planck and Einstein, of the early days of the quantum theory; the Berlin in which the first steps to the twentieth century revolution in theoretical physics were being taken. And here Rutherford met the men who were forging that revolution. From Berlin he went on to Leyden, in Holland, where he visited the great low-temperature laboratory of Kammerlingh Onnes. Not only were the lowest temperatures yet achieved being regularly produced in this laboratory, it was also the first example in the world of a really large laboratory, the first place where "Big Science" was practised. Onnes was not only responsible for a whole series of dramatic scientific "firsts", such as the discovery of superconductivity and the liquefaction of helium, he was also the first physicist to apply the power of twentieth-century engineering to producing experimental physics results. The vacuum pumps, the refrigerating machines that he used were on an industrial scale; he even founded schools of glass-blowers and instrument-makers to supply his needs for laboratory staff. In Leyden too, Rutherford met Hans Lorentz, another world leader in theoretical physics, who was working on the same problems as the leading Germans. Lorentz rapidly became a close friend and ally of Rutherford in the new arena of international scientific politics, and the alliance remained close for many years.

Lorentz had, quite unknowingly, influenced Rutherford greatly before they had ever met. Just a month earlier, on November 24th, 1908, the Dutchman had written to congratulate him, not so much on his Nobel Prize, as on his "beautiful" alpha-particle counting experiments. He had added "The close agreement of your value of 'e' with that found by Planck, which on theoretical grounds I think to be very reliable, is a truly admirable result, which like so many others we owe to you, could not have been dreamt of when you began your investigations." The nature and importance of this result of Rutherford's will be more fully examined when we consider the science of his Manchester period. But the fact that the great experimentalist accepted and used the revolutionary theory of the quantum developed purely theoretically by Planck, and accepted it without demur and without his usual roarings, can most probably be accounted for by his knowledge that his own value of "e", the single electric charge carried by the electron or the hydrogen atom, a value which was very considerably higher than that obtained by other experimentalists such as J.J., was supported by Planck and his theory. The impatience which one would expect Rutherford to have shown with the German theoreticians therefore never emerged, and from the start of his international "career" he was on good terms with Planck, Einstein and the others – relationships which were almost immediately to prove important.

The early part of 1909, following immediately upon his return from Stockholm, was largely occupied by accepting British plaudits. He had written to his mother in his usual dull and unemotional way that "we had a great time – indeed the time of our lives". He was livelier in accepting the congratulations of his friends and peers. From McGill, Eve had written congratulating him on the "splendid" prize "which the gods have shaken into your most deserving lap. You have certainly been sailing with a full sail, and a brimming tide." In February Manchester University paid its official tribute with a great dinner in the Whitworth Hall. J.J. concluded his speech, "Of all the services that can be rendered to science, the introduction of new ideas is the very greatest. A new idea serves not only to make many people interested, but it starts a great number of new investigations . . . There is nobody who has tested his ideas with more rigour than has Professor Rutherford. There can be no man who more nearly fulfils the design of the founder of the Nobel Prize than he does."

J.J. had sent his personal congratulations before the official prize-giving and had included some rather more worldly comments: "I was afraid when I nominated you that it would be thought that they ought to give other than English-speaking physicists a turn this year (for they attach a good deal of importance to giving each country a turn) and that you would have to wait for a year or two. I rejoice I was mistaken. No one ever deserved it more." In his formal reply to the official congratulations from the Cavendish, Rutherford paid his debt: "I feel I owe a great deal to the Laboratory and to you personally for whatever success has attended my work. It was in the Cavendish that I was first innoculated with the spirit of research and it was there I made my first investigations under your guidance. The work for which the prize is awarded was begun in the Cavendish Laboratory and I recall that the alpha-particle was born and christened in one of the lower rooms." Which may have sacrificed a little accuracy to gracefulness but was a reasonable and grateful version of the facts.

But soon it was back to battle and this time, for the first time, the opponent was the British government. The issue was whether Britain should send an official delegation to an international congress to set up a "Radium Standard", and whether the country would accept such a standard. The problem was one which had been growing more and more urgent among scientists for the past two or three years – and it was not a simple one to solve. It was impossible to manufacture a sample of pure radium metal by extraction from the ore; furthermore, since the radium was steadily disintegrating, as Rutherford had proved, into other related daughter elements, it would never be possible to obtain any sort of internationally-accepted piece of metal to which experimenters could refer, in the way that there were internationally agreed metal objects which were the standards for weight and length. Yet it was vital that there should be something by which scientists in different laboratories could compare their measurements of radioactivity – and it was clear that this standard must be of the type which specified a certain amount of radioactive energy being released by a certain amount of radioactive material. Once such a standard had been agreed, a "unit of radioactivity" could also be agreed and used by everyone.

There was, however, no obvious commercial motivation for setting up such a standard, no motivation such as there had been for standardising international units of length and weight to facilitate trading. The scientists would have to do it themselves, and the only scientists interested in such an operation were the very small number of radioactivists, a tiny proportion of the world's small total number of scientists. The whole project only became feasible when the Belgian government, possibly seeking some international prestige, offered to act as host for the required international congress. In the late summer of 1909 the Belgian Ministry therefore invited some twenty-five countries to send official representatives to a "Congrès Internationale de Radiologie et d'Electricité" which would be held in Brussels a year later – in September 1910. Rutherford, naturally, was approached by the Belgian organising committee to be Acting President for the Congress in Great Britain – and those with experience of launching new movements will know that this means he was asked (and he undertook) to "get the thing off the ground" in Britain.

He completely underestimated the understanding and interest of the British government. He had no concept of the depth of ignorance of the British ruling class of the time, as represented both by senior civil servants and by Ministers of Asquith's Liberal government. The Belgians had sent out their official invitations on September 27th, 1909. A month later there was still no reply from the British government and Rutherford decided that the time for action had arrived. He wrote to Sir Edward Grey, the Foreign Secretary, on November 4th, 1909, drawing his attention to the situation, and pointing out that the Congress would almost certainly result in the setting up of a permanent international committee to discuss and manage an international radium standard. The French had already agreed to send a six-man delegation to the

Congress, but the Belgians had received no reply from London. Six days later Rutherford received a stately reply from the Foreign Office saying that "the matter is still under consideration by the departments directly concerned". There is some evidence from a later stage in the story that three departments, the Foreign Office, the Home Office and the Board of Trade were shuffling the papers around, each trying to make the others responsible for the business which they did not even understand.

Rutherford therefore started to try to manipulate the levers of power. He brought the problem to the attention of Larmor, the Secretary of the Royal Society. Sir Joseph's reply must have made him realise the inertia he was up against:

> The British government needs very strong reasons for acting, and if it [the Royal Society] asks for this it will refuse something more governmental. My memory says that three years ago the Foreign Office consulted the RS about a similar request from the Belgian Minister, and that the RS could not offer any case for adhering to it. – But anything unofficial that can be done to back you up would no doubt be done with pleasure.

There is no evidence as to what Rutherford wrote in reply to this, but obviously he must have said that he would take the responsibility himself – a move that was to pay him off well in many future negotiations when he met unwillingness to take a risk. Larmor certainly replied by asking what he could further do to support Rutherford. His letter, dated November 13th, 1909 and cautiously marked "Unofficial", continued, "It is very noble of you to step in when others neglect", and went on to ask if it would be any use the Royal Society appointing delegates to the congress itself. "I don't see what the RS can do. It will not apply to the Foreign Office, as far as I can understand the temper of the Council, as if it did it could court a refusal," Larmor's letter continued, and he went on to point out a typical official difficulty; the Royal Society could of course only address itself to its equals in other countries and "Observe that your congress is not under the patronage of the Institute of France nor of any other principal Institute of Science of any country."

However, Rutherford's efforts had obviously caused some movement in the machinery of government, for two weeks later he received another letter from Larmor, marked, this time, "Private". It revealed that "The Board of Trade has asked whether this country should send delegates to your congress. I propose to ask the Council to say <u>yes</u> and to suggest you as the official representative . . . In the normal course of events they [the government] would pay expenses, but if too many are nominated they will be glad to evade that."

The battle was not yet won, however. Nearly eight months later Rutherford had to write to the Foreign Office again. It was now August 1st, 1910, yet the British government had still not sent any reply to the Belgians and the Congress was due to start on September 12th. Twenty-three other governments had already nominated their delegates since the Radium Standard "will be of considerable importance commercially as well as scientifically". And Rutherford ended with a paragraph which shows that he had been very quick in picking up the tone of officialdom, both in the slight threat of embarrassment and in offering the best way out of the problem. It is in fact a little masterpiece of official letter-writing:

> It would certainly be an anomaly that the one country which has done so much pioneer work in the recent development of the science to be discussed at the congress should be the one country not represented officially. I may mention, I hope without indiscretion, that the Royal Society has already agreed, that if asked by the government authorities, to recommend two of their members as delegates – viz, Sir J. J. Thomson and myself.

Of course he got his way, and got it in time to set off for a fortnight in Munich with his friend Boltwood before the pair of them went officially to Brussels. Boltwood had been working in Manchester at Rutherford's invitation, and it was undoubtedly at Boltwood's suggestion that they went to Munich, for the American was a great admirer of Germany, German methods and German beer – indeed it was Boltwood's rather pro-German attitude in 1914 that caused a temporary coolness between the two men. Rutherford could well afford the holiday, for he was rapidly learning the techniques of "management". All through the summer of 1910 he had been corresponding with Otto Hahn, who naturally was the German most interested in the outcome of the congress, with Stefan Meyer of the Vienna Institute, still the world's leading source of radium, and, above all, with Marie Curie in Paris, the discoverer of radium, the personification of radioactive studies in the eyes of the public. In May he had written to Madame Curie asking for one of her radium standards to compare with his own "empirical standard". She replied promptly, promising to send him a special standard which she would prepare for him herself. She hoped to see him soon in Paris and asked him whether he thought it would be worthwhile going to the Brussels Congress. Rutherford, of course, encouraged her to go and thus began a correspondence with her which lasted until her death in 1933.

At the same time he was discussing various problems with Hahn which they could both foresee were likely to arise at the first congress on the new subject. In June 1910, for instance, Hahn asked Rutherford "privately" for his suggestions on various terminologies, adding, "It is better to agree privately what to propose." They both wondered what Madame Curie would request, and they prepared, with their friends, to soften her by a little flattery, in that they all agreed that whatever new unit of radioactivity should be set up it would be called a "Curie". Just before the Congress, as they were all gathering in Brussels, Stefan Meyer was able to warn Rutherford that Marie Curie "wishes to use the term 'Curie' for the amount of emanation in equilibrium with ONE GRAM of radium . . . So it would be best for the committee to omit the two sections on radium and the Curie and report only on the main radium standard." From this we must conclude that the agenda and reports were ready long before the Congress met, and that the Congress was called to confirm what the inner circle of experts had already decided – this is indeed the only way to run an international congress successfully.

On the face of the record there is a paradox here, for the Congress developed into a public shambles, due to the poor organisation by the Belgian authorities. The final plenary session on the last day was described by Rutherford with considerable amusement to several of his correspondents. The Belgian chairman completely lost control as dozens of resolutions poured in from all over the auditorium in at least three different languages. Then came amendments, and amendments to the amendments. Finally a German scientist, W. L. F. Hallwachs, who had the advantage of possessing a very loud voice, managed to take control, "and expressed and moved the opinion of the Congress and it was carried". The outcome was that an international committee was set up to organise the official International Radium Standard. The committee consisted of Boltwood, Marie Curie, her colleague Debierne, Eve, Geitel, Hahn, Meyer, Rutherford, Evan Schweidler and Soddy. So from Rutherford's point of view the desired result was achieved. As he wrote to Meyer when he returned to Manchester "with an asymmetric face and a bad cold . . . On thinking it over I consider that our proposals were the best possible under the conditions." Three really important decisions had been taken: that Madame Curie would prepare a radium standard containing about twenty milligrams of the substance; that this standard would be kept in Paris and would only be used for comparisons with national secondary standards while it remained under control of the committee which would reimburse Madame Curie; and that the international unit of radioactivity would be called the Curie and would be based on an equilibrium of radium and its emanation.

The second critical factor in achieving this result was Rutherford's personal relationship with Marie Curie – a relationship quite unique in his life and one of its most admirable and attractive features. Rutherford seen from Marie Curie's point of view, the Rutherford one meets when reading a biography of Marie Curie, is so totally unlike the man his English-speaking scientific memorialists remember as to be almost unrecognisable. Marie Curie's Rutherford is a calm, gentle, supportive friend, almost fatherly, although he was four years her junior: a man to be turned to in trouble or sickness; the man who was called on by his fellow scientists to "handle" the cold and prickly Franco-Polish woman who was always so much on the aggressive-defensive as a woman and a scientist. The whole situation appears even more extraordinary when it is realised that Rutherford and his immediate circle of scientific allies did not hold the Curies in high regard as scientists, much though they admired the French couple's intense pioneering labours. On the Curie side, in the same vein, must have been the realisation that Rutherford had come to dominate the science they had founded. Furthermore several of Rutherford's friends actually disliked Marie Curie, while, by the time of the Brussels Congress, Pierre Curie had been tragically killed in a street accident and Marie, we can now see, was suffering from radiation-induced ill-health, if not precisely from radiation sickness.

The Rutherford–Marie Curie relationship had begun as far back as 1903 in personal terms, though naturally they had known of each other before then in scientific terms. But in 1903, when Rutherford came from McGill to England to defend or propagate his radioactive transformation theory, he took a short trip to the Continent and found himself in Paris with his wife. There, on June 25th, a postcard from Soddy caught up with him after a long journey to Geneva in his wake; the postcard said that Madame Curie had sent a message hoping that Rutherford would call on her. Immediately he went to the shabby laboratory where the Curies worked, only to find that it was locked and unoccupied. Marie Curie was at that moment facing three eminent examiners and achieving her Doctorate in Physical Sciences of the University of Paris with the distinction of "très honorable" – a remarkable breach of a male fortress. By coincidence the chairman of the examiners was Gabriel Lippmann who was to appear with Rutherford in the Nobel Prize list five years later.

The Rutherfords met the Curies at a celebratory dinner that evening, the dinner having been arranged by Paul Langevin, who was not only one of the Curies' very few friends but also a friend of Rutherford's from Cambridge days. The only other guests were Jean Perrin and his wife. Marie Curie's biographer says:

> Rutherford immediately took to Marie Curie, and after that first encounter always had a soft spot for her – even when some of his closest friends had developed different sentiments. He liked her no-nonsense way of dressing . . . Clearly Marie Curie's simplicity appealed to him just as his enthusiasm seemed honest and worthwhile to her. Once he turned to her and, in a way which she among few others could fully understand, said that radioactivity really and truly .was "a splendid subject to work on".

She was to remember these simple words to the year of her death. She knew that he treated women very much as equals; this must have flattered her. She did not find his bluntness repulsive. He was, at that time, one of the few physicists who was actively encouraging women to pursue scientific careers in his laboratories. It is more than likely that Rutherford was indeed attracted to Marie Curie precisely by her lack of sexual threat – it had taken her sister to persuade her to buy a new dress, black of course, for the day's official function. But the two had many things in common – their complete lack of commerciality; their similar attitudes to science as "pure research", and their proprietary attitude to radioactivity work as well as the feeling of jointly pioneering into unknown territory where many "establishment" scientists were unwilling to follow them. There was also their devotion to work; Marie Curie would never have expressed it Rutherford's way – "You know it must be dreadful not to have a laboratory to play around in" – but she might have been forced to admit that this was indeed her own attitude.

Certainly the dinner party was a resounding success, and apparently the scientific discussions aroused no animosity. The Curies' position on Rutherford's radioactive disintegration theory was one of non-commitment – as he put it in his first book a year later, they had "not put forward any definite theory". But they had already published some of their evidence to show that the first radioactive substance they had isolated, polonium – might not be an element after all. (It is, we know now, an element, but it was also "Radium-F" in the disintegration sequence that Rutherford's school established.)

These stimulating scientific subjects led them all to migrate into the garden at eleven o'clock that warm evening. Then Pierre Curie brought out from his pocket a tube containing radium and coated with zinc sulphide. The demonstration of brilliant luminescence caused by the radium fascinated them all. But, Marie Curie's biographer reports, it was by this light that Rutherford could see that Pierre Curie's hands were so raw and inflamed that he even seemed to find difficulty in handling the little tube. The terrible irony here is that Rutherford had only been able to perform his alpha-ray experiments, which clinched the disintegration theory, when he received a sufficiently powerful radioactive source which the Curies had sent to Canada for his use. Since there is no trace anywhere in Rutherford's life or papers that he ever suffered one iota of damage from his radioactive materials he might have rejoiced at the fact that his sources were never so powerful as those the French scientists had access to, though in fact he usually bemoaned his radioactive poverty.

Less than a year later, however, Bumstead was writing to Rutherford, "I haven't seen [the] Curies' last paper yet; I am never very eager to see them for I usually find that I have read the same things about a year before in one of your papers." And there seems no doubt that the Curies lacked boldness and imagination in developing theories that would account for radioactive phenomena on a broader scale than their own devoted chemical work. After Pierre Curie's death, Marie, not unnaturally, showed even less public confidence, though her devotion to work was unabated.

In 1908 she proved most uncooperative when Boltwood wished to compare his radium standard with hers, and he complained to Rutherford, "The Madame was not at all desirous of having any such a comparison carried out, the reason, I suspect, being her constitutional unwillingness to do anything that might directly or indirectly assist any worker in radioactivity outside her own laboratory."

In the following year Eve had been reading the Curies' collected works which led him to comment: "It is astonishing how slow [they were] to accept your conclusions . . . I believe if it had not been for you the whole subject would have been a grotesque muddle to this day." In the privacy of his letters to Boltwood, Rutherford was not particularly excited about Marie's enormous *Treatise on Radioactivity* published in 1910: "Altogether I feel that the poor woman has laboured tremendously, and her volumes will be very useful for a year or two to save the researcher from hunting up his own literature: a saving which I think is not altogether advantageous." And when they all met at the Brussels Congress Stefan Meyer privately expressed his belief that her attacks of nervous exhaustion which caused her to leave platforms and committees at crucial moments, were often brought on to suit her own purposes.

Rutherford however believed her to be genuinely ill, and he had medical advice to guide him. To his mother he wrote on October 14th, 1910: "Madame Curie looked very wan and tired and much older than her age. She works much too hard for her health. Altogether she was a very pathetic figure." Perhaps unconsciously he was comparing her with himself, for he told his mother in the same letter that the four days of the congress had consisted of "talking about 18 hours out of the 24", adding that he himself had spoken to most of the main papers and "had a good deal to do with starting the committee for fixing a radium standard".

Rutherford was right about her health. There was to be no more work for Marie Curie for several months. Throughout the winter of 1910–11 she was in hospital or convalescing. Her illness was described as tuberculosis and it was discovered she had kidney lesions as well. Rutherford wrote to Stefan Meyer in April 1911, that "Madame Curie has been seriously ill and threatened with tuberculosis . . . she is now better and returned to Paris . . . I am very sorry to hear about it as I am sure she consistently overworks." And so they were all due to meet again in November 1911, for the second radiation congress with very little progress achieved towards a realisation of a radium standard. Just before it began, he confided in Boltwood, "I am sure it is going to be a ticklish business to get the matter arranged satisfactorily, as Mme Curie is rather a difficult person to deal with. She has the advantages and at the same time the disadvantages of being a woman."

Whatever happened on the platform and in public at this second Brussels conference, the vital matter was settled in long, patient and private talking between Rutherford and Madame Curie. He wrote about it all to Stefan Meyer as soon as the conference was over. He had found that Curie had prepared the precise standard they required a month or two before, but had been keeping it in her laboratory to check the gamma-radiation level. Rutherford broached the possibility of the International Committee buying this standard from her but "for sentimental reasons, which I quite understand, she wished to retain this standard in her own laboratory. I pointed out that it would be quite improper for the committee to agree to the retention of the standard by a single institution." After some discussion it was suggested that a duplicate of the standard could be stored in the French National Bureau, which could, from the committee's point of view be considered as the primary standard. Madame Curie apparently fell in with this idea for she asked Rutherford to approach the Vienna Radium Institute to provide supplies of radium for the duplicate standard and whatever would be required by other governments and institutions. "This would be an excellent thing as it would simplify the costing."

For her part Marie Curie wanted the weight and purity of each preparation to be officially recorded so that she would have some check on her own standard and methods of measurement. This again, Rutherford felt, was "very sensible", for the scientific world would need to make every possible check to prevent mistakes. Then there was the question of money, and Marie Curie suggested that governments should be charged 1,000 francs for the preparation plus the cost of the original radium. Again Rutherford agreed: "This I think is reasonable as it is quite right to charge a good fee as some recompense for the time and trouble required." He had to be firm with her on some points, however: "I stated very definitely to Madame Curie that every duplicate standard that was required must be sent in the name of the International Committee, and not of herself as an individual. I was not quite sure whether she had not an idea of merely giving a personal certificate."

So the business of the International Radium Standard moved slowly forward, largely as a result of Rutherford's tactful treatment of the woman with whose name the public most closely associated the word radium. But Meyer was also working to the same ends, and he persuaded the Austrian government to sell radium at reduced prices for standards purposes – for which Rutherford thanked him and congratulated him on his success before the end of November was reached.

For Marie Curie that was the most dramatic month of her dramatic life. It was on November 4th, while she was still at the Brussels Conference that the French press "broke" the story of the scandal in her private life – the fact that Langevin, Rutherford's old friend, had taken the place of the dead Pierre Curie in her life. The French popular papers of that time do not make pleasant reading – quite apart from a scandal involving an international heroine, there were strong overtones of xenophobia against a Polish woman in that era of "Dreyfusards" and "anti-Dreyfusards", and there were even anti-Semitic attempts to hint that Marie Curie had Jewish origins. In the middle of all this, on the very day that Rutherford was writing to Meyer of his success in persuading Marie Curie to cooperate, she received a telegram informing her that she had been awarded the Nobel Prize in chemistry for the year.

It must have been the breaking of the scandal that led her to leave the Brussels Congress suddenly – but not before she had written Rutherford a little note – appreciating that he had spoken of her *"en termes sympathiques"*, and saying she was touched by all the attentions he had shown during the Congress; she would have liked to shake his hand before leaving but she was ill and could not do so – hence this note. He continued to keep a sympathetic watch over her in her troubles. The worst of public exposure was avoided by fairly clever tactics. Some publicity was inevitable when Madame Langevin sued for an order of separation from her husband, but Jean Perrin was able to write to Rutherford eventually (an undated letter from the Laboratoire de Chimie Physique, Faculté des Sciences, which must have been written some time in January 1912): "The news is good . . . Madame Curie *qui nous inquietait beaucoup vu son état d'épuisement*. The order of separation between Langevin and his wife makes no mention of Madame Curie but the damages are awarded against Langevin (who had not taken the precaution of having witnesses for those things which he could call up against his wife without bringing into court the name of Madame Curie)." He then gave details of the court orders about the disposition of the four Langevin children and the visiting rights. Langevin was granted their *"direction intellectuelle"*, and finally "Now I hope we shall all be able to get back to work again." But there was a postscript – Langevin wished him to say that he had been *"très sensible à votre amitié"* and Madame Curie has also been very touched by Rutherford's attitude. Marie Curie, not surprisingly, had another terrible attack of illness with high fever and kidney trouble; it even seemed possible that there might have to be an operation, but eventually the crisis passed. Poor Langevin, who never was successful at dealing with money, had therefore taken the blame, picked up the bill, and lost the woman, for in the social conditions of those times he and Madame Curie could only meet thereafter as academic colleagues.

Because of her illness she had to let her colleague Debierne manage the final stages of the formal acceptance of her radium standard by the International Committee. Rutherford went over to Paris for this strange cross between a scientific experiment and a ceremony in the spring of 1912. He was in fact relieved that Marie Curie was not able to handle the final details herself, and he considered Debierne a very "sensible fellow". Soddy was also in the party which found to everyone's relief that the Austrian standard prepared by Stefan Meyer and the Curie standard agreed to a fairly close tolerance of accuracy. And so Marie's radium sample was finally installed at the International Bureau of Weights and Measures as the world standard of radioactivity.

This was by no means the end of the strange, fond but sexless, relationship between Marie Curie and Rutherford. Nor was it the end of Rutherford's battles on the matter for he now had to get the British government to set up a national standard, calibrated, of course, by the International Standard, but kept at the National Physical Laboratory for internal reference. The Austrians, under Meyer's persuasion, had agreed to provide the radium, but there was throughout this period the competing demand for radium for medical purposes. It was this latter demand for scarce radium that had interested King Edward VII, and he had applied polite, unofficial, but powerful and royal, pressure on the Austrians for a supply of radium for medical use. A ludicrous mix-up followed. The Austrians supplied the radium to Sir Ernest Cassel, the millionaire who was a close personal friend of the King's, and very probably he paid for it personally, but the radium he received had been intended by the Austrians for the scientific standard. When Rutherford discovered this he wrote to Cassel, who felt he could not give up the radium, for he thought it was for medical use. The only thing Rutherford could do was to get some more radium from the Austrians. The Austrians, due to his close relationship with Meyer, were perfectly willing to oblige, but the question of paying for the scientists' radium was not so easily solved. Eventually Sir George Beilby, who was by this time Soddy's father-in-law, agreed to pay for it out of his own pocket.

All through the second half of 1912 this semi-comic affair dragged on, but it was precisely in this imbroglio that Rutherford seems to have won his spurs as a "wheeler-dealer", pushing here, soothing ruffled feelings there, taking the responsibility himself when official bodies and ministries were haggling over demarcation lines. Much of this correspondence was with Stefan Meyer. By May, Meyer had got the Austrian government to send out all the official documents, and Rutherford accepted the idea that all negotiations for radium should be between governments "to be sure that institutions are not attempting to get radium from your government on the cheap". He would inform Glazebrook – the first Director of the National Physical Laboratory – "quite informally" and then Glazebrook would be the proper person to initiate formal British moves. By September little had happened, and though Beilby had offered to supply the money for the British radium the whole thing seemed to be held up on committees.

Rutherford took things into his own hands – he asked Meyer to start preparing a British standard and "in the absence of a formal request you may hold me responsible in the meantime". In October Rutherford discovered that the radium had in fact been sent and was being held by Sir Ernest Cassel.

Eventually however a compromise seemed possible – Lord Rayleigh (president of the Royal Society) arranged that Cassel would hand over the purified sample for the standard if the medical interest could have a further twenty milligrams of radium which need not be in such purified form. "I gather it would be best for a definite application to be made," and this was duly done. But this left Rutherford and Glazebrook with an embarrassment at the start of 1913 – they were now due to get one sample of radium from Cassel and a second from the Austrians by official request, though Rutherford had already asked Meyer to prepare it unofficially. On January 8th, 1913, Rutherford had to ask Meyer to send the radium he was purifying direct to Glazebrook, once it had been got to about seventy-five per cent and then "we will arrange the matter".

It still took four months of further work to get the secondary British standard finally established at the National Physical Laboratory, and throughout 1913 and 1914 Rutherford received letters from G. W. C. Kaye who was put in charge of the radium at the institution, asking for advice on the equipment he needed, complaining that things were "chaotic", that Glazebrook "has small sympathy with the newer physics or with anyone interested in it," and finally questioning him whether he himself might become contaminated with radium, "so upsetting the measurements".

Nevertheless it would be wrong to picture Rutherford as toiling wearily through these troubles and ineptitudes. Several of the letters mentioned in the preceding paragraphs were written from Carcassonne or the Pyrenees which he toured in his motor car immediately after his trip to Paris for the official confirmation of the International Standard. He then returned to Paris accompanied, as he had been throughout his holiday, by his old friend W. H. Bragg, who had now become Professor of Physics at Leeds University.

There had been other trips to Brussels, too. In June 1911 Rutherford had received his invitation – needless to say marked "Confidential" – to the first Solvay Conference, described as "*Conseil scientifique international pour élucider quelques questions d'actualité des théories moléculaire et cinétique*". Lorentz was to be the president, and Einstein and Planck suggested that the conference consider the limits of movement around a rest position in the theory of radiation. The idea was to get everyone of great influence together, even when they were not specialists in the particular field, because *"nous nous trouvons en ce moment au milieu d'une évolution nouvelle des principes sur lesquels était basée la théorie classique moléculaire et cinétique de la matière"*. As well as Rutherford, Larmor, Schuster, Rayleigh, Jeans and J.J. were invited from England: they were joined by Einstein, Planck, Rubens, Sommerfeld, Wien, Marie Curie, Langevin, Poincaré, Perrin, Kammerlingh Onnes and van der Waals.

It may well have been during this conference that Willy Wien explained the theory of relativity to Rutherford at a pavement café. Rutherford had started it all by "twitting" Wien; the reply was an earnest discussion of the changes that Einstein insisted should amend the normal equations of relative motion, which Wien concluded by declaring, "But no Anglo-Saxon can understand Relativity." Rutherford's roaring reply was, "No! They have too much sense." This was a perfect example of the image Rutherford chose to present to the world, the bluff no-nonsense, colonial-farmer image – the English commonsense attitude to the wildly theoretical foreigners. His professional approach was totally different. Relativity was simply a theory, but at that time it did not conflict with anything which Rutherford or anyone else could prove. When the theory could be tested, by the apparent position of the planet Mercury as it "appeared" from behind the sun in 1919, no one was more interested than Rutherford. He was one of the first to hear the results of the astronomical expedition to the South Atlantic to observe the phenomenon, which confirmed Einstein's hypothesis, and he spread the good news as fast as he could among his friends. Whatever jokes he might crack about relativity over the lunch-table, as a scientist he treated it professionally, just as he treated Planck's quantum theory, except that quantum theory had an impact on his own work, whereas relativity had no direct bearing on his atomic physics.

The first Solvay Conference brought together probably the greatest gathering of scientific brainpower that one hotel room has ever collected. Yet all but two or three were men whom Rutherford knew and with whom he corresponded. The man who made it possible, the Belgian industrial chemist Ernest Solvay, not only entertained his distinguished guests, he also allowed them a thousand francs apiece for expenses. The conference did not achieve anything – it was not intended so to do. But it probably did perform a useful function in gathering together the people who were bringing about that revolution in physics, a revolution in our concept of the basic laws and materials of the world in which we live, and helping them to understand rather more about each other.

There was a second Solvay Conference in 1912, and by this time Ernest Solvay had endowed an International Physics Institute with a million francs, the proceeds of which were to pay for the annual "get-together" and to provide what we should now call research grants. For the next few years Rutherford was in regular touch with colleagues such as Marie Curie and Lorentz about the award of these research funds, and he was able to use his influence to obtain a proportion of the money for British research. He had been told about the benefaction before any public announcement was made, and in the same letter from Lorentz (in April 1912) he was invited to be the "English member" of the first Conseil Scientifique Internationale, which would have the task of disposing of the 18,000 francs a year which would emanate from the foundation, in research grants.

Marie Curie was pleased with the first tranche of research grants. She was enough recovered by October 1912 to write to Rutherford that she thought that the grants "*ont été très bien placées*". But more than that, she had discovered that during her absence from the scientific field Sir William Ramsay had invaded her territory and made the sort of claims for precedence which had earlier so infuriated Rutherford and his immediate co-workers. Ramsay had been working on a new determination of the atomic weight of radium, one of the figures worked out by Marie Curie when she was establishing, against considerable scientific opposition, the fact that radium was a true element. Now Ramsay had published his figures which were the same, in overall value, as Marie Curie's, yet were less consistent from one experimental run to another. She wrote, "*Malgré cela il conclut que son travail est le premier bon travail sur ce sujet!!! J'avoue que j'ai été étonnée.*" Furthermore Ramsay had gone on to make some "malicious and incorrect" comments about the Curie experiments on atomic weights. There was, of course, nothing that Rutherford could do about this except to sympathise and add Marie Curie to the world-wide list of those who complained about Ramsay and his public utterances. This "determination of the physical properties of radon", from which the weight of radium was found, was however, virtually the last scientific work that Ramsay undertook; but it is still interesting that his biographers appear to be quite unaware that he had so upset Marie Curie, or even that he had written anything which might be regarded as denigratory of her work.

Marie Curie's friendship with Rutherford was if anything

strengthened by his quiet support in her trials. She described to him the medical use of radium in Paris hospitals which had shown "*très bons résultats*", and in 1913 she discussed with him the possibility of setting up another international committee to coordinate methods and results. In 1914 she sought his support for a Solvay grant for a radiology laboratory in her native Poland where "*en raison de la situation politique en Russie, les carrières scientifiques sont fermées aux savants polonais*". And she nominated Rutherford to be President of the Congress of Radiology which was scheduled to be held in Vienna in 1915, to carry on the work of the Brussels congresses – it was, in fact, aborted by the world war.

No one could accuse Rutherford of regarding women as sex-objects. His attitude to fashions and beauty was disdainful – he wrote to his mother in 1911 describing a stroll around London in the spring with his friend Eliot-Smith and remarked that he had been "highly edified at the ladies' hats and dresses especially in Hyde Park. It was as good as a pantomime to see the women in hobble skirts and big hats." He continued to encourage women to work in his laboratory where he treated them strictly as colleagues. He was annoyed with Harriet Brooks for failing to come to Manchester and take up the post he offered her because she had decided to get engaged and therefore give up work. There is a rather brutal story (recorded by Andrade in the Rutherford Centenary celebrations) of the European lady who was working in the Manchester laboratory – a lady who is not named but who is described as an ardent feminist, something of a man-hater, and not particularly outstanding as a researcher. Rather than ask anyone to help her to get the lid off a bottle of sulphur dioxide she had tried to open the container by wedging the top in a cupboard door; she had been successful to the extent that she released a rush of gas which left her unconscious inside the cupboard, where fortunately she was found and revived. Rutherford heard of the incident – he heard of everything – and summoned the lady to his room. "What's this I hear? You might have killed yourself," he reproached the lady who replied rather sulkily, "Well, if I had, nobody would have cared". To this Rutherford's response was, "No, I daresay not, I daresay not, but I've no time to attend inquests."

Rutherford had no deferential respect for anybody. This was one of his hidden political strengths and rather than attributing it to his pioneer upbringing, it must have sprung from his profound, and humble, knowledge of his own worth and capabilities – a paradoxical self-evaluation which was not linked with vanity, and which was noted by all his friends. In the summer of 1912, the Royal Society celebrated its 250th anniversary. Rutherford had recently been elected to the Council of the Society, a considerable honour in itself, and as part of the many junketings attended by scientists from all over the world, there was a garden party at Windsor Castle where he and other luminaries were to be presented to the King and Queen. He described the occasion to his mother in a letter on July 25th, 1912:

> The ceremony was very simple. We went past and shook hands with the King and Queen and then made a beeline to find our women folk who were congregated at a suitable point. The day was not too hot and the King and Queen and family made a procession through the crowd of visitors, over 7,000, including everyone of importance in London. I understand the man who called out the names as we were presented got our group of delegates mixed up and every man was given his neigh-bour's name. It did not matter, as I don't suppose the King knew the names of any of them. After a scramble for tea we walked round and made for home about 5.30. It was a rather tiring function and I am quite content if I never go to a Windsor garden party again. I had to buy a silk hat and got quite accustomed to wearing that monstrosity. The ladies were all attired in their best, and the ultra-fashionable had piebald stripes, for example one half of the dress red and the other blue, the line of demarcation from one shoulder to the opposite foot. It was very amusing watching some of these queer get-ups.

The same letter describes functions at which he met the Archbishop of Canterbury, the Prime Minister, Asquith, the "leader of the Labour MPs", Ramsay MacDonald, and the Duke of Northumberland. But Rutherford's chief interest seems to have been meeting old friends from Canada and Australia.

His political attitudes, such as they existed at all, were plainly of a liberal trend, though the liberal is spelt with a small "l". He wrote, for instance, in 1911 to his mother: "You will have seen about the strikes in England which have apparently ended for the time being. No doubt the wages are in many cases too low. It appears fairly certain there will have to be an increase all round in wages during the next few years." Certainly his neighbours seem to have regarded him as liberal in his views, and especially on the subject of women and votes, coming as he did from a Dominion in which women had votes. He was invited by a local suffragette group to give a short talk on "Suffrage at Home" and in particular "What the effect is and how little it disturbs family life or women's 'womanliness' so called".

Rutherford's home life, with his wife and dearly-loved only

daughter figures hardly at all in the story. There were no letters between them, and Rutherford barely ever mentions his family in his letters even to his closest friends. There is the very rare occasion when some slight worry over his daughter's childhood ailments crops up; as for his wife it is often impossible to tell from the correspondence even whether she accompanied him on some foreign holiday. Undoubtedly the Rutherfords offered a great deal of hospitality in their home especially to visitors from abroad such as Boltwood, and to foreigners working in Rutherford's research team. Bohr and Geiger both write of the warmth of hospitality they received, and Rutherford, just once, mentions a Christmas spent quietly at home entertaining many of his colleagues. Hospitality in the Rutherford home, however, may not have suited many men quite as well as the serious and sober Bohr and Geiger. Mary Rutherford remained an extremely fierce opponent of alcohol all her life, and she was distinctly careful with her money – once she consulted a friendly neighbour asking advice about the cheapest possible way to give a party, and received a Lancashire answer to the effect that if she felt that way about it, it would be better not to give a party at all.

The life and activities that have been described in this chapter would have been a description of a full life for any young professor. Long before he left Manchester, before even the world war broke out, Ernest Rutherford had received every honour a man could reasonably want – he was on the Council of the Royal Society, he had won a Nobel Prize, and he was knighted in 1914. In scientific terms these were the rewards for the work he had done in Canada, for the work which had not only discovered entirely new phenomena and caused a revolution in the established paradigms of physics and chemistry, but which had also established on a firm basis an entirely new branch of science, the science of radiochemistry. The life described in this chapter seems a natural career fulfilment of those discoveries.

It is therefore difficult to realise that the background to all that has been described here was a continuous flow of new major discoveries; the chief achievements of Rutherford's scientific life were realised after he had achieved all the honours and international acclaim we have noted – or, it might be more precise to say, at exactly the time that these honours were being received. In this chapter, recording so full a life, there has been no mention of the even fuller life he was living in the laboratory. It is the essence of Rutherford that he could, and did, work very much harder than most other men. And he enjoyed this to the full. It was in these days that he said to H. R. Robinson, "I am sorry for the poor fellows that haven't got labs to work in." And many years later he was to write reminiscently to Geiger, "They were happy days in Manchester and we wrought better than we knew."

10

The Atom

When he arrived in Manchester in 1907 Rutherford's scientific reputation rested on his studies in radioactivity. At Manchester, radioactivity became his tool, his probe, with which he revealed the nature and the main structure of the atom. Manchester was also the arena in which he developed a new institution and a new style; it was here that he showed the effectiveness of the research "school", the large team of researchers, mostly working on parallel or converging problems clustered round one central theme or objective, a truly international team exchanging young postgraduates with foreign centres.

Around 1913 someone must have persuaded Rutherford to think about writing his autobiography, for there is an undated page of notes in his own handwriting, which does little but demonstrate once again how tongue-tied he became when writing about himself. But he does note that when he first came to Manchester the laboratory staff consisted of Petavel, Hutton, Beattie and Geiger, and we know that Schuster continued to work there on an occasional basis as well. Petavel moved to the National Physical Laboratory, Hutton went to Cambridge as Professor of Metallurgy eventually, and Beattie was soon placed as an independent Professor of Electro-Chemistry. The department was reasonably well-equipped, though not up to the standard of McGill and not prepared for radioactive work. There seem to have been three research students, who probably also acted as demonstrators. Within three years Rutherford had increased the number of his team to twenty-five, and there are photographs still extant showing them grouped stiffly in front of a gas-lamp and the arched doorway to the physics building. There are five foreign students and one woman in the group.

How this was managed financially is difficult to determine. Rutherford had only the Readership in Mathematical Physics financed personally by Schuster and the John Harling Fellowship at his disposal. Some of the foreign students certainly paid for themselves, and the younger men eked out a living by working as demonstrators, supervising the laboratory exercises of undergraduates. By 1914 Rutherford's official staff on the university payroll consisted of only nine men, who were paid in total £1,600 a year, the lowest paid earning only £125 a year. Rutherford's own salary was very high – £1,600 a year for himself. The departmental budget for supplies, almost exactly as Schuster had promised, averaged £420 a year from 1908 to 1914. In addition he squeezed two grants out of the government for £100 each during these years, and it is suggested by contemporaries that he may have got external finance for some major pieces of equipment. Quite frankly young researchers expected to be poor and were so – it was estimated that one could scrape by on £100 a year.

It was significant that when Harry (H. G. J.) Moseley was trying to get a place in Rutherford's team, money in the ordinary cash sense of the word is not mentioned. What the ambitious young physicist was looking for was something quite different, the making of a reputation which would ensure him a good job in the years far ahead, as it seemed in those stable pre-1914 days. So Moseley's consideration was the balance between the teaching necessary to earn himself enough to live on and the research opportunities to make his name. If there was to be too much teaching and not enough research, he wrote to his mother, "the advantage of contact with Rutherford is diminished but the mercenary advantage that he is an excellent stepping stone remains".

Moseley was undoubtedly the most brilliant of the research students attracted by Rutherford, but he was also one of the few upper-middle class men to be attracted. His father, a Professor, had died young, and Moseley's own relationships and performances seem to have been affected by this. It is difficult, seventy years after the event, to know how to appreciate a young man who could address a letter to "My dearest of small Mothers", but we know that he said of himself that he was better at research than at exams. The Oxford view of Manchester was that Moseley was "too good to waste on Manchester demonstrating"; the view was expressed by the Reverend Paul J. Kirkby, the senior demonstrator at the Oxford physics department whose head was J. S. Townsend, Rutherford's old Cambridge friend. But Kirkby's view did not impress Moseley because Kirkby himself was "about to take a country living in Norfolk . . . [after] . . . several quiet years of exploding mixtures of hydrogen and oxygen". Furthermore, "As an Oxford man he looks down on all things outside, whereas it is a fact that recommendations from Oxford are apt to carry less weight than from places where science is treated as a more serious profession; and a year or two teaching in a very large, well-organised laboratory at Manchester count for more than several years of the easy-going and less responsible demonstrating at Oxford."

Moseley joined Rutherford at the start of the 1910–11 academic year. He found to his dismay that he had to lecture on electricity to second-year Honours students and had nine to ten hours a week teaching physics to engineers. This prompted a letter which would be unthinkable in our times: "On Saturday I was disgusted to find a large proportion of coloured students with the thickest heads doing lab work. These seemed to include Hindoos, Burmese, Jap, Egyptian and other vile forms of Indian. Their scented dirtiness is not pleasant at close quarters."

Yet Moseley was the one man in Rutherford's team who openly expressed his disapproval of the leader's masculine and risqué conversation. The American biographer of this brilliant young man (and we will deal with his remarkable scientific work in due course) describes Rutherford as "the son of a New Zealand flax farmer, who possessed neither languages or culture, nor wealth nor mathematics, and who regarded with indifference most of the many branches of physics which Schuster had encouraged". He goes on to comment that after Rutherford had been in Manchester for three years, "The number of research students had increased somewhat while the range of research topics had conspicuously narrowed; Honours candidates took their general exams at the end of the second year, and then, after a brief initiation by Geiger, launched into full-time research, usually on a little radioactive subject picked by Rutherford."

The charge that he narrowed the curriculum receives support from an important quarter; from his close friend and co-worker, H. R. Robinson. In the Rutherford Jubilee celebrations Robinson said:

> The main degree examination in physics took place at the end of the second year, and a year later there was a paper on modern physics, and more especially radioactivity, for which students were expected to prepare mainly by the reading of current periodical literature. This scheme of undergraduate training, as I experienced it, had the grave disadvantage of leaving wide gaps in our knowledge of some important sections of classical physics – and this is putting it euphemistically, for the word gap normally connotes at least something more or less solid in which a gap can exist. On the other hand, under Rutherford the scheme did give to all students the chance of seeing physics as a living and growing science; the best students did, as they always will, get hold of their physics somehow, and the system produced a steady stream of men who could, under Rutherford's and Geiger's supervision, play useful parts in radioactive investigations.

It was at this time that Rutherford started drawing up at the beginning of each year a list of researches he wanted to get done. Often these were only scraps of paper covered with his scrubby pencil scrawls. There is one of these preserved, undated, but clearly from his early Manchester days. There are no fewer than twenty-three "projected researches", and only the last three are directly described as "radioactive projects" though the subjects are very far-reaching and include his own proposed work on "No. of alpha-particles from Ra". But at least six of the other projects involve using the high-pressure apparatus developed by Petavel, although two of the six concern the effects of pressure on radioactivity. Rutherford boasted to the end of his life that he "had never given one of my students a hopeless problem" and his admirers considered that this was equivalent to the Napoleonic claim of being interested only in "lucky" generals.

The complaint may still be made, and justifiably, that Rutherford's conduct of a big physics department was élitist; that it served admirably the needs of those who were naturally good at the subject and would become physicists themselves. But at Manchester and again at Cambridge, it was nothing like so satisfactory for those who turned out not to be gifted at the subject.

However, account must be taken of less formal contributions to history, such as some reminiscences recently (1981) published in the Manchester University student newspaper. These memories come from Mary Haworth, born in 1899, a local Lancashire girl, who was educated at a small grammar school and made her way to the university to study chemistry. She had to take some physics lectures and thus came into contact with Rutherford: "Some of the medics with us were taking their Physics exams for the fifth time and he had no need to bother with us at all, but it was a mark of his character that he was quite willing to do so . . . He was a fatherly figure, very approachable and generous, and when he described his experiments it was always 'we' and never 'I' . . ." She also remembers him setting up for the undergraduates one of his famous atomic experiments, using the original apparatus. There can be no doubt but the man had the ability to inspire with his own enthusiasm.

Whatever view one may take of Rutherford's attitude to the needs of students in a physics laboratory, his policy was a conscious one. He expressed it very clearly in a public speech he made at the opening of the Peters Electrical-Engineering Laboratory at the University of Dundee in 1910. The duty of any scientific department to the university and the country, he said, does not end with providing routine instruction for the student, and no science

271

lab should be content with handing on the knowledge gained in the past, "but it should be a centre, and an active centre, for the acquisition of new knowledge by original investigation". He put the point even more strongly:

> No university . . . is worthy of the name that does not do everything in its power to promote original research in its laboratories. It is the duty of the university to see that its Professors and teachers are not overburdened with routine teaching, but are given time for investigation and provided with reasonable lab facilities and the necessary funds for this purpose. The opportunity for investigation should not be confined to the teaching staff but should be extended whenever possible to the advanced student who shows promise of capacity as an original investigator. In this way a laboratory becomes a live institution instead of a dull, if somewhat superior, high school.

Rutherford's laboratory was certainly a live institution. There are many descriptions of the place and its atmosphere given in reverence by his former students and colleagues. To pick just one there is the letter from Hans Geiger to *Nature* written shortly after Rutherford's death, written when England and Germany were about to go to war:

> I see his quiet research room at the top of the physics building, under the roof, where his radium was kept . . . But I also see the gloomy cellar in which he had fitted up his delicate apparatus for the study of the alpha-rays. Rutherford loved this room. One went down two steps and then heard from the darkness Rutherford's voice reminding one that a hot-pipe crossed the room at head level and that one had to step over two water pipes. Then, finally, in a feeble light, one saw the great man at his apparatus and straightway he would recount in his own inimitable way the progress of his experiments and point out the difficulties that had to be overcome.

The other side of the man, too, Geiger saw:

> I always liked to recall another little episode which occurred at the time when much work was being done in the laboratory with sources of radiation consisting of extremely thin glass tubes filled with emanation. It was necessary to exercise great care lest any of this emanation should escape, for it spread rapidly throughout the building, and by virtue of its activity made experimental work an impossibility for periods of many hours. In his typically drastic manner Rutherford had threatened the severest penalties for offenders in this matter. One day I noticed that it had become impossible to use an electroscope in my room, where I had fitted up the first counting experiment for Rutherford, and

before long other research workers emerged from the neighbouring rooms with the same sad story. We were not long in discovering that the emanation had come from Rutherford's own laboratory where at that moment he was actively engaged with his experiments. As luck would have it, he came into my room shortly afterwards to inquire how my work was proceeding . . . I was brief and to the point. I told him that once again it was futile to attempt to do any work as the whole building was full of emanation, all of which came from his room. Rutherford looked surprised and replied, "Well, there you have further proof of the power inherent in this emanation." With this remark he left me; but he soon returned suggesting that I must be somewhat upset and that I needed a little fresh air.

And thereupon he took Geiger out for a ride in his new car, discoursing all the while on all the problems of physics. "In spite of the minor provocation, I would be loath to part with the memory of such a day, spent in fellowship with a master-mind," Geiger concludes.

It was in these Manchester days that Rutherford developed the habit of walking regularly round his domain at frequent intervals, certainly daily, and talking to all his "boys", discussing their results, chewing over their problems as he perched on the nearest lab stool. It was at this time, too, that his habit of intoning, tunelessly but audibly, "Onward Christian soldiers" when all was going well was first noticed. There is no doubt, equally, that everyone was fairly frightened during his black glooms, for his short outbursts of rage could be terrifying, though they would quickly blow over and be followed by apologies if he felt he had really hurt his victim.

Rutherford could be very severe, even in public, on pomposity and pretentiousness, and he hated anything that smacked of the "intellectual". He was not tolerant of the fool, but always seems to have respected the man who could bring him up with a sharp, well-placed retort. Similarly "Crimes against apparatus he kept in a special category; for such crimes he had little forgiveness and an uncomfortably long memory . . . Although Rutherford was, at any rate after the first few seconds, eminently reasonable about any accident that could fairly be ascribed to ill-luck, his view of a co-worker was apt to be coloured for many years by recollections of a bad case of carelessness," according to H. R. Robinson, one of the early Manchester recruits. Robinson describes how he was allowed access to the laboratory's high-tension electrical supply which consisted of a large bank of small and very fragile accumulators:

Geiger delivered a little homily: I was never to touch the battery connections while I was standing on the concrete floor; I must always keep a dry wooden board to stand on while making adjustments, and I must always hold one hand firmly behind my back while touching any part of the battery so that there could be no risk of the circuit being completed through my body. Before I had any chance of expressing surprise at, or gratitude for, this solicitude, he went on with obviously complete solemnity and singleness of mind: "You see if you get a bad shock you may kick out before you realise what you are doing, and the Prof would not like it if some of the cells got broken."

The same old friend records his honest friendliness, not marred by any self-consciousness or seeking after popularity – "his oldest friend, if he arrived inopportunely, was more likely to be received with a grunt than a smile". And it was Robinson who produced the immortal phrase (copied by many), "Few men can have made more friends, or lost fewer, than he did." Nevertheless it is admitted that the great man was a little inclined to crow over minor misfortunes afflicting close friends, and was always delighted to note if any of them was losing his hair more quickly than he.

Rutherford's true friendliness was emphasised by his complete lack of the English vice of snobbishness or over-awareness of class distinctions. Robinson again:

He had in a very high degree that friendly and companionable spirit which is a notable characteristic of people who have spent much of their lives either in New Zealand or in Canada. This undoubtedly had its effect in keeping the laboratory working as a more than commonly united family. Rutherford had, moreover, what is not typical either of England or of any Dominion, a curious insensitivity to differences of academic or social standing, within fairly wide limits, and it probably was very good for the morale of a young research student to see himself treated with as much – or as little – respect as an emeritus professor.

It must have been these personal characteristics, as much as his scientific ability that enabled Rutherford to build up his research team – and to build it so fast. Although J.J. in Cambridge had shown the way, it must be emphasised that the institution Rutherford formed was something quite new to the world of science – indeed new to the world in general. Firstly it was international to an extent previously unknown in a world where jet-travel was still unthought of; there were Russians, Poles, Austrians, Hungarians and Germans, mixing with Japanese, Americans, Canadians, Australians and New Zealanders. Secondly, the home-based core of the group was completely free of class distinction; there were those of the academic upper-middle class such as Moseley and Charles Darwin (grandson of the author of *Origin of Species*), but equally there were boys from the industrial area around Manchester, from families with little money and no academic tradition, such as Marsden, Chadwick, A. B. Wood and Royds. Robinson came as a "raw student" who was for many years quite satisfied with the tiny £125 stipend. He was personally devoted to Rutherford, and also the man who most often contradicted him directly. A. S. Russell, one of the chemists of the group, also came as a raw young Scot from no powerful background. Since so many of these youngsters ended their careers as knights or professors (or both) it is perfectly reasonable to look on Rutherford's laboratory as a machine for social mobility, and this characteristic was to be even more heavily emphasised when he eventually moved to Cambridge. Finally the Rutherford physics department at Manchester was, in its time, the most powerful and intellectually distinguished research team that had ever been gathered together: Rutherford, Bohr, Chadwick, von Hevesy, Moseley, Darwin, Russell, Fajans, Geiger, Marsden, Wood, Andrade, Robinson, to give only the names that immediately appear on a list of published papers.

They were more than a research group. They were a tribe with mutual loyalty, working on different projects with different partners within the tribe. Rutherford was the tribal leader, not only in the serious business of deciding what research should be done (and not always getting his own way against the stronger members of the tribe such as Bohr and Moseley), but setting the tone in small matters such as conversational style. True, the lesser members referred to him as "Papa", using a particular intonation they had picked up from the popular Manchester music hall; but they delighted in his conversation and remembered his style. They joined him in deriding outsiders with Rutherfordian phrases: "That man is like a Euclidean point; he has position without magnitude" – or of someone who had shown personal vanity, "I couldn't hear what he was saying for looking at it." Rutherford himself was certainly not above conversation that was deemed coarse by the standards of the time – he once described a very powerful magnet he had seen as "strong enough to draw the iron out of a man".

The men in the Manchester physics laboratory came to love, admire and revere Rutherford – the words are theirs, freely used by them in public speeches; there is no record of any of them disapproving of the leader. But the cement which bound them together was science, and the knowledge that they were, quite simply, the best in the world. If the atmosphere was that of a tribe, it was also the feeling of the dominant tribe. And the leadership they granted to Rutherford was not to his position as Professor, the formal head, nor to his undoubtedly appealing character, but was homage paid to his pre-eminence as scientist.

Taking his scientific ouput from his Manchester period, at least up to the outbreak of the world war, the rate of work, as measured in published papers, was almost equal to that of his Montreal period. And this was despite the fact that in Manchester he had a department to run, and his public position to keep up, whereas in Montreal he had been free to concentrate on research alone. Furthermore (as Feather has pointed out) the proportion of papers in joint authorship with one of his assistants, as compared with papers of which he was sole author, remains very similar. The important difference in Manchester was that so many major papers were published by men in his group without his name appearing at all.

The seven years immediately preceding the war can be divided, scientifically, into two periods. In the first Rutherford completed the work on "his" alpha-particle, finally establishing its weight, charge, and nature as essentially a helium atom carrying a double positive charge (it could not be said at the time that it was the nucleus of a helium atom stripped of its normal pair of electrons, which would be our modern description). In the second and greater period, Rutherford turned his alpha-particles into probes with which the interior structure of the atom could be investigated. With their aid, he produced the picture, or "model", of the atom which has stood all the tests of the past seventy years and which stands unchallenged as our present view of the basic nature of everything we see and hear and feel around us.

He started his Manchester work with the equipment and apparatus that were available when he first walked into his new laboratory. His first paper is a letter to *Nature*, dated only six days after his arrival in England, reporting long-term observations, lasting eighty or 110 days or 120 days, concerning the growth of radium in actinium preparations. More precisely the results were recording the fact that no radium appeared to grow directly from actinium, for the gist of the paper was that actinium could not be the parent element of radium. Although it is not mentioned directly, these results could only have come from samples and solutions he had brought to England from Canada with him. The time-periods quoted seem quite arbitrary, but long-term observations of samples of radioactive materials were a regular feature of Rutherford's, and all other, "radioactive" laboratories in the attempt to sort out the tangled relationships between active elements and their daughter products. Rutherford cautiously concludes in his first Manchester paper:

> I think we may safely conclude that, in the ordinary commercial preparations of actinium, there exists a new substance which is slowly transformed into radium. This immediate parent of radium is chemically quite distinct from actinium and radium . . . It is not possible at present to decide definitely whether this parent substance is a final product of the transformation of actinium or not. It is not improbable that it may prove to be the long-looked-for intermediate product of slow transformation between uranium-X and radium but with no direct radioactive connection with actinium. If this be the case, the position of actinium in the radioactive series still remains unsettled.

The final triumph of isolating and identifying the parent of radium was Boltwood's. It was achieved in that summer of 1907 and Rutherford gave it something very like an "imprimatur" in his third paper from Manchester, another short letter to *Nature*, dated October 27th. Although he did not like Boltwood's chosen name, "ionium", he accepted it and dealt some stern rebuffs to an objector, a Mr N. R. Campbell, who had proposed that every radioactive substance should be given a name which firmly showed its place in the radioactive family tree: "This system is very excellent in theory, but I have found it extremely difficult to carry out in practice. The continual discovery of new products in very awkward positions in the radioactive series has made any simple permanent system of nomenclature impossible." If names did not worry him greatly, he had shown that remarkable "nose" for the eventual scientific truth in the first of these papers and was very nearly completely right. In these two short contributions he had confined himself to guiding the detailed work which he left others – Boltwood, Hahn and still, to some extent, Soddy – to carry out, for he would admit they were better chemists, better at separating, filtering and crystallising, than he. In between he had recorded the results of his work with Petavel and his bomb, showing that very high temperatures did not materially affect the rate of radioactive disintegration of radium-C and the radium emanation. This was reported to the Leicester meeting of the British Association in September 1907.

But greater significance at Leicester was a discussion session that Rutherford was asked to lead. The subject was "The Constitution of the Atom". This shows that Rutherford's main interest lay in this direction, as he had declared in his Silliman Lectures. This, in his view, was the main task for physics as the twentieth century developed. However, he also delivered a review paper at this same meeting of the British Association – "The Production and Origin of Radium" – a review which he dated September 20th, and which therefore does not include Boltwood's discovery of ionium, which was announced on September 26th. (One may question the ordering of Rutherford's papers by Sir James Chadwick – there seems no reason why this review should be placed after letters to *Nature* clearly post-dating it.)

The real work which he initiated as soon as he settled down in his new lab at Manchester was a collaboration with Geiger on what must have seemed to his contemporaries an almost impossible task – the counting of alpha-particles emitted by radioactive substances. This literally meant counting, enumerating one by one, the invisible alpha-particles.

The reasons for setting out on this course were simple. "The determination of the true character of the alpha-particle is one of the most pressing unsolved problems in radioactivity, for a number of important consequences follow from its solution," said Rutherford in January 1908, speaking at the Royal Institution just as his experiments were getting under way. The problem was that he needed to find the exact charge carried by the alpha-particle in order to determine its exact weight – or he could do it the other way around. But only by discovering the weight could he decide if it really was a helium atom. "Unfortunately," he went on, "a direct experimental proof of its true character appears to be very difficult unless a new method of attack is found." There was a possibility of studying the scintillations produced by alpha-particles when they struck a zinc-sulphide screen – an effect noted many years before by Crookes. But "apart from the difficulty of counting the scintillations, it is very doubtful whether more than a small fraction of the alpha-particles which strike the screen produce the scintillations." Theoretically the ionisation produced by a single alpha-particle should just be measurable; but "the electrometer or electroscope used for measurement would require to be extremely sensitive, and under such conditions it is known that small electrical disturbances are very difficult to avoid".

What was needed was some method of "amplifying" the electrical ionisation effect of an alpha-particle so that the arrival of even one of them could be spotted. And this, with a great deal of help from Geiger, was exactly what Rutherford managed to do. He based the work on an effect first studied by his old friend J. S. Townsend, who had shown that a single ion moving in a low-pressure gas, with a strong electric field applied, produces several thousand more ions by colliding with the molecules of the gas. An alpha-particle can produce about 80 thousand ions in a gas before it loses its power of ionisation. So Rutherford fired his alpha-particles into a gas at low pressure with a strong electric field applied, and the amplifying ionisation effect was strong enough to swing an electrometer. In the actual experiment a small mirror was attached to the electrometer from which a spot of light was reflected on to a graduated scale. The instrument could be made sensitive enough to swing past 300 divisions of the scale by the application of one volt; Rutherford found he could get a swing of fifty divisions from the arrival of a single alpha-particle when everything was working well.

The apparatus was, in accordance with Rutherfordian tradition, simple. A plain brass tube about sixty centimetres long had a wire running along its axis which was connected to an electrometer. The gas pressure inside was reduced to as low a value as he could conveniently achieve with the primitive vacuum pumps which he had available, and an electric field of about 1,000 volts was applied between the wire and the brass tube. An extension of the tube, about another seventy-five centimetres long, was also evacuated, and in it he placed a small plate on which was a thin film of radioactive matter. This gave a pencil beam of alpha-particles which was allowed to enter the measuring tube through a tiny hole, two millimetres across, which was covered by a thin sheet of mica. The hole was adjusted until only some "six to ten alpha-particles a minute" entered the measuring tube. The arrival of each alpha was noted from the swing of the beam of light reflected by the moving mirror on the electrometer. By calculating back, the proportion of alpha-particles entering the measuring chamber compared with the total output of alphas from the active deposit could be determined. And thus the total number of alpha-particles emitted by a given quantity of radium could be found. In earlier experiments in Canada Rutherford had measured the total electrical charge emitted by a given quantity of radium, and both he and the French scientists had measured the heat given out by radium which is a measure of the energy of the alpha-particles.

From a whole series of experiments of this sort, Rutherford and Geiger were able to publish in the summer of 1908 a long paper packed full of detailed figures. The main qualitative result is that "we may conclude that an alpha-particle is a helium atom, or to be more precise, the alpha-particle, after it has lost its positive charge, is a helium atom."

But much more important to science as a whole were the numerical results. Because of his new ability to count individual atoms, Rutherford was able to establish two values of basic importance to science. The chemists had for many years believed in the existence of a number called Avogadro's Number, a very large value which gave the number of molecules in a cubic centimetre of gas at standard temperature and pressure; this number was, for sound reasons, believed to be the same whatever the gas, but up to this time no one had been able to give any very good evidence for exactly what the number might be. Rutherford calculated it must be 2.72×10^{19}, and this is, to all practical intents, the presently accepted value. He and Geiger also demonstrated that the basic unit of electric charge, that is the charge carried by a single hydrogen atom, the lightest of all atoms, when it is involved in electrical activity, is 4.65×10^{-10} electrostatic units. (Nowadays we might say that the charge carried by the hydrogen atom was the single positive charge of its nucleus, a single proton; and the charge on a single electron is of the same size but is negative.)

The great significance, for science in general, of these results was that Rutherford the physicist, the man who had apparently attacked the atomic theory of the chemists, had provided the best evidence to this date of the existence of single atoms as material realities, and had confirmed, in a way that no chemist had achieved, the reality of the concept of Avogadro's Number.

In rather more bounded terms the figure Rutherford provided for the unit electric charge was of greater importance. Rutherford's figure was considerably larger than that given by J. J. Thomson, the acknowledged master of this field, with his electron measurements. R. A. Millikan, the American, who was eventually to establish what was virtually the final figure by measuring the charge on a single electron, had provided a measurement greater than J.J.'s but still not as high as Rutherford's. Perhaps for this reason Rutherford wrote a note at the bottom of the appropriate page of his paper pointing out that Planck had deduced a value almost the same as his "from a general optical theory of the natural temperature – radiation". It was Larmor who drew Rutherford's attention to this agreement with the founder of quantum theory during the period when he and Geiger were actually writing up their results. This agreement strengthened the very different cases the two men were putting forward in their very different spheres. It may well provide a reason why Rutherford was so willing to accept the otherwise highly "theoretical" views of Planck. And looking at it from Planck's point of view, we find Thomas Kuhn writing, "Ernest Rutherford is the only scientist known to have been [converted] to 'the general idea of a quantum of action' by the special accuracy of Planck's computation." Though his interest proved consequential (it enabled him "to view with equanimity and even encourage Prof Bohr's bold application of the quantum theory to explain the origins of spectra") it was apparently unique. Kuhn continues, "By the time additional measurements of the electric charge unequivocally demonstrated the accuracy of Planck's prediction, his theory had won widespread acceptance by other means," and here he is apparently referring to Millikan's virtually final determination of the charge on the electron in 1911.

The relationship between Rutherford and Planck is fascinating. The two men were never in close personal relationship or correspondence, but their professional influence on each other was immense. It is not susceptible to proof, but Rutherford would have been justified in 1908 in being swayed in favour of Planck's theory by the fact that it predicted a value for the electronic charge that agreed with his experimental measurements. But at this stage, as Kuhn makes clear, Planck had not accepted the most important implication of his own work – the idea of discontinuity, the idea that at atomic or sub-atomic levels energy does not flow continuously but is moved in separate, discontinuous packets, the quanta. It is this discontinuity that is the truly significant concept of the revolution in physics wrought by Planck and his quantum mechanics. Planck came to accept discontinuity in 1910, and the Solvay Conference in Brussels in 1911 (referred to in the previous chapter) had discontinuity as its chief topic, although it was discussed in terms of the theory of specific heats rather than of "black body radiation" or atomic structure. And it was in 1911 that Planck himself wrote that his hypothesis might explain why alpha- and beta-rays could apparently only be emitted with certain predetermined velocities and concluded that "radioactivity seems effortlessly to fit the hypothesis of quantum emission".

Kuhn gives this quotation and the accompanying discussion of the historical growth of the concept of discontinuity without mentioning that here again it was primarily Rutherford's work which provided the experimental basis which confirmed Planck's theories – indeed in his whole book he scarcely ever describes a physical experiment, he simply describes a "problem-solving community" of theoretical physicists on which actual measurements obtrude only to confirm one view or pose another problem. Planck however does not seem to have downgraded the experimentalist in the way that modern students of his work have done. Rutherford certainly accepted quantum theory from the start, and continued to allow it to influence him and the work of his laboratories throughout his life, notably in accepting Gamow's theories in the 1930s which led to the building of Cockcroft and Walton's first accelerator with which the first artificial transmutation of matter was achieved.

Rutherford's 1908 atom-counting experiments with Geiger, as

they have been described, had an obvious methodological flaw – how did he know that it was indeed a single atom affecting his measuring apparatus? How did he know that he was only allowing six or eight atoms a minute into the measuring chamber? The basis of his certainty was that he also employed the scintillation method of counting, the method he had scorned in his Silliman Lectures, as a check. The usefulness and accuracy of the scintillation method had been improved by E. Regener in Berlin using a better quality zinc-sulphide screen and a low-power microscope, with a fairly large field of view, which enabled individual scintillations to be counted. Rutherford was told of Regener's first successful work, and his paper on the subject, by his faithful correspondent Otto Hahn, very shortly after its publication.

On January 27th, 1908, Rutherford wrote to Hahn congratulating him on his latest discovery of a new intermediate product in the radioactive decay of thorium: "You appear to have a private hunting ground in thorium." Then he started describing his new work with Geiger: "We have detected a single alpha-particle by an electric method" – and emphasising the practical value of being able to count the particles one by one – "I can then obtain the charge on each and hope to settle whether the alpha-particle carries two charges or not." On February 22nd Hahn replied with the news of Regener's latest experiments which had been reported "last night". He described the scintillation methods used in Berlin and how Regener "measures the current produced by the polonium in air and in vacuum; thus he gets a check on the mass of the alpha-particles . . . [and finds] a great probability that the alpha-particle carries two charges". Hahn knew, of course, that Rutherford had considered this to be the case for years, and that he was aiming at conclusive scientific proof of the matter, but Hahn obviously thought that "the value of the alpha-particle is now settled from different observers by different methods".

But this was not Rutherford's way. He got better zinc-sulphide screens himself, and started checking scintillations against electric countings. It seems fairly certain that he and Geiger approached Regener's method with scepticism and got a considerable shock when they discovered that scintillation counting was every bit as good as electrical measurement.

In the introduction to their paper they declare, "In considering a possible method of counting the number of alpha-particles their well-known property of producing scintillations . . . at once suggests itself. No confidence can be placed in such a method of counting the total number of alpha-particles . . . until it can be shown that the number so obtained is in agreement with that determined by some other independent method." But at the end of the same paper they are forced to admit, "The result, however, brings out clearly that within the limit of experimental error, each alpha-particle produces a scintillation on a properly prepared screen of zinc sulphide. The agreement of the two methods of counting alpha-particles is in itself a strong evidence of the accuracy obtained in counting the alpha-particles . . . by the electrical method." They conclude that there are now two independent and distinct methods of counting individual atoms.

The irony here is that Rutherford virtually dropped the electrical method of counting for the remainder of his life, and concentrated on the scintillation method for a whole string of experiments on many topics. Indeed the typical memory of Rutherford preserved by so many of his colleagues is of the darkened room, the periods of acclimatising the eyes, the difficult spells of counting tiny flashes of light, which the great man himself undertook and demanded of his colleagues. The electrical method of counting, dropped in 1908, was perfected by Geiger many years later. It is the basis of the "Geiger counter" which is perhaps the archetypal machine of our own nuclear age.

On the face of the evidence Rutherford continued his scientific work in a logical and determined way throughout 1908 and 1909. The list of his publications shows a continued set of results given about alpha-particles and radium emanation. But an examination of his laboratory workbooks shows that the real story is rather more complicated. The sequence is quite clear throughout his early Manchester days, beginning with the experiments using Petavel's "bomb". These seem, however, to have gone on rather longer than might have been thought, for there are notes showing that some were conducted as late as January 1908.

There are some changes in style, which obviously correspond with Rutherford's acquisition of a host of new, young collaborators. Quite often there is a formal statement of the experiment to be performed and the methods to be used in Rutherford's own hand, written in black ink. The following tables of figures and results are usually in pencil and in other hands.

The importance of the fresh supplies of radium which he obtained from the Vienna Academy despite Ramsay's manoeuvres is shown by the notation against Saturday, February 14th and Sunday, March 17th (both clearly 1908 though it is not stated) – "Experiments with Austrian Radium". The working papers show that Rutherford had an intense burst of work in early April, when there are recordings for every single day from April 2nd to April 10th. And there are all those delightful Rutherfordian remarks which make his notebooks so lively, showing that it was in his laboratory that he really lived – "best result", "too small to be reliable", "U tube had been accidentally filled with Hg. [mercury]" or "again, repeated 3 times, repeated He. yellow, again".

The April burst of work continued with many entries for May and June. This was a really massive effort and it does not take much reading to realise that Rutherford was deeply occupied in disproving Sir William Ramsay's extravagant claims which had been published that year. There are several pages of notes jotted from Ramsay's paper, especially concerning the production of neon from radium emanation. At one point there is a direct quotation from Ramsay written out in full: "We must regard the transformation of emanation into neon as indisputably proved, and if a transmutation be defined as a transformation brought about at will by a change of conditions, then this is the first case of transmutation of which conclusive evidence is put forward." On August 4th and 5th, according to the notebooks, Rutherford finally proved Ramsay was wrong. Alongside the results for these days he jotted "This neg. Ramsay's result", and in the margin he notes that Ramsay's mistake was probably due to neon having got into the apparatus with air.

In this same period of intense work Rutherford, with T. Royds, performed two important sets of experiments, while working in parallel with Geiger and the particle-counting operations. First they provided the first correct spectrum of radium emanation. This was completed by early July, provisionally published as a letter to *Nature* in that month with the enigmatic comment that "it would be interesting to compare" Rutherford's result with those of Ramsay. Rutherford, as we have noted, did not enter into controversy with Ramsay any more – he simply published his own results and allowed the reader to see that they differed from Ramsay's.

The second of the Rutherford–Royds experiments, probably performed at the start of the 1908–9 academic year, is my favourite Rutherford work. He had believed for many years that alpha-particles were essentially helium atoms – all the evidence pointed that way, and there were many compelling observations such as the Ramsay–Soddy experiments of as far back as 1903. But Rutherford felt some need to convince people beyond the possibility of doubt, and he got his chance because he recognised the skill of the glass-blower, Baumbach. Baumbach's premises were near the university and naturally most of his custom came from the laboratories even though in those days young scientists were expected to make most of their own glassware on the bench. Later, when the war came, Baumbach was to become a minor embarrassment to his family and his university customers as he insisted on making fiery and jingoistic pronouncements about impending German victories so that he got himself interned. But in 1908 he was at his peak and highly valued. He produced for Rutherford a device of great elegance – a glass cylinder with quite exceptionally thin walls, barely one-hundredth of a millimetre in thickness, and it was contained within another glass cylinder which was so fitted with taps that it could be evacuated or filled with any desired gas. The central, thin-walled cylinder was filled with radium emanation and the outer container was evacuated. Then the alpha-particles from the disintegration of the radioactive atoms of emanation penetrated the very thin glass and collected in the outer container while the emanation was held inside. Quite soon after this experiment had been set up it was possible to detect the spectrum of helium, and helium only, in the outer container, and this spectrum, the certain evidence of the presence of helium, grew stronger every day. Of course all sorts of other measurements had to be made under different conditions to show that nothing else could account for the effect, but the demonstration was absolutely conclusive. Thus Rutherford proved finally his long-held belief about the nature of the alpha-particle – it was in essence an atom of helium, which was thrown out of the nucleus carrying a double positive charge; this charge was neutralised as the particle picked up its normal two negatively charged electrons by collision with other atoms or molecules and it ended as a normal atom of helium.

The alpha-particle remained Rutherford's favourite for the remainder of his life, not for any sentimental reason, but simply because it was the most energetic of all the radioactive particles, with its high speed and large mass. Once he had settled its nature he used it as his main tool, the probe with which he could investigate the interior of all other atoms. The importance of this concept is underrated today. Rutherford should be acknowledged as the father of all the nuclear accelerators, the gigantic atom-smashers of our own times. It was he who first saw that the interior of the atom "must be the seat of very large electric forces" and who realised that extremely energetic particles of atomic size or smaller would be needed to investigate the structure of the atom. It is also interesting to realise that even at the time of writing (1981) new experiments with alpha-particles are still being set up in the great nuclear accelerators of the world.[1]

With his experiment finally proving that alpha-particles were helium atoms, Rutherford also experimented with a new method

[1] See for instance *CERN Courier*, May 1981, recording what was at the time the world record collision energy.

of publication. The first announcement of the result, in a brief and rather tentative form, was made to the Manchester Literary and Philosophical Society, in the form of a short address. The paper, "The Nature of the Alpha-Particle", was later printed in the *Memoirs* of the society, where it is marked "received and read, November 3rd, 1908". This society holds an important place in British scientific history. Firstly, its very foundation in the late eighteenth century was symptomatic of that moving of scientific endeavour away from the Royal Society in London, and away from the Universities of Oxford and Cambridge, into the hands of the practical middle classes of provincial Britain. The movement was an important feature of the rise of capitalism and the subsequent Industrial Revolution. Secondly, the society had won a place in history by being the first to hear of John Dalton's atomic theory nearly a century before Rutherford started to patronise its meetings. Whether Rutherford regarded the society as a good sounding board, or whether he regarded it as a place for early publication where he might out-manoeuvre Ramsay's sprints to the Royal Society, is not known. But certainly this publication in 1908 led to the society's third claim to historical fame, three years later, when Rutherford again used it to make his first preliminary claim to establish the existence of the nucleus of the atom.

It is also typical of Rutherford that he gave full credit to the glass-blowing skill of Herr Baumbach: "The walls of this glass tube were sufficiently strong to withstand atmospheric pressure but thin enough to allow the greater part of the expelled alpha-particles to be fired through them. After a number of trials, Mr Baumbach succeeded in blowing a number of such fine tubes for us." A very similar description of his experiments formed one of the high points of Rutherford's Nobel Lecture, when he went to Stockholm to collect his prize in December 1908. The formal full-length publication had to wait for the February 1909 edition of the *Philosophical Magazine*, and again Baumbach received his credit.

In the midst of the terrific output of scientific effort that marked 1908 Rutherford wrote to Bumstead, in the late summer, one of the most interesting letters of his life. "I have now got through the stress of my first year and have grown fat on it. At the same time I have never worked so hard in my life," he remarked and went on to describe the alpha-particle counting experiments: "Geiger is a good man and worked like a slave. I could never have found time for the drudgery before we got things going in good style. We had lots of difficulties, mainly in avoiding natural disturbance and an effect of scattering which took us some time to get down on . . . The scattering is the devil." He then goes on to describe the arguments he was using to establish the true value of the electronic charge, "e", by finding the charge on the alpha-particle and establishing whether this was really twice the value of "e" – his "smell for numbers" convinced him that the alpha carried "2e". "There is no rigging about this result. I am sure it is very nearly correct . . . J.J.'s result is not worth a cent." (It will be remembered that Rutherford's value for "e" was very much higher than J.J.'s and all previous results except Planck's theoretical calculation.) The letter goes on to report some gossip from Cambridge which had been relayed by H. A. Wilson: "H.A.W. says J.J. gave only his <u>two lowest</u> experimental values for "e" to come near Wilson's value and left out the rest. It is highly comic. J.J., however, is a little unhappy about it." There is then a description of the physical problems of counting the dim scintillations of alpha-particles, and he repeats, "Geiger is a demon at the work and could count at intervals for a whole night without disturbing his equanimity. I damned vigorously after two minutes and retired from the conflict."

Apart from the insights into scientific methods and personalities the letter is important on two counts. Firstly it shows that Rutherford, who had already assaulted and overturned the chemists' "establishment", was now preparing to attack his own masters and seniors, the leaders of the physicists. Secondly, there is the repeated mention of "scattering" – the deflection of alpha-particles from their true line of flight as they passed through matter. We have seen that Rutherford had noted this phenomenon when he got slightly "fuzzy" photographic images from beams of alpha-particles passing through mica sheets during his McGill experiments in 1906. In 1908 the scattering was a technical problem to be overcome – but, as with so many other of Rutherford's great leaps of scientific imagination, when the experiment was over he asked Geiger to look into scattering as a phenomenon in its own right. And from this fascination with the small anomaly great results were to be achieved.

Alpha-particles were not, however, Rutherford's only interest. The work of the Manchester physics department can clearly be seen to have had a broader strategic pattern – the attack on the great problem of the constitution of the atom. He organised it in the way he always publicly proclaimed such problems should be dealt with: by a broad attack using every approach that the scientific technology of the time allowed. A study of the publications of the other members of his rapidly-growing team shows that he was steadily directing his brightest youngsters into the investigation of beta- and gamma-rays and encouraging those who had a solid professional grounding towards the study of spectra, since it was widely believed in the laboratories that the spectra of the elements must give clues to the structure of all the different atoms.

Everyone is familiar with the spectrum of colours from violet through blue, green and yellow to red which we can see in a rainbow. Most people know that this range of different colours can also be obtained when we shine white light through a glass prism. What is rather less known is that if we heat an element, say sodium, until it is hot enough to give out light, and pass that light through a prism, we do not get all the colours of the spectrum, but only thin lines of yellow light. Correspondingly, if we shine white light through sodium gas before we pass the light through the prism we get all the colours of the rainbow except the yellow light that is typical of sodium – at the exact wavelength where sodium produces yellow light it also absorbs yellow light and there appear two black lines breaking up the full spectrum of white light. This makes it quite clear that the sodium atom must have some structure which emits light at very precise frequencies when it is excited, as by heating it, and which absorbs exactly those frequencies when white light is shone upon it. The techniques of spectroscopy had been very largely used by chemists for identification purposes. Once the spectrum of an element had been found, the presence of that element in any situation – including its presence in the sun or distant stars – could be immediately confirmed if either its emission spectrum (the bright lines of light) or its absorption spectrum (the corresponding dark lines in a bright spectrum) could be found. To Rutherford and his colleagues, however, the spectrum meant something different – it could give clues to the structure inside the atom, for there must be some reason why one atom would absorb light of one frequency while other atoms absorbed or emitted light at entirely different frequencies.

A simple examination of the papers published by the other members of Rutherford's Manchester group shows clearly that spectroscopy became a very important part of their work towards the end of the first decade of the century. Royds, working directly with the Professor, established the spectrum of radium emanation; E. J. Evans and R. Rossi did much work on establishing the spectra of metals, especially the rarer ones such as tellurium. That this was the general approach is shown by the nature of the work on beta- and gamma-rays, which was to establish that the emission of these rays was similar to spectra – that is to try to show that the rays tended to be emitted at specific energies by specific substances. Kovarik, Makower, Florance and Andrade all worked on this type of experiment. The results, of course, were not published in 1908; results often take a considerable time to come and the work is reflected mostly in papers published from 1910 up to the start of the world war. The mathematical physicists attached to the department – at first H. Bateman and later C. G. Darwin – also worked on the theory of spectra. But while it is true that the bulk of Manchester physics in these years was either radioactivity or spectroscopy, a quick examination of the titles of all the department's publications is quite sufficient to demolish the claims of some recent American scholars that Rutherford's colleagues were expected to drop all other subjects in favour of radioactive work.

All this is, however, but a prelude, the laying of a ground-base, to the greatest achievement of Rutherford's Manchester years, arguably to the greatest achievement of his life. To call this achievement simply "the discovery of the nuclear atom" both distorts and underrates the magnitude of the feat. Rutherford's own, more simple words are better – he discovered "what the atom looks like". More than that, his immediate colleagues, Bohr and Moseley, immediately worked out the main implications of his vision, linked it with the most advanced work being done elsewhere, notably by the German theoretical physicists and the Braggs' work on X-rays, and established the whole concept of atomic number which explained the nature of the chemical elements and their differences. Geiger and Marsden provided the experimental evidence to justify Rutherford's claims and Darwin provided the mathematical substructure. In three years' work the Manchester team, led by Rutherford, provided us with a new view of the nature of matter, a view which is still valid and which we use today. Furthermore it was a view which provided a common base for much other science. Nor was this view confined to the world of science; it was a revolutionary view, for it said that most of our universe is empty, void; it said that solid matter, the tables and chairs and stones around us, as well as ourselves, are mostly nothingness; that the basic unit of matter, the atom, really consists of a few particles gyrating in what are, by the particles' own standards, immensities of space. And all the infinite variety we perceive about us, red books and green leaves and blue sky and white pillows, is but the effect of increasing or decreasing the number of tiny particles linked to make units, one by one.

Rutherford's new vision of the nature of matter did not shatter any apparently well-established theory – it was the discovery of a new layer of reality, a new dimension of the universe which opened up a galaxy of new sciences. The process by which he reached his discovery contains little mystery except for the final leap of the imagination which no one can ever understand. The first stage, we know, was puzzlement – mental dissatisfaction over the fuzziness of the images of his beam of alpha-particles, a fuzziness which was greater when the particles passed through matter than when he had his strongest magnetic or electric field switched on. The second stage was plainly irritation – anger when the scattering of alpha-particles interfered with his counting of them. It was clearly at this stage that Rutherford decided to look further into what caused scattering, bearing in mind that alpha-particles, though unimaginably small, were travelling faster than any imaginable bullet or shooting star. Alphas had a velocity that was an appreciable fraction of the speed of light, and possessed what was, for their size, colossal energy, so that no one could envisage what cataclysm within the atom had expelled them. Furthermore, the alphas were present in vast numbers – Geiger and Rutherford had designed their entire experimental arrangement to reduce the numbers of alphas to a number that humans could count. The results of these experiments must therefore have been affected by statistical probability, as all counting operations of a small sample out of a large population are affected.

So the third stage of Rutherford's enquiry was to go back to class. At the start of 1909 he enrolled as a student to attend the elementary lectures on probability given by Professor Horace Lamb. This was no gesture or gimmick. Rutherford's laboratory notebooks bear witness to his attendance at these lectures and to the fact that he took extensive notes like any first-year student. As an example of simple and direct approach it can have few parallels in modern Academia.

One of the central features of the Rutherford–Geiger counting apparatus was a long metal tube, evacuated as far as their primitive pumps would go, down which the beam of alpha-particles was fed. It was inside this tube that "scattering" was first a problem and then the subject of experiment. Geiger, in his plain and unemotional prose, wrote about it thirty years later:

> In the electric counting of alpha-particles it was seen that the small residuum of gas in the four-metre long tube . . . influenced the result. We attributed this to a slight scattering of the alpha-particles. Later on I examined scattering quantitatively in several experiments. The most important observation was the appearance of isolated instances of extremely large angles of deflection, which were far outside the normal variations. At first we could not understand this at all.

These "most important observations" were made by the young Ernest Marsden – "a callow youth from Blackburn" according to his own account. He had been called on to the scene in one of the best-known "Rutherford stories", a story that Rutherford himself often repeated and perpetuated: "One day Geiger came to me and said, 'Don't you think that young Marsden whom I am training in radioactive methods ought to begin a small research?' Now I had thought so too, so I said, 'Why not let him see if any alpha-particles can be scattered through a large angle?' "

The result was quite extraordinary. Marsden found that a countable number of alpha-particles actually bounced back from a thin sheet of gold foil. Of course the majority of the particles in his pencil-beam went straight through the gold and were only slightly scattered. Rutherford always declared it was the most surprising result he had known, and he coined a graphic phrase which, again, he often used: "It was as though you had fired a fifteen-inch shell at a piece of tissue paper and it had bounced back and hit you."

The paper published by Geiger and Marsden in the *Proceedings of the Royal Society* in June 1909 says exactly the same thing but in words more appropriate to a learned society. "On a Diffuse Reflection of the Alpha-Particles" begins by pointing out that many observers have seen beta-particles bouncing back off a metal plate, though some people still regard these as secondary particles ejected from the metal by the impact of the primary beam of betas. Then, still in the first paragraph, it gets down to its surprising business:

> For alpha-particles a similar effect has not previously been observed and is perhaps not to be expected on account of the relatively small scattering which alpha-particles suffer in penetrating matter. In the following experiments, however, conclusive evidence was found of the existence of diffuse reflection of alpha particles. A small fraction of alpha particles falling upon a metal plate have their directions changed to such an extent that they emerge again at the side of incidence.

The rest of the paper continues in this simple and easily understandable style which Rutherford favoured. All the counting of alpha-particles was done by the scintillation method, and the experimental apparatus was so simple as to require no more than a sketch plan to show the positions of the radioactive source, the reflecting plate, and the scintillating screen with the low-power microscope for counting the number of impacts of alphas. By using reflecting plates of eight different common metals they showed that the number of particles reflected back increased roughly as the atomic weight of the metals increased – thus lead with an atomic weight of 207 and gold with an atomic weight of 197 reflected back about twice as many particles as silver whose atomic weight is 108, and twenty times as many as aluminium where the atomic weight is 27. They showed that roughly one alpha-particle was reflected back for every 8,000 shot at the metal. And they demonstrated that the amount of reflection depended on the volume of the metal in the reflecting screen and not upon its surface, though this relationship was not a simple one. Here they permitted themselves their only comment: "If the high velocity and mass of the alpha-particle be taken into account, it seems surprising that some of the alpha-particles, as the experiment shows, can be turned within a layer of six one-hundred thousandths of a centimetre of gold through an angle of ninety degrees and even more. To produce a similar effect by a magnetic field the enormous field of a billion absolute units would be required." In two brief references to the work they did in checking against errors they state that about one reflection a minute seemed to come from nothing more than the air in the apparatus – an interesting result that may well have stuck in Rutherford's mind for later use.

But for the moment Rutherford was not prepared to go any further than his colleagues. During the summer of 1909 he was preparing a lecture that he gave at the annual meeting of the British Association, which was held at Winnipeg in Canada. While his main reaction to this unusual travel was delight that it gave him such an opportunity for a holiday and meeting with old friends from his Canadian days, Rutherford always took such lectures most seriously, and in 1909 he was President of Section A – the physics section. His lecture was therefore one of the presidential addresses. Towards the end of it he referred to this very recent work by Geiger and Marsden but confined himself to saying, "The conclusion is unavoidable that the atom is the seat of an intense electric field, for otherwise it would be impossible to change the direction of the particle in passing over such a minute distance as the diameter of a molecule."

When the new academic year started in Manchester in October 1910, the matter was left in abeyance. Geiger, who remembered "we could not understand it at all" went back to studying small-angle scattering, which resulted in two papers in 1910. The first, written with Rutherford, was titled "The Probability Variations in the Distribution of Alpha-Particles" and clearly showed the effects of Lamb's lectures. The second – "The Scattering of Alpha-Particles by Matter" – referred back to his work with Marsden and pointed out that ". . . a simple calculation, assuming the ordinary probability law, shows that the probability of an alpha-particle being scattered through an angle exceeding ninety degrees is extremely small and of a different order from what the reflection experiment suggests. It does not appear profitable at present to discuss the assumption which might be made to account for this difference." Feather suggests that one interpretation of this opaque remark is ". . . if Rutherford is baffled it was certainly not a profitable use of time for anyone else to discuss the significance of the facts."

Marsden went off temporarily to entirely different work – in fact to the meteorological station on the Pennines where he produced a paper on the electrical state of the upper atmosphere. "Rare indeed are those whose first observation in research have given a clue and stimulus to a conception which changed the whole face of modern physics," wrote Chadwick in an eightieth birthday tribute to Marsden in 1969, when he had become Sir Ernest Marsden, a great "scientific statesman" who had spent many years building up the New Zealand government's scientific organisation. The young Marsden was to return to the scattering of alpha-particles and his work was to do as much to establish the reality of Rutherford's concept of the nuclear atom as it had to stimulate it. He was in fact to become one of the great "stars" of the Manchester lab. When Andrade joined the team in 1913 he found "Marsden was at the height of his research activity and his stimulating vigour pervaded the lab", and had at least six young men working under his direct guidance. Rutherford himself was to write in 1915 of his "very high opinion of Marsden and the large amount of important work he has already done in radioactivity".

Rutherford himself brooded and thought, and carried on doing so for a full eighteen months. His output of scientific work fell from fourteen publications in 1908, to only six in 1909, and the same number in 1910 and 1911. Furthermore, many of these were only brief notes and letters, though it has to be remembered that these were also the years in which he was deeply involved in the international negotiations over the radium standard.

No one can ever tell how an inspiration, a discovery, finally comes. Rutherford seems to have immersed his mind in the world of individual atoms until he was so familiar with the size of the forces involved, and the dimensions of the bodies he was dealing with, that finally he emerged with a concrete picture of what was going on – he "knew what the atom looked like". He has left some picture of the way his mind worked – given twelve years after the event in an address to the Science Masters' Association in Cambridge in 1923. He asked his audience – in what for him was an unusually imaginative passage – to travel back in time to Manchester in 1911 and to think about the work with alpha-particles; particles with such "enormous" energy that it could be expressed as a potential difference of 4 million volts – "much greater than we can impress . . . on a particle . . . in a vacuum tube by the potential available in our laboratories".

On account of its great energy of motion the alpha-particle is not only able to penetrate but to pass through the structure of the atoms in its path. If it meets with a powerful deflecting field in the atom it is diverted from its rectilinear path or, in technical terms, "scattered" . . . This scattering is in general small, and, as Geiger showed, increases with the atomic weight of the material, but even in the case of a heavy element like gold, only amounts to an average deflection of a few degrees. Another very unexpected observation on scattering has been noted by Geiger and Marsden. When a pencil of alpha-rays falls on a thin sheet of gold foil or other metal, a few of the alpha-particles are scattered backwards and are deflected through an angle of more than ninety degrees. The number of these backward particles is enormously greater than was to be expected if the small angle scattering of the main beam is to be ascribed to the resultant effect of a multitude of small random deflections. It seems clear that in passing through matter, the alpha-particle occasionally encountered such an intense field in the atom that it was turned back in its path as the result of a collision with a single atom. When we consider the mass and energy of the alpha-particle this is a most astonishing observation, indicating that the atom is the seat of gigantic deflecting fields. Supposing that the forces involved in such collisions are of the ordinary electrostatic type, it can readily be calculated that in order to produce such a large deflection of the alpha-particle in an atomic encounter, the atom must contain a massive charged centre of very minute dimensions. From this arose the conception of the now well-known nucleus atom, where the atom is taken to consist of a minute positively charged nucleus containing most of the mass of the atom, surrounded at relatively great distances by a distribution of electrons equal in number to the units of resultant positive charge on its nucleus. On this novel theory of atomic structure, the alpha-particle in collision described a hyperbolic orbit around the nucleus and the fraction of the alpha-particles scattered through different angles varied according to a simple but definite law. The accuracy of the law of distribution for a number of elements and for angles between five degrees and 150 degrees was verified experimentally in an elaborate investigation by Geiger and Marsden.

That was Rutherford's own hindsight – but there was a good deal more to it than that. Firstly there was a rival concept – the "plum pudding" atom which had been suggested by none other than J.J. and which had been supported by a number of other major scientists. This visualised the atom as a sphere of positive electricity in which the negatively charged electrons were stuck like plums in a pudding. J.J., however, was already working on more subtle patterns which solved some of the difficulties inherent in both his and Rutherford's concept. Furthermore J.J. had people working on the scattering problem in his own laboratory, and a paper by one of his men, Crowther, became of crucial importance in the battle between the two concepts of the atom. It is, however, too often ignored that Rutherford's superior concept of atomic structure also involved the overthrow of his master's model, and the victory of Rutherford's atom also meant victory for Rutherford and established him as the new dominant figure in British physics.

It is clear that Rutherford reached his momentous conclusion in the first days of December 1910. Both Darwin and Geiger have given accounts of "the birth of the atom", but can only place the date at "just before Christmas, 1910". Rutherford however had written to Boltwood on December 14th:

> I think I can devise an atom much superior to J.J.'s for the explanation of and stoppage of alpha- and beta-particles, and at the same time I think it will fit in extraordinarily well with the experimental numbers. It will account for the reflected alpha-particles observed by Geiger and generally I think will make a fine working hypothesis. Altogether I am confident that we are going to get more information from scattering about the nature of the atom than from any other method of attack.

It must have been just after this that there came a memorable evening. Darwin remembered:

> One of the great experiences of my life was that on one Sunday evening the Rutherfords had invited some of us to supper, and after supper the nuclear theory came out. The main principle was that the whole deflection must have been done in a single operation. At first he talked about it as simple scattering and then he realised that this was ambiguous and I can still recall the satisfaction in his voice when he hit on the name single scattering since nobody could mistake what that meant.
>
> He assumed a central charge in the atom – it was indeed a year or two before it was renamed the nucleus – repelling the alpha-particle according to the ordinary laws of electricity so that it should travel in a hyperbola. He worked out the law of scattering, that is to say the number of particles that should be found at any angle of deflection. Rutherford was not a profound mathematician, and I would doubt if he had thought about the properties of a hyperbola since he had been at school, but he had remembered just exactly enough of them to serve his purpose and he told us the trigonometrical formula that should give the number of alpha-particles deflected through any angle. Indeed, as

far as I can remember on that very evening he asked me to check over his work, which was of course perfectly correct.

I remember him also saying that there was nothing in the experiments to show whether the force was repulsive or attractive, though it was natural to suppose that all nuclei had a positive charge, so that it should be repulsive. [A positive nucleus would repel a positively charged alpha-particle, whereas a negative nucleus would attract the particle.]

I also recollect that even on that first evening Rutherford was already speculating how small the nucleus might be. He had worked out the distance of closest approach of the alpha-particle, and he was hoping that some observations would conflict with his formula, so that he might say that the law of repulsion between the two charges altered when they got within a certain distance of one another. In this hope he was, of course, disappointed because the nuclei are so small that, at any rate in those days, no alpha-particle could get near enough to show any departure from the behaviour of a point charge.

Now comes a little point I would like to mention which illustrates Rutherford's generosity. In his work he had been thinking about the nucleus of a gold atom, which is much too heavy to be shifted by an alpha-particle. It occurred to me at the time that interesting things might happen if one studied the collision of an alpha-particle with a hydrogen atom, because then one might expect that it would knock the hydrogen nucleus forward at a higher speed than the alpha-particle itself had had before the collision. Rutherford was much interested and said that had not occurred to him. It is this that makes me think that it could not have been many hours before our supper when he had actually worked out his theory, because it was an obvious fact that nobody could have missed for long. Not long after this he came into my room at the laboratory and told me to write a paper about it. One has heard tales of some really great men who have had suggestions given to them by their juniors, and in thinking them over they have incorporated them into their own train of thought to such an extent that they forget their original source. This point was really obvious, but even so Rutherford did not forget where it had come from . . .[1]

The immediate result of this Sunday supper-time session was a visit to Geiger's room in the laboratory – very possibly the next morning. Geiger recalled, "One day Rutherford, obviously in the best of spirits, came into my room and told me that he now knew what the atom looked like and how to explain the large deflections of the alpha-particles. On the very same day I began an experiment to test the relation expected by Rutherford between the number of scattered particles and the angle of scattering."

Darwin was not justified in his belief that Rutherford managed

[1] This account of the "birth of the nucleus" was given by Sir Charles Darwin at the Rutherford Jubilee conference in Manchester in 1962.

to remember schoolboy mathematics in order to suggest that the path of the alpha-particle as it approached the nucleus was a hyperbola. Feather found Rutherford's undergraduate copy of the first 111 sections of Newton's *Principia*, edited by Frost, and marked "E. Rutherford, Trinity College". On pages 220 and 221 there are marginal notes in Rutherford's hand opposite the discussion of the hyperbola and the action of the inverse square law. This law is the one which Newton used in his great formulation of the Law of Gravity as it acted between heavenly bodies. Unfortunately the notes cannot be dated and cannot be stated definitely to be part of Rutherford's working-out of his nuclear theory. It seems, in his own terms, statistically unlikely that there should be no connection. It is surely more likely that Rutherford, considering the movement of one body near or around another under the influence of attractive or repulsive forces, should have referred to one of the great classic works on the subject and found there a treatment of his problem which he could then adapt to the precise situation.

From about this time there is a collection of thirty-five loose sheets of paper – including some left-overs from the McGill University physics department – entitled "Theory of the Structure of the Atom". "Suppose," it begins, "atom consists of +ve charge 'ne' at centre and −ve charge as electrons distributed throughout sphere of radius 'b'." There is then a very rough sketch of this. "Suppose," it continues, "charged particle 'e' and mass 'm' moves through atom so that deflection is small at perpendicular distance from centre = 'a'." There are then two and a half pages of mathematics ending with a complicated formula against which he notes, "This is three times average deflection as found according to J.J.'s theory of +ve sphere (ne) and −ve electrons distributed." A little later there is consideration of the calculus of the chances of a collision – presumably again derived from Horace Lamb's lectures.

Next he considers "deflection of alpha-particle through a large angle", and here there is a rough pen sketch of the hyperbola situation. From this he goes on to calculate what the deflection of the incoming particle would be if the centre was attractive rather than repulsive – a calculation which takes another page of mathematics, ending, "Therefore deflection somewhat greater than for repulsive force but would not be appreciable difference except for small values of 'n'." A different handwriting now appears – there are tables of rough values which are definitely not in Rutherford's thick scrawl and some other supporting mathematical notes – which may have been Darwin's instant checking of Rutherford's maths. But the whole thing finishes with some calculations about the size of the nucleus which are plainly Rutherford's own.

Geiger's first sets of results seemed quite satisfactory enough for Rutherford to decide on an early publication. Again he chose the Manchester Literary and Philosophical Society for his stage and on March 7th, 1911 he read a short paper on "The scattering of the alpha- and beta-rays and the Structure of the Atom". It is not two pages long – but it goes immediately to the heart of the matter. "There seems to be no doubt that these swiftly moving particles actually pass through the atomic system and a close study of the deflections produced should throw light on the electrical structure of the atom. It has usually been assumed that the scattering observed is the result of a multitude of small scatterings." He then describes the J. J. Thomson atom and the support given by Crowther to that model which demanded that the number of corpuscles in an atom is about three times its atomic weight in terms of hydrogen. But this sort of theory cannot account for the large scatterings seen by Geiger and Marsden – only single atomic encounters can account for this phenomenon. "In order to explain these and other results, it is necessary to assume that the electrified particle passes through an intense electric field within the atom. The scattering of the electrified particles is considered for a type of atom which consists of a central electrical charge concentrated at a point and surrounded by a uniform spherical distribution of opposite electricity equal in amount." From this virtually theoretical basis he deduces his law of scattering, adding, "This law of distribution has been experimentally tested by Geiger for alpha-particles and found to hold within the limits of experimental error." Without giving any evidence, but simply "from a consideration of general results on scattering by different materials", he suggests that the central charge of each atom is very nearly proportional to its atomic weight: "The exact value of the central charge has not been determined but for an atom of gold it corresponds to about 100 unit charges" – a figure which we now know to be very much too high. There are a few sentences stating that Crowther's results can be explained just as well by single large-angle scatterings as by large numbers of small deflections, and a final promise that Dr Geiger is making further examinations of the implications of the new model atom, and that is all. A very cautious statement, not yet including the word nucleus, not even suggesting whether the central charge is positive or negative and certainly not containing the suggestion that the mass of the atom was concentrated in the tiny central body.

It is worth emphasising at this point that though most people will see an immediate analogy between the solar system, with planets orbiting around the sun, and Rutherford's atom model, with electrons orbiting around a massive nucleus, this was not the way Rutherford saw the structure. Essentially the proportions are very different, the nucleus in this sense being much smaller and heavier than the sun, while the space, the amount of emptiness, is much larger.

A fuller statement of the nuclear atom theory was put forward, under exactly the same title as the Manchester statement, two months later, in May 1911, in the *Philosophical Magazine*. The word nucleus had still not found its way into the vocabulary and there was still no decision as to whether the central charge was positive or negative. The paper is in fact a properly presented version of the "Theory of the Structure of the Atom" notes referred to above, some of the phraseology even being the same, and the hyperbola diagram is also included. It does, however, contain one odd and irrelevant statement in its final paragraph: "It may be remarked that the approximate value found for the central charge of the atom of gold (100e) is about that to be expected if the atom of gold consisted of 49 atoms of helium each carrying a charge of 2e. This may be only a coincidence, but it is certainly suggestive in view of the expulsion of helium atoms carrying two unit charges from radioactive matter." The idea that all atoms were made up of a selection of much smaller units was very much "in the air" at the time – papers had been written suggesting the existence of "protyles", four possible basic units of matter which in various combinations could make up the distinctive atoms of all the elements. Rutherford must at least have been playing with the idea that his favourite alpha-particles might be a basic unit of all matter.

Certainly his main nucleus theory was put forward simply as a hypothesis. He had devised a model of the atom which, as far as he could see, fitted the few experimental facts. Now in strictly professional scientific terms it remained to be seen whether further facts would prove or disprove its correctness. In formal terms Rutherford hardly referred to the matter again for nearly two years. Many commentators have pointed out that in his book *Radioactive Substances and their Radiations*, which he wrote during 1912 and which was published in 1913, he gives only a few lines to the matter, although he does introduce the word "nucleus" for the first time, before dropping it again to revert to "central charge".

In his mind he "knew what the atom looked like" and he wrote in a more uninhibited fashion to his friends and private correspondents. At this time (and for many years to follow) W. H. Bragg was one of Rutherford's most intimate friends; their relationship, which had started with the older man's letters from Adelaide to McGill virtually seeking Rutherford's approval for his first efforts in radioactive research had blossomed into a great friendship when Bragg came to Leeds as Professor of Physics in 1909.

Bragg badly needed this friendship during his first months in England, for his family found the industrial grime of Leeds hard to bear after the sunshine of Australia, and Bragg himself was fighting a long losing battle against his contemporary, Barkla, over the nature of X-rays. Rutherford's first letter to him on the subject of the nuclear atom seems to be no longer extant, but must have been written within the first few days after he had conceived his great idea, for Bragg, writing on December 21st, 1910, and inviting the Rutherford family to come and stay in Leeds for the last few days of the year, says, "I would have liked to have seen you before this to hear about the new atom . . . The atom sounds very fine. My boy wants to know if he may hear about it too, as he has been going to J.J.'s lectures . . ." The "boy", a Cambridge undergraduate, was later to become Sir Lawrence Bragg, Rutherford's successor to the Chairs at both Manchester and Cambridge. These two friends, in a succession of letters, were particularly bitter about the paper by Crowther, which appeared to support J.J.'s atom structure, and discussed the new atom at length.

On January 5th Bragg criticised Crowther's work in detail but added, "I think you will have to put some sub-centres in your atom to account for the X-ray effects." On February 8th Rutherford told Bragg that Geiger was working on "large scattering". His preliminary results "look very promising for the theory. I am beginning to think that the central core is negatively charged, for otherwise the law of absorption for beta-rays would be very different from that observed . . . I have thought a good deal about the possibility of accounting for X- and gamma-rays but have made no definite progress." Bragg, writing on the same day, continues the arguments for or against negative and positive centres, and also discusses whether there is just a central charge or a "big centre". The following day Rutherford replies that he has worked out all about the scattering from thin plates – the number of alpha-particles bouncing backwards should be proportional to the thickness, at least when the plate is very thin. He goes on, "I have looked into Crowther's scattering paper carefully, and the more I examine it the more I marvel at the way he made it fit (or thought he made it fit) J.J.'s theory . . . Altogether I think the outlook is decidedly promising." Two days later he had more to say about Crowther's paper: "I am quite sure the numbers of the earlier part of the curve were fudged," and he writes of Crowther's "scientific imagination".

On March 7th, the day Rutherford first spoke publicly about his new atom at the Manchester Literary and Philosophical Society, Bragg wrote to him that "Campbell tells me that Nagaoka once tried to deduce a big positive centre in his atom in order to account for the optical effects. He thinks Nagaoka, but it was a Jap anyway. Time about 5 or 6 years ago, when Schott and others were on the subject."

Professor Nagaoka, it will be remembered, was the Japanese who had visited Rutherford in Manchester and had commented on the simplicity of his methods. Rutherford lost no time in writing to him – a long letter dated March 20th, 1911, in which he began by outlining J.J.'s position and Crowther's support which "apparently confirmed experimentally" the view that all alpha-particle scattering was the result of many small deflections. But by considering large deflections of alpha-particles Rutherford had come to the conclusion that a single encounter between one atom and an alpha-particle tended to play the main part. "I have devised an atom which consists of a central charge 'ne' surrounded by a uniform spherical distribution of opposite electricity, which may be supposed, if necessary, to extend over a region comparable with the radius of the atom as ordinarily understood," he wrote, and went on with the other details such as the central charge being roughly proportional to the atomic weight, the path of the alpha-particle being a hyperbola, and the support from Geiger's first results. He explained that thus alpha- and beta-particles must be considered as passing "right through the atomic system" and he went on, "You will notice that the structure assumed in my atom is somewhat similar to that suggested by you in a paper some years ago. I have not yet looked up your paper but I remember that you did write on that subject." And in his full, May, paper Rutherford did make proper reference to Nagaoka's "Saturnian atom" – but pointed out that this was essentially disc-like rather than spherical. It is also true that Nagaoka's atom was very different from Rutherford's in that its central body was very large, whereas in the Rutherford conception the nucleus is very small, though very heavy, and the atom consists mostly of emptiness compared with the Japanese scientist's suggestion.

The letter to Nagaoka contains other news, emphasising that Rutherford was still taking his atomic theory as a hypothesis which was by no means occupying all his time or that of his laboratory. The letter contains the first announcement that the Manchester team had found that radium-C in its radioactive disintegration could break up in two different ways – "a side branch at this point constituting only a small fraction of the total". He added, ". . . I have long supposed that such effects must be present, although it is a very difficult matter to get complete experimental proof. For example, it seems fairly certain that actinium is a side branch at some point of the uranium-radium series" – another example of Rutherford's supremely accurate "scientific nose" telling him, long in advance of experiment, what would ultimately be proven.

But Rutherford was writing to many other people about his theories – and they were not necessarily influential people. On March 8th, 1911 he wrote to John Madsen, a young research worker at the University of Sydney in New South Wales. Madsen was a protégé of Bragg's who had also taken up work with beta-particles and their scattering, and Rutherford expounded the new atom to him at equal length and in very much the same words as he had used in his formal publication. He made it rather clearer, however, that the actual results obtained by Crowther could equally well be explained by "large" scattering as by multiple small scatterings. "I may mention that the theory of large scattering will hold equally well if instead of one large central charge one supposed the atom to consist of a very large number of smaller charges distributed throughout the atom. It can be shown however that, on this view, the small scattering should be much greater than that experimentally observed. It is consequently simplest to consider the effect of a single point charge." And Rutherford arranged, through Bragg, to send out a radium sample that would give Madsen a good source of beta-rays for his future work.

It may be added that at the same time Rutherford was also writing yet another book – *Radioactive Substances and their Radiations* – yet another "grand tour" of the ever-expanding field of radioactive studies which was widely acclaimed when it appeared. (There is a rather tattered remnant of Rutherford's primitive financial system which shows that he obtained 4,000 marks' advance in 1912 against a German translation of this forthcoming work; the advance was agreed by his publishers, the Cambridge University Press, who paid him £42.18s.0d. in royalties for that year, of which 7s.6d. came from his first book, *Radioactivity*.)

Looked at from the viewpoint of a continental physicist, although one who was devoted to Rutherford personally, this stretch of Rutherford's life was described thus:

Atomic structure . . . is the outcome of experiment and affords a measurable proof of his accuracy and profound individuality. The emptiness of space penetrated with matter, the concentration of mass and the electrical charges in nuclei of minute dimensions, the play of forces between such nuclei are ideas which in his hands became condensed into figures and equations . . . The nature of the atom in its characteristics is thus perceived; with surprising rapidity and with the happy cooperation of the younger generations of all lands it soon becomes an undisputed possession of science.

The words come from Geiger, the man who did most to establish the correctness of Rutherford's "figures and equations", but written many years later, as one of the tributes on Rutherford's sixtieth birthday in 1931. The importance of this statement by Geiger is that it was made at exactly the moment when Rutherford's atom was being reinterpreted by the great European theoreticians, when Schrödinger, Heisenberg, de Broglie and Max Born were showing that particles could be considered as waves, and waves could be considered as particles, when Rutherford's original "sphere of opposite electricity" was shown to be as valid a concept as that of particulate electrons in physically defined orbits around the nucleus.

Not many months after he had published his theory of the atom, at a time when it aroused very little interest in the world of physics, Rutherford confided to Bragg after his return from one of the Brussels conferences, "I was rather struck . . . by the fact that the continental people do not seem to be in the least interested in trying to form a physical idea of the basis of Planck's theory. They are quite content to explain everything on a certain assumption and do not worry their heads about the real cause of the thing. I must say I think the English point of view is much more physical and much to be preferred." A little later Bragg remarked, after starting from his problems with the nature of X-rays:

It is curious to reflect that Newton rejected the pulse theory [of light] for wrong reasons and Huygens the corpuscular theory for reasons also mistaken. It is even more curious to consider how little their mistakes affected their work. Their theories were no more to these men than familiar and useful tools. Much of the heated argument in which we occasionally indulge arises from the failure to recognise that hypotheses are in the first instance made for personal use. We really have no justification for demanding that others should adopt the means which we find most convenient in the modelling of our own ideas.

The mathematical physicists of Europe were to be able to prove their point of view; they were able to show that it is impossible to "form a physical idea of the basis" of the atom, and indeed they showed that there was no "real cause" in Rutherford's physical sense. Yet Rutherford's physical "model" has remained the "familiar and useful tool" with which the atom is approached when we wish to use its properties in chemistry, or electronics, or nuclear energy.

Rutherford, of course, believed in experimental proof for a theory. In the early summer of 1911, therefore, he lured Ernest Marsden back to Manchester from a post he had taken at Queen Mary College in London. Marsden describes it as "Rutherford kindly invited and made it possible for me to return to Manchester", but it is clear that what actually happened is that Rutherford obtained a grant from the Royal Society, for the paper that resulted ends with the words, "We are also indebted to the Government Grant Committee of the Royal Society for a grant to one of us, out of which part of the expenses has been paid." This sort of sentence at the end of a paper is so commonplace nowadays that no one notices it – but in 1911 it was virtually unprecedented.

Marsden and Geiger then settled down to a year's unremitting labour testing the theory of the new atom – or more correctly testing the correctness of the scattering law that Rutherford deduced from the hypothetical structure of the atom. This law predicted that different numbers of alpha-particles would be scattered at different angles and that the scattering would vary proportionally to the atomic weight of the metal from which the reflections were coming. Marsden remembered thirty-five years later:

> The complete check was a laborious but exciting task. I remember Geiger making a calculation that in the process of the work we counted over a million individual alpha-particles. When I look back and consider the apparatus used and the nearness of our heads and bodies to the large sources of radium emanation used and the time of the exposure to the radiation, I marvel that it did so little physical harm to us. On modern safety standards and using the same method of counting scintillations the apparatus would need to have been almost unworkably complicated for the purpose. Our work did, I think, sufficiently establish the theory on a firm basis. We had checked the effect with angle over a range of 1:250,000, with velocity 1:10, and with atoms from carbon to platinum, all the results being in accordance with theoretical expectation.

When Geiger and Marsden published their results in English in April 1913 in the *Philosophical Magazine* they made it quite clear that "Professor Rutherford has recently developed a theory" and that their object was to check it. There was still no decision whether the central charge was positive or negative. In this paper the number of alpha-particles they counted was given as 100,000 – not the million suggested by Marsden, and there are a number of mentions of the difficulties involved. Towards the beginning of the paper they state, "It may be mentioned in anticipation that all the results of our investigations are in good agreement with the theoretical deductions of Professor Rutherford and afford strong evidence of the correctness of the underlying assumption that an atom contains a strong charge at the centre, of dimensions small compared with the diameter of the atom." But at the end they get much firmer in the declaration, ". . . we have completely verified the theory given by Professor Rutherford". In particular they claim the results much strengthen the suggestion that the central charge is proportional to a figure of approximately half the atomic weight of the different elements. But they admit their results are still only "approximate, probably correct to 20 per cent", and apply only to elements heavier than aluminium.

The great investigation from which Rutherford had hoped for so much is seen, in retrospect, and contrary to the opinions of many memorialists of Rutherford, not to have provided firm proof of his nuclear atom theory, but to have done little more than verify his scattering law. Note Marsden's caution, quoted above, "Our work did,. I think, sufficiently establish the theory on a firm basis." Rutherford himself later claimed very little more than "this verification of the law of scattering". In fact the work of Geiger and Marsden was completed by July 1912, many months before the usually quoted date, which is put at the English-publication date of April 1913. However a very neat piece of detective work by Dr Trenn has shown that the English paper is no more than an "emended English version" of an original paper in English which was translated into German by Geiger and published in Vienna in 1912, clearly dated from Manchester in July 1912. Geiger had, in fact, left Manchester at the end of the academic year of 1912 to take up a new job in Berlin, and he oversaw the publication of the paper in Vienna. Presumably this choice of Vienna was based on the use of Viennese radium sent to Rutherford by Meyer at the start of the Manchester period. Dr Trenn points out that "this first edition actually contains a more complete and corrected statement of the results", but it is marginally less firm in its support of the nuclear atom theory. Rutherford himself started using the word "nucleus" only about August 1912, and at about the same date seems finally to have decided that the central charge is positive.

Trenn suggests that the publication in Vienna "may also to some extent reflect a divergence of viewpoints" between Geiger and Rutherford–Marsden over the significance of the results. And he makes the most interesting point that the Vienna paper, although preceding the English paper by six months, is never cited in the scientific literature, nor is it even mentioned in semi-official bibliographies of the output of Rutherford's Manchester laboratory: "The lack of citation of this first edition . . . is a further indication of a minimal concern for priority." But the most important point made by Dr Trenn in this study is when he suggests:

> . . . it was not the Geiger–Marsden scattering evidence, as such, that provided massive support for Rutherford's model of the atom. It was rather the constellation of evidence available gradually from the spring of 1913 and this, in turn, coupled with a growing conviction, tended to increase the significance or extrinsic value assigned to the Geiger –Marsden results beyond that which they intrinsically possessed in July 1912.

This seems to be a very accurate summary of the situation. Rutherford's nuclear atom did not prevail because of direct evidence in its favour – it prevailed because of its extraordinarily successful explanatory power. It had the mark of the truly great hypothesis: it inspired a great amount of highly successful new work. Invented to explain puzzling results in the physics laboratory, the nuclear atom theory was shown to provide satisfactory explanations for large areas of problems in chemistry, particularly regarding the nature of the elements and the regularities and differences between them. It gave explanations for the physical phenomena of spectra, and its development in the hands of Niels Bohr led to that great outburst of discovery, the new physics, which is the intellectual crown of the first three decades of our century.

It was not the verification of the rather primitive scattering law that established the nuclear atom theory – it was the suggestion that the central charge of the atom was proportional to about half the atomic weight that was fruitful and convincing. Rutherford had drawn this proportionality from the original, 1909, scattering work of Geiger and Marsden, and it was somewhat strengthened by the 1912 work. Rutherford had throughout this year, 1912, been off chasing what turned out to be hares, in the shape of beta- and gamma-radiations, hoping that they would provide evidence of the structure of the atom that would support or illuminate the evidence of the alpha-particles. He did this work with Robinson and Andrade, and it was good solid work, published in a series of papers throughout 1913 and 1914, but it led to no very profound conclusions in itself. He also directed a number of other colleagues in the laboratories to the same or similar subjects, and it is in this sphere, in the running of his laboratories, in his ability to attract and select the greatest of his younger contemporaries, that he must be given full credit for the brilliant series of results which were produced in Manchester on the basis of his nuclear atomic theory.

The years 1911 and 1912, which saw Rutherford creating the hypothesis of the nuclear atom, also saw him recruiting Niels Bohr and Harry Moseley; if the director of a modern research laboratory had recruited two such men in a life-time of work he would be regarded as a genius for that alone.

11

The Atom in Action

To understand the power of explanation provided by the Rutherford atom, one must understand the problems faced by the scientists of 1910. Most of the ninety-two naturally occurring chemical
elements were already known then, and the weights of the atoms of
these elements were also known. It was understood that the
differences in the elements could be accounted for by the differences in their atoms, and the Russian Mendeleev had spotted that,
if the elements were arranged in increasing order of atomic weight,
then certain regularities emerged in the arrangement. This was
called the "Periodic Table" and was widely considered to be a step
towards an explanation of the nature of the elements, since it made
it possible to group elements which exhibited similar qualities – for
instance the metals, or the newly discovered noble gases – in
regular patterns of occurrence. But there were some who considered the Table an artefact of chance. J.J. had suggested that his
atomic models provided some explanation in that the characteristics of the different elements could be accounted for by their having
different numbers of electrons in different orbits or "shells". The
difficulty he faced was that electrons were so small that he had to
imagine atoms containing hundreds of electrons to account for the
known masses. There were also certain widely acknowledged
problems, or anomalies, in the regularity of the Periodic Table. All
these problems were eventually solved when Rutherford's team
came up with the suggestion that what mattered was the atomic
number of the element and not its atomic weight. They proposed
that the smallest atom, hydrogen, atom number one in the Periodic
Table, had a central charge, positive, of one unit and therefore had
one negatively charged electron in the space around the nucleus.
Helium, atom number two in the table, the second lightest, should
have two units of charge on its nucleus (and this of course is
Rutherford's alpha-particle), and two electrons circling around.
And so on up the Periodic Table until we come to the heaviest
natural atom, that of uranium, now known to be number ninety-

two, with ninety-two units of charge on its nucleus and ninety-two electrons in many different orbits and shells.

It is a beautifully simple idea; it very soon resolved all the technical problems surrounding the Periodic Table; it has provided the basic view of atomic structure for both physics and chemistry ever since; in simple words it was right and everybody saw it was right. And it was this application of Rutherford's original idea that convinced the world of science of the reality of the nuclear atom. Simultaneously Niels Bohr applied the concept of the quantum to the structure of the atom and removed the greatest objection to the nuclear theory – and we shall examine that development in greater detail a little later.

Rutherford, looking back ten years after the event, put these two developments in the reverse order. He told the Association of Science Masters:

From general evidence it seemed certain that the nucleus of hydrogen had only one charge and that of helium two, and it was suggested by van den Broek that the observations [of Geiger and Marsden] were not in disaccord with the view that the nuclear charges in fundamental units might correspond to the atomic number of the element, i.e. to the original number of the element arranged in order of increasing atomic weight. This relation between nuclear charge and atomic number seemed very plausible and was used by Bohr in his theories before its correctness had been established by the work of Moseley . . .

Moseley . . . had already shown his power of attacking successfully difficult problems. I remember very well a discussion with him as to the next problem to take up after his work with Darwin on X-ray spectra had been finished. As a result Moseley decided to attack the fundamental problem whether the properties of an element were defined by its atomic weight, as in the periodic classification of the elements, or by its atomic number or nuclear charge. On the nuclear theory, as we have seen, the motion and distribution of the outer electrons are controlled by the attractive forces from the charged nucleus . . . On this theory the mass of the nucleus should exercise only a very secondary influence on the motion of the outer electrons. In order to throw light on this subject, Moseley examined the X-ray spectra of a number of successive elements. It was found that the K-spectrum was very similar for all elements consisting mainly of two bright lines, and that the frequency of vibration corresponding to these two lines varied by definite steps from element to element . . . If the atomic number of an element were a measure of its nuclear charge such a relation was to be expected on the Bohr theory of spectra. Moseley concluded that this was the case . . .

This work of Moseley was a great step in advance, for it not only brought to light the remarkably simple relation that existed between the X-ray spectra of different elements, but showed that the nuclear charge of an element was in numerical units given by its ordinal (atomic) number. In addition he was able to assign the value of nuclear charge to all the elements and to predict with certainty the position of possible but missing elements. This relation found by Moseley is of unexpected simplicity, for it could hardly have been predicted that, with few exceptions, all atomic numbers between hydrogen, 1, and uranium, 92, would be represented by known elements in the earth. The proof that the properties of an element depend on the whole number representing the nuclear charge is of fundamental significance and has formed the basis of all subsequent work on atomic constitution. The law of Moseley, as it may be termed, in my opinion will rank in importance with the discovery of the periodic law of the elements and of spectrum analysis, and in some respects is far more fundamental than either.

The first firm statement on paper of anything approaching the theory of atomic number seems to be in a memorandum written by Niels Bohr for Rutherford's consideration in the middle of June 1912, though this paper is more usually studied for its demonstration of Bohr's initial thoughts about the application of quantum theory to atomic structure. Bohr himself in later life recalled that some of the inspiration came from a remark by George von Hevesy, the inventor of the concept of radioactive tracers, who was also working in Rutherford's laboratory at the time, on the apparently different subject of isotopes, a concept which had then just been put forward by Soddy. Bohr had been attracted to the subject by his consideration of, and criticisms of, Charles Darwin's latest theoretical paper on the mathematics of alpha-particle scattering.

Darwin remembered: "By the end of 1911 the existence of the nucleus was firmly established. In Manchester we all knew it was very important indeed and we had the idea that all the elements must have atomic numbers . . . However, perhaps we did not quite realise how tremendous the consequences were going to be." He pointed out that "there was no proof that the nuclear charge would even be a whole number, though everybody thought so, and there were problems with the known gaps in the Periodic Table. The result was that everyone in Manchester was rather cross when the Dutchman, van den Broek, put forward the full hypothesis of atomic number" in a German physics journal in 1913. Darwin wrote: "I can recall that we felt a little annoyed at this, because it was based on the Manchester work, and it ran exactly on the lines we had all been thinking. We rather felt that an opportunity had been missed of stating an almost obvious fact."

Rutherford, in 1910, had seemed to favour the idea that one element could only disintegrate in one way, though very shortly his own laboratory was to prove that this was not so. So at first Rutherford postulated that all beta-particles from one element left the nucleus with the same speed or energy, but were affected by the electron structure of the atoms in different ways, so as to emerge with differing energies, having produced gamma-rays in their interactions with the electrons. This implied that the energies of the gamma-rays should be mathematically related to the energies of the beta-particles, and Rutherford did manage to produce a simple law which seemed to relate the energies of the beta and gamma emissions from radium-C.

Rutherford published this idea in the *Philosophical Magazine* for October 1912. The paper is titled "The Origin of Beta- and Gamma-Rays from Radioactive Substances" and is dated "Manchester, August 16th, 1912". Moseley must have seen the proofs of this paper when he returned from his summer holiday, for in a letter to his mother dated October 14th, 1912 he wrote, "Today I was surprised to find a sad blunder in Rutherford's latest paper, in which he gives a new theory of beta-rays. I fear all his calculations are wrong, but when I demonstrated it to him, he philosophically acknowledged his error and declared that, even if the calculations did no longer fit the theory (which was made to suit them), he is sure the theory is right all the same."

Rutherford's "blunder" was to misunderstand the new mathematics demanded by the theory of relativity, for in this case the beta-particles travelled at such a significant fraction of the speed of light that the effects of relativity had to be taken into account and a special mathematical formula – the "transformation" invented by the Dutchman Lorentz and incorporated into the theory of relativity by Einstein – had to be used. The blunder caused Rutherford to write the only full public retraction that he ever had to make in his long career – it appears as a letter in the December 1912 issue of *Philosophical Magazine*. As always he gave credit where it was due: "Mr Moseley drew my attention to the fact, which I had overlooked, that according to the Lorentz-Einstein theory . . ." He ends this letter of retraction in the same philosophical, but unyielding mood, that Moseley reported: ". . . It is of great importance to know accurately the distribution of the beta-rays from active products both as regards velocity and number, for it is only with the help of such data that we can hope to explain the origin of the remarkably complex beta radiation from active substances and its connection with gamma-rays." Which is very much the same as saying he was sure the theory was right all the same.

Moseley had attained the right to correct Rutherford by solid experimental work, achieved against all the difficulties of pre-world-war apparatus.

My work is going through much tribulation. On Thursday I at last induced my apparatus to stay at a pressure of . . . a three-hundred-thousand of an atmosphere after plastering many suspected leaks with a red sticky stuff which resembles butter. Then I started my experiment, but a glass tube of thickness much less than tissue paper, filled with radium emanation, chose to break off its stalk inside, and everything had to come to pieces to get it out. Feeling for a thing which breaks at a touch was too risky to be tried, for to let emanation loose upon the laboratory is a capital offence. The method of getting a vacuum I use is delightfully simple. Heaps of charred coconut shell in a bottle are sealed into the apparatus and then cooled by liquid air. The charcoal sucks up almost all the gas in a remarkably short time, and so long as it is kept cool, keeps the pressure down.

This was Moseley's description of his work to his sister in a letter of November 1910.

Darwin's memories of working with Moseley are rather less sympathetic:

Working with Moseley was one of the most strenuous exercises I have ever undertaken. He was without exception the hardest worker I have ever known. Even before the time of our collaboration I used to go into his room sometimes to see what he was doing, and I would find him obviously in the last stages of exhaustion, having probably been up all night. When I told him he ought to be at home in bed, he would answer that when he was feeling well he wanted to be out walking in the country, and that it was only in this condition when he was tired out that he felt inclined for laboratory work.

There were two rules for his work. First: when you started to set up the apparatus for an experiment you must not stop until it was set up. Second: When the apparatus was set up you must not stop work until the experiment was done. Obeying these rules implied a most irregular life, sometimes with all-night sessions, and indeed one of Moseley's expertises was the knowledge of where in Manchester one could get a meal at three in the morning. In spite of these irregular hours it was extraordinarily stimulating, because all the time we were discussing the implications of the current experiment either to fundamental theory or to the next experiment we should try. Here there arose dangers from yet another of Moseley's habits. He was always ready to take the whole apparatus to pieces and set it up again if he could see any possible improvement to be hoped for. Often he was quite right but some of his proposals were really rather trivial and I can remember occasions when I opposed him, suggesting we had better give the current set-up a trial before spending a lot of time in changing it, and sometimes we found that the existing set-up gave all we needed.

It may well have been as a result of Moseley's successful exposure of Rutherford's blunder in the beta-ray work that he had sufficient "leverage" to persuade Rutherford to allow him to switch to work on X-rays. Certainly there was fierce discussion about the shift and the exact experiments to be done. Darwin recalls that Rutherford at first rather discouraged Moseley, "saying that we in Manchester had no familiarity with X-rays, and that we would be at a disadvantage in a new field which was better understood in other places". But Moseley persisted and in the end won the argument by showing that his tremendous work rate could make up for any initial disadvantage.

X-rays were one of the foremost points of controversy and development in physics at this time, some fifteen years after their first discovery. Were X-rays a form of light though of much higher frequency than light? Were they waves, like light waves, or were they particles like alpha- and beta-rays? The puzzle was made the more difficult by the fact that Einstein's work on the photoelectric effect (the production of electrons by the action of light on metal surfaces) demanded that light be regarded as corpuscular, consisting of massless particles called photons, in opposition to the classical view that light was a wave-motion, a view held since Young's famous slit experiments of the eighteenth century and supported by Maxwell's electromagnetic field equations. Naturally physicists turned to spectroscopy in their efforts to solve these problems, asking whether X-rays gave spectra in the same way as light, and they found that X-ray spectra could be obtained by reflecting the rays at very small angles from the surface of materials. (Incidentally, another line of attack on the problems was a continuation of Rutherford's early work on the ionisation of gases caused by X-rays.) Some years earlier Barkla, who clashed with W. H. Bragg so seriously over the nature of X-rays, had shown that different elements did produce lines in the X-ray spectrum which were characteristic for each different element. These spectral lines were named K, L, and M lines for no very definite reason, and no one was very clear as to how or why they were produced.

In 1912 a small group in Munich, under M. von Laue, came up with a most exciting result – they shone X-rays right through a single crystal and photographed an absolutely regular pattern of black dots and marks which resulted. Unfortunately for the Germans, they did not understand just what they had done. To Moseley and to the two Braggs (W.H. and W.L., father and son) it was quite clear that this proved that X-rays were waves and that the regular patterns of dots were interference patterns caused by the X-rays being diffracted by the layers of atoms that made up the crystal. Moseley was called on to lecture to his colleagues in Manchester on the meaning and significance of the German result. He confided to his mother that he had been very nervous, not least because Bragg, "the chief authority on the subject", had been present.

> I was talking chiefly about the new German experiments of passing the rays through crystals. The men who did the work entirely failed to understand what it meant and gave an explanation which was obviously wrong. After much hard work Darwin and I found out the real meaning of the experiments and of this I gave the first public explanation on Friday. I knew, privately, however, that Bragg and his son had worked out an explanation a few days before us, and their explanation, although approached from a different point of view, turns out to be really the same as ours. We are therefore leaving the subject to them.

"Leaving" an area of research to another team by mutual agreement is not common nowadays, to put the matter mildly. But in the first decades of this century, with physics in a revolution, and with so many fewer scientists at work, it was a sensible division of labour and well-suited to the British "gentlemanly" tradition. The Braggs had been working on von Laue's results with ever-increasing excitement throughout the family summer holidays. The explanation to which they were led came largely from the son's experiments conducted when he returned to Cambridge in October 1912, which meant that the father had to give up most of his long-held views on the nature of X-rays. It was in this same winter of 1912 that further experiments by the son showed the father the way to the invention of the X-ray spectrometer, a machine which is now standard equipment in both physics and molecular biology laboratories. And for their joint work the Braggs duly won a Nobel Prize.

The elder Bragg's conversion to his son's theory is explained in two letters to Rutherford, written in the first months of 1913, mixed with long discussions as to whether he should take a post in the new University of British Columbia. In January he was still puzzled: "Supposing the identity of X-rays and light to be established, the supposition is this, I take it. The energy travels from point to point like a corpuscle: the disposition of the lines of travel is governed by a wave theory. Seems pretty hard to explain but that is surely how it stands at the moment." This is an interesting comment historically, for it pre-dates the Bohr theory of complementarity and the work of Schrödinger and de Broglie by more than ten years, yet shows that physicists were already accepting some form of wave/particle duality long before the theoreticians expressed this duality in mathematical form. Two months later Bragg felt much more comfortable. "I think I have got the whole thing more or less. It is against my corpuscular theory in part, but also against the pulse theory in part. However that can be argued separately. I will just give you the facts." And he laid out a set of experimental results for Rutherford to consider.

Having left what we now call "X-ray diffraction" to the Braggs, Moseley had, however, plainly seen, somewhat against Rutherford's earlier conviction, that X-rays must give information by their reflections about the constitution of the atom. His X-ray lecture had been given in the first days of November 1912. Where the Braggs were looking into crystals with their X-rays, Moseley concentrated on the reflection of X-rays from different elements. First he had to establish that what was reflected was still an X-ray, and the obvious way to do that was to see whether it still ionised a gas according to the old Rutherford method. But in the first months of 1913 he was as puzzled as W. H. Bragg by the problems of waves and particles.

To his beloved sister, Margery, Moseley explained in a letter of February 2nd:

The X-ray problem is becoming intensely interesting and lots of people are working at it, so we staked out our claim in the conventional way by sending to *Nature* a letter which, after some delay, has now been published. I feel therefore that there is no longer a terrible hurry needed as we no longer fear someone else claiming a monopoly in the subject. [The letter, written jointly with Darwin, was called "The reflection of the X-rays".] We have now got what seems to be definite proof that an X-ray, which spreads out in spherical form from a source as a wave through the ether, can, when it meets an atom, collect up all its energy from all round and concentrate it on the atom. It is as if when a circular wave on water met an obstacle the wave were all suddenly to travel round the circle and disappear all round and concentrate its energy on attacking the obstacle. Mechanically of course this is absurd, but mechanics have in this direction been for some time a broken reed.

There is some most mysterious property of energy involved which the Germans have for some years been groping after but which we see no immediate hope of comprehending.

Throughout that spring of 1913 the puzzlement about the nature and behaviour of X-rays continued. It was a problem which occupied the minds of many scientists: "In Germany . . . there is no laboratory there which comes up to England in the experimental side of modern physics. They run away after strange theories and experiment gets neglected. Also their national bigotry is rather a serious obstacle in the path of any foreigner. In that respect the French are a hundred times worse and their suspicion that their pet ideas will be stolen seems in some cases to be such that conditions become intolerable." This was Moseley writing to his mother when he was beginning to wonder where he should go when his spell at Manchester came to an end in a few months' time. It confirms the extraordinarily xenophobic attitude of this well-educated and well-brought-up specimen of the British upper classes of the time, and shows a simplistic and jingoistic patriotism which was to lead to Moseley's early death when he ignored all possibilities of war-work appropriate to his shining scientific talent. But in May he suddenly began to get glimpses of something further. It was on May 18th, 1913 that he wrote:

> The whole subject of X-rays is opening out wonderfully. Bragg has of course got in ahead of us and so the credit all belongs to him, but that does not make it less interesting. We find that an X-ray bulb with a platinum target gives out a sharp line system of five wavelengths which the crystal separates out as if it were a diffraction grating. In this way one can get pure monochromatic X-rays [i.e. it was possible to obtain a splitting up of X-rays into bands, each of which had a precise wavelength]. Tomorrow we search for the spectra of other elements. There is here a whole new branch of spectroscopy which is sure to tell one much about the nature of the atom.

It seems likely that Moseley must have obtained the first rough experimental clues to the regular separations between the X-ray spectra characteristic of a few elements in May and June of 1913. Furthermore he must have talked with Bohr and everybody else in Manchester where the idea that atomic number and atomic charge meant the same thing was steadily becoming clearer. And it seems likely that the conception of his final experiment was formed during the long summer vacation. Certainly his experimental apparatus was completed and his really conclusive results had started to flow some four weeks after the new academic year started in October 1913.

The apparatus he devised, though not so simple as those used by Rutherford himself, had a Rutherfordian simplicity in its conception. It consisted of a row of small trucks – like toy railway trucks – running on a track; on each truck was mounted a specimen of the elements he wished to examine, and usually these were a series of elements adjacent to each other in the Periodic Table. The little train on its track was all fitted into a glass vacuum tube, with a string coming out at either end so that the specimens could be moved along. Each specimen in turn could be moved into the beam of X-rays so that the spectrum of each element could be examined. By Sunday, November 2nd, 1913, he could write to his "Dearest of Mothers":

> Since Wednesday it has been astonishingly successful . . . I can now get in five minutes a strong sharp photograph of the X-ray spectrum, which would mean days of work by the ionisation method. In the last four days I have got the spectrum given by Tantalum, Chromium, Manganese, Iron, Nickel, Cobalt and Copper and part of the spectrum of silver. The chief result is that all the elements give the same kind of spectrum, the result for any metal being quite easy to guess from the results for the others. This shows that the insides of all the atoms are very much alike and from these results it will be possible to find out something of what the insides are made up of.

A week later he wrote further good news:

> My work has turned out so extremely interesting and important that I will go on with it for a long time to come and the only question is when to break off, publish and start afresh at Oxford. I feel a very selfish fellow over it, as it is so very easy and so little trouble and gives so rich a return for a minimum of work that I should like to keep it all for myself. If I publish, a horde of hungry Germans will be down on it directly, and if I delay perhaps someone will get in ahead. So for once I feel thoroughly commercially-minded over the whole thing . . . I hear from Rutherford that the Solvay Institute has given me a thousand francs which will come in very useful for getting apparatus in Oxford.

This last comment was hardly grateful to his Professor, for Rutherford was at this time on the committee which disposed of the funds of the institute, and must therefore have been influential in getting Moseley this useful grant at a time when the young man was leaving Manchester and going back to Oxford, which move was made early in the New Year.

At the same time, November 1913, Moseley was seeking Bohr's advice:

> I am puzzled by this apparent simplicity and by the wonderful way in which "alpha" fits the simple formula. Of course they show that it is well worth while to examine every element in this way, and until this is done I do not see the explanation clearly. I feel however that they lend great weight to the general principles which you use, and I am delighted that this is so, as your theory is having a splendid effect on physics, and I believe that when we really know what an atom is, as we must within a few years, your theory, even if wrong in detail, will deserve much of the credit.

Bohr's reply credits Moseley for "most interesting and beautiful results" and agrees they are "suggestive", but he goes forward cautiously: "As to the detail interpretation, I must confess that for the present I cannot offer any valuable suggestions. I hope very much that your further investigations shall be able to throw light on the problem . . . For the present I have stopped speculating on atoms. I feel that it is necessary to wait for experimental results."

Rutherford by this time was showing an intense interest in Moseley's work – according to Eve the Professor "caught the prevailing X-ray fever and began bouncing gamma-rays (which are the same as X-rays but of a higher frequency) off a beautiful crystal of potassium ferrocyanide". What was clearly showing from Moseley's work even before he left Manchester was that the clearest spectral line (the "alpha" of the letter to Bohr) from each element differed in frequency from the chief line of the preceding element in the table in a completely regular way. Moving up the list of elements was like moving up a regular staircase in terms of the X-ray spectra. It is worth repeating that the elements had, up to this date, been arranged in an order which depended solely on their atomic weight, and there was no regularity in the difference in atomic weights – no regular steps up the table. Moseley's results showed that there was, in fact, a difference of one "something" between each atom and the next in the table; indeed it showed that the Periodic Table did have a reality, that the atoms of the elements were arranged in this order due to some essential difference between each one and its neighbours.

Bohr's theory was that this essential difference was a difference of one unit charge on the nucleus of each atom, and hence a difference of one more electron for each element. For the moment Moseley's results did not prove that this was the real difference between atoms; all they showed was that there was some real regular difference between one element and the next. When Bohr was, shortly afterwards, able to advance an explanation for all spectra, Moseley's results fell into place as the strongest evidence for the Rutherford atom as spelled out in detail by Bohr.

At the start of 1914, however, Moseley's work was not completed. He had found the spectra for most of the metals and a number of the other elements, but there was much work to do when he moved to Oxford at the beginning of that year if he was to complete the examination of all the elements then known to science. One of the problems he faced was that some of the elements gave only very weak spectral lines in the range of frequencies where the metals gave strong and clear lines. But by March 1914 he was able to send Rutherford another list of twenty-four elements with their atomic numbers, all clearly showing the regular differences. Indeed he could now show that there were certain gaps in the complete list of elements, where science had so far failed to find the substance which fitted into the vacant place in the periodic table – all these gaps have since been filled with substances that fitted Moseley's predictions. However Moseley was not infallible, any more than his master, when it came to mathematics. The day after he had sent Rutherford this list of twenty-four more elements safely placed in their positions in the table, he had to send off a panicky post-card. "I find a slip has thrown out all the values of N [the atomic number or its equivalent, the number of units in the nuclear charge] from yttrium onwards by one unit. The (alleged) element between strontium and yttrium is therefore a myth. All else stands."

And Moseley's work has indeed stood the test ever since. There was little more of it, for he was in the army before that year was out and dead before the next year was finished.

Moseley's arrangement of the elements would have encountered one major problem if there had not been a contemporaneous development, which was also largely the work of Rutherford's colleagues and pupils. The placing of the ninety-two elements from hydrogen to uranium on Moseley's scheme left, as we have seen, a few gaps for as-yet-undiscovered substances. But it did not provide anything like enough room at the top, or heavier, end of the table for the many new elements produced in radioactive decay families – and it had been, of course, an essential part of Rutherford's original Canadian work that such substances as the emanations of radium, thorium and actinium were true elements in their own right. Yet, since that time, many further radio-elements had been discovered. It is an old and worn-out controversy whether the Law of Radioactive Displacement was discovered by Soddy, by A. S. Russell or by Kasimir Fajans, and it is customary nowadays to award the prize to all three without distinction. But there is no doubt that it was Soddy who first clearly advanced the idea of isotopes. What Soddy suggested was that all the atoms of any element were not absolutely identical – that there could be two sorts of sodium atoms, say, identical in their chemical behaviour as sodium, but differing in their atomic weights. He was compelled to this conclusion by his own and other chemists' discoveries that certain products of radioactive decay, which by his own and Rutherford's theory of disintegration must be elements, were not separable by any chemical method he could find from certain other radioactive elements, although it was known that the atomic weights of the inseparable elements were different. The number of isotopes known to science has steadily increased, particularly since the start of the nuclear energy programmes of the 1940s. It is now known that virtually all the elements have isotopes, more or less stable. What used to be regarded as the "normal" form of the element is now understood merely to be the most stable, or the most common; many of the less common isotopes are radioactive. The Displacement Law brought some order to the burgeoning list of radioactive decays known in the first decade of the century. What it said was that an atom which decayed by emission of an alpha-particle became an atom of the element two down in the Periodic Table, because the alpha-particle carried away two units of nuclear charge. An atom that decayed by emission of a beta-particle, known to be an electron, moved one place up the Periodic Table, for by the loss of a negative charge from the nucleus, the atomic charge must increase by one positive unit. The Displacement Law has been stated, if simply, in modern terms. It was described somewhat differently when first proposed, but it is clear that from the start it meshed in closely with the Manchester view of the nature of the atom, and provided considerable support for Rutherford's new ideas. Rutherford's correspondence throughout 1913 and 1914 contains many letters, especially from Kasimir Fajans, who had spent two years in Manchester in 1910 and 1911, about the rival claims for priority over the Displacement Law but Rutherford continually urged the contestants to refrain from public controversy, and eventually he got his way, and the honour is still shared.

Russell, with the tolerance of old age when he had become a don at Christ Church, Oxford, and was speaking at the Rutherford Golden Jubilee celebrations, had some interesting comments to make: "Why were the inorganic chemists so slow in the years immediately preceding 1914 in arriving at the conceptions of isotopes and atomic number? When we realise the importance of these conceptions in the nuclear physics of today, as well as in inorganic chemistry, we cannot but criticise the casual manner in which the earlier chemists eventually arrived at them." He pointed out that Crookes had made the first suggestion of isotopes as far back as 1886, but the idea had not been followed up. Twenty years later, in 1906, the evidence began to gather when Hahn discovered radiothorium by its radioactivity but found it could not be separated from thorium. More and more cases of "non-separability" accumulated where the different radioactive behaviours showed that there were two elements present in a mixture, but no chemical procedure could separate the two, even though they were known to have different atomic masses.

Russell continued:

With the fact of nonseparability, with about forty radio elements somehow to be explained or fitted into the Periodic Table, and, as we now know, only ten places above the position of lead available, it must strike you as odd that the chemist hesitated for so long to distribute the radio elements in the available places of the table. That would have given us at once the idea of isotopes. It would have given us also the displacement laws, and from the displacement laws it would have been seen that the number of positive charges carried by an atom was more important than its mere atomic mass. Note that it would not have given us the absolute values of the atomic numbers of thorium and uranium; that had to wait for the work of Moseley and of Bohr, but it would have given us the atomic numbers of all the radio elements relative to that of one of them arbitrarily chosen – and this as early as 1911, a year before Rutherford and his co-workers were evolving the idea of atomic numbers from their experiments on the scattering of alpha-particles by elements like gold and platinum, and two years before Moseley's classical work on the X-ray spectra of the elements. In defence of his apparent lack of insight, or even courage, the chemist could have advanced two points. He could argue that it is unwise to erect into a principle the mere fact that some analytical chemists had failed to effect certain separations. Inability to do something that seemingly ought to be done is surely better met with "the more fool you" retort than with the acceptance of a new principle. He could argue also that in 1911 you could not, in fact, "trust" the Periodic Classification. As a classification it was obviously not Periodic. The rare earth elements demonstrably did not fit in. There were difficulties over the atomic weights of argon and potassium and other pairs of adjacent elements and, worst of all, there was no criterion for limiting the number of elements that might be heavier than lead or lighter then hydrogen.

The real difficulty with the inorganic chemists in that period, however, was not that Mendeleev's Periodic Classification in its development had never kept pace with discovery (so that it always appeared to critical minds more a help to the student or a mnemonic than a true principle in embryo) but that the climate of opinion in inorganic chemistry then was adverse to any kind of speculation or even theorising. Chemistry, we were always being told, was an experimental science. No good ever came from pontificating on the ways of nature from the comfort of an armchair. The laboratory bench, not the sofa, we were sarcastically told, was where truth would be found. And so no inorganic chemist of eminence, not Ramsay, Crookes, Madame Curie, Soddy, Hahn, Marckwald or Welsbach, took the final and decisive step in 1911 . . . Valency was the property insisted on. Position, now so simple and obvious an idea, was not then thought of. ·

So the weight of evidence in favour of Rutherford's nuclear atom built up – and Russell for instance was working in Manchester through these crucial years as one of the 'tame chemists". But there was an apparently fatal flaw in the Rutherford atom that was all too clear to physicists. It was essentially a mechanical flaw, for a simple application of the mechanics developed by Newton showed that Rutherford's atom could not work. By Newton's Laws the electrons, which were presumed to orbit around the nucleus in some undefined way, must of necessity eventually give up their energy by radiating it away and must collapse into the massive central body, the nucleus, which attracted the negative electrons by its positive charge. There were plenty of people to point this out to Rutherford – perhaps the most eminent was Sir Oliver Lodge, Professor of Physics at Birmingham university. He wrote in July 1913: "The astronomical view of the structure of the atom, with a minute, positive, massive nucleus and a neutralising number of electrons subject to the inverse square law is in many ways attractive. But there are difficulties about it, chief among which is a question of stability . . . I suppose you have your own ways of getting over these difficulties."

There can be no doubt Rutherford understood this quite well himself; ten years later he wrote:

At this stage it was clear on the nuclear theory that the properties of an atom depended mainly on its nuclear charge, for the latter must not only define the number of the outer electrons, but their arrangement, on which the ordinary physical and chemical properties of an atom depend. Nothing, however, was known about the constitution of the nucleus or of the arrangement of the outer electrons. On the ordinary views of electrodynamics, the outer electrons could not be permanently stable in an atom of this kind but should fall into the nucleus. In addition it seemed certain that the hydrogen atom contained only one electron and the helium atom two, and the movements of these electrons must in some way account for the very complicated spectra of these elements.

The problem of the mechanical instability of orbiting electrons was, in any case, perfectly well-known to "atom-builders" and had been for a number of years. J.J. had grappled with the problem and had indeed come up with a mathematical arrangement of electrons which could be stable: it involved eight electrons all in the same orbit, but it could not solve his problem for he needed hundreds of the lightweight electrons to give his atom sufficient weight. "Atom building" was a popular speculation of the time – in Soddy's annual review of radioactivity for 1911 he gives considerable space to a theory by J. W. Nicholson, "a structural theory of the atomic weight of elements . . . which has a definite theoretical foundation and is certainly in surprisingly good accord with the best chemical determinations". In this theory the whole mass of the atom was regarded as electromagnetic and the positive charges were carried in small units, very massive and smaller in size than the negative electrons. But all atoms were made up of four types of "protyle" and the theory demanded the existence of two elements, coronium and nebulium, one of which was lighter than hydrogen, and Soddy judged that the view of J. J. Thomson that the positive charge was spread throughout the volume of the atom "must be retained". In the same summary of the year's work Soddy remarked on the possibility of Rutherford's alpha-particle scattering experiments "giving an experimental means of learning something of the nature of atomic structure." He pointed out that the passage of alpha-particles through matter was "the only known phenomenon for which the dictum that two particles of matter cannot occupy the same space at the same time is not true".

Plainly Soddy kept up very well with all the literature on this fascinating subject, and so his is as good a comment as any in his 1913 report, where he describes "Rutherford's Theory of Atomic Structure" without mentioning the word nucleus and states that "the mass of the atom is associated with positive electricity, about one unit of charge per two units of mass." He goes on, "This atom is, of course, not stable according to ordinary electrodynamical laws, for nothing apparently operates to prevent the dispersion of the extremely concentrated central positive charge", which is not the reason usually given for the instability of the model. Soddy however retrieves the situation: ". . . but it is now recognised that these laws require modification. The model has been used with very considerable success in conjunction with Planck's theory of quanta, and leads to results . . . in striking accord with experimental determination." Soddy then goes on, without saying that there is any direct connection between the two theories, to give the evidence which perhaps swayed him and many others in favour of the Rutherford model. He gives the van den Broek theory of atomic (or nuclear) charge corresponding to atomic number; he adds the radioactive displacement law and his own new definition of isotopes, which he adapts to the new theories as "Isotopes are elements for which the algebraic sum of opposite charges is the same, but the arithmetical sum is different". That may be neat for the academic, but is very confusing for the layman. He rushes on, however, to give what must have been the latest speculation in scientific circles, for it runs far ahead of the evidence. He writes of the assumption that both alpha-particles and beta-particles (electrons) are expelled from the nucleus of the atom and not from any outer ring structure, and he goes on to ask whether the nucleus contains electrons or whether it consists only of helium atoms with one unit of charge to two units of mass which pick up their second unit of charge in some unexplained fashion, or whether the hydrogen nucleus is also a constituent of all other nuclei. These were problems which would continue to occupy physicists, including Rutherford, for the next twenty years, indeed they are problems to which we still do not have final all-embracing solutions.

But Soddy had got the essence of the matter *right*. Rutherford's nuclear atom model had had its main problem, the problem of instability, solved by calling Planck's quantum theory into play. The man who had done this, the man who had shown that Rutherford's atom was correct, was Niels Bohr.

Of all the gods in the scientific pantheon Niels Bohr is most like the Christian saint, and the feeling of love and reverence for him persists to this day when those who knew him, family and friends, will even stage street protests against a commercial film which they feel distorts his record. But Bohr's methods and approach were unusual. He started much of his work from spotting apparently small mathematical defects in other people's work, and reached momentous conclusions by developing the lines of thought that arose from these criticisms. The development of his thought was often achieved in lengthy conversations with friends and pupils and these talking sessions could often last far into the night and could sometimes turn into long monologues. For English-speaking colleagues these long conversations were often peculiarly difficult, for Bohr spoke slowly and his command of the language was imperfect, though the results were often delightful. But Bohr, who had been a member of the Danish Olympic football team, had been taught by his father to love things English, and especially the science of J. J. Thomson's school in Cambridge. So when he completed his doctoral thesis on the electron theory of metals, which explained the physical properties of metals in terms of the activity of free electrons in metal structures, he elected to take a year's study abroad, and he chose to go from Copenhagen to Cambridge. This was not, perhaps, the wisest choice, for J.J. had given up the electron theory some years previously and was then working on "positive rays", while the greatest authority on electron theory was Lorentz at Leyden.

When Bohr arrived at the Cavendish in 1911, he was delighted with J.J.'s friendliness and his promise to read his thesis, which had been roughly translated into English. The thesis contained some criticism of J.J.'s work, pointing out that in one of the master's calculations he had failed to take account of the time involved in electron collisions. But it seems unlikely that J.J. ever worried about this, for he never read the thesis, leaving it on his untidy desk; nor did he ever let himself get involved in one of Bohr's lengthy discussions. Not that he was at all unkind or unwelcoming; it was just not his way. In a series of letters to his brother and his fiancée, Bohr describes the failure of communication. First, on September 29, 1911:

> I have just talked to J. J. Thomson, and I explained to him as well as I could my views on radiation, magnetism, etc. You should know what it was for me to talk to such a man. He was so very kind to me; we talked about so many things; and I think he thought there was something in what I said. He promised to read my thesis and he invited me to have dinner with him next Sunday at Trinity College, when he will talk to me about it . . .

But a month later he wrote:

> [Thomson] has not yet had time to read my thesis and I still don't know if he will agree with my criticisms. He has only chatted about it a few times for a couple of minutes, and that was on a single point, my criticism of his calculation of the absorption of heat rays . . . Thomson said he could not see that the collision time could have so great an influence on the absorption: I tried to explain and the following day gave him a very simple example (an example corresponding to his calculation of the emission) which showed it very clearly. Since then I've talked with him for only a moment, and that a week ago. I think he thinks my calculation is correct, but I'm not sure he doesn't believe that one can design a mechanical model to explain the law of heat radiation on the usual electromechanical principles . . .

This letter clearly shows that Bohr was already thinking in terms of having to apply some new principle (possibly quantum methods) to the solution of sub-atomic problems. His doctoral thesis had already suggested such an approach. However, at this time Planck was only just producing the second and more powerful version of his theory, and was indeed only just managing to convert himself from the classical physical view to an acceptance of discontinuity which his own work forced upon him. It was precisely at this time in the autumn of 1911 that the Solvay Conference first discussed the new idea and from that conference acceptance of the validity of quantum theory slowly started to spread outwards.

The problem at Cambridge for Bohr was not that Thomson or others would not accept the new theory. Bohr continued to regard J.J. as "the genius that showed the way for everybody", but found that the man "would break off in the middle of a sentence, after a moment's conversation, when his thoughts ran on something of interest to himself". J.J. did set Bohr a small laboratory problem in the area of positive rays (in itself a valuable development which led J.J. to the first discovery of a non-radioactive isotope a few years later and which also led to Aston's development of the mass-spectrometer), but the particular problem turned out to be trivial and in any case experimentation was not Bohr's métier. The basic problem was that "the two men . . . differed profoundly in their approach to physics". In interviews later in his life Bohr explained that to Thomson, models were mere analogies and fundamental problems of little interest. Thomson did not demand consistency among the different models he employed nor did he worry about quantitative agreement between experiments and calculations based on the models; indeed, in Bohr's words "things needed not to be very correct, and if it resembled a little, then it was so" as far as J.J. was concerned.

Bohr was not by any means unhappy at Cambridge, for he had many letters of introduction provided by his father. One of these was to a former pupil, the physiologist Lorrain Smith, then working at Manchester. It is probable that during a visit to Lorrain Smith in November Bohr first met Rutherford. Certainly in December Bohr saw Rutherford at one of the annual Cavendish dinners, where visiting Europeans were so astonished at the spectacle of the most famous English physicist, red in the face, singing silly songs or linking arms in "Auld Lang Syne". It is recorded that on this occasion Rutherford was introduced as the one who, among all the young physicists who had been trained in the laboratory, could swear most profoundly and most fluently at apparatus that refused to do his bidding. This caused the great boom of Rutherford laughter to dominate the room. Bohr was apparently impressed.

Certainly in January 1912 he wrote to Rutherford asking to come and work with him in Manchester "in the end of March". He explained that he had spoken to J.J. about this and would like to move as soon as he had finished some theoretical work. A week later Rutherford accepted the young Dane and offered to arrange for him to "do some preliminary work in getting practice in radioactive methods". The following day, January 28th, Bohr replied that he would be in Manchester by March 20th, "to see you before you leave for the vacation about the work you desire me to do in April". So Bohr moved to Manchester and was immediately set to take Geiger's preliminary course in radioactive laboratory work.

Rutherford soon set Bohr to work on a small piece of research which turned out as irrelevant as the task he had been set in Cambridge. Experiment was not Bohr's way; talk and discussion was. In Rutherford's laboratory, although experimentation was the rule, there was also a vast amount of talk and discussion so that even those of a primarily mathematical turn of mind, such as Darwin and Bohr, were mixed with the experimental specialists. Bohr insisted that it was a chance remark by von Hevesy about the impossibility of separating radium-D from lead that gave him, independently, the idea of isotopes and set him thinking about the Rutherford model of the atom in such a way that he formed the idea of nuclear charge being the equivalent of atomic number. And it was a paper by Darwin, on the subject of alpha-particle scattering, which contained errors in the mathematical assumptions, that gave Bohr the starting point from which he could advance. The crucial time was in the first weeks of June 1912. He wrote to his brother on the 12th . . .

> a couple of days ago I had a little idea for understanding the absorption of alpha-particles (the story is this: a young mathematician here, C. G. Darwin (grandson of the right Darwin) has just published a theory about it and I thought that not only was it not quite correct mathematically (a rather small thing however) but also very unsatisfactory in its basic conception, and I have worked out a little theory about it, which even if it is not much in itself, can perhaps shed a little light on some things concerning the structure of atoms . . . In recent years [Rutherford] has been working out a theory of atomic structure which seems to be quite a bit more solidly based than anything we've had before. And not that my work is of the same importance or kind, yet my result does not agree so badly with his . . .

Just a week later he wrote, again to his brother Harold:

It could be that perhaps I have found out a little bit about the structure of atoms. You must not tell anyone anything about it, otherwise I certainly could not write you this soon. If I'm right, it would not be the indication of a possibility (like J. J. Thomson's theory) but perhaps a little piece of reality. It has all grown out of a little piece of information I obtained from the absorption of alpha-particles (the little theory I wrote you about in my last). You understand that I could still be wrong, for it's not completely worked out (I believe it's not, however); nor do I believe that Rutherford thinks it's completely mad; but he is the right kind of man and would never say that he is convinced of something that is not entirely worked out.

It seems that Bohr proceeded from criticism of Darwin's paper for having omitted to consider the effect of the electrons on an incoming alpha-particle to start considering what the effect of the electrons must be – and electron theory was, of course, his own specialised subject. At some time towards the end of June, or possibly in early July 1912, Bohr produced for Rutherford a "memorandum" which contained the essence of the ideas that made Rutherford's atom into the Rutherford–Bohr atom which is virtually the model of the atom that science still uses today.

Bohr's letters to his brother and to his fiancée show clearly that at this time his relationship with Rutherford was that of a fairly timid junior colleague to a loved and respected senior. The letters between the two men over the next two years are apparently purely scientific in content, the slow hammering out of the new and better model of the atom; but they also show a complete change in the personal relationship, as Bohr steadily emerges, through a combination of scientific brilliance and personal doggedness, into an equal partner, a man accepted by Rutherford as a fully independent colleague. The relationship developed into an immensely strong friendship, albeit with a flavour of a "mutual admiration society" as the years passed: years in which the families got to know each other and enjoy each other's company; years in which both men showered on each other all the scientific patronage of lectureships and formal openings of institutes that were at their individual command.

But the letters throw another light on Rutherford, especially the letter of June 12th. The sentence "Rutherford . . . is the right kind of man and would never say that he was convinced of something that was not entirely worked out" speaks loudly of Rutherford's scientific caution, his ability to distinguish very clearly between speculation and well-supported belief. He was by no means averse to speculation – his nuclear atom was clearly put forward as speculation, no more than a useful hypothesis – and he welcomed

Bohr's hypothesis of the new electron structure. But these things were at first no more than speculation, and they had to be tried and proven against every argument that could be put against them; above all they had to stand the quantitative tests of experimental measurement applied to the best mathematical theories that could then be provided. It has often turned out that more sophisticated mathematical treatments of the phenomena have superseded the comparatively simple rules – the theories of dispersion or scattering – that Rutherford used, but his grasp of the comparative size of the quantities involved has kept all his greatest intuitions intact through the years.

The essence of Bohr's proposal for the electron structure of the atom was that the electrons had to be in strictly defined orbits, each of which corresponded to some definite level of energy. The electrons could change orbits but only in definite steps or jumps, each of which corresponded to a change of energy in definite units or quanta. When energy was put into an atom in the shape of a certain number of quanta an electron would jump up into a higher energy orbit; when an electron went down into a lower energy orbit it emitted energy in a certain number of quanta. This explanation, which was produced to get over the difficulty of the lack of stability in the electron structure of Rutherford's original atom, immediately provided a possible solution to all the spectroscopic problems concerning the precise and exact wavelengths at which different atoms absorbed and emitted light and X-rays. It has been one of the major tasks of both physics and chemistry in the seventy years that separate us from Manchester in 1912, to work out all the details of all the possible electron energy changes in all the different permissible orbits in all the different atoms. And during this same period quantum mechanics has taught us that the reality cannot be seen as a small solid satellite orbiting round a large planet. But the essence of the model remains as our working vision of what our universe and ourselves are made of.

At the start Bohr had only seen "a little piece of reality" – the remainder of his life was to be taken up with the expansion and refinement of his vision. Furthermore, in the summer of 1912 he had another preoccupation on his mind – he was getting married. In July 1912 he left Rutherford and Manchester and atoms and electrons for Denmark and matrimony. He was soon back, for he changed his honeymoon plans so that he could call in on Rutherford on his way to Scotland. But the next stage of the development of the Rutherford–Bohr atom was conducted by correspondence between Copenhagen and Manchester in the early spring of 1913. Bohr had spent all the winter toiling at his problem, when not distracted by the problems of his new job as Assistant Professor of Physics. He had started, as his memorandum to Rutherford shows, in his usual way from a criticism, a criticism of the "impossibility" of J. J. Thomson's atom. But Rutherford's atom, too, suffered from the instability of its rings of electrons:

> In the investigation of the configuration of the electrons in the atoms we immediately meet with the difficulty (connected with the mentioned instability) that a ring, if only the strength of the central charge and the number of electrons in the ring are given, can rotate with an infinitely great number of different times of rotation . . . In the further investigation we shall therefore introduce and make use of a hypothesis from which we can determine the quantities in question. The hypothesis is, that there, for any stable ring (any ring occurring in the natural atoms), will be a definite ratio between the kinetic energy of any electron in the ring and the time of rotation. This hypothesis, for which there will be given no attempt of a mechanical foundation (as it seems hopeless) is chosen as the only one which seems to offer the possibility of an explanation of the whole group of experimental results which gather about and seem to confirm conceptions of the mechanism of the radiation as the ones proposed by Planck and Einstein.

Letters to Rutherford in January, and to von Hevesy in February 1913, show that he was still working primarily on the problems of stability and the relationship between electron numbers and observed chemical differences between the elements. It has been argued, and the argument seems strong, that it was the influence of the "atom-building" of Nicholson, who founded much of his speculation on spectroscopic evidence from the sun and stars, which suddenly interested Bohr in spectroscopic evidence. It is certainly true that his great trilogy of papers, the formal announcement of his theories of atomic structure, start with the spectroscopic approach and relegate the problem of stability to later consideration.

On March 6th Bohr sent the first draft of the great paper to Rutherford with a covering letter:

> Enclosed I send the first chapter of my paper on the constitution of the atom. I hope that the next chapters shall follow in a few weeks. In the latest time I have made good progress with my work and hope to have succeeded in extending the considerations used to a number of different phenomena; such as the emission of line spectra, magnetism and possibly an indication of a theory of the constitution of crystalline structures . . . As you will see, the first chapter is mainly dealing with the problem of emission of line spectra, considered from the point of view sketched in my former letter to you. I have tried to show that it

from such a point of view seems possible to give a simple interpretation of the law of the spectrum of hydrogen, and that the calculation affords a close quantitative agreement with experiments . . . I hope that you will find that I have taken a reasonable point of view as to the delicate question of the simultaneous use of the old mechanics and of the new assumptions introduced by Planck's theory of radiation. I am very anxious to know what you may think of it all. As you will see, I have by the consideration of the first chapter, been led to an interpretation, different from that generally assumed, of the origin of some series of lines observed in stars, and also recently by Fowler in a vacuum tube filled with a mixture of hydrogen and helium. Instead of ascribing them to hydrogen, I have tried to give reasons for ascribing them to helium.

He goes on to ask whether an experiment to confirm his view might be done in Manchester, since the equipment was not available in Copenhagen, and he suggests a short visit to Manchester when he has finished his paper.

To this Rutherford replied with speed and some enthusiasm. On March 20th he wrote back:

I have received your paper safely and read it with great interest, but I want to look it over again carefully when I have more leisure. Your ideas as to the mode of origin of the spectrum of hydrogen are very ingenious and seem to work out well; but the mixture of Planck's ideas with the old mechanics make it very difficult to form a physical idea of what is the basis of it. There appears to me one grave difficulty in your hypothesis, which I have no doubt you fully realise, namely how does an electron decide what frequency it is going to vibrate at when it passes from one stationary state to the other? It seems to me that you would have to assume that the electron knows beforehand where it is going to stop.

There is one criticism of minor character which I would make in the arrangement of the paper. I think in your endeavour to be clear you have a tendency to make your papers much too long, and a tendency to repeat your statements in different parts of the paper. I think that your paper really ought to be cut down and I think this could be done without sacrificing anything to clearness. I do not know if you appreciate the fact that long papers have a way of frightening readers who feel they have not time to dip into them.

Rutherford goes on to offer to give detailed study to the paper; offers to send it to the *Philosophical Magazine*; offers to correct the English; offers to do the same for the later papers; welcomes Bohr's proposed visit to England; and offers hope that E. J. Evans, in his laboratory, will carry out the spectroscopic experiment requested by Bohr.

Before this letter could reach Bohr there was another on its way from Copenhagen to Manchester. On March 21st Bohr wrote, "Since I have sent my paper to you I have worked some more on the subject with the result that I have found it necessary to introduce some small alterations and additions. These which I have introduced in the copy enclosed are, however, only of a formal character." Several of these additions and alterations are based on consideration of Nicholson's theories, but some of the later ones refer to Rutherford's views on the origins of alpha- and beta-rays and Bohr goes on:

> My considerations do not give much more than expressing the results of experiments in new words. They suggest however, a possible, very simple way of accounting for a number of facts, and further the most beautiful analogy between the old electrodynamics and the considerations used in my paper. In the very last days I have seen that the theory of emission and absorption sketched may allow a very simple interpretation of Planck's formula of radiation (being without books, as the library is closed I have however not been able to finish this point; if it will turn out right I have thought to introduce a remark about it in the proof). I am now busy working on the next chapters. I hope very soon to be able to come to Manchester and should be exceedingly glad to get the opportunity to hear your opinion about different questions.

The reply from Manchester was scientifically encouraging but editorially firm; dated March 25th it read:

> Dear Dr Bohr, I received this morning the amended manuscript of your paper which I have read again. I think the additions are excellent and appear quite reasonable; the difficulty is, however, that your paper is already rather full and long for a single paper. I really think it desirable that you should abbreviate some of the discussions to bring it within more reasonable compass. As you know, it is the custom in England to put things very shortly and tersely in contrast to the Germanic method, where it appears to be a virtue to be as long-winded as possible. I should consequently be glad to hear what parts you think might be jettisoned or cut down. I think it would not be difficult to reduce the paper by one-third without sacrificing any of the essential points. As I mentioned in my last letter it is very desirable not to publish too long papers, as it frightens off practically all the readers.

Bohr admitted, many years later, that this "brought me into a quite embarrassing situation" and he set off to Manchester as soon as he could.

I therefore felt the only way to straighten matters was to go at once to Manchester and talk it over with Rutherford himself. Although Rutherford was as busy as ever, he showed an almost angelic patience with me, and after discussions through several long evenings during which he declared he had never thought I should prove so obstinate, he consented to leave all the old and new points in the final paper. Surely both style and language were essentially improved by Rutherford's help and advice and I have often had occasion to think how right he was in objecting to the rather complicated presentation and especially to the many repetitions caused by reference to previous literature.

Rutherford was indeed very busy at the time writing one of his more important books, *Radioactive Substances and their Radiations*, as well as doing further experiments on his favourite alpha-particles. The difficulties he saw in Bohr's new ideas, the problems of mixing classical and quantum treatments, the absence of a clear physical model of what was going on, and the query as to how the electron would know what its future state was to be, were all accepted by Bohr as "very pertinent, touching on a point which was to become a central issue in the subsequent prolonged discussions".

Bohr's position was that a revolution, an abandonment of the paradigm, was absolutely necessary – whatever the outcome.

My own views at the time . . . were that just the radical departure from the accustomed demands on physical explanation involved in the quantum postulate should of itself leave sufficient scope for the possibility of achieving in due course the incorporation of the new assumptions in a logically consistent scheme. In connection with Rutherford's remark, it is of special interest to recall that Einstein, in his famous paper of 1917 on the derivation of Planck's formula for temperature radiation, took the same starting point as regards the origin of spectra, and pointed to the analogy between the statistical laws governing the occurrence of spontaneous radiation processes and the fundamental law of radioactive decay formulated by Rutherford and Soddy already in 1903.

Despite his concrete and quantitative turn of mind, Rutherford came easily to accept that precise physical models could not be formed of the new quantum theories – a remarkable flexibility of thinking that enabled him to keep in the forefront of physics in the years ahead when most men of his age would have been fighting for a more conservative view. His question about the electron knowing what it was going to do had far greater depth and perspicacity than is usually allowed him. It was ignored or forgotten for many years, but now in the second half of the century and in a different guise it has reappeared as one of the great and still unsolved problems of the new physics which takes quantum mechanics as its foundation. But in the early summer of 1913 Bohr's theory was no more than an exciting speculation, and one of little interest to the vast majority of physicists outside Manchester and Berlin. The long evenings of discussion and Bohr's "obstinacy" had, however, a personal effect. Bohr came out of the state of pupillage, and Rutherford came to accept him as an equal, though he still felt protective. Rutherford does not yet seem to have given final acceptance to the new theory; he was certainly not willing to pin any scientific colours to this particular mast – yet. But he plainly felt considerable commitment to Bohr and his ideas.

In June Bohr came back to Manchester for yet more help with the second of his trilogy of papers. This contained more emphasis on the relationship of nuclear charge to atomic number, with the electron structure corresponding to the numbers of the elements. By going into the question of the numbers of electrons in different orbits and by introducing the idea of electron "shells" in a manner earlier tried by J.J., Bohr started the explanation of why there were groups of similarly behaved elements in the Periodic Table. At the same time Evans in the Manchester laboratory was working on the experiments with the spectra of hydrogen and helium which eventually turned out to support Bohr's ideas.

Again in September Bohr was back in England, seizing the opportunity to see Rutherford at the meeting of the British Association, which was held in Birmingham in 1913. There was a very high-powered discussion on radiation with Rayleigh, Larmor, Madame Curie and Lorentz participating "and especially Jeans, who gave an introductory survey of the application of quantum theory to the problem of atomic constitution. His lucid exposition was, in fact, the first public expression of serious interest in considerations which outside the Manchester group were generally received with much scepticism." It was at this meeting that Larmor solemnly and publicly requested the great Lord Rayleigh to give his views on the new developments only to receive the reply, "In my young days I took many views very strongly, and among them that a man who has passed his sixtieth year ought not to express himself about modern ideas. Although I must confess that today I do not take this view quite so strongly, I keep it strongly enough not to take part in this discussion." Rutherford cherished this piece of wisdom.

Back in Copenhagen after the Birmingham meeting Bohr wrote to Rutherford and privately attacked the Thomson atom model which was still in the field. There were similarities in the results to be deduced from the two rival models, but "this agreement has no foundation in the special atom model used by Thomson, but will follow from any theory which considers electrons and nuclei and makes use of Planck's relation". Furthermore many of Thomson's other values came out wrong by four times or twice too small, and the model suffered from its "mechanical system". Following publication of Bohr's three papers in the *Philosophical Magazine* during the winter months of 1913, his model began to hold sway. More than that it began to interest a wider field of physicists. When the German, Stark, reported a doubling of the spectral lines when a strong electrical field was also present, Rutherford rapidly wrote to Bohr and suggested that he should write something about this and the Zeeman effect which had long been known to produce a similar doubling with a magnetic field (the two effects turned out to be essentially very different). The correspondence between Rutherford and Bohr in the spring of 1914 mainly concerned these new effects, and in the March issue of the *Philosophical Magazine* they both had papers on the constitution of atoms. Rutherford's gave his broad support to Bohr's theory but with the rather grudging "there may be much difference of opinion as to the validity of the underlying physical meaning of the assumptions made by Bohr".

But at exactly this time the evidence of Moseley and the realisation of the meaning of isotopes and the radioactive displacement law were bolstering up the Rutherford–Bohr atom and bringing more scientists to accept it – at least as the working hypothesis which brought order into both physics and chemistry. Bohr's version of the outer structure of the atom made it clear that radioactive decay and the emission of alpha-particles must be an entirely nuclear phenomenon. Hence the nucleus must have structure and contents too. As Feather put it:

> Once Rutherford accepted the cogency of this [Bohr's] conclusion he can have had no more doubts. For nearly twenty years he had devoted the major part of his energies to the problems of radioactivity. Almost incidentally, in the process, he had discovered the nucleus. Now it was clear that the property of the spontaneous emission of particles was a property of the nucleus itself. It was indeed clear that his latest discovery had opened up a new realm of nature, and he accepted the challenge of its exploration without demur.

The importance of the work of E. J. Evans in checking by spectroscopic methods the theories of Bohr, has, however, been much underrated, for its impact on the thinking of scientists at the time was very considerable. The situation was the classical one of the verification of a hypothesis – Bohr's theory made a prediction that certain spectral lines were emitted by helium, and Evans' measurements showed that this was so. Immediately after the Birmingham meeting of the British Association, von Hevesy, who was later to work with Bohr for many years in Copenhagen, went on to a scientific congress in Vienna, where he met Einstein. He reported to Rutherford that there was more knowledge in Vienna, but far more ingenuity in Birmingham, and he went on:

> Speaking with Einstein . . . we came to speak on Bohr's theory, he told me that he had once similar ideas but he did not dare publish them – "Should Bohr's theory be right so it is from the greatest importance". When I told him about the Fowler spectrum [i.e. the result of Evans' experiment], the big eyes of Einstein looked even bigger, and he told, "Then it is one of the greatest discoveries". I felt very happy hearing Einstein saying so.

The welcome for Bohr's ideas had apparently not been so great at Birmingham, for in the same letter (October 14th, 1913) von Hevesy told Rutherford of some sharp words that had been exchanged: "J. J. Thomson has, in spite of his greatness, in some ways a distinctly demagogueous behaviour; he tries to be witty, ingenious and popular and does not trouble at all about being very fair. The general appearance was that he told something highly ingenious and Bohr something very stupid – just the contrary was the case. So I felt bound to stick up for Bohr . . ." One can hear a faint murmur here of the typical Cambridge arrogance in conflict with the slow and ponderous English of Bohr and the excitable Hungarian English of von Hevesy.

In a much more domestic sense the spectroscopic confirmation of Bohr's theory was important too. A. B. Wood, then a very young research student in Manchester, remembered the effect: "This brought a new lease of life to the optics division of Rutherford's laboratory, which included the remnants of Sir Arthur Schuster's laboratory in the person of E. J. Evans. Sure enough Evans found Bohr's predicted new lines in the hydrogen and helium spectra, and our opinion of the somewhat eclipsed optics research laboratory was much enhanced."

Rutherford himself, as we have seen, was not willing to give immediate full-blooded support to Bohr's theories. His lectures to the Royal Institution, a series of six given in the summer of 1913, were full of the problems with which Bohr dealt. The expulsion of beta-rays from radioactive materials was "an exceedingly complicated phenomenon" which nevertheless must eventually "throw light on the modes of constitution of vibration of the interior structure of the atom"; gamma-rays and X-rays probably come from "vibrations of rings of electrons", and it is significant that as early as May 31st, 1913, in the second of these lectures, Rutherford is either groping himself towards a quantum explanation, or has already been influenced by Bohr's thinking. For instance, he said, "The beta-ray in escaping from the atom sets the rings in vibration and the energy absorbed is a definite multiple of a unit of energy which is constant for each ring. One or more of these units of energy may be abstracted from the beta particle in passing through each of the rings of electrons."

His considered view of the achievement of his Danish colleague was given ten years later:

> A radical departure from accepted views seemed essential if progress were to be made. The departure made by Bohr consisted in a novel application of the ideas underlying the quantum theory of radiation of Planck which had already shown its power of successful explanation in several departments of physics. In Planck's view, radiation is emitted only in definite quanta . . . We must bear in mind that in the initial stages of the Bohr theory, the quantum theory was applied to the atom in such a way as to be concordant with the known series relations of the spectrum of hydrogen and other elements . . .
>
> The electron in the hydrogen atom, for example, is supposed to be capable of moving for more or less indefinite periods in any one of a series of stationary states or non-radiating orbits in which, except for the absence of radiation, the classical laws of dynamics are obeyed. Under suitable stimulus, however, the electron can change from one stationary state to another, with the emission or absorption of monochromatic radiation of which the energy and frequency are connected by the quantum relation. This theory of the origin of the spectrum of the hydrogen atom appeared at the time to be wildly revolutionary and aroused much opposition, for no physical mechanism seemed capable of this behaviour. It was not realised then, as it is realised today, that we cannot and have no right to expect to construct a purely mechanical model of the atom.
>
> But the test of any theory is its power to suggest new relations, and in this respect Bohr's theory was triumphant from the first.

He went on to explain how Bohr's hypothesis was shown to provide a perfect explanation of the spectra of hydrogen and helium. It is eminently obvious that Rutherford made a clear distinction between what could be admitted officially and publicly in the formal literature of science, and what he was prepared to accept privately at the same time.

While in the March issue of the *Philosophical Magazine*, as we have seen, he was still cautious about the lack of physical explanation of Bohr's work, a month earlier he had received a suggestion from Schuster, who was then Secretary of the Royal Society, that there should be a Royal Society discussion meeting on the constitution of atoms. On February 2nd, 1914, Rutherford replied:

> I agree with you in thinking it is an excellent subject of discussion on which our knowledge is moving very rapidly. I think it is a question where there is not much room for dogmatism and consequently room for interchange of opinion among all types of people interested in this sort of topic. I should be very glad to introduce the discussion, but of course you know that I have promulgated views on which J.J. is, or pretends to be, sceptical. At the same time I think that if he had not put forward a theoretical atom himself, he would have come round long ago, for the evidence is very strongly against him. If he has a proper scientific spirit I do not see why he should hold aloof and the mere fact that he was in opposition would liven up the meeting. I think I told you that I went for him on this problem with a good deal of momentum at the Brussels Committee and he really took it very well. As a matter of fact I believe he knows in his heart that his own atom is not worth a damn and will not do the things it has got to do.

Rutherford obviously summoned up the forces supporting "his" and Bohr's atom quite carefully, for on March 4th, in a letter containing important details about the atomic numbers of various elements he had just measured, Moseley remarks, "I imagine I have to thank you for the invitation to discuss atom structure at the Royal Society . . ."

J.J. did, bravely, turn up for the meeting on March 14th – and that is virtually all we know about this momentous discussion, for the Royal Society records give no details of what was said. The meeting is minuted, J.J. is reported to have been present, and that is all. Nevertheless the meeting can be seen as a watershed. No other atom model, except modifications and the clarification of this one, has ever commanded any serious scientific support, and the Rutherford–Bohr atom is still our view of how the universe is built.

12

Rutherford at War

Rutherford was knighted in 1914. J.J. had been knighted only three years before, but Rutherford was still under forty-five years of age and the significance of the honour was soon to emerge when the First World War broke out and Sir Ernest Rutherford was a much more powerful figure than plain Professor Rutherford, for England was then, even more than it is now, a "deferential society". One of the most important changes wrought by the 1914–18 war was that its pressure forced the foundation of "government science" – for the first time official committees and research councils became part of the average scientist's world. It was during the war that the first partnership between government and science was formed, however imperfect and unsatisfactory that partnership may have been. No man was more important in forming the partnership than Sir Ernest Rutherford; and the men who worked with him were virtually all his friends and members of his immediate circle: J.J., the Braggs, father and son, Eve and Langevin. Out of their experience came the improvements of the inter-war period when science came to be more readily acknowledged as a national resource; and through their vast circle of contacts the ideas of government science were spread to the USA, Canada, Australia and the main European countries.

To Rutherford himself all these developments would appear as battles won or half-won. They were battles fought, not for himself, but for science, or at least for a growing vision of what science ought to be able to do for the community. We have to accept that this altruistic attitude was indeed his true motivation, for never did he draw any personal benefit from his activities in the field of "public science", never could he be caught manoeuvring for personal aggrandisement, and it is manifestly clear that he regarded all these matters as peripheral to his own career which lay in academic science. Furthermore, his own personal delight continued to be in his own work at the laboratory bench.

His acceptance of his knighthood was entirely in this vein. As he repeatedly told his former pupils and colleagues when they wrote to congratulate him, "I am very glad to feel that my old students are pleased with this recognition of my labours in the past. I think you know that I do not lay much stress on such forms of decoration for they have obvious disadvantages in the case of a scientific man like myself. However I am getting quite used to the change of title and trust it will not interfere with my scientific activities . . ." That was his reply to Otto Hahn; to von Hevesy he added:

> It is of course very satisfactory to have one's work recognised by the powers that be, but the form of recognition is a little embarrassing for a relatively youthful and impecunious Professor like myself . . . Eileen was very pleased at the news and was greatly excited on New Year's Day with the succession of telegrams. She is of the opinion that neither of her parents has the "swank" and natural dignity for such decorations . . .

It is somewhat unfair to his attitude to himself and his honours to quote from his letters to Hahn and von Hevesy as I have done, for his words about this subject are no more than a few remarks in long epistles devoted to scientific matters. He was principally concerned to describe to both these men his own work (done with Andrade) on gamma-rays for, as he told von Hevesy, "I have been hard at work all the Christmas vacation determining the wavelengths of gamma-rays, and have made such good progress that I see my way pretty well clear to its conclusion." He was also most relieved by a recent result of Soddy's which confirmed the results of a man called Antonov on the transformation products or radium-D. Antonov had produced this work in 1909 and 1910 in Rutherford's first years at Manchester, and the work had been criticised by others who had not been able to get the same results. "I am very glad that things have turned out all right for I should have been very sorry to feel that the laboratory had one piece of bad work to its discredit. I think Antonov has scored heavily in the matter, for he never wavered in his certainty that he was right," he confided to von Hevesy.

Rutherford's attitude to his knighthood was but one facet of his indifference to the standard attitudes of middle-class British life. In particular he did not share the simple jingoistic nationalism which prevailed so strongly at the start of the First World War – so much so that he has been described as "totally lacking in public spirit": he did not regard himself, or other scientists, as being subject to the normal pressures of their society. To those of us who have lived through the "total" war of 1939–1945, in which the whole national community was involved, the private unconcern of Rutherford, his obvious feeling that science and scientists should be truly international, may come as strange, though admirable. This is not to say that he did not firmly believe in, and work for, victory for his own side. But throughout he kept in touch with his Austrian friends in the Vienna Academy, and he followed the fortunes of Geiger with as much interest and concern as he followed Russell, Andrade, Florance and many other of his former students. News of his German friends came through neutral countries, often through American colleagues, or sometimes directly from Chadwick, who was caught in Berlin by the outbreak of war, interned there, and helped to while away the years by setting up an experimental laboratory in the prison camp through the assistance of Geiger and other German scientists.

In early 1915 Rutherford reported to Marsden in New Zealand that the lights in Manchester were darkened "but apart from this and the occasional sight of troops marching about, it is difficult to believe that war is going on".

Rutherford had in fact only just returned to England from Australia when he wrote these words. In May of 1914 he had been invited to Washington to give the Hale lectures, and he had had to obtain leave of absence from Manchester University to fulfil this function. The contents of these lectures were important scientifically and will be dealt with at greater length in the next chapter. Politically this visit to Washington was important in that it enabled him to strengthen his American connection, confirming old friendships and making new contacts which were to prove valuable in the war years to come. There was at that time no feeling of war in the air. In July he wrote to Stefan Meyer in his usual scientific style and slipped in a phrase of condolence over the recent "murder of your Archduke". He was more interested in the arrangements for the meeting of the British Association in Australia in September 1914, and he had been in correspondence with his old professional foe, the chemist Henry E. Armstrong, over the problems of choosing which ship to travel in. Rutherford confided, "Frankly I want to avoid Dixon," and he admitted that there had been a "hitch" in booking cabins for one particular ship "as not enough had declared for her". This again was to have long-term results, for one of the spare places on Rutherford's ship was given to a young chemist, Henry Tizard, who thus met for the first time the man with whom he was to work so closely and successfully in the arena of government science.

In Australia Rutherford gave a number of public lectures as well as taking part in the meetings of the British Association, and a lecture in the Lyceum Hall at Sydney (used because the town hall was engaged for the Lord Mayor's reception and ball) was a noteworthy occasion because it resulted in one of the very few public criticisms of him that were ever recorded. The fearless reporter of the *Melbourne Age* printed on August 25th, "Though a great scientist Lord Rutherford is hardly an ideal lecturer, at any rate to a popular audience. He is fond of using specialised terms that convey nothing to the majority of his hearers, while he frequently drops his voice as though soliloquising in front of the screen. Nevertheless he told a marvellous tale of the achievements of science investigating the extraordinarily minute." Plainly Rutherford had succeeded in confusing this reporter for, later on, he wrote, "The so-called transmutation of radium into helium, with some general views on the question of chemical elements and the future possibilities of changing elements from one form into another was referred to." The *Sydney Morning Herald* gave a much shorter and more polite report. After referring to the very large audience and interlarding the whole with several exclamations of "(Applause)", it merely recorded that Rutherford felt that "the atomic theory had now attained a very high degree of probability through recently made measurements and one must admit a granular structure in matter".

It seems likely that what confused the Melbourne reporter so badly was a passage on isotopes, referring particularly to lead. Rutherford was by now convinced that lead was the final product of the three great radioactive families, although the case was not yet proved. The concept of isotopes provided the solution to the otherwise puzzling results that had been obtained, and Rutherford declared, "There may be two pieces of lead which look exactly the same and yet their physical qualities may be quite different. That may not be believed now, but it will be later." He has, of course, been proved correct and there are now known to be at least four radioactive isotopes of lead, which can also be called, radium-B, radium-D, thorium-B, and so on, and at least three stable isotopes of lead which can also be regarded as radium-G, actinium-D, and thorium-D.

Immediately after the British Association meeting many of the English visitors returned home from Australia because of the outbreak of the war. Typically Moseley was among them and he volunteered for the army as soon as he was back. Rutherford, showing his lack of public spirit, went on to his home country, visited his parents and relations in the North Island of New Zealand, returned to Christchurch to receive a civic welcome and give a public lecture on "The Evolution of the Elements", and then went on across the Pacific, travelling slowly, visiting his Canadian friends en route and returning to England only in January 1915 despite the worries about submarines haunting the transatlantic shipping lanes.

Back in Manchester his concerns were only with university and scientific matters. He confided to Schuster that various of his laboratory people were already in the army but "we shall however with a little rearrangement be able to carry on the work temporarily all right". Most of his letter was concerned with getting a £100 grant from the Royal Society in order to provide a supply of radium to Marsden, who had just taken up an appointment as Professor of Physics at Victoria College in Wellington, New Zealand. Indeed even in May of 1915, Rutherford seems to have been chiefly interested in organising the annual meeting of the British Association which was due, that year, to be held in Manchester. He wrote again to Schuster about the problems of inviting foreigners to this meeting:

> I daresay a few might come over from France and Russia, but I am a little doubtful of neutral countries, for there are special reasons why they might not feel inclined to take obvious sides. However, they can only refuse. There is to be a discussion on Friday about this matter amongst some of us. Whom would you like to suggest to invite? . . . If we write at all it will be worth while seeing that physics is not left out in the cold.

Those of his own staff who had joined up were almost apologetic. D. H. Florance actually excused himself and added, "I felt sure you would have no objection to my offering my services to the common cause. I was anxious to carry out the work you left me to do and so I delayed as long as possible, but I felt the call for the strong and vigorous was so urgent that I could no longer hold out and consequently applied for a commission." He joined the Royal Field Artillery but in January 1915 he was stationed in Ireland, "still waiting patiently for guns, horses and other necessaries", which can hardly have helped Rutherford to excuse his departure. However, Florance's letters may have turned Rutherford's mind towards another aspect of the relationship between scientists and the war, namely the possibility that they might have a special role to play. His young protégé wrote, "We need scientific men in the artillery, and some of the officers who do not happen to know that the exterior angle of a triangle equals the sum of the two interior and opposite angles find themselves tied up."

In July Eve wrote to Rutherford suggesting that Moseley would be more valuable "if he were set to solve some scientific problem" rather than acting as a Brigade Signals Officer, although the problem was that Moseley "would naturally be very much incensed" if he was simply brought home for his own security. There was an immediate reaction to this and Rutherford wrote to Sir R. Glazebrook to see whether suitable scientific work could be found, but in August Moseley was killed. Florance, who had written, "In the old days you would never have thought you had so many desperate swashbucklers in your lab. I think our lab compares favourably with any of the other departments of the university for turning out hefty men to do men's work," and whose letters were always full of simple, cheerful, patriotic phrases, was as shocked as any at Moseley's death. "I thought at the time he went to the Dardanelles that it was a colossal mistake as a man doesn't always get a sporting chance in that death trap." And Rutherford wrote his tribute to the brightest of his stars in *Nature* on September 9th, 1915. But in addition to his scientific tribute he voiced public criticism:

> Scientific men of this country have viewed with mingled feelings of pride and apprehension the enlistment in the new armies of so many of our promising young men of science – with pride for their ready and ungrudging response to their country's call, and with apprehension of irreparable losses to science . . . It is a national tragedy that our military organisation at the start was so inelastic as to be unable, with a few exceptions, to utilise the offers of services of our scientific men except as combatants in the front line. Our regret for the untimely death of Moseley is all the more poignant because we recognise that his services would have been far more useful to his country in one of the numerous fields of scientific enquiry rendered necessary by the war than by the exposure to the chances of a Turkish bullet.

This was brave stuff in a society where young women could give white feathers to non-uniformed men in the streets and where there could be a public outcry against men such as Schuster because of their Germanic names.

Rutherford would have nothing to do with such simple herd behaviour. He had received letters from Geiger, already wounded once in the German artillery, from Meyer and Hess in Vienna, where they were working still "as in the deepest peacetime" and actually still had an Englishman and Godlewski, the Pole from Rutherford's Canadian school, working beside them in the laboratory. From his American friends – perhaps the neutrals who would not want to be seen taking sides – he heard different stories. W. E. Hale admitted that he was "privately" very anti-German, and Kovarik, born in the Austro-Hungarian empire, but nationalistically Czech in his sympathies, was pro-Allied. Rutherford's relationship with Boltwood cooled considerably during the war, as

Boltwood continued to argue the German point of view, while Lorentz, from Holland, reported that the German physicists felt that they were indeed "defending" their country.

The apotheosis of Rutherford's disinterested attitude to the war came right at the end in what is probably the most famous of all Rutherford stories. It is related – and the story seems to come from American sources such as Dr Karl T. Compton and Bumstead – that Rutherford either missed, or was late for, a joint Allied meeting on anti-submarine warfare. He sent a message of apology stating that he would be delayed by the necessity of completing certain laboratory experiments in which he thought he had succeeded in splitting the atomic nucleus, and ending with, "If this were true, its ultimate importance is far greater than that of the war." And indeed towards the end of the war he did largely withdraw from his anti-submarine activities and concentrate on his own work. But in between he had contributed a great deal to the war effort.

The Board of Inventions and Research, BIR, was an ill-fated, short-lived body, brought into existence for mixed and irreconcilable objectives. Yet it was an important initiative, the first conscious attempt to harness the power of scientific investigation to the needs of a nation at war, it planted the seed from which many more important organisations sprang. It must be said, however, that the short history of the BIR is one of the most disgraceful episodes in the history of the Royal Navy, and fully justified the charges made by modern American naval historians of the technological backwardness of the navy, caused largely by the social snobbery which so seriously afflicted the senior service in the years leading up to, and including, the First World War.

BIR was the creation of A. J. Balfour, who became First Lord of the Admiralty in May 1915, when Winston Churchill was dismissed because of the failure of the Dardanelles expedition. But shortly before Churchill's departure the navy had also lost Lord Fisher, the outstanding officer of his generation, the man who had introduced the Dreadnought, and who had sought to modernise both the tradition and the equipment of the service. Fisher had resigned during an argument with Churchill, but at least half the navy rejoiced at his going, for the service was split between Fisherites and anti-Fisherites in a bitter internecine war involving many issues and personalities, including Fisher's conflict with Admiral Beresford. Giving Fisher the job of running the BIR would give him a task for which both his enormous energies and his technological bias were suited, and would bring back into the navy its most outstanding officer. Unfortunately his appointment aroused all the old enmities and BIR, to many a naval officer, and especially to the powerful in the Admiralty, was translated as "Board of Intrigue and Revenge".

Scientists had been pressing their potential usefulness in the war effort upon a government which contained very few men who understood what science was about. J.J. had led this pressure from his position as President of the Royal Society, and, as we have seen, Balfour had close links with the Cavendish. It was therefore largely to Cambridge that Balfour turned when he sent an invitation through a letter to Sir Joseph Larmor for scientists to help the BIR. The response was immediate, and Fisher, as President of BIR, was given a central committee of J.J., Sir George Beilby (Soddy's father-in-law), and Sir Charles Parsons. By the end of the summer they had acquired a panel of scientific experts which included not only Rutherford, but his friend Bragg, senior scientists such as Sir William Crookes and Sir Oliver Lodge, and the ill-fated founder of aeronautical research, Bertram Hopkinson.

Before Fisher was appointed to head BIR, the Admiralty had welcomed the idea of such a body of scientists. The Admiralty's interest in it was as a sieve through which they could pass the thousands of suggestions for war-winning devices which had poured in from citizens who were convinced – as H. G. Wells was convinced – that the Germans had coopted scientific invention into their war machine far more successfully than Britain. The Admiralty wanted the scientists to examine these suggestions, filter out any that might be of any use, and pass these on to the service, thus at the very least avoiding the charge that the war was being lost through failure to take advantage of British inventiveness. This job of sieving any gold that might be present in the flow of inventions was indeed taken on by the scientific panel of BIR. J.J. estimated that they had considered 100,000 ideas by the end of the war. Only thirty were of any use at all, in his considered opinion, but the political effect of satisfying the public was "very considerable". Some of the ideas they received were of a gorgeous dottiness: train cormorants to peck at mortar between bricks by putting food on it, then release numbers of trained cormorants in the Ruhr to peck down the chimneys of the Krupp factories; train seagulls to expect food on periscope-like objects, and then the gulls would spot any German submarine periscope and swoop on it, betraying its presence to escort vessels; train sea-lions to hunt underwater noises as a food source and they would lead ships to submarines, a proposal that led to much work because the circus sea-lions first employed showed that they could indeed hear underwater and would go long distances to an underwater dinner-bell, though unfortunately they were easily distracted and travelled only three or four miles an hour. High flying balloons carrying long cables smeared with bird-lime to entangle Zeppelins was a further suggestion which had sad echoes in the Second World War when the proposal for stopping German bombers by a barrage of parachuting aerial mines on long cables brought Tizard and Lindemann into a political conflict which harmed them both.

The scientists did, indeed, sift through this strange medley of ideas, but they were not content to leave it at that; it was their nature, as scientists, to consider the problems of the navy in a broader context, and they were urged on to this more active role by Lord Fisher. At the first meeting of the board, on July 29th, 1915, he declared that their objective was not only to "sift" but to "promote" invention. Far from being "facile dupes" or "servile copyists" the board would initiate, with imagination, with "big conceptions and quick decisions" and "with Nelsonic bravado". The scientists seem by instinct and professional practice to have steered a course between the Admiralty's intentions and Fisher's desires; they set up committees and panels on the various subjects where it seemed to them that the navy might benefit from scientific help – committees on submarines, aeronautics, naval construction, marine engineering, oil fuel, anti-aircraft, noxious gases and ammunition.

Rutherford was put on to the anti-submarine committee. He flung himself into the work with another of those bursts of almost superhuman energy which astounded all his contemporaries. He found himself in a scientific vacuum, an area where no work had been done, where there were almost no established quantities, yet within three months he had outlined the chief features of this unknown ground and correctly identified where the main immediate "attack" must be. His work in this field is little known, and was rarely mentioned by his contemporaries and memorialists, yet it is probably the clearest single example of his greatest genius, that feel for physical realities, that instinctive grasp of the comparative quantities, in an area of intellectual adventure which few had entered previously. His work here also demonstrates another quality of mind with which he is very rarely credited, a clear vision of the state of technological advance in the instruments which will be necessary for conquering the new area. Another feature of his irruption into this new field, a feature which was appreciated by those closest to him in the work, such as A. B. Wood, was the extraordinary mental flexibility which allowed him, as he approached the age of fifty, to switch his subject from radioactivity and atomic structure, where he had worked for more than twenty

347

years, into the totally different specialism of underwater acoustics.

In his first three months on anti-submarine work Rutherford produced at least three "Secret" reports which the secretary of the BIR sent immediately to the Admiralty. It was in these three reports that he drew up the "map" of underwater warfare which has remained unchanged to the present day. What Rutherford said about submarine-hunting in 1915 remains true in the 1980s, and it is only within the last decade that technology has come anywhere near providing devices which could satisfy some of his longer-range requirements. Rutherford started by considering the basic ways in which a submarine could be detected under water according to the known laws of physics. He found four characteristics of the submarine which could be "spotted": the sounds it makes, the heat it gives off, the electromagnetic disturbance caused by the movement of a large steel body in the earth's magnetic field, and its ordinary visual characteristics as seen by light. The possibility of spotting a submarine by the use of underwater laser beams is just at the edge of technology now, though as far as is publicly known there is no current laser system which will see a submarine at a militarily useful range. Detecting submarines by their thermal emissions using infra-red detectors, possibly mounted on satellites, is probably even further beyond the limits of our present technology as long as the submarine is underwater. There is serious consideration now being given to the detection of submarines by magnetic detectors known as "squids" mounted on satellites which use the properties of superconducting electric currents in metals kept at temperatures near absolute zero. But no one has yet thought of a method of detecting submarines other than those covered in Rutherford's papers. His advice that the only practical way of detecting submarines by any technology in use in 1915 was by sound remains equally valid in the 1980s. Acoustic detection was the way ahead, said Rutherford, and the Royal Navy was forced to accept his advice as soon as a few experiments by other enthusiasts in other fields were reported by BIR to have produced no practical results.

It may seem to us nowadays that this first "overview" of submarine detection was terribly simple, that anyone could have done it. The point at issue is that no one else had ever done it before, no one had applied the mind of a physicist to the problem, and no one else has yet managed to enlarge on his view, though many thousands have laboured to fill in the details of the map he drew of this territory.

In September 1915 Rutherford went to see the only work that was then being done on the detection of submarines by sound. At

Granton, on the Firth of Forth, a lone figure, Commander C. P. Ryan, was working to develop hydrophones, which can be regarded, simply, as underwater microphones. His small group was termed the Hawkcraig Experimental Establishment, but it was run on totally non-scientific lines – Ryan did not for instance carry out experimental measurements to compare one device or system with another, but was interested only in trying to make systems that would work.

It is indeed extraordinary, and a condemnation of the Royal Navy's attitude towards investigations of almost any sort, that virtually no work had been done towards combating the submarine menace, although underwater vessels had been in service since the beginning of the century. What little underwater work had been done was directed towards the problem of signalling to and from submarines, and this again had been done mostly by "freelance" officers who had developed an interest in some aspect of the subject, and it had been done on an unscientific basis. There had also been some work by commercial companies in both Britain and America on underwater signalling, using bells and steam-driven sirens and similar noise-producing devices. The navigation authorities, Trinity House in Britain and the Lighthouse Board in the USA, had made occasional incursions into the same territory, seeing the possibility of using underwater sound as a warning device when visibility was poor. The hydrophone was developed essentially by extending the ideas of Bell's telephone, and had been tried by the Royal Navy as a navigation device just at the turn of the century, and then dropped. Underwater signalling and navigation aids were reconsidered some ten years later, when a committee was set up consisting of the captains of the Navigation School, the Torpedo School and Submarines with the Superintendent of the Signals School. Some forty ships and submarines were fitted with receivers of various sorts to take underwater signals, and it was believed, according to intelligence reports of 1912, that German submarines could signal to each other underwater.

The American Submarine Signalling Company was probably the leading authority and producer of equipment at this time, and interest in the subject of underwater detection was further aroused by the sinking of the *Titanic* in 1912, which led a number of scientists to look into the problem. Professor Fessenden made a striking advance for the American company when he produced his oscillator in 1913 – this consisted of a large steel plate, called the diaphragm, which was vibrated rapidly (540 times a second) by an electric induction motor. The diaphragm could transmit its vibrations through water with very considerable efficiency, and could be made to produce the dots and dashes of the Morse code. Furthermore, it was the first example of what we now call a transducer, in that the diaphragm could receive as well as transmit such signals, though it was rarely used in this way. Nevertheless Fessenden detected an iceberg at a range of two miles in April 1914, and demonstrated that his apparatus was a practical signalling device as well. Following reports from the British Consul at Boston, who had witnessed successful trials by Fessenden in June 1914, the Admiralty bought one of the sets, and was satisfied with its performance in October 1914. But the subsequent plan to fit the device to large numbers of ships was shelved by the pressures of war and it was eventually fitted only in a small number of submarines.

The medium of sea water was little investigated since the first accurate measurement of the speed of sound in water by J. D. Colladon and J. C. F. Sturm in Lake Geneva in 1826, showing that sound travelled at 1,435 metres a second in this medium, that is more than four times the speed of sound in air. A few further measurements showing that the speed varied slightly with variations in temperature and salinity had entered scientific literature in the remaining years of the nineteenth century. So some of the BIR's first actions were to consult Lord Rayleigh about the propagation of sound waves in water, to set Professor Sir Horace Lamb to provide a theoretical mathematical treatment of the same subject, and, on Rutherford's part, to set up a number of water tanks in his Manchester laboratory in which he set about measuring various basic quantities, such as the natural resonating frequencies of metal diaphragms in water. Later he was to have most of the basement of the laboratory turned into a very large concrete water tank in which more and larger measurements could be made.

Rutherford's visit to Captain Ryan's establishment at Hawkcraig was made on September 15th and 16th, 1915. He was accompanied by Sir Richard Paget, Sir Richard Threlfall, and the Duke of Buccleuch. Unfortunately the Duke could not get there on the first day, and all had to be repeated again on the second day when he did manage to turn up. Nevertheless it may have been as well that such a socially superior person as the Duke was able to join the party, for the Royal Navy was much infected with snobbery, as will be seen later.

Rutherford wrote the official report, giving hearty approval to Ryan's work and pressed that it should be supported. He pointed out that no submarines had been made available for the scientific visitors to listen to, as "all were on active service". He urged the "importance that a ship should be placed immediately at Commander Ryan's disposal in order to carry out practical experiments" on listening to submarines and other ships, and he complained that Ryan was "only allowed to spend £1 on any one experiment". Instead of this niggardliness "Commander Ryan should be given all necessary financial and scientific assistance" and, if and when he required it, "the services of two or three trained physicists should be placed at his disposal". Ryan was developing two different systems, Rutherford reported. One was a row of hydrophones tethered near the shore, each connected by telephone wires to a listening post. With this detection system listeners should be able to hear a submarine approaching, say, a harbour, and could then detonate mines by electrical signals when the sub was within range. This idea was eventually developed into a fairly formidable coastwise sound-ranging system which performed some useful work. But Ryan was also working on a more important "drifter" hydrophone set, which could be dropped over the side of a ship to detect the sound of any nearby submarine. The problems of this system were to occupy much of Rutherford's time in the next few months, and were to keep a number of people busy until the end of the war. It was also noted that Ryan had started experiments with "Broca tubes", invented in France, and other sound-collecting devices which he had invented himself. Such devices were also being tested by Rutherford in his lab.

A much longer "secret" report by Rutherford was produced two weeks later. Dated September 30th, it was entitled "On methods of collection of sound from water and the determination of the direction of sound" and it formed, according to Sir Richard Paget, the basis of a "scheme of acoustic research" which the BIR proposed formally in October, and which it carried out over the remaining years of the war. Plainly Rutherford intended such a scheme to be founded on his report, for the whole thing is written in the plainest of plain English, obviously intended to convince non-scientific readers. It begins: "The problem of collection of sound from water has received very inadequate attention in scientific journals," but he admits it has been the subject of practical work and adds, "In addition many important experiments have been and are being made in the navy, not only on the general problem of submarine signalling, but on the more difficult subject of detection of ships at a distance by their natural sounds communicated through water." He then explains the basic science – since the speed of sound in water is 4.3 times greater than the speed of sound in air, so a single sound wave is 4.3 times longer in water. But the amplitude of the motion of the wave in water is only one-sixtieth of the amplitude in air, since water is more dense, and therefore for equal energy at a given frequency the pressure of the sound wave in water is sixty times greater than it is in air. Hence thick steel diaphragms can easily be vibrated in water and sound waves coming from outside will be transmitted through the steel sides of a ship to any listening device inside the hull.

There are, therefore, three avenues of research – firstly the generating of sound waves of great intensity in water and the methods of detecting them at considerable distances; secondly, the detection of ships in motion by the natural sounds of their machinery communicated through the water; and thirdly, the detection of the direction of sound in water.

It is of considerable interest to note that Rutherford does not, at this stage, seem to have a clear distinction in his mind between what we now call "active" and "passive" sound-detection systems. In active systems the searcher sends out a sound beam of his own and seeks to detect reflections of this sound wave bouncing back from the submarine he is looking for; in passive systems the searcher tries to hear the noises made by the object of his search. Because the main objective of those who were trying to develop underwater signalling was to produce a very powerful source of sound the issue was clouded, and the earliest efforts to use "active" sound detection were unsuccessful attempts to find the "shadow" of a submarine which was being "illuminated" by a powerful source of sound. Nevertheless it was clear to the scientists that the ideal method of searching for a submarine was to develop a narrow beam of sound that could be used like a "searchlight" (J.J.'s words) in the water. They had no technology that looked remotely like producing such a thing, but from physical first principles they guessed that it would be found, if at all, at very high frequencies. This report by Rutherford says that his sub-committee has "discussed the possibility of a system of secret signalling by the use of sound waves of frequency beyond the limit of audition". This is the first hint anywhere of the system that would eventually become our modern sonar – at first it was referred to as "supersonics", and the navy's word for the devices in the Second World War was ASDIC. One reason why so little is known about Rutherford's major role in this development is that the whole subject was one of the greatest official secrecy at the time of his death, just before the Second World War, and it could not therefore be detailed in his obituary notices and official biography.

Apart from the problems of generating strong sound waves – the first subject of his discussion – there is the related subject of collecting the signal. The likeliest solution here, he writes, is the "tuning" of the receiver to the frequency of the sound generator, so that the receiver will more readily "pick up and reinforce" the signals. Trials of such a signalling system were then going on at HMS *Vernon*, the headquarters of the navy's School of Signals. Rutherford, in search of enemy submarines, therefore concentrates on the second of his main divisions, the detection of ships at a distance by listening to the noises they make – what we should now call "passive sonar". He starts with a disappointing result which he had himself obtained in his first experiments in the Firth of Forth: that there is no natural frequency in the noise made by any ship – the frequency varies according to the speed of the ship, it varies from one ship to another, and if there is any predominating frequency generated by the ship's screw, say, it will be masked by all the other noises of the ship's auxiliary engines and the sound of its motion through the water. There is, therefore, no hope of "tuning" a receiver so that it can pick up the special frequency of a particular type of ship such as a submarine. Some of this basic information had come from the observations of Sir Richard Paget, a man who possessed "perfect pitch" – the ability to detect the exact frequency of a musical note. Sir Richard had been hung over the side of a boat with his head in the cold waters of the North Sea while Rutherford and others had hung on to his legs and he had reported on the frequencies he could hear underwater.

There were, Rutherford wrote, three devices for picking up the noises made by ships: the simple collector, a stick or tube with a diaphragm at the end which could be held to the ear; hydrophones, where the vibrations received by the diaphragm actuated a microphone which produced sounds as in a telephone; and magnetophones, which were similar to hydrophones but used magnetic induction instead of a microphone – essentially the Fessenden system. By his own experiments with simple collectors such as the Broca tube, Rutherford felt that he could "judge that with such a tube the sound of a submerged submarine in motion would certainly be detected under favourable conditions for a distance of about one half-mile or possibly more". Two tubes, one for each ear, extended the range of detection. But in practical terms one needed to increase the intensity of the sound received by at least twenty times, and a further practical difficulty which Rutherford himself had experienced was that such systems only functioned if used from the side of a ship at rest and in fine weather: "Even in a light breeze the effect of the wind on the ears and the splash of the waves on the side of the ship seriously interfere with the picking up of weak sounds." Those are the words of a man who has tried it out for himself.

But there was still some hope of using the direct listening method if the observer could be in a closed room inside the ship with the outside diaphragm linked directly to a tank of still water inside the ship which would transfer the sound to the ears of the observer through something like a Broca tube. (This method was indeed followed up, with receivers on each side of the ship to get a sense of direction. A register was made of blind people, preferably musicians, who might be trained as special "aural observers".) Rutherford also did much work on the possibility of using very large wooden trumpet-like receivers to collect more of the sounds, rather like primitive hearing aids. This suggestion came from a Mr S. G. Brown, and was one of the few inventions which the BIR was able to take seriously.

Nevertheless Rutherford showed most interest in this report in the use of hydrophones. Again he praised Ryan's work highly. He was "strongly impressed by the value of the work", referring to the string of moored hydrophones, and "the use of submerged hydrophones as a means of detecting ships at a distance by their characteristic sounds has reached a practical stage and should be still further developed . . . [They] can detect and recognise a submarine with certainty at one mile range." He described the Ryan hydrophone as a vessel containing a manganese-steel diaphragm, 7.5 centimetres in diameter, and one millimetre thick, to the centre of which a microphone was attached. But there were problems, he admitted. One was the variability of the performance of the hydrophones – a matter to which Ryan did not pay much attention but which worried the scientific mind – and even Ryan had admitted that a ship searching for a submarine with a hydrophone would only be successful if it was at rest and in fairly quite weather.

Magnetophones had been tested and found to be too insensitive; they could "only hear a submarine at about 100 feet".

Next, Rutherford discussed his third main problem area – methods of finding the direction from which sound was coming. He outlined four possibilities: Method 1, to use the instinctive directional sense of the human ear; Method 2, to use some device like the long trumpets in the water; Method 3, to use interference methods; and Method 4, to compare the sounds heard on opposite sides of the listening ship. "The writer has tried some experiments on (1)," he continued, but admitted that "no very reliable results were obtained". Commander Ryan had been trying experiments of Method 2 but they had not proved "very practical". Method 3 would only work if the submarine was found to have some note of a definite frequency, but it had been shown that a sub generated many frequencies, and in any case it would be very difficult to convert such a laboratory technique to a practical scale – "observations of this kind are difficult to carry out from the deck of a ship except under very ideal conditions in calm weather". And so Method 4 seemed the most likely to succeed and "in my opinion it is important that experiments along these lines should be tried on a practical scale". He went on to propose the use of a "swift sub-hunter" which would stop her engines at intervals to listen for submarines, and there were additional possibilities in trying to neutralise the disturbance of "own ship's sound".

With hindsight we can see that Rutherford's touch was less sure when he considered, as a separate subject, the "Detection of submarines by reflected sound". His comment was: "While it is difficult to make definite calculations of the amount of sound returned in this way, there can be no doubt that it is very small, and would be difficult to detect except at short distances." Fessenden's success at iceberg spotting by reflected sound at ranges between one and a half and two and a half miles was only relevant in that the icebergs were probably a hundred times larger than a submarine, and therefore a submarine would only be detectable by this method at a range of about a quarter of a mile. To this Rutherford wisely added "and only in deep water where there is no reflection from the bottom". "It appears to me that it would be more practicable at such short ranges to detect the submarine by its own characteristic sounds when in motion." He also pointed out that if a submarine was "lying doggo" on the sea bed it would always be difficult to find because of the problems of bottom reflection. In this understanding of the difficulties caused by reflections of sound from the sea bed (and from the underside of the sea surface), Rutherford showed again his magnificent grasp of physical reality, for any ex-navy man who hunted subs in the Second World War can recount stories of losing submarines in the shallow waters of the Continental Shelf when the same submarine could easily be detected as soon as it moved into the deep water of the oceans.

This clear and important report, the foundation of subsequent anti-submarine warfare, ends by suggesting seven lines for further investigation, all of which were followed by BIR scientists. But it also ends with demands: "It is of urgent importance that an experimental ship and station with a trained personnel should be placed at the disposal of the committee . . . without this, [the] work of the committee will be slow and difficult." To strengthen his argument he points out that Professor L. V. King, of McGill University, had already been lent a ship by the Canadian government when he had started determining the depth of water by measuring the time of sound reflection from the bottom "at his own initiative and expense". This work was the start of our present echo-sounding technology, just as Rutherford's report was the first demand for what was to become the Royal Navy Scientific Service. As to Rutherford's cautious approach to "active sonar", the detection of submarines by reflected sound, success here would only come after the development of a successful electrical amplifier, using valves to capture the very small amount of sound reflected, and the development of the piezoelectric method of producing very high-frequency sound, ultrasonics. And in this Rutherford played a major and heretofore quite unrecognised role.

For the moment however, at the end of 1915, the most important feature of his report was the demand for ships and experimental stations manned by trained scientists. This demand was exactly parallel to demands coming from other sub-committees of the BIR – for instance, in December 1915, J.J. submitted to the First Lord of the Admiralty a demand for a "Naval Engineering Experimental Station". The BIR, he wrote as Chairman of the Scientific Panel, was "unanimously of the opinion" that such a station was "a real necessity if HM Navy is to be kept in the forefront of the great advances in modern engineering which are continuously taking place". The best way to start such an urgently-needed project was to appoint a "suitable well-qualified gentleman" as the head, give him sufficient staff and ample funds and find any convenient place where they could begin work. A half-hearted start was made in this direction by setting groups to work at the City and Guilds Engineering College in South Kensington, and at Finsbury Technical College, in North London. More important is the conclusion of J.J.'s submission: "There is a strong opinion on the part of this board, that a permanent general research department is much required in connection with the navy, and that it might be well to consider the establishment of such a department at the same time. There is little doubt that such a department would soon save the amount of its cost and upkeep."

Fisher had written to Balfour: "N.B. Man invents, Monkeys imitate . . . Noxious gases made us send Professors to study German asphyxiation! German mines and submarines have walked ahead of us by leaps and bounds although many years ago we were in a position of apparently unassailable superiority." Fisher wanted his BIR to invent, but here were the scientists pointing out a much more serious weakness in the whole structure of the navy, the lack of research, the lack of a technological or scientific approach to all the problems. Such criticism was not what the navy had envisaged.

356

Then came another of Rutherford's "secret" reports. Again running through all the possible ways of detecting submarines, it concludes: "Of these the only method which has shown itself capable of detecting a submarine at ranges of more than 100 or 200 feet is that of detection by the sound emitted by a sub when in motion." Under the best conditions the acoustic method could detect the fleet at sea at a range of thirty or forty miles and a single submarine at ten miles. It was now known to be easy to distinguish between turbine and reciprocating engines. Submarines made more noise when submerged than when on the surface and a trained observer could tell the difference between electric and turbine motors. It was even possible, under favourable conditions, to point out the direction of the submarine to within a few degrees.

All these findings were not just matters of opinion – they had come out of experiments conducted by Commander Ryan and Rutherford and his team. And there was a straightforward account of the practical difficulties – the hydrophones and listening devices were badly affected by any rush of water past them, by the noises of other ships, by the noise of the waves on the surface and even by any vibrations in the receivers themselves. All these problems affected fixed arrays of hydrophones such as Ryan had developed first, but there were further complications of "own ship's noise" from machinery and motion if the listeners went afloat to search for submarines. Therefore "the most favourable conditions for listening . . . are undoubtedly to be found in another submarine submerged and having its engines temporarily stopped . . . There has not yet been opportunity for testing to what extent sound reception in a submarine submerged is interfered with by its own engine noises, etc." If submarine-hunting was to be conducted from surface ships "only really by stopping . . . can you listen". This would obviously expose the hunter to torpedo attack. So Rutherford suggested that either the hunters should be structurally protected (i.e. by extra armour or nets) against torpedo attack, or else ships of very light draught (i.e. very shallow) should be used. One might use motor-torpedo boats, or even seaplanes, he suggested.

Finally he elaborated his earlier direction-finding proposals of having observers isolated inside ships with hearing devices on both sides.

These broad views over the whole field of anti-submarine warfare – what we might nowadays call "overviews" – however much physical intuition they might show, were also the result of a great deal of extremely hard work in the Manchester laboratory water tanks and many visits to Hawkcraig. The conclusions about what was practicable were based on a series of technical reports Rutherford made to BIR. These covered the use of Broca tubes – "air-filled collectors" – in the BIR's official report on its work at the end of 1916; the design of new types of directional listening apparatus; fundamental work on the vibration and resonance of diaphragms in water which led to Rutherford designing diaphragms ten times more sensitive than those in current service use; and a careful study of electromagnetic underwater signalling. There is also a short report from him on his earliest experiments with piezoelectric crystals, work which was to have the most profound results in later years.

So much for the official record. But a brilliant illumination of what this work actually involved has been shed by the fortunate discovery of a previously unknown series of letters written by Rutherford to Albert Beaumont Wood. Wood and H. Gerrard were sent as the first two young scientists to work with Commander Ryan at Hawkcraig. Albert Wood was a typical Rutherford discovery and protégé. He came from a solid, but totally unscientific background in the Pennine village of Shuttleworth, and followed the route through local schools to the University of Manchester and the physics department. Here Rutherford and his team from 1910 to 1913 made him into a fairly typical research student working in radioactivity and he went on to a junior post at Liverpool University. He corresponded with Rutherford even after he had left Manchester, obtaining help in the provision of radioactive sources and support from his teacher when he successfully applied for an Oliver Lodge Fellowship. Furthermore, Rutherford put Wood in touch with what was then Rhodes College, at Grahamstown in South Africa, where there was a vacancy for a physics lecturer, but the young man did not pursue this opportunity.

In his report on Ryan's work at Hawkcraig in September 1915, Rutherford had suggested that Ryan should be given the support of two or three young physicists. Action on this suggestion followed rapidly and on October 7th, 1915 Rutherford wrote, rather peremptorily, to Wood: "I want to see you on Saturday morning, preferably between 11 and 12 at the laboratory, about an important matter." The letter is in Lady Rutherford's handwriting, for she worked frequently as her husband's amanuensis, though the names of recipient and the signature are in Rutherford's own hand. Wood reported immediately and three days later received a scribbled note from Rutherford: "Matters of which I spoke are moving rapidly and your services may be wanted possibly within a week. Consult Wilberforce [the Professor of Physics at Liverpool] and make tentative arrangements for your departure. The Admiralty will arrange pay – but I am sure it will be quite satisfactory. I hope to see Gerrard on Monday in connection with the same matter." There is an addendum: "Give my regards to Prof. Wilberforce and tell him I am sorry to interfere with his department but hope he can make suitable arrangements to take over." And in the top corner is another addendum: "If you are appointed you had better give up university salary for the time you are away." This is indeed a buccaneering way of getting staff – Rutherford at his most forceful. For neither was Gerrard one of Rutherford's own men, though he worked in the adjacent electro-chemistry department. However, it must be said that Rutherford's own department had already been stripped of all its fit, younger, men by the demands of the forces. Certainly there was no protest from the robbed Professor Wilberforce and he remained on friendly, though not close, terms with his more powerful colleague in the years ahead.

Rutherford wrote to Wood again the following day, October 11th, 1915. It is the clearest possible evidence of the fact that science, and radioactive matters in particular, took absolute priority in his mind over all other subjects, even over his war-work with which his hours were mainly filled. This letter starts with the sentence, "You will have got my letter I wrote yesterday – there is nothing more to report on that subject," and then discusses nothing else but the significance of some results Wood had obtained concerning the different ranges of alpha-particles emitted by thorium-C. The letter ends: "Please examine this point if you have time," which was perhaps somewhat unsympathetic towards a man who had already been ordered to make all his arrangements for leaving his laboratory, but which clearly shows the priorities accorded.

Three days later another note, handwritten, asked Wood to come to Manchester "early on Saturday morning to do some experiments on sound", and added that his name had been sent to the Admiralty and a decision should soon be received. Rutherford had to learn that Admiralty arrangements could not be made to work as briskly as his unofficial network of scientists could be made to respond, and the next scribbled letter, again three days later, deals with nothing other than ranges of alpha-particles emitted by thorium. Pinned to the single-page note is a small sheet of graph-paper on which Rutherford has plotted a theoretical set of points and the observational points found by Wood.

There is no trace of Rutherford's method of finding out what was going on, though presumably he worked through BIR headquarters. But he must have been pressing hard, for after only three more days he explains to Wood that the delay in getting him to work at Hawkcraig is because "the delivery and work of the ship has been rather delayed". The ship eventually allocated to Ryan's station – and the allocation had been one of Rutherford's demands on Ryan's behalf – was the fishing trawler, technically a "drifter", called *Heidra*. But in this delay in allocating her to Hawkcraig there is the first warning of the trouble to come. However, basic work on the acoustics of the sea could still go ahead and Rutherford, who never worried about weekends either in peace or war, invited Wood to come over on the Saturday when "we can get some more work going". A false alarm that the Scottish work could begin in the next few days led to a telegram from Rutherford to Wood on October 26th: "Delay final arrangements until you get official notification." And Wood and Gerrard did not finally get to Hawkcraig until November.

When the two young physicists arrived at the naval establishment they became a sort of front-line outpost of Rutherford's Manchester laboratory, rather than helpers of Commander Ryan. Rutherford himself visited the station fairly regularly, and frequently went out in the ships themselves. (Another vessel, the *Tarlair*, was later added to the strength.) But basically Wood and Gerrard were carrying out the full-scale experiments which sprang from Rutherford's laboratory work. They therefore had to be given detailed instructions as to what he wanted, and these are contained in a series of forty letters written between December 1915 and April 1916. Much of the content of these letters is engineering detail, and the making of arrangements for experiments and the movement of material. The first, dated December 6th, 1915, for instance, starts off abruptly: "I am hoping to send up today a new Broca, which should be more sensitive than the last two for notes above 1,000. As soon as you can, please compare its range with the last two I sent up, and the multiple."

But there were several different and parallel lines of research that Rutherford kept going. Further on he writes: "I have been doing a good deal of calculation with the effect of water on one and both sides of the diaphragm on the natural frequencies . . . I want this matter pushed through as rapidly as possible, as I believe that by this arrangement we shall greatly increase the efficiency of the electromagnetic methods [of underwater signalling]." Then come further instructions: "I want each of the 7 diaphragms tested, say with the *Tarlair* if a submarine is not available, (a) when a Brown telephone is attached to each diaphragm, (b) with water in the tank and a Broca placed close to each vibrator as receiver, (c) with a Brown 6 inches, or 12 inches, trumpet placed close to the vibrators with water in between . . ."

360

It is with sheer enthusiasm, rather than threat, that the letter ends: "I am getting several other pieces of apparatus constructed, and we shall keep you very busy after this in testing," for Rutherford was a good leader and did not expect his troops to do what he would not do himself. "I am off to London today but shall return tomorrow evening. I get very little time to do much work of my own," he concludes.

But Rutherford and his two young assistants had to learn that neither the navy, nor the element in which the navy worked, the sea, were as amenable to control as laboratory experiments. In his next letter, four days later, he is "sorry to hear that the weather has gone wrong and that you have not got any further with the experiments". Likewise "your little device may be quite useful if it works well in practice. I would certainly show it to Ryan." No more is ever heard of the "little device" and it soon turned out that showing things to Ryan was not very easy, for the physicists were rapidly becoming aware that the navy administrative machinery was finding young civilians difficult to deal with. It was not until the February of 1916 that the two men received "Admiralty Indispensability Certificates and also Navy Armlets" and then they were advised that "you are exempted from the provisions of the Military Service Act, under which the Government have power to grant exemptions to those in the employ of HM Government. Should you be approached by the Local Recruiting Authorities, it will be sufficient to exhibit the Indispensability Certificates."

The following month there was further rather angry correspondence with the secretariat of the BIR over the knotty problem of whether they counted as "officers" and whether they were therefore allowed first-class rail travel when called to London. This matter ended in the hands of the Accountant General, no less, because the BIR was bound on such important questions by Admiralty regulations. And although the physicists were granted the facilities due to officers when they were at Hawkcraig, it was made quite clear to them that they were still not socially acceptable as the real equals of commissioned naval officers. Still less were their families, when they moved to the area later in the war, accepted or even politely welcomed by the navy community.

In the middle of December Rutherford turned his attention to trumpets for collecting sound, and tried to find the best materials for making them. Apart from the odd paragraph discussing how he and Wood should publish the results and theoretical work on alpha-particles from thorium-C, the letters for the rest of this month are all about water noises in the Broca tubes and:

I have been very busy at work with paraffin trumpets and comparing their efficiencies with different diaphragms, etc. I believe I can make a substantial improvement on the wooden ones. I intend to examine next time I come up, whether a paraffin trumpet cannot be used in place of a Broca for direct listening, and I am hopeful that the water noise will be much reduced . . . I think the frequency outfit is nearly ready, and Beattie tells me that the source telephone is much louder than the one we use . . .

The remainder of the letter deals with the arrangements that Rutherford is making directly with Ryan for the two young men to have some Christmas leave. But Christmas, too, always took second place to science and there is a brief note to Wood, dated Christmas Day, asking him and Gerrard to call and stay to lunch in a couple of days. It ends: "Bring details of experiments on scintillations, diameter of wire and area under microscope."

Nor was New Year's Day any time for relaxation. Plainly some contretemps had occurred as soon as the two got back to Scotland, for Rutherford's letter of January 1st starts that he "quite understands the situation" and that he is coming up himself that week and has asked Ryan to make *Heidra* available for various seagoing experiments, for which he asks Wood to make preparations.

On this trip the party must have experienced the January weather in the Firth of Forth at its most bitter, for in Rutherford's next letter he advises, "For such experiments it will be desirable to wait for a quiet day, for I found there is too much noise in the listening room when the ship is in motion in rough and windy weather." But most of this three-page letter is a long list of the exact experiments he wants carried out with various diaphragms. There is also a long list of work to be done with the paraffin trumpet, which turned out to be disappointing, and demands for further work on neutralisation experiments and valve experiments. Yet only three days later, on January 13th, 1916, he writes to Wood suggesting further ideas which stem from his own experiments and from the first results that Wood has sent him. In addition there is a rough sketch of a new method of mounting diaphragms and listening equipment in the hull of the ship.

The rate of work was quite fearsome. Only two more days had passed and Rutherford had been considering the implications of Wood's results, so he wrote: "The different law of decrease of sound must be due either (1) to some direct effect of the trumpet or diaphragm or (2) that waves of different frequencies are unequally absorbed by water." But then he added a further thought: "Or (3) greater screening by hull of higher frequencies." He went on: "I expect you will have tried (1) by now, but if (2) is correct it is very interesting but very disturbing . . . On the other hand this idea is hardly borne out by the fact that the diaphragm in the trumpet gave about equal effects on the valve with a battle fleet about five miles away . . ." Then another afterthought: "After reflection I think (3) is the more likely explanation." Jammed on to the top of the second page is yet another rough sketch for a way of mounting the diaphragms and the letter also contains the first mention of direction finding by observing maximum and minimum effects as the diaphragm is rotated. The letter ends: "We are hopeful that some suggestions we have made to the Admiralty will make work at the station in general more satisfactory and rapid in the future."

The letters continued to pour out from Manchester, one every few days through January and February. Sometimes they were several pages long, typed and full of advice about the best ways to do experiments. Sometimes they were scrawled in pencil, containing engineering details of the diaphragms and other devices he sent up for trial, and there were graphs of his own laboratory results. The whole drive was to find the basic laws of the conduction of sound by seawater and the development of accurate, reliable and strictly comparable devices with which to measure the basic quantities involved. On February 7th, a tiny, but important, practical detail was involved, which is very revealing of the care that Rutherford put into the work that gained him the reputation of being the greatest experimentalist of his time. At the end of another long typed letter there is a handwritten postscript:

> We have found in testing diaphragms it is very important to avoid grease on them and make certain that bubbles of air are not attached to the diaphragm. We had been noting variations in resonance points and finally traced them to Powell's industry in carefully greasing the diaphragms to prevent rust! A clean surface is essential if results are to be repeated. Grease layer usually lowers the frequency of resonance while bubbles raise it. Keep your eyes open on this point. It would be worth while running a clean sponge over the surface of a diaphragm in position in the water to get rid of the bubbles. Verb sap.

In the middle of February Rutherford himself paid another short visit to the station, showing Professor Broca, over on a visit from France, some of the work being done with his listening tubes. By the end of the month work had also started on the project of towing hydrophones in some sort of streamlined casing outside a ship in order to avoid the problems of "own ship's noise". This work went on until the end of the war. The same letter also mentions the first work on "a valve" for magnifying the effects received in the hydrophone – in other words a primitive version of what we should now call an amplifier. Rutherford admits that he has no experience himself with working with valves but he has been enquiring among those pioneers who had recently begun to use them. He advises Wood and Gerrard to concentrate on the vacuum pressure inside the valve and suggests the use of a heating coil to adjust this value. He has also found out that a magnification of about twenty times the received signal should be possible so he writes: "It will make a great difference to all our experiments with telephones if we have a magnification of 20 on tap. There are a number of other experiments on magnification I have been enquiring about and we shall possibly use them later." The upshot of all this was that he encouraged Wood and Gerrard to cooperate with Ryan's work on a valve, although this was really Admiralty work. This letter also contained a mysterious hint: "I am working on another method of detection altogether and shall probably make experiments with it the next time I come up."

But progress was slow in March – the valve magnified the unwanted ship and water noises as much as anything else and did not give a greatly increased range of detection; also, the towed body would not proceed in a straight line behind the ship. The inability of the navy to provide a submarine for experiments was plainly beginning to become very annoying. Rutherford paid another visit to the station in March: "Left early on Sunday morning and arrived in Manchester after 12 hours' cold journey via York and Leeds." The rather irritable letter ends: "I hope we shall be able to make suitable arrangements for carrying out our work far more expeditiously in the future, but that will depend a good deal on the Admiralty," and he added a list of nine major experimental routines he wanted performed. Two weeks later the physicists had proved that an idea of Professor Bragg's for underwater signalling by use of a bell would not work, but little progress had been made with the Admiralty. On March 26th Rutherford had to write: "I note what you have to say about the ship question. Negotiations are at present in progress with the Admiralty and we are awaiting their decision in regard to future arrangements. I hope that this will not be long delayed so that we shall know exactly where we are."

By now Rutherford's drive had moved on to the question of finding the direction from which the sounds came. He sent a series of newly-devised direction finders up to Scotland to be tried out, because he was getting "puzzling" results in his small laboratory tanks. In the last days of March he told his young men "privately" that negotiations were now "at a critical stage". But there must have been some further trouble from the navy, for the postscript to

Rutherford's letter of March 31st reads: "I received your letter this morning and had quite anticipated the information. It is difficult to believe we are in the midst of a great war!"

Nevertheless it was at just this moment that Wood and Gerrard were allowed for the first time to test out their hydrophones and other devices on an actual submarine. It was immediately apparent that the sound of the sub was not at all similar to the sounds emitted by the *Tarlair*, the ship on which they had done most of their observations, for the loudest noises emitted by the sub were at about 700 to 800 cycles per second, whereas the *Tarlair* and similar surface vessels emitted their peak noise energy at about 1,000 cycles. Rutherford's reply to this information, in a letter of April 1st, is another masterpiece of physical insight:

. . . I suggested some time ago that the origin of sounds from submarines was probably mainly due to the vibrations of the propeller blades set up by their motion through the water. These motions may be either transverse, like a tuning fork, or lateral . . . If this point of view is substantially correct for a submarine running on electric motors, we shall anticipate a good deal of sound over the whole spectrum; but it would obviously be best to have a diaphragm tuned to one of the higher stronger harmonics to reduce water noise. Probably the propeller blades of the *Tarlair* have different natural frequencies from the submarine, and I should imagine are relatively more rigid for their size. On this point of view the sound of a submarine is generated by the propellers; part is directly communicated to the water; and part through the water and shaft to the body of the submarine, which acts like a sounding board. Of course with the petrol motors in motion you get an addition of the sound due to them. If my point of view is substantially correct we shall anticipate that the relative sensibility of a diaphragm should be independent of the speed of the propeller. I should anticipate the number of beats should be equal to the number of blades multiplied by the number of revolutions. Possibly Paget may go up soon to test the notes of a submarine in motion. This is at present only a working theory, but I am of the opinion that it has much to commend it, for, as far as I can see, there is no evidence that contradicts it.

Two days later Rutherford wrote to his two men to inform them "privately" that "a decision has been reached about the future of our work at the station, and that Bragg will take the position I outlined to you during my last visit. I think matters will be definitely straightened out and we shall be able to get down to solid work under reasonable conditions."

This was in fact the solution to the problems that BIR tried to impose – sending Professor W. H. Bragg to Hawkcraig to set up and control a separate and purely scientific unit working alongside Ryan, and cooperating over the use of naval facilities. Bragg had been associated with Rutherford on the Anti-Submarine Committee since the first days of BIR, but had confined most of his work to his own laboratory, where he had been helping in the study of optical glass, the supply of which from Germany to Britain had naturally been cut off. Bragg's work in the first months of BIR's existence had also been cut down by the fact that he was in the process of moving from his position at Leeds to the Professorship of Physics at University College, London. In February 1915 he had written to Rutherford . . . "You know most things and I expect you are quite aware that the University College are asking me to take Trouton's place . . ." He had gone on to explain how he had been bargaining with University College for better research facilities and had now decided to accept their offer. His position had been much strengthened by the recent award of the Nobel Physics Prize, which he had received, jointly with his son W. L. Bragg, for their work on X-ray diffraction.

Bragg, quiet, shy, gentlemanly, with none of Rutherford's extrovert brashness or push, was nevertheless a powerful man, achieving much by persistence and solidity of character. He was also much more firmly realistic about relationships between science and the outside world of industry and war. For instance he wrote to Rutherford on April 25th, 1915, asking his advice about possible successors at Leeds. What was wanted, he said, was a man with the

> ability to take up research on lines that would directly benefit this district. There are a number of physical problems in the textile and other industries which ought to be attacked. I dare say that – if we may anticipate – one result of this war will be a better appreciation of scientific research in industrial circles. So there may be hope that the manufacturers about here will lend a more willing ear than in the past. Anyway the one thing to do here in this university is to convince people that research is some use.

In like vein Bragg wrote on July 23rd, 1915: "If we could ourselves get in touch with the fighting units and know what they want, or sort out the inventors and put the sensible ones among them into touch with the fighting men, that would be the thing. And we might be able to help the real people to work out their ideas in our own labs." Bragg had lost one of his sons on the Western Front, but the other was making a great reputation for himself by running a sound-ranging unit which used scientific methods to pinpoint German guns behind the trench lines, and several of Rutherford's

Manchester scientists eventually joined Bragg or other units run on the same lines. Bragg himself worked closely with Rutherford on their war problems – a letter of his to A. L. Rogers, an instrument-maker, in December 1915, remarked: "I see a lot of Professor Rutherford. He came over yesterday to discuss some new theories of his. We agree closely about things and are almost in opposition to the Cambridge people."

Bragg arrived at Hawkcraig in May 1916. His biographer writes: "A. B. Wood became his right hand man, they had the same modesty and integrity and worked comfortably together under the difficult conditions." It was undoubtedly a strange station. On Ryan's staff was a young RNVR officer, a Lieutenant Harty who was later to be Sir William Hamilton Harty, the conductor of the Hallé Orchestra, but who was meanwhile employing his musical talents striking diaphragms with a little hammer to sort out the higher from the lower pitched; there was also Ryan's dog which regularly stole people's dinner from the table until the scientists attached a high-voltage battery to a piece of steak. The mixture of scientists and naval officers attracted many visitors: Beatty and Jellicoe from the battle fleets stationed nearby, parties of Allied officers from other countries, Rutherford and other members of the BIR, and Sir Richard Paget who used his musical talents not only to discover the note of a submarine, but to improvise an oratorio from an advertisement, and to sing a wordless, two-part song as he accompanied himself on the piano.

One visiting party illustrates well the naval attitudes which made work with any other body of men apparently impossible. The visitors were a delegation of French scientists and officers, but early on the morning that they were due to see the work of Hawkcraig, Bragg received an urgent signal from the Admiralty, "on no account show them anything". The visitors were so well entertained at lunch, with the help of Beatty's Flag Lieutenant of the Battle Cruiser Squadron, that they were in no condition to notice how little they were shown of the anti-submarine work. The French were extremely anxious to collaborate with the British on scientific developments and were thoroughly prepared for a truly "two-way traffic" on inventions. The French were at this time considered the most inventive of European nations; it was held for instance that they had invented most of the devices which made this war different from other wars – the breech-loading rifle, the machine-gun, the aeroplane were examples; it was part of the accepted wisdom of the time, not least among the French, that they invented, but the Germans and the British turned their inventions to profit.

The French had first raised the possibility of scientific cooperation at a meeting between Painlevé and Lloyd George at the Ministry of Munitions in 1915. In 1916 they invited two British delegations – Captain Cyprian Bridge, R.N., led the first in April 1916 and Professor W. J. Pope went over in July of the same year. The French Ministry of Inventions revealed all they had to show in the way of microphones and hydrophones for use on land and in water, for towing behind ships and for detecting submarines – the BIR's official report admits to the French offering eight major lines of research, in many of which they were equal to, if not ahead of, their British counterparts. In aerial photography they were particularly far ahead thanks mostly to a new camera developed by Gaumont. In August 1916 the French authorities asked, through Professor Pope, whether a BIR representative might be permanently stationed with them at the Inventions Ministry, and they would like to send a representative to Britain to liaise with BIR. The Admiralty totally refused to have a French officer aware of their inventions and this led to considerable, and not unnatural, "soreness" on the part of the French. The Foreign Office also got involved and raised the point that if the French had access to new British ideas all the other Allied governments might claim similar rights. Soon other services and ministries accepted the French proposals for liaison. The "resentment at the aloofness of the Admiralty" grew on both sides of the Channel.

Eventually Pope and Bridge wrote a joint letter to Vice-Admiral Sir Richard Peirse, the official Admiralty representative on the Central Committee of BIR, asking that

> the advantage to be gained by the navy [be] open to reconsideration in view of the highly important and interesting matters already communicated to us by the French . . . It may be pointed out that practically all the most important inventions of recent years which had any great bearing on the present war have been originated and first developed in France . . . The French, as a nation, are more imaginative and consequently more inventive than we are and the study of scientific matters has probably received more encouragement in France in the past than it has here.

It was thus in September 1916 that the French party was allowed to visit Hawkcraig, and shortly afterwards the navy capitulated when it was told that the French had selected the Duc de Broglie, who held the rank of Lieutenant de Vaisseau, as their scientific representative. Once again the navy seems to have been impressed by a duke.

It was, of course, a very difficult time for the Royal Navy, the months surrounding the Battle of Jutland, when it had not yet become clear that they had won a strategic victory over the German High Seas Fleet, and when British losses seemed unaccountably high. The scientists had listened on their hydrophones to the underwater sounds of the battle fleets steaming out to combat, and shared the bewilderment and loss when they returned. J.J., in London, recalls Fisher's extreme dejection the morning after the battle, with the great, office-bound Admiral storming about the room saying, "They've failed me, they've failed me! I have spent thirty years of my life preparing for this day and they've failed me!"

Perhaps because of the technological inferiority of the British fleet, so frighteningly demonstrated at Jutland, naval officers were now beginning to think that the scientists had not invented enough or speedily enough. Jealousy and mistrust abounded. One day Bragg asked the skipper of *Heidra* to perform a certain manoeuvre and when the ship returned to Hawkcraig Commander Ryan promptly punished the skipper with fourteen days' confinement to his ship for disobeying Ryan's orders. Wood recorded: "Prof. Bragg went to see Ryan and apologised for having asked the man to do what Ryan had previously told him not to do, and accepted all the blame, but the sentence had to stand." On another occasion, in Bragg's absence, a length of hydrophone cable which had been laid at his request for use as a telephone line, was taken up by Ryan's people without explanation to the scientists, simply "by order". By the end of the year it had become quite obvious that no further progress could be made at Hawkcraig, so bad were relationships between Ryan and the scientists.

The whole matter was taken up to Balfour's level, and his first suggestion was that Ryan should be posted elsewhere. Bragg refused this offer, whether on the stated grounds of fairness – that Ryan had been there first and should be allowed to stay – or more probably on his own belief that scientists should work more closely with the fighting units who had the real problems. So early in 1917 the scientific unit moved to Parkeston Quay at Harwich, close beside Admiral Tyrwhitt's base, where his light forces and submarines were engaged in aggressive operations against their opposite numbers across the North Sea. Here it stayed until the end of the war, expanding and achieving far more useful results, and eventually coming under the command of Rutherford's friend, A. S. Eve. Eve had enlisted in a Canadian Highlander regiment of which he became Colonel, before coming to Rutherford's aid in the anti-submarine war, and therefore claiming the distinction of being the only Highlander Colonel in command of Naval Units.

Before leaving Hawkcraig Bragg summed up its work in a letter to J.J., writing on December 10th, 1916:

. . . You expressed regret at the Panel meeting at the delay due to the unfortunate situation at Hawkcraig. It is not perhaps so bad as you think. The work of BIR has been the cause of useful advance in two directions. In the first place the BIR has pressed forward the plan of fitting listening apparatus to submarines which is now being adopted and used successfully. It is true that there has been much delay in fitting all the boats but the delay is mainly due to the reluctance of higher placed naval officers than those at Hawkcraig. The second advance has been the evolution of a practical direction finder which may be of real use. This was finished – in the main – last July; and again the delay in introducing it is due to officers at the Admiralty . . . Perhaps even if Ryan had been in full sympathy with BIR, we should not have succeeded any better in bringing about the thorough trials of the instrument which ought to be made. It is not so much to these things that the need for our transference to Harwich has arisen. It is more because we are not likely to get any further in a place where we are practically cut off from all contact with the navy except such part of it as is hostile to BIR. I do not know that we should have got much further with the acoustic problem we were sent to Hawkcraig to investigate if we had been entirely on our own: after all we have learnt a lot from Ryan. But there must be many practical developments of the problem and many other problems which we shall have a real chance of getting in touch with in our new circumstances. I believe the direct contact between the men using the submarines and the physicists with their workmen will be really inspiring and fruitful . . .

It was Rutherford of course who had pressed for the introduction of listening apparatus into submarines, as we have seen. It was he, too, who had invented the first practical direction finder. And the remainder of his letters to Wood and Gerrard, during the period when Bragg was in charge, are concerned mostly with the direction finding problem.

Back in 1916, on April 8th, he told them that Bragg and Paget had been able to listen to submarines at Portsmouth and had completely confirmed his view about the nature of the sounds received from submarines. His next letter, however, showed that the strain had been telling on him – he announced that he was going for two weeks' holiday in the Lake District, adding, "I feel the need of a little rest as I have not taken a day off for six months." But before he went away he sent up further pieces of apparatus to be tested, and his letters include a most interesting sketch, as usual in his stubby pencil-scrawl, of a rotating direction finder, consisting of two listening tubes mounted on a horizontal bar so that they were approximately half a typical wavelength distance apart. The whole was rotated by a vertical tube mounted halfway between the two active diaphragms, so that the whole apparatus had the shape of an inverted "T". His covering note concluded: "Troubles will probably arise due to possible differences of phase of diaphragms near resonance point, but try the matter out thoroughly on shipboard. Tank experiments are no good for this kind of thing."

This device was not, in fact, particularly successful. But at the same time Rutherford had sent up to Hawkcraig another direction finder – they called it the "little direction finder" – which provided a successful solution to the problem. This was the extremely simple device of fixing a metal diaphragm inside a metal ring which could be rotated, and clamping a microphone directly to the centre of the diaphragm. At first it hardly worked at all, until in the middle of April 1916, Rutherford suggested fixing a dummy, exactly equivalent to the microphone, on the opposite side of the diaphragm. This worked immediately and it turned out that the diaphragm picked up sound excellently when it was "broadside on" to the source, and detected almost nothing when it was "edge on". By May 6th Wood reported extremely successful results. And, when Bragg arrived a few days later, practical development in streamlining the apparatus went on apace. It is interesting to note that the successful design arose from a suggestion made by Rutherford in March for mounting a single diaphragm inside an iron tube in order to get a directional effect as the tube was rotated. Experiments by Wood and Gerrard in March had shown some directional effect but terrible trouble with water noises. The length of the iron tube was reduced in efforts to get rid of this problem and it was found that the directional effect increased as the tube was shortened, until the simplest mounting in a rigid metal ring was achieved.

In May, several weeks after his direction finder had been shown to work well, Rutherford heard from both Bragg and Wood that Commander Ryan had discovered a slight directional effect in a device he was developing which contained two diaphragms. The actual development of the two devices illustrates the naval attitude, as it was reported at a meeting of the full BIR on March 29th, 1917:

> As an illustration of this lack of cooperation, the following case may be brought to notice: in May 1916 the Board of Invention and Research developed and produced a submarine direction finder. Its capabilities were, at the time, fully made known to the Admiralty Departments concerned, and it was inspected and tested by the then Assistant Director of Torpedoes and the Chairman of the Submarine Com-

mittee. Two orders for 50 have been given, but apart from these, which were chiefly required for demonstration purposes, no more were ordered, and nine months have thus been wasted in providing instruments which are now regarded as being of considerable value. A few weeks ago a Mark II apparatus was completed by Captain Ryan (Admiralty Experimental Station, Hawkcraig), and on the strength of his opinion – based upon the behaviour of a single instrument – an order for 700 was eventually given by the Admiralty, while at the same time an order for 200 Mark I (the pattern developed by BIR – approximate cost £20 each) which was about to be placed, was cancelled, although this latter type had been tested at sea and shown to be capable of important results. This decision was taken without any comparative trials having been carried out to ascertain the relative merits of the two instruments.

This example was quoted by an independent committee of non-scientists later in 1917, the Sothern Holland Committee, which we shall look at more closely later in this chapter.

From June 1916 a different note creeps briefly into the exchange of letters between Rutherford and Wood, for Wood sought his Professor's advice on whether he should marry in view of the unhappy and possibly impermanent nature of things at the Hawkcraig establishment. Rutherford was encouraging, though he could not offer certainty of financial security in the troubled years of war; later letters offered congratulations and a wedding present – a small silver sweet dish which is still preserved by Wood's widow. Furthermore the frequency of the letters decreases in the summer of 1916; the reason is that most communication now went directly to Bragg. In any case Rutherford's interest was now switching heavily into a new field: in September 1916 he wrote, "I have got a new scheme for determining the actual amplitude of diaphragms in very weak vibrations. These methods may ultimately prove useful when proper investigations are made on the energy of the sound picked up in water. I have also a new, and, I think, accurate method for determining the minimum sound for audibility."

Rutherford was referring here to his growing interest in what were called "supersonics". No man can claim to be the "only begetter" of what we now call "sonar", and which was long called "ASDIC" by the British, but Rutherford was certainly one of the two main inventors of the idea: the other chief claimant is his old friend from Cambridge in 1895, the Frenchman, Paul Langevin. The two men were working absolutely parallel, though separately, from 1915 to 1917, and in 1917 the French revealed their progress through the scientific exchange scheme which the Admiralty had so long resisted.

It had been clear to the scientists, from a consideration of physical principles in 1915, that a "searchlight" of sound could only be found by generating sound at very high frequencies, far higher frequencies than were generated by the Fessenden electromagnetic vibrator working mechanically on to steel plates. Some early work was therefore devoted, by Rutherford among others, to increasing the frequencies generated by Fessenden methods, but these proved unsuccessful. It was Rutherford who first looked at piezoelectric crystals as an alternative way of producing high-frequency sound. Wood records that even before he went to Hawkcraig, i.e. as early as the summer of 1915, Rutherford "was hopefully, if not very optimistically, scratching small pieces of quartz crystal (with a telephone headpiece connected) to discover if the piezoelectric effect of quartz was likely to prove useful. The result of this was inevitably disappointing as no means of amplification was available at that time."

When Wood came to contribute his memories of Rutherford for Eve's biography in 1937, the very existence of ASDIC and all its details were official defence secrets, carefully guarded. Nevertheless Wood wrote to Eve: "Rutherford initiated in the Manchester laboratory what ultimately developed into the most effective method of detecting and locating submarines. It is of course not possible to disclose the nature of this work but the thanks of the navy are due to him for his untiring efforts in the anti-submarine war." Eve did not include this passage in his life of Rutherford. By 1963, when Wood contributed a memorial to the *Collected Papers of Lord Rutherford* he was able to say: "Rutherford played an important part in the early development of the piezoelectric quartz ASDIC which ultimately became the most effective method of detecting and locating submarines." But there is a persistent story, still told by those who knew Rutherford personally, that his friends at one time urged him to make a public claim to be the inventor of ASDIC; to which Rutherford replied "If Langevin says he did it first, that's good enough for the public, let Langevin have the credit", or words to this effect.

The piezoelectric effect was a little-known curiosity of physics, discovered by the brothers J. and P. Curie in 1880. (This P. Curie was Pierre Curie who is now more famed for having given his name to his wife, Marie, which again emphasises how very small the world of physics was at the start of our century.) The Curie brothers discovered that there are certain crystals, and natural quartz is one of them, which give off an electric field when they are physically distorted by being strained, compressed or struck. And in reverse these crystals change their shape slightly, or distort, if they are subjected to an electric force. This means that if such crystals are subjected to a rapidly changing electric current they should vibrate at the frequency of the current. If they are in water this should be transferred to water as a sound wave. Thus while the hydrophone work and the Fessenden oscillator and the natural frequency of ship noises were all concerned with frequencies of about 1,000 cycles a second, the first experiments on piezoelectric vibrators were conducted at 10,000 cycles per second and frequencies of 100,000 cycles a second were being discussed and attained by 1917.

Rutherford certainly continued his piezoelectric experiments in the small water tanks in his Manchester laboratory throughout 1916, and used the phenomenon to provide an accurate measurement of the natural frequencies of his steel diaphragms. The French revealed "Professor Langevin's . . . proposals for the production of supersonic vibrations for the detection of mines, submarines, etc." and this was immediately examined by a small research team that Rutherford set up under R. W. Boyle. According to BIR's official list of reports received, Sir Richard Paget, following up the general visits to France by Captain Cyprian Bridge and Professor Pope, brought back the first details of Langevin's work on June 8th, 1916. By September 9th Boyle reported lack of success in generating supersonic vibrations by direct electromagnetic methods, and on the 28th of the same month Rutherford personally supplied a report on his own work with piezoelectric crystals for various purposes and particularly for "generating high-frequency vibrations". By October 1916 they had completed an examination of the electromagnetic method of producing these supersonic waves in quartz crystals and were going on to examine electrodynamic and electrostatic methods of driving the crystals. One month later Boyle and his team were already reporting their first work on the application of special high-frequency microphones for detecting supersonic vibrations in water. At this time Bragg launched a search underwater to see if ships themselves generated supersonic vibrations, but found nothing.

It was in February 1917 that Langevin submitted his own first full report to BIR. He proudly pointed out that in experiments in Toulon harbour in the autumn of 1916 he had achieved a range of three kilometres for signalling and had detected the echo from a sheet of iron at 100 metres. "This work, although not in itself of great practical importance, marked a definite advance," Wood comments, and Boyle's team, which had been working for several months on supersonic techniques, was soon shifted to Parkeston

374

Quay to begin serious development. Langevin had undoubtedly achieved several important advances – by using electrostatic methods of distorting his quartz he had achieved frequencies of more than 100,000 cycles; he used whole plates of quartz rather than single crystals; he was the first to use quartz plates as receivers of the ultrasonic signals; he had developed methods of practical mounting of his quartz plates which are the basis of modern sonar equipments; above all, he had applied electrical valve amplifiers, developed by the French army wireless services, to his system. Another of Wood's comments is relevant here: "Although the possibility of using quartz was seriously considered in 1916, nothing of importance was done about it as we had no knowledge then of high gain amplifiers. Langevin was the first to try a plate of quartz as a receiver. Used in conjunction with a valve amplifier this gave very promising results. At Parkeston Quay good progress was made when French amplifiers became available." In this respect many of Rutherford's comments in his letters to Wood, quoted above, show remarkable foresight.

In BIR's list of official reports the work of Boyle's team, throughout the spring and summer of 1917, now begin to dominate, but though the first ASDIC sets were fitted to navy vessels just as the war ended, ASDIC was not operational until after hostilities ceased.

The official record bears out Rutherford's unmade claim to be at least the co-inventor of sonar. His correspondence with R. W. Boyle makes his claim even stronger. Boyle has appeared earlier in Rutherford's story, as probably the best of his Canadian students, and as seeking Rutherford's advice and support when he took up a physics post at the new University of Alberta. In January 1915 he wrote to Rutherford offering to come to Britain and do war work for no payment – "Why is there no attempt to mobilise Empire scientific effort?" he asked. By the late summer of 1916 he was in London leading the small team which Rutherford had asked to look at "active" or "beam" methods of submarine location, and, incidentally, Rutherford lent him £15 on a personal basis because his "supplies" had not come through from Canada. By the middle of October 1916 Boyle is telling Rutherford that he is having some success with a receiver circuit that can work up to 30,000 cycles a second, and what he needs is a good, tunable, transmitter but there are "limitations to the production of supersonics by the mechanical method".

Letters from Rutherford to Boyle and vice-versa at the beginning of December 1916 show clearly that, at that time, Rutherford was deeply involved with his own experiments on piezoelectric devices and was directing the work of Boyle's small team on the same subject. Pieces of equipment such as voltmeters were being interchanged, and Rutherford wrote on December 2nd, "I have thought about the effect of the end pieces on the quartz, but find it difficult to calculate their effect. The trouble is that I am not quite sure what are the forces per unit extension due to the electric field." This can only mean that he is trying to apply basic science to actual devices produced by Boyle, and in the same letter he mentions briefly his own experiments "with the piezoelectrique" which is one reason he needs a particular voltmeter. On December 12th Boyle writes that he is trying to measure the minimum amplitudes that could be picked up by his high-frequency detection microphone "using your quartz piezoelectrique to give us small measurable amplitudes", implying that Rutherford has provided the device. This same letter contains lengthy numerical results with piezo techniques which Boyle wants to discuss with Rutherford in Manchester. In January there were further long letters about "your quartz piezoelectrique" and the first mention of a quartz device which would both transmit and receive supersonics; this suggestion definitely came from Boyle but he asks Rutherford whether such an experiment has ever been carried out. Through the latter part of January and most of February the correspondence concerns Boyle's move to Parkeston Quay and the delays in achieving it, but Boyle offers to test one of Rutherford's "vibrators" towed by a ship as soon as he gets to his new base. And in March, when he has arrived at Harwich, Boyle reports enthusiastically on the idea of using "valve after valve" to amplify the received signals.

This correspondence shows that in the last half of 1916 Rutherford and his collaborators were working exactly in parallel with Langevin and were quite as far ahead technically, with their eyes fixed perhaps more firmly than the French team on the use of supersonics for sub detection rather than signalling. Rutherford's notebooks fully confirm the vast amount of work he performed with "piezo". Nothing can detract from Langevin's successful development of practical devices which made sonar a reality, however, and it is probably typical of Rutherford that he began to lose interest in taking part in the development personally when the line of research began to change into a line of technological practicalities.

There was however a much more practical reason why Rutherford withdrew from the front line of "supersonics" in April 1917 – he was sent to lead the official British scientific mission to the USA on anti-submarine and naval matters. He undertook this task only after he had received a telegram from J.J. The telegram is no longer extant but we can deduce its contents because Rutherford wrote to J.J. on May 4th, thanking him for it, and remarking, "I wanted to be quite sure the mission was worth undertaking and not entirely of a complimentary nature." The story of Sir Henry Tizard's mission to the USA early in the Second World War, taking with him the secrets of British radar superiority such as the magnetron valve, and the early designs of the Whittle jet engine, is well known. It is rarely appreciated that Rutherford, who was Tizard's guiding light and political support during the 1920s and 1930s in government and defence science, had done almost exactly the same thing, indeed had set the precedent, soon after America joined in the First World War. With his experience of American conditions and his wide circle of American contacts, Rutherford was the ideal man to lead the mission. He took with him Commander Cyprian Bridge, who had performed much of BIR's early reconnaissance of French science; and there was a much larger French delegation travelling and working with them.

They set off for the USA on May 19th and stayed until July 9th, and apart from extensive discussions in Washington they visited Boston and the newly-built anti-submarine experimental station at Nahant; they went to the laboratories of Western Electric and General Electric; and Rutherford went to see L. V. King's echo-sounding work at Montreal. But, before starting, Rutherford had visited Paris to coordinate his approach with his French friends. There he had gone to the Ministry of Inventions and discussed submarine matters. He had met Langevin again, and Perrin and Debierne, and, of course, Marie Curie, "looking rather grey and worn and tired. She is very much occupied with radiology work, both direct and for training others," he told his wife.

Rutherford's report on his return from America was for many years covered by the Official Secrets Act, and was apparently not available to Eve and other early writers on his work. It is a fascinating document, written with that great clarity which Rutherford was invariably able to summon up when dealing with a non-scientific audience whom he wished to impress. This is the secret of the very great political, or governmental power that Rutherford was able to achieve in his later years; he could communicate very well and very powerfully with non-scientific civil servants and administrators. His report states:

> At the outset we think it may be said that as regards the acquisition of information we found very little of material importance beyond certain specific devices and proposals detailed in the Appendices . . . As

regards [giving all possible information and assistance] we are of opinion that the visit of the mission was of very great value to the American authorities by reason of the information on scientific and technical questions which we were able to give and also on account of the advice and assistance of the whole mission in the general scientific organisation for war purposes in America. From the above, it would appear that the immediate results of the visit of the mission to the United States are somewhat one-sided, but we are strongly of the opinion that the organisation of the scientific resources of the American nation, in which we were able to advise and cooperate, should produce results at an early date of great value, not only to America, but to the Allied cause in general. The French mission are in agreement with this opinion which is based on the knowledge of the enormous and hitherto unutilised scientific and technical resources which are available in America to an extent at present quite unobtainable in England or France. These resources comprise not only large numbers of highly-skilled scientists and assistants with numerous large and well equipped laboratories, but also practically unlimited mechanical assistance for the manufacture of experimental apparatus. In addition special ships, stations and submarines have been detailed for anti-submarine research purposes, and, if required later, excellent manufacturing facilities are available for the output of apparatus in quantity for general service use.

If there is some implicit criticism of the Admiralty performance in the last sentences that would only be clear to those who knew the inner history of the BIR. The remainder is a well-written statement of success achieved, which could be understood by politicians, ministers, civil servants or naval officers. What is even more fascinating is that in both phraseology and content it is a direct precursor of the reasons for sending the Tizard mission twenty-four years, just one generation, later, and its claims for success are exactly the same as those made in histories of the Second World War.

In both his official report and in his private letters to J.J., who was technically still his "boss" at BIR, Rutherford stressed how "opportune" the mission turned out to be. From the Hotel Powhatan in Washington on June 4th he told J.J., "The meetings were, I think, highly successful and we gave them a fairly complete general outline of the work done in England and France. I think our American friends had not been led to expect we had covered so much ground . . . They all state we arrived at a most opportune time . . ." He goes on to tell J.J. about American reports of work done with the Fessenden system by the commercial companies such as Submarine Signalling, Westinghouse, and General Electric: "As a matter of fact they get some results at a distance but they do not know themselves to what it is to be ascribed . . . We feel it well worth while coming so far to give the information they want here if their energies are not to be wasted in research on war problems." And so they went to the research stations "to see for ourselves".

At a different level in his official report he writes:

> We found that the American authorities were very conscious of the great potential value of science in its various applications to war, and that certain organisations of a scientific character were already formed and in process of formation . . . but it was soon apparent that they were much handicapped owing to lack of knowledge of the practical conditions and complete want of information regarding the scientific work already carried out in England and France . . . In this connection the arrival of our mission was most opportune . . . Although we found subsequently that they had then only progressed to a point which we reached a year or eighteen months ago, many of them were evidently of the opinion that they were already ahead of us. As a result of long conferences . . . in which we gave most detailed information of the work already carried out . . . this opinion was entirely corrected.

No impression should be left, however, that he criticised the Americans – he only regretted the previously poor flow of information. He wrote:

> In scientific matters . . . we found the American authorities most anxious to receive and act upon the advice we were able to give, and we feel convinced that the same spirit of cooperation exists in the other directions . . . and that definite expert advice is of great importance in order to avoid further delay in taking advantage of the great resources of the American Republic. We do not wish to imply . . . that the American authorities are lacking either in energy or initiative. Very much the reverse is the case.

Probably the biggest single result of Rutherford's mission was the setting up of a new naval research centre at New London. Rutherford claimed on his return: "We were also instrumental in the formation of a second experimental anti-submarine research station . . . at New London." He said this would lead to the reduction of congestion at the existing station at Nahant, but he made it quite clear that he thought the investigations of the Fessenden devices by the commercial electrical companies had come to a sticky end: "We arrived at the conclusion, with which the French mission were in complete agreement, that there was nothing to prove that the method proposed has so far produced any practical results of value, and we are very doubtful if it will be capable of practical development in the future. Our opinion on this matter was shared later by many of the American scientific men . . ."

In favour of the New London station he wrote: "Advantage will also be taken of the great university laboratory resources available at convenient distances. In the whole organisation for the attack of the submarine problem the closest possible understanding exists between the scientific representatives of the National Research Council and the officers of the Navy Department, and all requirements are supplied by the latter to the former with the least possible delay." Rutherford was certainly not above criticising the Admiralty and putting forward his own views by implication in the same way as the author of *Gulliver's Travels* had done. But he also provided a great deal of practical advice to the Submarine Committee of the National Research Council, which was headed by Millikan, the man who had finally measured the charge on the electron, about the best infrastructure to set up, about the grouping of problems for investigation in fields which Britain had already explored, and about countermeasures to quieten their own ships.

Not the least notable feature of Rutherford's mission and report was the excellent cooperation he achieved with the French, and perhaps the most important outcome of the mission was the cooperation he initiated with the Americans both at an official and an individual level. His American friends turned to him for advice after the war when the question of setting up a peacetime naval and national scientific system had to be faced.

Behind the success of the Rutherford mission to the USA lay the colossal physical and mental effort he put into it. The whole seven-week trip was another of those great outbursts of his working energy. He was, of course, unusually young to be the leader of such a mission, still under fifty years of age. But his letters to his wife tell the tale of a schedule which would have defeated many men. He arrived in New York late in May 1917: "Dinner party in the University Club by Dr Pupin and met a number of scientific men engaged in submarine question." In Washington on June 1st: "A very busy day . . . now hard at work", the hotel "not much of an hotel but Washington is so crowded that it is difficult to get a room". Visited the Secretary for War and the Secretary for the Navy; met Hale and Millikan; lunch at Army and Navy Club and reception by French Ambassador; "slept like a top under a sheet for the weather here is pretty warm". Next day, met his old friend Jacques Loeb by chance, and officially Dr Taylor in the Food Commission; dinner with Hoover, the Food Controller. Then "we had a great meeting discussing phases of the submarine question with the American committees – five hours in all today and more tomorrow. I took a prominent part in the question so have had a full and rather tiring day as the weather has been distinctly warm." Dinner with the French Commissioner and then an official scientific dinner given by Hale, plus "a long all-day committee on Saturday".

> We had breakfast in pyjamas on Sunday morning – grapefruit and tea – and then discussed organisation with Millikan and Co. At 12 we called on the British Ambassador . . . we go to lunch there on Tuesday after a meeting with Fabry and Co. Today we pay special calls and go in the afternoon to the Bureau of Standards . . . The weather is getting pretty warm but I sleep well in pyjamas without a sheet and generally my constitution stands well the strain of so many lunches and dinners.

A day later he is admitting the heat is steamy but they went to a music hall in the evening after dinner with some British naval officers. The following day lunch at the Cosmos club with Professor Stratton: "I met there a number of old scientific friends." Then there was the official dinner with American scientists and British and French Ambassadors, at which Rutherford had to speak. "I felt pretty weary this morning but had a long meeting with admirals, scientific, naval and military men and had to speak for three parts of an hour. I felt very sleepy and tired but we had three hours of it in all. I was bored stiff by the last hour which was very uninteresting. I tried to get a little rest this afternoon, but with no success, for the telephone bell summoned me to another interview and further work." On June 11th he went to New York and then on to Boston, after two more official dinners in Washington and a motor trip to the Civil War battlefields. He was still "fairly fit and well considering our late hours and continuous talking".

In Massachusetts he met old friends and students, Bumstead and Zeleny, Kovarik and Boltwood, who had "changed over his views completely and is now busy working on anti-submarine devices". He received an Honorary Degree from Yale, and visited the new scientific anti-submarine base at New London. Back to New York to visit Thomas Edison, in the inventor's laboratory at Orange – "received very well by the old man who was as enthusiastic as a schoolboy over his ideas". Dinner with a man from Bell Telephone, up to Albany to spend three days at the General Electric labs at Schenectady. Three days in Montreal and then back to New York to see Western Electrical and thence home. His full report to BIR was delivered seven days after his return to England.

Whether he appreciated his mother's comments on this adventure is not known. She wrote congratulating him on meeting

Edison and on getting his Honorary DSc from Yale which "make pleasant reading for proud parents, for though your father does not say much he is glad and proud of the distinctions won by you . . . You cannot fail to know how glad and thankful I feel that God has blessed and crowned your genius and efforts with success. That you may rise to greater heights of fame and live near to God like Lord Kelvin is my earnest wish and prayer . . ."

But the receipt of Rutherford's report was one of the Board of Invention and Research's final formal acts. It had become quite clear that an informal gathering of scientists, who were mostly Professors still working officially in their universities, could hardly continue to run an organisation with quite large resources such as the establishment at Parkeston Quay, and could hardly operate at their best in the face of continued obstruction by the Admiralty. Sir Robert Sothern Holland and Sir H. Ross Skinner were deputed, by the Admiralty, to investigate the activity of BIR and to "ascertain what can be done to give more expeditious and practical effect than is at present the case to the adoption of new methods and inventions", according to the wording of a letter to Lord Fisher from Sir Eric Geddes. It took these two eminent civil servants rather less than a month to visit some dozen naval establishments and interview some forty scientists and naval officers.

The result of their work – the Sothern Holland Report – is an extraordinary condemnation of the Admiralty and the attitudes of the Royal Navy of the time. Their conclusions – and they were neither of them scientists – were exactly what the scientists of the BIR would have wished. Men of the greatest scientific knowledge were not being used to the fullest extent and were being wasted on committee work; BIR was not in close enough touch with the Admiralty and vice-versa; the scientists were not in and among the problems they worked on; the BIR was not constituted to allow of essential driving and coordinating power; and the Admiralty was multiplying experimental establishments, leading to dissipation of forces and confusion. They proposed, therefore, that BIR should be abolished, and that the Admiralty should establish a powerful Director of Research and Experiments, with a new central Establishment of Research and Experiments, and a panel of scientists to advise the Director.

The rest of the Sothern Holland Report consists of some twelve pages of continuous condemnation of the Admiralty, beginning with: "The navy up to quite a recent date has possessed no research institutions and such small establishments as now exist have grown up in spite of little encouragement and the absence of any general plan." Then we have: "Owing to the serious menace of the submarine the Admiralty has suddenly to mobilise scientific men to aid the navy in the attempt to find means for defeating the submarine . . . But why was the navy without suitable instruments for detecting the movements of submerged craft by sound?" and it is pointed out that Leonardo da Vinci had suggested how to do this. Furthermore, "Even at this late stage of the war it is not considered that the problem is now being grappled with sufficient earnestness or with sufficient vigour."

In the remainder of the report facts and figures are given proving the claims of duplication and dispersion of forces among the navy's establishments; giving evidence of the Admiralty denying information to the scientists who were supposed to be helping it; showing how the scientists were arbitrarily denied facilities and equipment. It is a sorry story, but only one paragraph in it refers directly to Rutherford by name. To illustrate their claim that scientists were denied both sympathy and information by the navy, the authors provide a footnote from the Minutes of the BIR Sub-Committee 2, dated March 27th, 1917, which reads:

> In the course of the discussion which ensued Sir E. Rutherford, Mr Threlfall, Professor Bragg and the Secretary referred to the feeling which was general among the scientific members of the committee that their utility to the navy in connection with the anti-submarine question was largely impaired by lack of information as to the actual service conditions and as to the nature of the methods already employed for the detection and destruction of enemy submarine depots. BIR has been informed for example that it was not possible to put them in touch with the anti-submarine depots. It was pointed out that in scientific research it was found to be essential that the researcher should have the widest knowledge and personal experience of the difficulties to be solved. Commodore Hall dissented and expressed the view that the only information necessary to be given was that the enemy submarines were in the sea, and that means were required to detect their presence.

The mind may boggle at the thought of Sir E. Rutherford's response to the Commodore, which perhaps fortunately is not recorded. But in the Commodore's defence it must be said that the same attitude was to be found in serving officers in the Second World War, and it was one of the fruits of scientific effort in the Second World War that the services became convinced that this attitude was not productive.

The recommendations of the Sothern Holland Report roused fury in the Admiralty. According to a modern historian, Gusewelle[1]:

[1] In *Naval Warfare in the Twentieth Century*, ch. 7.

Jellicoe's approach was to challenge the honesty of the Holland Committee and to blame every possible person and/or circumstance for whatever failures or errors of judgment the Admiralty may have committed. He went on to make the preposterous observation that the BIR scientists were interested only in personal gain both material and professional. There is absolutely no evidence to indicate that the distinguished scientists of the BIR misused their trust, and in fact they made considerable personal and professional sacrifices in order to do war work.

Certainly the Admiralty managed to avoid implementing the recommendations of the Sothern Holland Committee. True they appointed a Director of Research and Experiments (Sir Charles Merz) but "His role was nominal. In effect complete supervision of scientific research was placed in the capable hands of the Admiralty technical department officers." No big central research establishment was initiated, though an Admiralty Research Laboratory was set up and a small Naval Scientific Service, split up into several stations – and incorporating such men as A. B. Wood – was kept up between the wars. The navy did not learn the lesson and at the start of the Second World War there was a repeat of the 1915 emergency call for scientists to help deal with problems such as the magnetic mine. The main, and most successful work, in the inter-war years was the development of ASDIC.

The demise of BIR did have one important result for Rutherford, however. From the ashes of the organisation and from the lessons learnt, and using many of the people involved in BIR, Balfour provided the Committee for Scientific and Industrial Research, which duly became the influential Department of Scientific and Industrial Research, with which Rutherford was to become deeply involved in the years of peace, and through which he was to exert much influence. And here again, as so often in Rutherford's life, we may find the pioneering wisdom of J.J. at work. In July 1915, when he first invited Rutherford to join BIR, J.J. had written: "I hope we shall be able to do something useful both for the prosecution of the war and for the introduction of a closer coordination between science and the navy. If we could get a good scheme for this it would be a very important thing as other government departments would probably adopt it, and the principle of making the fullest possible use of scientific help would I think spread to most of our industries and business." J.J.'s son, Sir George Thomson, told Balfour's biographer of his father's attitude to the BIR: "This was one of the first examples of government science. I don't know that it had any influence on the war, and my father complained that the work they did on the detection of submarines was ignored until there was a sudden panic, when it was too late to develop it properly. However, I think it probable that the existence of this body helped in making science known to the governing classes."

When BIR ceased to operate, a number of local committees of scientists and businessmen continued to help in the anti-submarine war. Rutherford was a member of the Lancashire Anti-Submarine Committee until it was wound up in 1919. He also continued to work on an inter-Allied high-level committee. It was at one of their meetings that he made his famous remark about having split the atom which he considered to be of greater consequence than immediate war work.

He certainly attended the inter-Allied conference on the super-sonic method of detecting submarines which was actually in progress when the war ended in 1918. He gave his wife a graphic description of a cross-Channel journey: "We crossed over in the regular boats used by soldiers on leave and they were crowded with Tommies and Officers. Everyone had to wear a lifebelt. Three to four fairly swift boats leave together and are escorted by destroyers and airships, the journey taking about 1½ hours." In Paris, where he again met Langevin, he found the French in very good spirits: "The Place de la Concorde (where the old guillotine used to work) was filled with a great number of captured German guns and there was always a big crowd looking at them. The statue of the City of Lille was decorated with flowers as I was there only a few days after its liberation . . . It is a very exciting time to live in but people here are very quiet and refrain from celebrations until our main enemy goes under."

Rutherford could appreciate and share in the joy at the ending of the war, but his deepest attention was, as always, upon his own scientific work, and his true interest had indeed passed on from submarines and acoustics and returned to the incredibly minute world inside the atom.

13

The Atom is Smashed

To those of us who live with the threat of nuclear weapons and the promise of nuclear power it must appear that the splitting of the atom was Rutherford's most significant achievement. But a deeper perspective presents this achievement as but one of several major discoveries which have given us our view of the material world. Radioactivity, under Rutherford's investigation, had shown that the atom must exist and that it was not indivisible. The theory of the nuclear atom had shown what the broad structure of atoms must be. The splitting of the atom – more precisely the splitting of the atomic nucleus – was but the first step in showing what were the contents and structure of the nucleus; it was the first discovery of sub-nuclear science, a penetration into a yet deeper layer of the knowledge of the universe.

It was also the classical example of the Rutherford method at work. First there is the noting of an anomaly, a strange result, which he was the first to realise might open up an entirely new "Tom Tiddler's Ground" – one of his favourite phrases. Secondly there was the slow and deep contemplation of the new area of knowledge. At the end of this it appears that he "knows" what the answer must be because he can "see" what is actually happening. Then comes the furious outburst or outbursts, of experimental energy and imagination, in which he eliminates every other possible explanation. And finally there comes the cautious, carefully phrased, almost timid, statement of a revolution.

In this particular case Rutherford did almost all the work himself, assisted only by William Kay, the Manchester laboratory steward. The work was done in his spare time, in odd moments when he could break away from the problems of anti-submarine warfare and the running of a university teaching department in wartime. Perhaps because it was all his own work his laboratory notebooks have survived to tell an unusually complete story of his successes and failures, the blind alleys up which he stumbled, and the successful elimination of every alternative he could conceive of to account for that original anomalous result.

The start of this story can be accurately dated to 1914. At the start of the summer term of that year Rutherford obtained exceptional leave of absence from Manchester University, and paid for replacement lecturers, in order to give the newly-instituted Hale Lectures to the National Academy of Science at the Smithsonian Institute in Washington. The title of these lectures was "The Constitution of Matter and the Evolution of the Elements". They were hailed by scientists as a great success and by the shorthand writers as "impossible to transcribe"; we must therefore take his scripts as being the true representations of what he thought about atoms and other matters at the time.

"The great majority of scientific men regard the atomic theory not only as a working hypothesis of great value, but as affording a correct description of one stage of the subdivision of matter," Rutherford declared at the start of his first lecture; he was of course referring here to the chemical atomic theory, and he added that only twenty years before there had been a limited revolt against that theory by those who held it "was of necessity unverifiable by direct experiment and should therefore not be used as a basis of explanation of chemistry". But radioactivity (and, by implication, Rutherford) had changed that: "It is somewhat remarkable that while the study of radioactive phenomena has clearly indicated that the atom is not always permanent and indestructible, it has at the same time supplied the most convincing proof of the actual reality of atoms and has provided some of the most direct methods of determining the values of atomic magnitudes."

The atoms whose existence he had demonstrated were, however, the seat of enormous quantities of energy, he said, and went on to discuss the meaning and problems of atomic energy at some length and with what seems almost superhuman foresight, tempered only by a sound appreciation of the technology of his time. "It can easily be shown that the energy emitted from a radioactive substance which expels alpha-particles is several million times greater than the energy emitted from an equal weight of matter in any known chemical reaction . . . During the life of the emanation the total energy emitted corresponds to an engine working at 128,000 horsepower for one day." Hence it followed that any appreciable amount of emanation would be an enormously concentrated source of power, "many million times greater than for an equal weight of the most powerful known explosive". The reason for all this, and for the non-availability of the energy, was that the energy "is in the atom itself".

". . . The atom is believed to consist of a large number of positively and negatively charged particles which are collected in a very small volume and held together by intense electrical forces";

". . . such a structure involves a large store of energy"; "the enormous reservoir of energy that must exist in the atoms themselves"; these are the phrases Rutherford used to impress his audience with the idea that the atom and energy had to be considered together, for, as he pointed out, there was every reason to believe that all the ordinary atoms contained this energy as well as radioactive atoms. But, here comes the practical man, "This store of energy does not manifest itself and is not available for use"; the energy is only liberated when "a drastic rearrangement of the atom resulting from an atomic explosion" occurs. And he ends this section with a remarkable paragraph. "There is at present no evidence to indicate that we shall be able in any way to influence radioactive changes, . . . we can only watch and investigate." He was very doubtful of the prophecies recently made by H. G. Wells that the atom's energy would become available.

> This will only be possible on a large scale if we are able to alter in some way the rate of radioactive change and to cause a substance like uranium or thorium to give out its energy in the course of a few hours or days, instead of over a period of many thousands or millions of years. The possibility however of altering the rate of transformation of radioactive matter or of inducing similar effects in ordinary matter does not at present seem at all promising.

The object of this discussion of the energy inside the atom was not to make predictions about a possible future use of atomic energy by man, it was simply drawing the map of the atomic world as Rutherford saw it at that time, and since the essence of Rutherford's grasp of the atomic and sub-atomic world was a perspective of the quantities and sizes involved, it was necessary to include the energies and forces involved. He went on to describe "the idea of structure within the atom" which was implicit in those energies. He pointed out that J.J. had proved the existence of the negatively charged electron, but that there was "no sign of a corresponding positive charge ever found associated with a mass less than the hydrogen atom" so it seemed doubtful "whether . . . a positive electron exists". Considering that the main drive of his scientific work at that moment involved demolishing J.J.'s concept of the structure of the atom, he was remarkably generous to his old master: "The Thomson atom proved for many years very useful in giving a concrete idea of the structure of the atom, and had the great advantage of being amenable to calculation."

Rutherford proceeded to give the history of the discovery of the "nucleus" atom and the evidence for believing that his own view was right, but he several times emphasised that it was only a hypothesis, a "supposed" structure of the atom. In a previous chapter this work has been described in detail; it is only necessary here to pick out a few remarks which show important new trends in his thought. He emphasised, for instance, that his nucleus atom was not the same as the Saturnian model, for in Rutherford's view the nucleus, although it contained almost all the mass of the atom, was, in physical size, smaller than the solitary electrons. He explained this apparent contradiction by calling to his aid the electromagnetic theory of mass -- the great mass of the nucleus is due to its colossal concentration of electric charge, and he pointed out that if charged particles such as electrons are closely packed together their fields must interact, and Lorentz had shown that in such a case the mass of a clump of closely packed electrons was not necessarily the same as the sum of the individual masses.

And so he pictured the nuclei of all atoms – "It is to be anticipated that the nuclei of all atoms are built up of positive and negative electrons," and the positive electrons, which are the nuclei of hydrogen atoms, are always in excess of the negative electrons so that the nucleus always has a positive net charge. So the nucleus of a helium atom would be likely to consist of four positive electrons and two negative electrons, giving the final double positive charge which was carried by helium nuclei, his beloved alpha-particles. Possibly this arrangement was a particularly stable one, which might well be a regular sub-feature, or component, of the nuclei of all atoms. But Rutherford was always honest and rarely, if ever, fudged an issue. He admitted here, for instance, that there was no evidence that negative electrons existed in any nucleus, but presumably they must be there for they were shot out as beta-rays in radioactive transformations which were certainly nuclear events. Here Rutherford shows that he was caught in a well-known philosophical trap; it is not logically necessary that what comes out of something must have been inside it before it came out. This fallacy was to lead Rutherford into a number of blind alleys in the years ahead, until the mathematics of quantum mechanics got him, and science in general, out of the impasse; and it remains true today that although electrons, as beta-rays, and helium nuclei, as alpha-rays, emerge from the nucleus, we still have no evidence of their independent existence as such within the nucleus. (Indeed, the phenomenon of the emergence of an electron from the decay of a neutron is the archetypal display of one of the four fundamental forces that control the universe, the so-called "weak" force which was undreamt of in Rutherford's time.)

With hindsight we can see that the most significant of the extra remarks that Rutherford made in this section of his lecture was a reference to some very recent work in his Manchester laboratory. Explaining how Geiger and Marsden's work on scattering alpha-particles had demonstrated his nucleus theory he added that "H-particles", separate nuclei of hydrogen atoms, which he considered positive electrons, had been observed causing scintillations on the little screen after travelling four times as far as an alpha-particle and with 1.6 times the velocity of an incoming alpha-particle – exactly the speed and distance which one would expect from elementary considerations of the conservation of momentum if an alpha-particle had struck a naked hydrogen nucleus and pushed it forward as one billiard ball strikes and pushes another. Rutherford claimed that this observation supported his general theory, but the significance from our point of view is that it shows his mind was already dwelling on this phenomenon and beginning to incorporate it into his personal internal map of the atomic world.

He concluded this first lecture by bringing his audience right up to date with the latest views on the structure of the atom and the evidence for it – namely he explained Bohr's refinement of the "nucleus atom". In Bohr's view, Rutherford explained, an electron could temporarily occupy any one of the number of orbits which were fixed in relationship with the nucleus, while in moving from one orbit to another (technically from one stationary energy state to another) an electron emitted radiation of a frequency which was connected to the difference in energy between the two orbits by Planck's constant. Using this theory Bohr had been able to account for the spectra of hydrogen and helium, and his theories had been supported by work in Manchester by Fowler and Evans. "While there is room for much difference of opinion as to the interpretation of the rather revolutionary assumptions made by Bohr . . . there can be no doubt of the great interest and importance of this first attempt to deduce the structure of simple atoms and to explain the origin of spectra," said Rutherford cautiously before ending on a rather higher note: "The agreement of the properties of such theoretical structures with the actual atoms is in several cases so remarkable that it is difficult to believe that the theory is not in some way an expression of the actual facts."

The second of the Hale Lectures was much more speculative – yet it shows how successfully Rutherford was capable of looking beyond his cautious attitude to real, proven, quantifiable results, towards what was likely to emerge. He was for ever creating, and was always willing to display, "glimpses of a possible universe". But more importantly, perhaps, this second lecture, "The Evolution of the Elements", shows the direction in which his mind was moving, the way in which he was probing towards the structure of the nucleus. Speculation had long been that it was "probable" that the many different elements were built up from some fundamental substance, and the two branches of science which had supplied suggestive evidence for this idea were astronomy and radioactivity. He called up Lockyer's observations of the spectra of hydrogen and helium in the sun and the hottest stars, and he referred to the discovery of the spectra of other elements in cooler stars. Perhaps this meant that as temperatures decreased the lighter elements, hydrogen and helium, combined to form the heavier elements – a speculation in which he was totally wrong quantitatively, yet with which he was plainly exploring in the right direction.

The evidence against the possibility of building up heavy elements from lighter ones was then produced: "No indication has yet been observed that the radioactive processes are reversible under any ordinary conditions", and "No new elements have ever been made by electric discharges", the last statement being of course a burial of Ramsay's claims to have achieved exactly that. Radioactivity showed the enormous energies that would be needed to build up new and heavier atoms by shooting the lighter atoms of hydrogen or helium into existing atoms, by the enormous energy of the particles they shot out, energies of a class far higher than anything that could be produced in the laboratory. Furthermore it was, by definition, difficult to obtain any direct evidence on the behaviour of such small objects as atoms. But Rutherford finally took the audience into his confidence: "I personally am inclined to believe that all atoms are built up of positive electrons – hydrogen nuclei – and negative electrons, and that atoms are purely electrical structures."

His peroration in the typewritten script concludes that "the fundamental problem which lies at the basis of physics and chemistry" was the problem of the structure of the atom. But fortunately the shorthand writers, who had been so baffled by most of the lecture, managed to get at least the last paragraph. What Rutherford actually said was:

> In regard to the actual structure of the atom to explain the physical properties and explain the chemical properties, etcetera, I will say, in fact, instead of these very simple atoms, it is obvious that there is going to be a very complex affair, and will not be done in this generation or the next and probably not completely for many years, if at all, or for many hundreds of years, because the constitution of the atom is of course the great problem that lies at the base of all physics and chemistry, and if we knew the constitution of the atoms we ought to be able to predict everything that is happening in the universe.

He underestimated his own powers and his own future successes; he underestimated the growth in science and the number of scientists who would attack his basic problem; and he underestimated the surprising results which Bohr's pursuit of the quantum theory would expose, such as Heisenberg's uncertainty principle.

But even while Rutherford was delivering these lectures in Washington the next stage of attack on the "basic problem" was being prepared, unknowingly, in his laboratory at Manchester. At that memorable supper party in his house when Rutherford had first produced the idea of the nuclear atom, Darwin had pointed out to him that all his calculations were based on considering a small alpha-particle approaching the nucleus of a heavy atom, a nucleus much larger than itself. It will be remembered that Rutherford was much excited by Darwin's idea that there should be interesting results when an alpha-particle met a nucleus lighter than itself, such as the nucleus of a hydrogen atom, and Rutherford had encouraged him to write a paper on this. Darwin's calculations showed that an alpha-particle hitting a hydrogen nucleus should propel the hydrogen forward at a speed 1.6 times that of the incident alpha and with a penetration or range of some four times greater than that of the alpha. This penetration calculation was based on the "law of absorption" which the elder Bragg had produced from his earliest researches on the range of different alpha-rays. When the main work of "proving" the existence of the nucleus by measuring alpha-particle scattering had been completed by Geiger and Marsden, and Geiger had returned to his native Germany, Marsden set about providing experimental evidence for this other form of scattering where the alpha-particles impinged on nuclei lighter than themselves.

It was immediately found that "H-particles", the rapidly moving nuclei of hydrogen atoms, could be produced by the impact of alpha-particles on a suitable target. These H-particles caused the zinc sulphide screens used in all these scattering experiments to scintillate in a way which was clearly different from the effect of alpha-particles. The H-particles produced a smaller, more point-like flash of light, and they could be observed at ranges of about ten centimetres in air, much farther than ordinary alpha-rays would travel in air without being completely absorbed. All this had been discovered in time for Rutherford to tell his American audience how the results confirmed his nuclear theory.

But as the summer of 1914 wore on Marsden began to have trouble with his experiments; "a serious anomaly showed itself". Under certain conditions of absorption in air, and especially if he examined H-particles emerging obliquely, Marsden found far too many of the H-particles appearing, far more than the impact of alpha-particles could account for. He assumed that he was suffering from hydrogen contamination in his apparatus, probably hydrogen in dust or grease particles contaminating his radioactive source. But whatever he did he could not get rid of it, he could not reduce the number of H-particles flashing on to the tiny zinc sulphide screen. By the time Rutherford was in Australia and the war had begun, Marsden was forced to the conclusion that H-particles were being shot out of the radioactive source itself. Under pressure of war conditions, complicated by the fact that Marsden himself was moving to take up his first university teaching job, out in New Zealand, the work was rather hurriedly wound up, written up as "The passing of alpha particles through hydrogen" which appeared in the *Philosophical Magazine* as two articles, one in 1914 and the other in 1915, and included a passing mention of the anomaly with the conclusion, "There seems a strong suspicion that the H-particles are emitted from the radioactive atoms themselves, though not with uniform velocity."

When Rutherford got back to his laboratory in 1915, he was faced with the exciting possibility that his radioactive atoms emitted something else as well as alpha-, beta- and gamma-rays, but he was also faced with a war and the submarine menace, and no Marsden. He wrote to Marsden in New Zealand, asking his junior if he would "mind" the Professor continuing the experiments himself. Marsden certainly did not "mind"; he was much more impressed with Rutherford's kindness in pressing the Royal Society to get a supply of radium sent to New Zealand for his use. And although Marsden insists that Rutherford agreed with the possibility that a few of the H-particles might come from the radioactive source, it is clear that no serious thought was really given to this possibility, except as one explanation which had to be eliminated.

Rutherford's laboratory notebooks from his return from Australia in January 1915 to his departure for Cambridge at the end of 1919 are remarkably complete but they are also a dreadful mess. All his time and work seem to be accounted for in these twelve volumes (NB 13 to NB 25); for instance there is a clear gap in the summer of 1917 when he was away on his American mission. But only some experiments are dated, and the notebooks represent faithfully a picture of the times. There are experiments and results which plainly are part of his war-work on submarine acoustics; these are interspersed with short bits of work on academic, primarily radioactive, subjects, where it is obvious that in his few moments of spare time he went back to his true love. Plainly he had about six of these notebooks "on the go" at the same time, and on any day might pick up any of them: Notebook 17, for instance, contains work from both 1915 and 1917; Notebooks 19 and 20 contain both acoustic work with piezoelectrics from 1916 and vital work on atomic constitution from late 1917.

This confused story becomes clear only when we know what the end result was. What Rutherford eventually showed was that the H-particles seen by Marsden were the chips flying off the atomic nuclei of nitrogen atoms when those nuclei were hit by alpha-particles. In modern terms the nitrogen atoms were turned into oxygen atoms by having a proton (the nucleus of a hydrogen atom, an H-particle) knocked out of their nuclei by the impact of the alpha-particle. Thus, in a sense, it was Marsden who first split the atom as early as 1914, but he had no knowledge of what he was doing, and it was Rutherford who saw what was happening, who understood and explained the process, who provided the evidence proving the case, and who was first, deliberately and knowingly, to smash an atomic nucleus. Furthermore, the notebooks show that he knew what he was doing as early as 1917, and that the last two years of this work, up to the formal publication date in 1919, were spent essentially in collecting all his evidence and eliminating every other possibility. It was the supreme example of the Sherlock Holmes dictum that when you have eliminated every other possibility, what remains, however unlikely, may be the truth.

Rutherford finally presented this work as a logical continuation of the main thrust of all his developments during the Manchester period, the outcome of Darwin's first thoughts when the idea of the nucleus was announced. There are four papers under the collective title of "Collision of Alpha-Particles with Light Atoms". The first three deal straightforwardly with what happens when alpha-particles strike hydrogen, nitrogen and oxygen atoms, and also cover briefly the effects of alpha impact on a large number of other light elements – lithium, boron, carbon, sodium, and so on. The fourth paper, apparently a mere addendum no more than five and a half pages long, is titled "An anomalous effect in nitrogen". In its very last section it concludes cautiously:

> From the results so far obtained it is difficult to avoid the conclusion that the long-range atoms arising from collision of alpha-particles with nitrogen are not nitrogen atoms, but probably atoms of hydrogen . . . If this be the case we must conclude that the nitrogen atom is disintegrated under the intense forces developed in a close collision with a swift alpha-particle and that the hydrogen atom which is liberated formed a constituent part of the nitrogen nucleus.

The whole mass of four years' work ends with the prediction: "The results as a whole suggest that, if alpha-particles – or similar projectiles – of still greater energy were available for experiment, we might expect to break down the nucleus structure of many of the lighter atoms." Today, more than sixty years later, scientists in specially built laboratories that have cost tens of millions of pounds are still vigorously following up that vision.

But Rutherford's published results do not tell the real story of what he did – as is the case with many scientific papers. The notebooks show clearly that right from the start of the four years' work involved he was "chasing" Marsden's H-particles. The start of this chase began with a number of different, and apparently disconnected experiments, rather like a hound casting around for a good scent. There are gamma-ray experiments in March and April 1915, but then no more is heard of this subject. In June 1915 there are "Measurements of current in H (Hydrogen) and He (Helium) experiment". Everywhere there are the very human remarks with which all Rutherford's notebooks are full: "Done very carefully", "No good, unsteady", "Easter Monday", "Sunday evening", "No good", "Stopped here for lunch . . . Repeated after lunch".

In three notebooks – Numbers 17, 18 and 19 – it is quite impossible to reconstruct the sequence of action, as the dates jump about so much. Sometimes they are even in the wrong order within one year – as when work of March 1917 directly precedes similar work on "H-scintillations" of January 1917. But it seems fairly clear that his attention throughout most of 1916 was occupied with acoustic problems and particularly his work on piezoelectrics. His thinking about matters atomic at this time must have been largely on the wrong lines, for it is equally clear that in that same year he was thinking that the H-particles might have revealed the presence of a previously unknown light gas. It must be remembered that Ramsay's discoveries of the noble and inactive gases argon, neon, krypton and xenon, and his isolation of helium were among the outstanding scientific results obtained during Rutherford's lifetime. Rutherford himself, with Soddy, had discovered new gases, the emanations radon and thoron.

But on December 6th, almost certainly of 1916, he set up a new experiment to "Test for the presence of light gas in small quantity by conductivity methods". It was a version of some of Ramsay's work, with a platinum heating coil sealed in complicated arrangements of glassware. There was a connected system producing hydrogen by electrolysis, and another platinum spiral in a water tank. At another stage helium was carefully purified. The changes of resistance with temperature were carefully measured in the first arrangement of glassware. Then there were also tests for "H" among the gases liberated by the rare mineral samarskite. The whole project came to nothing and very soon afterwards, early in January 1917, the whole complexion of the notebooks begins to change – the hound had picked up a promising scent, Rutherford's scientific "nose" had given him a lead, and the pace and concentration become more intense.

But the idea of an unknown light gas continued to haunt him. One of his famous lists of projected researches for the immediate post-war period includes a proposal for a student of his to search for such a gas. In the same notebook as his own experimental results on the subject, there is a section devoted to speculations about the sun's atmosphere, where helium had been found. The notes are plainly a discussion between Rutherford and himself; they read at one point: "This discussion is based on the assumption that radioactive matter exists in the sun's outer envelope. For simplicity suppose this is uranium plus radium . . . It is reasonable to suppose that the outer limit of the sun's atmosphere is mainly hydrogen with which radioactive matter is mixed in small proportions."

He goes on to theorise about the impact of alpha- and beta-particles on hydrogen, and among pages of rough calculations and "theoretical physics" he considers the work of C. T. R. Wilson on the "potential gradient of the atmosphere" and the "penetrating power of the aurora and constitution of the earth's upper atmosphere" apparently considering the possibility that light gas atoms might reach the earth from the sun. There is a date of September 22nd, 1917 on one of these pages and this is unexpected because later notebooks show him hot in pursuit of his pure target throughout that month with quite different work in progress on the 19th, 20th and 27th. It is therefore tempting to speculate that he had put the wrong date on his thoughts about the sun and its atmosphere, and that these notes belong to September 1916. Their place in the disturbed sequence of the notebook would justify this redating, and it can be shown in other places that Rutherford sometimes got his years muddled. However the light gas remained a possibility in the back of his mind throughout the rest of his life, and in the end he was justified, for his final, and much-neglected triumph, was to demonstrate the existence of tritium, hydrogen with an atomic weight of three units, the fuel of hydrogen weapons and neutron bombs in our days, and the light form of helium, helium three, yet another light gas.

1917 was the crucial year – though we have to go backwards in the notebooks (to Numbers 17 and 18) to find the work that was so important. At the very start of the year – January 4th is the first certain date – Rutherford started a careful study of the "Range of H-atoms". He immediately met experimental difficulties and noted: "Preliminary observations failed due to inadvertence in not protecting alpha-rays due to handling brass apparatus." It seems that he had devised the idea that the anomalous H-particles might be produced in the thin sheets of metal or mica which were used to absorb the alpha-rays and thence to measure velocities. Four days later he is puzzled: "Difference on wrong side if H atoms produced in Mica," he scribbled.

The notes for February 28th mark a crucial step forward. They show a crude little drawing of the equipment with which he finally solved the problem. A battered little brass box, with pipe attachments for introducing the required gases was all he needed; inside is a movable mounting arm on which he could place his radioactive source at a measured distance from an exit hole at one end. Across this exit hole he placed thin foils of gold, aluminium or mica to absorb the alpha-rays produced from the source; they allowed H-particles to emerge, however, so that they could be counted through a microscope observing a zinc sulphide screen mounted immediately outside the exit. The remains of the apparatus are still preserved at the Cavendish Laboratory, and it is difficult for anyone who has visited the enormous "atom smashers" of our time, great circular concrete tubes a mile or more in diameter, to realise that it was in something so small and crude and simple that the atom was first split.

The notebooks show that in March 1917, he was still pursuing the idea that the H-particles were expelled somehow from the metal foils. "Experiments on H-scintillations March 13th, 1917," he writes, and then notes, "Absorption of H-particles from active matter – Decay of H-atoms from active matter." It was a good day's work, for he scribbled at the side: "Continuation of the day's expts. Good and reliable curve. Fall off rapidly at first and then more slowly." But puzzlement creeps in. After "Important results on production of particles in aluminium and mica", we find "Aluminium effect ? to be verified by another experiment." And later, when using gold foil, there comes: "Experiments not very good, probably hydrogen not pure." Eventually there comes a "Summary of results" showing even worse puzzlement – there were more H-atoms from platinum foils than from brass; there appeared to be more in one plane than at right angles. There was "no effect in this case of production of scintillations from aluminium or mica", and so finally: "It would appear large number of high-speed

H-atoms must be repeated; results quite contradictory from old experiments."

Parallel with these results there is a whole notebook full of calculation and rough theories. There is only one date in it, March 8th, 1917. But though it is boldly marked on the outside "H-Atoms", and it opens firmly "Production of H-atoms and other high-speed atoms", the puzzlement through which Rutherford was passing is clearly marked. He starts with Darwin's original calculations on the energy of light atoms hit by alpha-particles, and applies these to atoms of hydrogen, carbon and lithium. Then he tries to work out which atoms should be detectable if the theory is correct: "Suppose H-atoms can be detected of one-sixth the energy of an alpha-particle from radium-C." This is followed by detailed working out of the scattering theory, with the remark: "This agrees with Marsden's experiments." But later there is trouble – a table of the relative energies of nine sorts of atom recoiling from alpha-particle impact has a firm "wrong" in the margin beside it. And next comes the remark: "Range therefore difficult to estimate unless change of effect of mass is known." By now things appear to have got somewhat out of hand, for in ink we read: "Therefore iron atom should be detectable" and there is an insertion of "not" in thick pencil. This is followed by a "return to square one" with references to Sir Horace Lamb's formulae and calculations on whether the inverse square law or the inverse cube law describes the forces acting on an alpha-particle approaching a nucleus. There are even scribbled reproductions of his own "hyperbola" sketch of an atomic impact, and several pages of scribbled calculations are marked "Good". But then come two pages marked "No good".

The very last pages of this notebook show, however, an interesting and significant change of the direction of his thought. He now starts considering the gases involved in his experiments. There are some notes on "Absorption of nitrogen atoms in hydrogen" and a mysterious "Numbers to be expected from nitrogen" and "Carbon from carbon dioxide not detectable."

The complete break from laboratory work during the summer of 1917, while he went on his American mission, shows not merely by the absence of any dated work between March 1917 and September of that year, but also in the character of the notes after September. From that point onwards there is a clear and enormously energetic pursuit towards the final conclusion. To use a Rutherfordian phrase, "it is difficult to avoid the conclusion" that he had come up with the right answer, the correct explanation of Marsden's anomaly, while he was in America; that powerful mind had been brooding on the problem and on those explanations which he had already excluded. He was left with but one reasonable explanation – whatever was happening was happening in the gas inside the experimental box. He opened a new notebook as soon as he got down to work in Manchester. "Range of high speed atoms in air and other gases, Sept. 8th, 1917," he wrote at the start.

It is quite probable that he already "knew" that the answer must be that the alpha-particles were splitting nitrogen nuclei, for on September 11th he was measuring "final particles after absorption of nitrogen atoms" and he noted that these final particles had "slow absorption like hydrogen, cannot be carbon". Then in "test to see whether new atoms come from air" (air consists mostly of nitrogen), he performed "not very satisfactory experiment but seems to indicate nitrogen atoms reduced in number . . ." In the following week he managed to show that helium under alpha bombardment produced no "long-range particles", but his experiment to show the same with oxygen failed: "No definite evidence drawn on account of contamination," he noted.

October and November 1917 saw one of the Rutherford outbursts of creative energy – for spells of days on end he did nothing but experiment, struggling against difficulties with his apparatus, with problems such as the irregularities in the thickness of the gold or silver or aluminium foils he was using. He had enormous problems in getting supplies of absolutely pure gases – contamination is a comment that appears with dreadful regularity. Furthermore, the actual counting of the scintillations on the tiny screen had all to be done by Rutherford himself or the faithful laboratory steward, Kay. Each set of observations is carefully marked with the name of the observers – "Kay observer" or "R obs". At one stage the gold foils reflected so much light that observation of the scintillations proved impossible. One day in October he noted, "Results very unsatisfactory due to strong luminosity," but on the next day, "Good – silver instead of gold foil." That remark came with "myself observer", followed by "Kay in afternoon, erratic and not so good". It was necessary to adapt the eyes to the dark before starting to count, and counting itself was such a strain, concentrating on the faint scintillations, that one man might only be able to manage a couple of minutes' work. On at least one occasion Rutherford had to note: "No observations because of poor eyesight."

It is now quite clear from the record of work in the notebooks that a long process of elimination has begun. On October 3rd Rutherford set out to "test whether effect observed beyond the range of alpha-particle comes from the gases, air, oxygen, carbon dioxide or the source itself". Even by the end of the first day he is able to note: "Indications are <u>all</u> in direction that effective particles observed come <u>mainly from the gas itself</u> and not from source." By October 18th he is concentrating on the difference between having air or pure oxygen inside his apparatus and is having to take great precautions: "Apparatus completely cleansed by washing in hot caustic potash to remove any grease that might lead to the generation of H. Atoms." But the result did not quite suit his theory, "Effect larger than previous observation compared with air," so he adds, "Requires retesting."

Then he had another idea. After a page of "Theoretical considerations" he jotted down: "What if natural scintillations are connected with anomalous change of rad-C observed by Fajans," referring to the little-known fact that very occasionally a radium-C atom disintegrated by a different process from the usual. He replied to himself:

> The general absorption and range indicate these may possibly prove atoms of beryllium which if they carry one charge should travel 13 millimetres and be half absorbed 8 to 10 millimetres. If this be the case it is to be interpreted new product should be noted through several centimetres of air . . . If there is anything in this hypothesis it is clear that a fair amount of recoil product should travel a millimetre or two without an electric field for collection.

The importance of these notes is that they show without any doubt that in October 1917 Rutherford was fully aware that he was making "new" particles and that these particles were atoms of an element that had not originally been present in the apparatus. In other words he knew that he was splitting atoms; what he had to prove was just which atoms were being split into what.

So he went back to first principles once again, and again his notebooks show the original considerations of an alpha-particle curving in a hyperbola as it approaches the nucleus of an atom. This time he was trying to work out the direction in which the results of such a collision would move away after the encounter. Then he, very significantly, calculates a general theory for the "Number of atoms expelled along the range of alpha-particles in oxygen, nitrogen and carbon". This involved him in an impassable maze of calculations to consider what might happen if an electron in the outer shell of an atom was "deformable" under the impact of the forces brought by an alpha-particle. The possibility of an electron deforming, remembering that in size the electron is larger than the much heavier nucleus, became of great importance ten years later when theoreticians such as Max Born were developing the theories of quantum mechanics. Rutherford had to admit to himself in 1917 that "Effect not so far calculable".

If the process of scientific discovery were strictly logical and rational, one would have to say that the crucial experiments in this process were performed on November 6th, 7th, 8th and 9th, 1917 – Tuesday to Friday – four days of continuous intense work. It seems likely that Rutherford already expected the answer he got on these four days, and this work simply provided the essential proof that he needed. Page 39 of Notebook 22 carries the vital information – but more important are Rutherford's scrawled comments. The entries for November 6th are headed "Absorption of H-particles" and the first sets of numbers carry the comment, "Results in fair agreement for small numbers," demonstrating that Rutherford had already formed a view of what was happening. He adds: "Important: it is seen however that the addition of three centimetres of air, instead of diminishing number increases it more than twice." In other words, the number of H-particles causing scintillations on the screen increased when there was more air between the source of alpha-particles and the screen; one would have expected the air to absorb the flying particles and reduce the number, but more air produced more scintillations and therefore the scintillating particles should come from the air; but very little hydrogen is found in ordinary air, which consists mostly of nitrogen; therefore logically it was most likely that H-particles came from nitrogen.

Wednesday, November 7th, was mostly taken up with eliminating the most obvious sources of contamination which might introduce hydrogen into the air. First sets of figures are headed "To test whether long range scintillations from air due to water vapour in air". The air was bubbled through sulphuric acid to get rid of any water vapour before being put into the apparatus. On that afternoon Kay took up the bulk of the observations and the possibility of dust in the air was removed: "It was thought possible dust in the air led to production of H-particles. For this purpose tube packed with cotton wool in path of air." Admittedly the readings were not all that great – "Rather small numbers" – and the filter was a bit simple, but the results were good enough to encourage the two men, alone in the darkened laboratory with a Great War going on outside. Rutherford muddled the dates of his next experiment, writing both "Thursday, November 7th" and "Friday, November 9th". So, on Thursday, November 8th, he introduced pure nitrogen, made by the chemical reaction of sodium nitrate and ammonia chloride, into his apparatus and fired alpha-particles through it. The results were highly satisfactory: "Good expt," he jotted beside the figures on Page 43 of the notebook. Then he added his conclusion: "It is clear from these experiments that chemical nitrogen gives long-range H-particles which produce scintillations at least as bright as H and have about the same range (to be tested accurately)."

On that same day he carried out identical experiments with carbon dioxide gas in his apparatus instead of nitrogen. The conclusion is marked "Important" and it reads: "Since introduction of carbon dioxide gives very small effect it is clear that no appreciable number of the particles come from carbon or oxygen."

Rutherford had therefore shown, however roughly, that if he shot alpha-particles, which are helium nuclei, into nitrogen gas, hydrogen nuclei came out the far end. This could only mean that the alpha-particles were splitting the nitrogen atoms and knocking hydrogen nuclei, the positive particles of the atomic nucleus, out of the nitrogen nuclei. On Friday, November 9th, he confirmed his results and cautiously noted again: "Increased effect due to pure nitrogen."

Now he had to prove his results beyond doubt, and the first problem was to show that it was hydrogen and nothing else that came out – in his notebook terms – "To settle whether the scintillations are nitrogen, helium, hydrogen or lithium." Hydrogen, helium and lithium are the three lightest atoms in that order, the likeliest chips to be knocked out of the comparatively large nucleus of nitrogen, and the possibility had also to be eliminated that the flashes on the scintillation screen were caused simply by nitrogen atoms being propelled out of the gas whole and undamaged.

It seems that Rutherford paused for a week to consider how to do this. The notebook takes up the story on November 16th with calculations showing that the range of hydrogen particles would be very much greater than that of nitrogen atoms, which gave a satisfactory method of eliminating nitrogen. There is also some consideration of the different appearance of the scintillations when they were caused by hydrogen particles hitting the screen; in particular there was an effect which Rutherford called "instantaneous doublets" and there are two special notes about this: "Kay considers average distinctly brighter", and "This observed before but requires careful looking into".

The rest of November 1917 was taken up with eliminating the possibilities that atoms of lithium, boron or beryllium were appearing. December was devoted to providing really solid results and measurements on "Long-range scintillations from nitrogen and air", but there were problems as, for instance, on December 12th where the record says, "R. observer", and "Special precautions taken to dry air – but counting uncertain as eyesight poor". And a few days later a new problem arose as the notes say, "Rather small, obviously special experiments required to make sure effects not due to possible water vapour or hydrogen or H-compound in air: to be repeated later."

The first fortnight in January 1918, saw the elimination of the possibility of the scintillations being caused by whole nitrogen atoms, but the possibility of an unknown "H-compound" being present loomed larger. This was really another version of the "light gas" probability, for Rutherford wrote, "Suppose long-range scintillations in nitrogen are due to an atom charge +e, mass M = 2 called 'x' " – plainly he was considering the possibility of a hydrogen atom carrying double the normal weight, and this of course is the atom we know now as "heavy hydrogen" (deuterium), which was not to be discovered for another fifteen years. He was to some extent floundering at this time and he also did a couple of small experiments to eliminate the possibility that either fluorine or uranium could be involved in any way. In the last week of January he found his way out of the difficulty by reverting to a technique which had been successful in his Canadian days. He proved that the particles which were causing his "long-range scintillations" were atoms of hydrogen by subjecting them to a magnetic field during their flight, and showing that they behaved exactly as known hydrogen atoms would behave under this force. The first results of the new method were triumphantly successful and he noted: "It is clear therefore scintillations are nearly all from source and are deflected by magnet-like positive particles. Curve [and he sketches it here] agrees well with H-atoms. It is clear natural scintillations *are* H-atoms." But he laboured all through February to provide publishable results to prove his point, going through the hard work that follows inspiration, firing his H-particles through the magnetic field, then reversing the field, accounting for the "doublets" he saw on the scintillation screen and so on.

For the remainder of 1918 the pace of work slowed down. It appears that Rutherford took a break in April and May, and worked much more slowly through June, July, August and September. Virtually the whole year's work was on H-particles, the laws governing the passage of alpha-particles through a gas of hydrogen molecules, the numbers and ranges of hydrogen atoms that could be expected to be shot forward by impact from the alphas, and the experimental demonstration of the conclusions. The work took him into many byways; the problem of the doublet scintillations was one of the most troublesome, involving, he thought, the possibility that some hydrogen nuclei might consist of two particles. There were problems, too, with the new "Coolidge-tubes", an American invention which he had come across when he went on his mission to the United States. They were the most powerful source of X-rays yet invented, and Rutherford eventually had to install his tube "in the next room" to avoid the light and interference it gave off. Possibly he was getting really tired as the war dragged on – a whole weekend's work in July was wasted. "This expt. unsatisfactory as number of alpha-particles appeared capriciously, probably due to gusts of emanation from the pump," says the dejected note. But he also finally disposed of the difficulty that some of the H-particles might come from the radioactive source itself, and in November, when the pace of work increased again, he built up the results which determined the general behaviour of nitrogen under impact from alpha-particles. The massive work on hydrogen made the first two of the four papers he finally published. The effects of alpha-particles on oxygen and nitrogen went together to make the third, and all these first three can be seen as contributions towards the development of his nuclear atom theory. The fourth paper, as we have seen, "An Anomalous Effect in Nitrogen", cautiously told the world that he might have made another major discovery.

His notebooks show that the first three months of 1919 were spent in "tidying up" the work, and there was a pleasant rounding off of the structure as though a good dramatist was at work, in the brief return of Marsden to the Manchester laboratory, now appearing as Major E. Marsden, of the New Zealand Division Signals Company. The last notebook of the series shows Kay and Marsden working together as observers in "Experiments on H-particles, Jan. 3rd, 1919 (Friday)". They did careful work with gold, silver, copper and paraffin wax screens, half filling the apparatus with hydrogen and then gradually leaking air in until there was half hydrogen and one-third air. There were also control experiments in which the alpha-particles were provided from a polonium source. But all the notes at the end of the experiments and much of the recording were still in Rutherford's hand. A few loose sheets of paper in the final notebook of his Manchester period record the laboratory work up to April 30th, 1919, but by then the paper announcing the splitting of the atom was on its way to the printing presses. Rutherford also had other matters on his mind: he had just been made Cavendish Professor of Physics in the University of Cambridge, the acknowledged leader of physics not just in Britain, but a post with an imperial role in the British Empire of those days.

It is worth realising how hard he had worked, and why he placed more value on the disintegration of the atom than on attending a committee meeting on anti-submarine warfare. There is a paragraph in the first of the four papers that were published in the June of 1919 where among the figures, graphs, diagrams and passive prose of the scientific paper, he strikes a personal note:

> In these experiments, two workers are required, one to remove the source of radiation and to make experimental adjustments, and the other to do the counting. Before beginning to count the observer rests his eyes for half an hour in a dark room and should not expose his eyes to any but a weak light during the whole time of counting. The experiments were made in a large darkened room with a small dark chamber attached to which the observer retired when it was necessary to turn on the light for experimental adjustments. It was found convenient in practice to count for one minute and then rest for an equal interval, the times and data being recorded by the assistant. As a rule, the eye becomes fatigued after an hour's counting and the results become erratic and unreliable. It is not desirable to count for more than one hour a day and preferably only a few times per week.

There is another remark which was to be of significance in Rutherford's attitude towards atom-splitting and the availability of atomic energy. Towards the end of the first paper he wrote:

> It is clear from the results given in this paper that a close collision between an alpha-particle and a hydrogen nucleus is an extremely rare occurrence. Only 1 in 100,000 of the alpha-particles passing through one centimetre of hydrogen . . . gives rise to a high-speed H-atom . . . Thus for every one thousand million collisions with the molecules, in only one case does the alpha-particle pass close enough to the nucleus to give rise to a swift H-atom.

But Rutherford knew the value of the work he had done. It was not arrogance that phrased his apology to the international anti-submarine warfare committee for his absence, "If, as I have reason to believe, I have disintegrated the nucleus of the atom, this is of greater significance than the war."

14

Cambridge and the Cavendish

In a very real sense it was the wartime work of the Board of Invention and Research, and the rumpus accompanying its demise that eventually led to Rutherford moving back to Cambridge. But the route was a very English one, depending on the oddities of ancient institutions. Trinity College, the largest of the colleges that make up the University of Cambridge, is unique in that the Master of the College is not elected by the Fellows but is appointed by the Crown, which means in effect by the Prime Minister of the day. Early in 1918 the Master, Montagu Butler, died, and the Prime Minister, Lloyd George had to recommend a successor. He chose J.J., and remembered, many years afterwards, his reasons for doing so: "His super-eminence as a scientist was known, even to a barbarian like myself who never had the advantage of any university training. As one of the War Directorate I knew what invaluable services Thomson had rendered in the conduct of the war . . ."

J.J. was installed as Master of Trinity on March 5th, 1918, with the full ceremonial of knocking on the locked outer gate of the college and having to pass the patent of his appointment through the barely opened wicket-gate to the head porter, although this official had once been his "gyp" (college servant). In rather the same unrealistic but gentlemanly mood, J.J. gave up his salary as Cavendish Professor, but seemed to intend to retain the power and control over the Cavendish Laboratory, and plainly he believed he could carry on his scientific work, being only sixty-one. Events, however, swept aside any possibility of gradual change at the end of his thirty-four-year rule over Cambridge physics. The war ended in November 1918, the government promised, and gave, liberal grants to those who wanted to complete their interrupted education at university, demobilised officers flooded back to Cambridge, and there were such additions as 400 naval officers sent on special scientific courses. There were even 200 American soldiers for the first term of 1919. So fierce was the pressure that the thirty-year-old controversy over the university's demand for some knowledge of Greek as a condition of entrance was settled finally, with only fifteen votes for retaining Greek against 162 who decided that an education need not necessarily be based on the classical languages.

J.J. was worried that a four-year war, a year longer than the average undergraduate's stay at university, might have broken the memories of traditional conventions and shattered the continuity of clubs and athletic activities. He soon found he need not have worried about tradition, but should have been concerning himself about the change in spirit which came with the euphoria of victory. The government – perhaps more correctly the governing classes – had realised what science and scientists could do for them in a war; perhaps, too, they had learned from the early German advantage what a good educational system and a wide knowledge of science could do for a country's industrial base in peacetime. Certainly government for the first time began to think seriously of supporting the universities and of expanding science teaching facilities in particular.

J.J. himself had dropped slowly out of the leading wave of physicists after his discovery of the electron in 1897, though he was still universally admired by young scientists as the man who had opened up the new physics. Lawrence Bragg, one of the brightest young lights of J.J.'s Cavendish, viewed him as "a curious link between the old and the new physics. It has been said of him that he opened the door to the new physics but never went through himself." The important advances made in the Cavendish while Rutherford was in Manchester – such as C. T. R. Wilson's invention of the cloud chamber, and Aston's early work on the mass spectrometer and the identification of non-radioactive isotopes, were not inspired by J.J. in the way that so many of the achievements of Rutherford's colleagues were inspired by Rutherford. As Crowther has remarked, it was Rutherford who went through the door that J.J. opened.

In addition to this, the Cavendish itself had run down during the scarcities and shortages of war, although J.J. had husbanded its finances magnificently so that there was something saved to begin the post-war redevelopment. In early 1919 he had the laboratory secretary, a Miss Pate, draw up a set of accounts. These showed that the laboratory had started the year 1918 with £1,000 in hand. The main income came as £2,633 in students' fees, along with £1,313 from the navy. The total on the credit side for the year was £5,763. Lecturers and demonstrators were paid £2,673, and technicians and lab boys earned just over a thousand pounds. Only £500 had been paid out for all the running costs of apparatus and materials – apparently no new instruments had been bought. J.J.

waived all his entitlement to fees and was able to put £500 in the laboratory's deposit account, leaving a balance of £841 to start the financial year of 1919. With literally hundreds of students demanding science courses, with the laboratory trying to accommodate double the number of people for which it had been designed, with the navy demanding special courses for young officers, the entire organisation had to be changed and it was difficult to see J.J., withdrawn and introspective as he grew older, changing it.

Rutherford was the obvious man to take J.J.'s place, but no one knew how to make this come to pass. There must have been "conversations", "soundings", murmured discussions in college Combination Rooms and Senior Parlours, probably also in London clubs such as the Athenaeum. But the electors to the Cavendish Professorship do not vote on candidates, they comment on the names put before them and if there is only one name then that candidate is elected.

J.J. was persuaded to resign the Cavendish Professorship at the beginning of March 1919. He was given, in return, a new personal research professorship, and guaranteed the use of certain rooms in the laboratory, with enough funds to pay for a personal assistant researcher and a technician. The first mention of the matter on paper came on March 4th, 1919, when Sir Joseph Larmor wrote to Rutherford in terms which make it clear that the whole matter had been discussed in conversation beforehand: "Our Vice Chancellor, who is a man of rapid action, had ordered us to elect a new Cavendish Professor on April 2nd. It appears to be tacitly assumed that there would be no prospect of attracting you back to the scenes of your earlier youth. I wish there were." Larmor, who was himself one of the Electors, and probably the most influential, went on to say that the stipend for the professorship was £850 a year, but there would also be easily £200 available for the Professor out of the fees paid to the laboratory by students. He explained the arrangements made for J.J. and emphasised that the new Cavendish Professor would be Director of the Laboratory in sole control. And he concluded, "I confess to a strong personal longing that you might be available under some conditions or other to help make this the Imperial University that it is expected to be in the new scheme of things, and if I get any encouragement at all, I will not let it rest." The letter was marked "Private".

The letters poured out almost daily after this. There seem to have been personal meetings, too. Rutherford appears to have been just as indecisive, just as much torn between loyalty and ambition, as he had been when invited to leave McGill for Manchester. At Manchester his salary was £1,250 a year, one of the highest-paid professorships in the whole of Britain, and he certainly did not intend to move to Cambridge at a personal loss. Furthermore, there were difficult shoals of propriety to be navigated, but Larmor set himself to reach the goal that everyone seemed to think was really the correct solution – that Rutherford should return to Cambridge. Larmor had talked to Rutherford personally immediately after his letter of March 4th, and two days later wrote to point out that a College Fellowship could certainly be made available, which would bring in an extra income and give Rutherford a powerful place in university politics. But "the Electors, or some of them, might take umbrage at an invitation to you to come up to talk the matter over if they were not concerned in it . . . One has to walk warily where people are so touchy as to their rights and influence."

Rutherford did the sensible and honourable thing; he wrote to J.J. himself on March 7th, addressing the letter from the Savile Club in London's Mayfair, and asking for J.J.'s "views and frank opinion". If Rutherford got the job, "I feel that no advantages of the post could possibly compensate for any disturbance of our long-continued friendship or for any possible friction whether open or latent that might possibly arise if we did not have a clear mutual understanding with regard to the laboratory and research arrangements." He went on to enumerate possible causes of friction: who should assign the research students between the two Professors? Would J.J. be working solus with his own team of assistants? And should the Director of the Lab have charge of all researchers and assign them their lines of work and their supervisors? Again the Director of the Laboratory must have responsibility for changes in the organisation and personnel of the lab – "with which changes you might not altogether concur". Finally there was the whole question of the finances of the laboratory and the division of them between the two Professors.

Meanwhile Larmor was lobbying hard; two days later he wrote that he had been "to see all the chief people concerned" and found that "if you could come you would have a unanimously enthusiastic welcome". In practical terms he was certain that Cambridge could get the total income for Rutherford "near to £1,200" and "nobody would wish you to come at a loss". But "the chief trouble is the criticism of the literary fraternity which is of late years becoming more and more persistent on principle against the greater share for scientific purposes". While Larmor thought Rutherford was wise to seek an understanding with J.J., the problem did not really lie in that direction: "He [J.J.] was always an isolated figure here, had very little contact with mathematicians or chemists who would have responded to contact. There are vast possibilities open in that direction." Larmor agreed that the Cavendish was too small, physically, to hold two Professors, but an extension of the laboratory was "practical politics". And he went on:

> The fame of the Cavendish ought to make it easy to beg for funds for an extension. Shipley [the Vice-Chancellor] and our other public men are past masters at importunate mendicancy, and would be available . . . I now see the possibilities of expansion that are practical if people here had a strong lead. They complain of J.J. that he not only did not give a lead, but poured cold water over projects. He possibly resented interference and interruption of his own way of life.

Others were pressing on similar lines, notably W. J. Pope, Professor of Chemistry, who had become one of Rutherford's friends and allies in the BIR. On March 11th he wrote:

> . . . I certainly think it very unlikely that J.J. will prove troublesome to the Cavendish Professor, first on personal grounds – J.J. concerns himself very little with other people's affairs – and on traditional grounds – it is not considered the thing here to interfere with another man's job . . . For myself I hope very sincerely that you will accept the post. The amount of freedom which is allowed a Professor here cannot be realised by anyone who has not been in such a position, and I am quite sure you will find more opportunities here than in Manchester.

He assured Rutherford that the university had a strong intention to expand all the physics, chemistry and engineering labs, and added, "Our classical friends have lost status during the last few years and all now realise that the future of the university is in the hands of the experimental science people." Furthermore a number of Cambridge chairs were occupied by very old men, "who must drop off within the next few years", leaving the "prospect of a young active professoriate within a short time which will make for progress". As for finance, Pope assured Rutherford that Cambridge was not an expensive place to live in: "The old custom of elaborate entertainment was on its last legs when I came here ten years ago and had been practically discontinued before the war." And he ended by urging that two Professors of Physics could work together "to the Glory of God and the Advancement of Physics and Chemistry".

Larmor continued beavering over the practical financial details. Letters flew to and fro at intervals of two or three days. The problem was that a Cambridge professor's income was made up from various sources buried deep in the constitution of the university – College Fellowships, university stipends, laboratory fees

– and no one was empowered to offer an actual salary or to say definitively what money a new professor would receive. This constitutional problem had Rutherford "a good deal worried". He complained: "As a business proposition it seems to me highly unsound for me to send in my name for a post at an official salary less than that for which I am prepared to come, while trusting in Providence for the necessary addition. Your constitution ought to be elastic enough to give special guarantees in special cases." This was not personal greed on Rutherford's part; he explained that he was particularly keen not to have to draw on laboratory funds for his own needs when it was not at all certain whether the lab fees would be adequate for paying staff and buying equipment. He could not see, he continued, just what body Larmor represented in making the financial arrangements: "Possibly what I have said may appear somewhat captious to you, but you must remember that I have not had dealings before with a body whose powers appear to be so restricted in practical directions."

J.J., however, was proving remarkably amenable. He replied promptly to Rutherford's letter: "I am very glad you are still entertaining the possibility of coming to Cambridge . . . If you do you will find that I shall leave you an absolutely free hand in the management of the laboratory." Everything would be settled easily except possibly the assignment of research students. Even that, J.J. thought, would "work out naturally" and he added in his off-hand manner: "If any took subjects on which we were neither of us very keen we might toss up." He repeated his promise to leave Rutherford entirely independent and emphasised that he would ". . . never dream . . . of expressing any opinion about matters of policy". And he ended with a postscript: "There is a very keen hope that you may see your way to come to Cambridge. Nothing would give me so much pleasure as to have for my successor my most distinguished pupil." In practical terms J.J. sent Rutherford the laboratory balance sheet, and organised a Fellowship for him at Trinity College, of which he was now Master.

Manchester University was necessarily involved in these negotiations as well, and made considerable efforts to keep its star. The physics department "is even now, with you in it, the best in the country – or even the world", wrote the Vice-Chancellor, Henry A. Miers, on March 22nd in a letter in which he asked Rutherford to name any improvement he might want in his stipend, in his laboratory equipment or in his staff assistance. There was "nothing that I would not ask the Council to do". Rutherford drafted in his own hand a long reply and sent it off the following day. The department was depleted by the war and needed reorganising from the foundations, he responded. In particular the newly-created PhD degree would involve providing for many more postgraduate students, and this would involve more equipment, more staff, and more laboratory assistants or technicians than they had had before the war. In any case they needed a new post of Professor of Mathematical Physics, and two or three readers who would be specialists in different branches of the science, such as chemical physics. Again there was no element of personal greed in these demands, for Rutherford explicitly stated his own willingness to carry some of the load of elementary physics lecturing. On the contrary, this was the sort of pattern he would recommend to Cambridge, when he got there, and it was the pattern he was recommending for big independent naval laboratories in both Britain and the USA at exactly this time.

Larmor's lobbying had obviously taken him far afield in the search for funds for Rutherford himself and for the Cavendish Laboratory, for there is a letter extant from Sir Charles Parsons, the famous engineer, and yet another of the influential men with whom Rutherford had come into contact through the BIR. From his London home at 1, Upper Brook Street, Sir Charles wrote to Rutherford that "to avoid any possibility of misunderstanding" what he had said to Larmor was that he "would certainly contribute something to the Chair. I have not said I would <u>guarantee</u> anything nor has any sum or recent contribution been mentioned by me." He went on to explain that his first responsibility was towards engineering and particularly the Engineering School at Cambridge, and this for the very good reason that his own work lay in this field and the money he possessed came from it, too. "Nevertheless I am fond of physics. Personally I should, as a member of the old university, be very pleased if you came there and our paths crossed oftener," he added.

It has often been written that Rutherford only applied for the Cavendish Professorship by telegram on the actual day of the election. But it appears that this is somewhat oversimplifying the matter. He telegraphed his decision to stand to Larmor on March 29th, some four days before the formal election on April 2nd. Larmor responded, "Your telegram to hand. Though I shrunk from the responsibility of pressing you, in my view you have beyond all question taken the right course and your mind will at length be at rest." And he was elected automatically as the only candidate. Nevertheless he insisted on a formal exchange of letters with J.J. on a personal basis giving a fairly detailed agreement between them on all the possible points of conflict that might arise, and this "treaty" was concluded and written out in Rutherford's hand in August 1919.

Rutherford's letter of resignation to Manchester was quickly written, and he agreed to carry on until the end of the summer term, the end of the academic year. Nevertheless, he wrote, it was with "feelings of great regret that I sever my connection with a university with which I have been proud to be connected for the past 12 years", and he added that it was "only after much hesitation and pressure that I felt it my duty to take up a more difficult task elsewhere". In his formal reply to the Council some months later Rutherford referred to Manchester as "the most progressive of our universities". It is interesting to see that, just as when he left McGill, the dismay at his departure was not so much at the loss of a brilliant research leader, it was more at the loss of a man who had been a wise counsellor in university affairs; the Vice-Chancellor wrote personally of his "regret that I am to lose the benefit of your advice and cooperation in so much university business". The significance of Rutherford's discoveries in the atomic field and the importance of his disintegration of the atom had not been realised; it was the loss of a powerful business leader that was felt.

Rutherford's letter to his mother a few days after he had been elected to the Cambridge post probably reflects his own feelings with some accuracy.

> It was a difficult question to decide whether to leave Manchester as they have been very good to me, but I felt it probably best for me to come here, for after all it is the chief physics chair in the country and has turned out most of the physics professors of the last twenty years. It will of course be a wrench, pulling up my roots again and starting afresh to make new friends, but fortunately I know a good few people here already and will not be a stranger in Trinity College.

Lady Rutherford, always practical if not graceful, had started househunting in Cambridge even before the election was formally announced – one of her very few letters to her husband was written from Trumpington Street in Cambridge, undated except for "Saturday". It described Larmor as "like a nice old hen with one chick to find a nest for" but added that he was "awfully kind and good about your job" and explained that he had not understood about Rutherford having to "drop the insurance" and so "had rushed up to London about it to raise more money as soon as he got your letter". She also passed on Larmor's assurance that "The electors merely sit in judgment on the names submitted – if there's only one he's appointed."

The Rutherfords soon found their home, Newnham Cottage,

right on "the Backs", which is Cambridge's name for the famous riverside where the Granta flows past the old colleges. It was to be their home until his death, though they only leased it from Caius College. Apart from its favourable situation the house was reasonably large without being vast and draughty, and it had a large garden with plenty of trees.

In describing Rutherford's negotiations over the Cavendish Chair it has been emphasised that his bargaining for a financial guarantee was not personal money-grubbing. This view is fully supported by the very recent discovery of a full inventory of the Rutherfords' possessions which was made when they moved into the house in 1922. The inventory was made for insurance purposes and was discovered when old files in a Cambridge solicitor's office were being cleared out. They owned nothing of any real value. The most valuable thing in their possession was Lady Rutherford's Bechstein grand piano, then worth £410. The remainder of their furniture all together was worth little more than £1,000, including "5 stone and 2 indiarubber hot water bottles in the parlourmaids' pantry". The house contained but two reception rooms – a drawing-room and a dining-room – and a study for Rutherford himself. Nevertheless it was a house designed to be run by a number of servants, since there was a pantry, kitchen, scullery and servants' sitting room. There were a great many smallish rooms upstairs, including bedrooms for "living-in" cook and housemaid.

Rutherford and his wife at this stage of their lives had separate bedrooms – Lady Rutherford had the main room with a single three-foot bed in it. Rutherford himself slept in "Sir Ernest Rutherford's dressing-room", again in a single-bed. They lived simply: Lady Rutherford's entire wardrobe was valued at only £200, her most valuable possessions were a moleskin scarf valued at £15 and a diamond and sapphire ring worth £35. He had a gold watch, but all their silver and plate together were worth only £200. Plainly, too, the décor was in the cluttered style we now consider "Victorian" – there were no fewer than twenty-three vases and statuettes listed in the drawing-room, while the dining-room contained what the valuers describe as "A Japanese bronze figure of woman with rat and holding aloft a candle nozzle. 10 inches high £3.10s.0d." Probably the most impressive part of the listing is the number of books – they included the standard English classics such as Shakespeare, Tennyson, the Waverley novels, and more than 200 general books, which hardly supports the picture of the ill-read scientist. And in Rutherford's own study there was an extremely large array of scientific works – all the papers of Rayleigh, Kelvin, J.J., Lamb and Heaviside, a large collection of scientific periodicals, and more than thirty dictionaries and works of reference. The inventory draws a picture of a comfortable, tasteless, academic home, lacking in grace or inspiration, run by three or four servants in the manner of the times, with a wife whose main interest was in her garden, for a husband whose main interest was in his laboratory.

The first task in Cambridge was to get the Cavendish Laboratory reorganised, expanded, and back into its position as the world's leading centre of experimental physics. Rutherford had been assured that the university was prepared to put money into the laboratory, and he had been told by several people, J.J. in particular, that there was a new attitude to science in government circles and that the government intended, for the first time, to support the universities financially. He therefore prepared a remarkable paper on "The History and Needs of the Cavendish Laboratory, 1919" which he duly sent off to the Vice-Chancellor.

"While the Cavendish Laboratory had, from the beginning, been a focus of research in physics, the beginning of a definite research school dates from the year 1895 when the University opened its doors wide to advanced students from other Universities," he wrote, forbearing to mention that he had himself been the first of those advanced students. He went on: "This new step led to a rapid increase in the research students in the Cavendish Laboratory . . . The output of important work grew rapidly and the Laboratory was soon recognised as the chief centre for research activity in physics . . . It may safely be said that a large proportion of physicists now holding important scientific positions in the country and also in our Dominions have at one time worked in the Cavendish Laboratory."

Of J.J.'s leadership and achievement he reports: "[The laboratory] has witnessed within its walls the proof of the existence of the electron and of the part it plays in the structure of atoms and molecules . . ."

Rutherford then presents the case for support of scientific research as a national need; this is the new thinking engendered by the harnessing of science to the war effort and the appreciation of scientific input to industrial advance. It is a case that has not yet been entirely victorious in Britain, and which must have been very "avant-garde" in the Cambridge of the 1920s where many of the dons still clung to the ideas and ideals of a university with the social and political values of the mid-nineteenth century. The idea that trained and qualified chemists were useful to industry was fairly well accepted, and industry, especially in Germany, had profited by supporting chemists. Rutherford had to impress on people the

415

parallel position of physicists, and he chose to do it through the example of something that had become familiar to everyone through the war – radio communications. "While discoveries of this kind [the electron] may appear at first sight as only of intellectual interest in widening the bounds of knowledge, yet experience has shown that they inevitably sooner or later become either of direct practical importance or lead indirectly to further developments of value to industry." And he instanced how Clerk Maxwell's theoretical work had led to the discovery of the practical possibility of radio communication, while J.J.'s research on the movements of ions in gases and on electric discharges had "been instrumental in leading to the development of modern oscillators and amplifiers" which were an essential part of all wireless and radio communication. He pressed on: "It is now generally recognised that active schools of research in pure science are essential if we are to make full use of the application of science to industry. Without the stimulus of new fertilising ideas and principles, technical research in industrial problems tends to wither and become ineffective." This is the classical statement of the case for supporting pure research as well as applied research – it is a case now universally accepted by scientists, but still, today, questioned by Treasury officials of every country in the world, and continuously ignored by the "hard-nosed" businessman.

What the Cavendish Laboratory needed, therefore, was more space and more money, and it needed these things for two reasons: to increase both the amount of teaching and the amount of research it could do. The Cavendish was now teaching 600 men, including fifty naval officers, in a building designed for half that number: "The classrooms and laboratories have been crowded to excess," wrote Rutherford, rather moderately. He pointed out that the lab had been built in 1877, extended in 1895 and 1904, and was now being further expanded by a new storey being built over the elementary laboratory. "This," he admitted, "will relieve the strain somewhat, but still further increase in space and increase in teaching power will be required if the laboratory is to be brought up to date as a great centre of teaching and research in physics." There was just as much need to turn out physicists as there was to produce new results in physics. The war had emphasised in a striking way the importance to the state of having a number of highly trained physicists available – and he instanced the location of guns (Bragg's work), the detection of submarines (his own work) and the development of radio signalling. Furthermore peacetime industry would need physicists for applied research and behind all this there was the need for physics teachers in the

Above left: On a Lake District holiday in the later 1920s.

Above right and below: On holiday at Burton Bradstock, Dorset, 1931.

Left: In conversation with J.J. The talk may have been about cricket rather than physics.

THE CAVENDISH

Right: The Cockcroft-Walton accelerator. Big Science has now entered the Cavendish. Note Cockcroft crouched in the counting chamber which was closed by a curtain for real work.

schools. In addition, it was now "the declared policy of our government to farm out to the universities the pure researches required for the army, navy and aviation". His final point was that "the rising prestige of this country and the eclipse of Germany will lead to an increase in the number of research students from neutral and Allied countries" – this was a point which was much exercising Balfour and other government leaders and which was to be important in the parallel struggle to reorganise the degree structure of the older universities.

In practical terms all this meant that the Cavendish needed £75,000 for a major new building and a further £125,000 as an endowment to provide income for maintenance, purchase of equipment and salaries for new teaching staff. Rutherford demanded three new, well-equipped laboratories for applied physics, optics and a subject he called "General Properties of Matter". To staff these developments he also required three new University posts of lecturer and another professorship in physics. He particularly foresaw a much-increased research effort into the problems of radio signalling from land and water, and he intended that this work, as well as the work on properties of matter, would be carried out in close cooperation with the Department of Engineering. He had in fact already made a small start in the radio research, he pointed out, and the site he needed for the new buildings would soon become available when the Engineering School moved into its new accommodation. All this would be necessary "if we are to play our part in the researches required by the state and in providing well-qualified research men for various branches of industry and for the scientific departments of the state".

This statement of what was, in fact, his plan for the Cavendish, effectively disposes of those critics of Rutherford who claim that he was only interested in radioactivity and atomic physics and that he concentrated the work of his laboratories virtually exclusively in these directions. His claim that the state would need the young physicists he produced was amply justified when another world war erupted and it was largely Cavendish men who provided the war-winning developments of radar.

At the same time that he was seeking out a route for the laboratory to follow in the future, Rutherford was also fighting battles on the university front to provide a structure into which his young workers would fit. Cambridge was one of the last universities to institute the Doctorate of Philosophy as the standard second degree – the PhD which is the objective of most modern research students at the end of a three-year course, the standard apprenticeship in research in any subject from physics to French

417

literature. The battle to get Cambridge to accept the PhD had begun in 1916, largely under pressure from Balfour and Haldane. The essence of the case, when it came up for public discussion on June 7th, on a report from the General Board of Studies, was stated by J.J.:

> I think it probable that after the war many students from neutral countries [and he meant, of course, the USA] will be unwilling to go to Germany for their postgraduate studies, and would much prefer to come to England. I have reason to believe that such students attach considerable importance to the attainment of the degree of doctor – a titular degree is all they require – and unless some such degree were obtainable they would not be likely to come in any considerable numbers.

He was supported by the Master of Emmanuel College who said he had received a letter from the head of one of the largest universities in America supporting the view J.J. put forward.

But there was fierce opposition. A Dr Hobson declared that the PhD "was a piece of goods made in Germany" and adopted in America, which had the supreme disadvantage that it used the word Philosophy "not in the sense used in Cambridge" (and there is no greater condemnation than this available to a Cambridge man). The whole proposal was "a piece of window dressing quite unworthy of the dignity of the university". A Dr Ridgeway pointed out the warning that had been given by the Rhodes Scholarship scheme at Oxford (which all the rest of the world regards as a success), but the good Dr Ridgeway managed to have a view that was quite exceptionally narrow: "Let them do nothing rash in order to secure more students. They wanted to get colonials, certainly, but let them think specially of their own students." There was one serious difficulty, namely the position of the potential Doctors in the university, for in Cambridge's very individual system a man (and there were certainly no women) had to obtain a Master of Arts degree in order to become a senior member of the university and thus be able to vote in the Senate. To obtain this M.A. degree it is not necessary to do any studying, to pass any exams, even to live in college – it necessitates the payment, however, of £5 three years after one has obtained a B.A. degree. It was hardly surprising then that Dr Ridgeway should remark that he "hoped the standard of their degrees would be untouched by any sudden desire to bring Americans or even colonials to the university".

But even the parochiality of Cambridge can eventually be

overcome. There was in existence an organisation of a somewhat tenuous nature which collected together "the Universities of the Empire", and which had held its first congress in 1912. It had planned to hold a further meeting in 1917, though this was postponed by the war. However Balfour, who was by now Chancellor of Cambridge, as well as Chancellor of the Exchequer, and his ally H. A. L. Fisher, who was President of the Board of Education, set up a small Standing Conference to represent all the Empire universities, and this group met in London in May 1917 and again in 1918. Balfour proposed that a delegation from this organisation should go to the USA in spring 1918, to find out what might be needed from Britain. And to the organisation the Conference of Canadian Universities reported in 1917 that "only by the establishment of doctorates that might be obtained within a reasonable time . . . can we hope that the stream of students which has of late set towards the USA will be diverted to the Universities of Britain". In another letter to this umbrella-body it was officially requested that Cambridge, specifically, should set up the PhD degree to attract Americans. In wartime Cambridge the battle went on, and by the end of 1917 the innovators were winning. Nevertheless it was not until May 1920, just in time for Rutherford's arrival, that by Royal Patent, the degree of Doctor of Philosophy was formally established at the university.

One of Rutherford's first official university positions when he arrived in Cambridge was to sit on the Board of Research Studies which was set up in April 1920. He immediately had to report a "large number of applications" from graduates of Cambridge and other universities "to pursue courses of research", and he launched a campaign for even further degrees to be made available as recognition of good research. He suggested degrees such as M Litt (Master of Literature) and MSc (Master of Science). These degrees would be needed for those who did satisfactory research but did not attain the PhD standard, also for those who only wanted to spend a year or two at Cambridge and did not want the full three-year doctorate. There would even be men who could not afford to spend a full three years, men whose scholarships or grants from home lasted a shorter time span – and surely he must have been remembering his own Exhibition Scholarship which brought him to England twenty-five years before but which did not last three years. He begged his colleagues to look at the matter from a broad angle: "If this university is to be one of the chief centres of postgraduate research not only for students of this country but for those coming from our Dominions and abroad, it is essential that they should have machinery to recognise good work on the research side". Apart from the problems of student finance there were occasions "when even good men found that a problem proved much more difficult than they imagined and so were unable in their allotted time to achieve positive results which would make a reasonable thesis".

In November 1920, J. Ellis M'Taggart, the philosopher who had given Rutherford kidneys for breakfast in 1895, and G. F. C. Searle, the doyen of physics instructors from Rutherford's own Cavendish Lab, made a formal announcement that they would vote *"non placet"* against the new lesser degrees when the matter came up for approval in the Senate. In the little broadsheet on which they published their decision they gave their reasons and particularly mentioned the "overcrowding of laboratories by students of little ability and resource" which could lead to "discouragement of ordinary university teaching". Rutherford however got his way; the new degrees were established, but they were not a great attraction.

The following year, presenting the report of the Board of Research Studies in February, Rutherford showed that there were seventy-two research students in the university preparing for the PhD degree. Fourteen of these were Cambridge's own products, six came from the USA and twenty-six from the Dominions, while the largest number, thirteen in all, were in his own physics department. Every year these numbers increased. In 1925 there were 248 research students; thirty-three had obtained PhDs, only eight had received the degree of MSc and but one was M Litt. There were a further 110 admissions that year, forty of whom were in the physics department, double the number of any other department except biochemistry, where Gowland Hopkins' Dunn Laboratory was becoming the second great centre of research science in Cambridge. Of these researchers, eighty were Cambridge graduates, totally disproving the fears of the parochialists, thirty came from the USA, forty-two were from the Dominions, and there were four whose first university was European, plus one Japanese and one Chinese graduate.

But whatever the victory of the modernists and scientists over the degrees, Rutherford did not get the money or support that he had requested for the Cavendish. Support from central government, apparently promised in the euphoria of victory, did not materialise until the ponderous machinery of the Department of Scientific and Industrial Research, the Research Councils and the University Grants Committee had been set up and had begun to work efficiently – a process which took so long that money for scientific research did not start to become available until the second half of the decade.

The absence of the money he needed had a profound, and negative, effect upon Rutherford and the Cavendish. Plainly Rutherford refused to adopt the "importunate mendicancy" which Larmor had told him was a strong point among the great men of Cambridge. Whether this was on account of pride, or because he was truly indifferent to money for himself and believed in simplicity in scientific method, or because he felt that people should only get what they were willing to pay for, or from a combination of all these emotions, it is impossible to say. But it is quite certain that he would not ask for money. Long after Rutherford's death Chadwick revealed that the Master of Caius College, R. K. Anderson, who had personal connections with the rich shipping family that owned the Orient Line, had arranged for a considerable sum to be made available to Rutherford for his own research work. But it was a condition that Rutherford had to ask for the money – and this he never did. Nor did he reveal to anyone, neither to Chadwick, his personal assistant, nor to any of his professional colleagues in the laboratory, nor even to his wife, that the money had ever been offered. The only reason why Chadwick ever knew about it was that Caius was his own college – indeed he had been offered a fellowship there out of a desire to help Rutherford and his men – and one evening, after dinner, Anderson asked him why Rutherford had never even troubled to ask for the funds that had been offered. Chadwick, of course, was completely ignorant of the negotiation although the whole business of buying equipment and arranging supplies for both teaching and research at the Cavendish had by this time been delegated to him. Yet this refusal to ask for private money must have been a matter of principle with Rutherford, for he was perfectly willing, indeed anxious, to obtain grants for equipment and staffing from official bodies such as the Department of Scientific and Industrial Research and the Royal Society in the immediately following years.

The problem of money at the Cavendish introduces Chadwick into the centre of Rutherford's life. James Chadwick had been an extremely poor boy, so poor that his family could not even let him take up a scholarship to Manchester Grammar School. He was, however, very bright, and found his way to Manchester University and into research in the physics department, at the age of eighteen, very shy, very immature, and terrified of Rutherford. He was set to work on the problem of the radium standard:

I saw a few snags about it quite early. I'd seen a perfectly obvious little snag in the arrangement that Rutherford had proposed as soon as I began work, but I was so afraid of him that I daren't mention it. Of course when I showed him some of the measurements he saw at once what the trouble was, and I think he was rather disappointed in me that I hadn't pointed it out . . .

In those early days in Manchester the only thing that mattered was getting on with the experiments that he was doing – the things he was particularly interested in himself and to other people to a slighter degree. And he was a pretty hard taskmaster. I was very definitely afraid of him and I was quite immature.

It took Chadwick a couple of years to establish himself in Rutherford's trust, and it was even then mainly as a man who could do some of the chemistry the Professor sometimes needed. In 1914 Chadwick had gone, with his master's blessing, to work in Berlin with Geiger. There he was trapped by the outbreak of war and interned in a prison camp at Ruhleben. Geiger and other German scientists kept him provided with some spare bits of equipment and Chadwick set up a small research laboratory in the camp, where he not only kept himself busy, but introduced a young, Regular Army officer, Charles Ellis, to the delights of science. At the end of the war, both men returned to England, but Chadwick was suffering, at least in his own opinion, from a permanently ruined digestion. The immature youth had hardened into a shy, dry, unenthusiastic, and rather dismissive man, but a man of strong character whose scientific expertise had not been blunted by the distractions of the war. He had, however, no job, no prospects, very few clothes and only £8 in money when he called on Rutherford in Manchester at the start of 1919. He was immediately put to work helping on the final disintegration of the nitrogen-atom experiments. And Rutherford arranged for Chadwick to come to Cambridge with him, where his Caius College Fellowship provided a living, and Rutherford organised a small scholarship to help him get started. The arrangement by which Rutherford took Chadwick to Cambridge was quite clear: Chadwick was to deal with the day-to-day running of the Laboratory and its research students.

Chadwick is, therefore, the best authority on the effect that the lack of money had upon the Cavendish, its people and its policy. The number of research students, as we have seen, was growing steadily, year by year, in the early 1920s, and Rutherford held strong views on the duty of the laboratory to produce and train physicists who would be needed by industry, and society generally. It became, then, a question of policy, rather than of science, whether the Cavendish should concentrate on what we now call nuclear physics. In an interview not long before his death Chadwick explained the problem:

> Rutherford had said some time before . . . that the question of the structure of atomic nuclei was one of the most important questions of physics at the present time but that it was very difficult to attack. He didn't see much way of doing it, and . . . he said it was a problem which could be left to the next generation. Well, of course in my view – I think in his too, but he didn't say definitely – it was his duty to do it. Who else was there who would go on those lines? . . . He did make one or two attempts in other fields but we hadn't the experience, we hadn't the equipment; we decided it couldn't be pursued any further – that we must concentrate.

So the Cavendish concentrated on nuclear physics.

This enforced narrowing of the laboratory's field of vision brought criticism upon Rutherford, mostly from those who held wider views about the purpose of education and the role of scientists in society. Chadwick said, "Occasionally he would come back from London and tell me so-and-so had been criticising the work of the Cavendish and the line which he was pursuing, and he was a little perturbed by it, perhaps only for the moment." The defence of Rutherford's policy – a defence which was perforce raised posthumously – is that when war came in 1939 it was largely his students from the Cavendish who went into radar research and development and proved their flexibility and the excellence of their training as physicists by turning radar into a war-winning weapon; and of course nuclear and atomic physics was rapidly developed from the most esoteric of scientific disciplines into the most powerful of industries in which Cavendish men such as Chadwick, Cockcroft and Massey played vital early roles in Anglo-American cooperative efforts.

Nevertheless Chadwick felt there was some justification in the criticisms, and especially those from other professional physicists: "It was very difficult to make any progress. One had to jump around and just do what one could. And two things added to the difficulties. One was the shortage of equipment which really was very serious indeed . . . We hadn't much money to spend and I had to be very careful and buy the cheapest . . . meters." Pumps were another device in short supply and Chadwick found "you could only buy a few at a time. When I say 'at a time', each year. And I spent the money during the course of the year but saving up some so that at the end of the year I should know better what was needed and then buy it. And then I should know how many people were coming into the laboratory and have a better idea of what they were going to do."

The combination of penury and lack of a clear view of the way to scientific progress led to an argument and even a short breach of relations between Chadwick and Rutherford in the years 1923 and 1924. Because of the necessity to "jump around" and follow any slight lead that might show a possible avenue towards solving the problem of the structure of the nucleus, Chadwick wanted to set a room apart from daily use by the ever-increasing number of research students. In this room he would keep certain items of equipment which could be immediately available to follow up any lead. This was not élitism on Chadwick's part – it was by his own ability to follow up very quickly some apparently inexplicable results from the Joliot–Curie Laboratory in Paris that he earned the distinction of being the first to establish the existence of the neutron. It was the Cavendish's slowness of reaction that, in his opinion, led to the Parisians being the first to establish the existence of artificially induced radioactivity. Nor was this conflict between economic demands and scientific achievement confined to the Cavendish; the failure of Ernest Lawrence's team in California to record significant scientific advances when they first constructed their cyclotrons in the 1930s was very largely due to the fact that the giant machines had to be used primarily for medical work, since the funds to pay for them had been raised by medical charities and appeals.

Chadwick described the situation thus:

> We had, I must say, a difference of opinion about how the laboratory should be run. I thought we were having too many research students, because we did not have the room for them and because certain things, some points, were not being looked into because they might give a negative answer, but in any case would not occupy a man, or sometimes two men working together, for a couple of years, and what I wanted to do was to have a few rooms set aside in which various pieces of equipment could be ready for any examination we wanted to make in a hurry, such as a cloud chamber, counting apparatus of different kinds, which only meant a few rooms, but it meant shutting off these rooms from the occupation of these research students. Also there was some difficulty in keeping the apparatus in order because we had very little technical assistance. And I suppose Rutherford was quite right in saying that we shouldn't do that. Well, anyhow, we didn't do that.

The invading horde of research students did not worry about scientific policy or the shortage of equipment, or even what subject they would be asked to concentrate on. "Their first desire was to work in the Cavendish under Rutherford, and in very few cases did they have any particular problem in mind," Chadwick explained; it was up to him and Rutherford to satisfy the students' needs for problems, and equipment to solve those problems and degrees to prove they could solve problems.

> He generally consulted me about what people were going to do because in any case I had to provide the apparatus. But we always had some problems in mind. And that is where . . . the PhD degree provided us with some difficulties, because the regulation was that at least two years' research should be spent on the PhD course. It was technically three, but a year could be allowed for previous work at research done at his own university . . . We didn't want him for three years, but we didn't want to have to start from the beginning . . . We had to think of problems that would occupy that time and which were big enough for them to get their teeth into and keep them busy and give some kind of positive and reasonable results. That limited us very much . . . Our procedure was to jot down during the course of the year various problems that we thought must be looked into. We did that independently. Then in the summer we knew what research students we were going to have although one or two might turn up later. But we were pretty certain by the middle of the summer what the composition of the laboratory was going to be. Then we sat down together and exchanged our views and decided what could be given, what would provide suitable research work for a student reading for the PhD degree. Well, that meant that quite a number of interesting things couldn't be done. We didn't have time or equipment actually.

Senior scientists of our own time recall with some pleasure such incidents as asking the laboratory head technician – the famous Mr Lincoln – for a piece of metal tubing and being told to cut the amount they required from an old bicycle frame. But the situation did not look so funny at the time even for the brightest of the youngsters. Chadwick recalled how Mark Oliphant

> was in real difficulty because he hadn't got the equipment to get a proper vacuum. He wanted a high vacuum [pump]. I said, "Well, I'm sorry, there isn't one." We had a bit of a talk and I went away, and, realising his difficulties, I saw that he really couldn't get on without a high vac, and I went round the laboratory and somehow or other I got hold of one. I think, as a matter of fact, it was one I kept for Rutherford's lectures. But I took a chance and I had it put on Oliphant's bench. He had gone to lunch almost in tears, so he told me much later. He came back and all was well.

Meanwhile one of the laboratory's most senior men, F. W. Aston, continued his pioneering work on making the original mass-spectrometer and establishing the existence of most of the rare, but non-radioactive, isotopes of the elements, in a cellar, making almost all the parts of the apparatus himself. C. T. R. Wilson continued the development of the expansion chamber, or cloud chamber, by himself, virtually without technical assistance, doing all his own glasswork – there is a fable that someone took his leave of Wilson one day watching him grinding a glass junction by hand, and returned after a trip abroad some months later to find him still sitting in the same spot grinding the same junction.

Wilson's cloud chamber was one of the Cavendish Laboratory's major contributions to science, for it is the ancestor of all the giant bubble-chambers which we nowadays use at the particle-accelerator laboratories to discover and track the sub-atomic particles which are the basis of matter. It was also one of Rutherford's favourite instruments, one which he never ceased to praise and proclaim largely because it enabled him to "see" electrons and protons and alpha-particles and demonstrate their reality to lay audiences. Wilson himself was of the same age as Rutherford and had been at the Cavendish when the New Zealander first arrived in 1895. A Scot, he was, like so many others there, a man of little wealth who had travelled to Cambridge via Manchester University, where he had obtained a First in zoology and converted to physics research with no formal training in that subject. The seminal period of his life had been a two-month spell as a meteorological observer on the top of Ben Nevis, where he had been fascinated by the cloud effects he could see. At Cambridge he followed up earlier work on clouds and found he could create a cloud in an enclosed chamber by expanding moist air. By combining this idea with the main line of J.J.'s work on the ionisation of gases, Wilson found that the ionising effect of a charged particle could be revealed, since water droplets (cloud) would first form, as the gas in the chamber was expanded, along the line of the particle's track where molecules had been ionised, and upon which the droplets would form. Thus the passage of the charged particle was shown by a line of water droplets in the chamber. This line could be photographed, and as the years passed the process could be speeded up, repeated at rapid intervals and even automated. Rutherford himself nominated Wilson for the Nobel Prize several times in the 1920s, writing in 1927 "the researches of Mr Wilson represent one of the most striking and important of the advances in atomic physics" which very often provided the *"experimentum crucis"*.

Rutherford's main problem in the years 1920–25, was, then, the management of the world's greatest experimental physics laboratory. Irving Langmuir, after visiting the Cavendish in 1922, wrote to its Director: "After visiting many laboratories in England and on the Continent I became thoroughly convinced that the Cavendish Laboratory, not only for the work accomplished, but for the stimulating scientific spirit derived from its leader, stands far above all others." The last years of this period, however, saw an important development in finance – the opening of a new channel through which official funds could be attracted. This was the newly-formed Department of Scientific and Industrial Research, the direct successor-machinery to the former Board of Invention and Research. In 1924 the DSIR agreed to pay Chadwick's salary in a new post of Assistant Director of the Cavendish Laboratory. The work he was to do in this new position was exactly that which he was already doing and was to continue to do – keeping an eye on the research students, controlling the apparatus, and so on. Salaries of even the most senior men at the Cavendish were only £400 or £450 a year at that time, and this was the size of the annual grant for Chadwick. But out of a total salaries' bill of £2,700 for the whole laboratory this amounted to a considerable contribution. Chadwick himself, in describing this development many years later, threw much light on Rutherford's attitude towards the sources of laboratory finance. Rutherford was discussing the supplies of radioactive source material for experiments and Chadwick had his usual tale to tell of a much greater demand than he could satisfy. So he remarked, "What a pity it is that somebody didn't give you a gram of radium as the women of America had given to Madam Curie." Rutherford's reply was, "My boy, I'm damned glad that nobody ever did. Just you think, every year I should have to think of the import of what I had done with a whole gram of radium, and I should find it impossible to justify it."

The reasons for Rutherford's attitude in this example and in the case of the grant which he refused even to request, were, in Chadwick's view, primarily modest:

> I believe he was essentially a very modest man. I think he was always surprised at what he had done. He gloried in it. Oh, he used to enjoy anything written in his praise. He or I would come across something, say in German publications he didn't read . . . so he would ask me to translate for him. And he enjoyed it. But one word which he enjoyed particularly was *Bahnbrechend* (pioneering). When that came he would repeat it and glory in it. But nevertheless . . . I knew him fairly well, because it wasn't only in the laboratory that I knew him. I knew him at home . . . I knew him in off-guarded moments when he was not being the great physicist . . . That, I think, was modesty. It was also a reluctance to be responsible to somebody outside the laboratory.

At the same time, from the end of the war throughout the early 1920s, Rutherford set about rebuilding the international position of science and of his own leadership in the experimental physics area. Niels Bohr, of course, got in touch with Rutherford as soon as hostilities ceased. His first post-war letter expresses his views on internationalism and the possible future leaders of Germany. It is the letter of an unworldly man, a letter which must seem naïve to us now, but which expressed the profound hope for peace of the typical quiet, retiring academic: "All here are convinced that there can never more be a war in Europe of such dimensions; all people have learned so much from this dreadful lesson." Rutherford's reply was the first of many in which he offered Bohr posts in England, both in Manchester and Cambridge. Bohr's reply to this and all following offers was polite, nay reluctant, but negative; he was already committed to his life's work of building an Institute of Theoretical Physics in his native Copenhagen.

Bohr invited Rutherford to visit Copenhagen in 1920, an invitation which was accepted, and in thanking Rutherford for his visit, Bohr described his recent first trip to Berlin and his meeting with Planck and Einstein: "I spent the days discussing theoretical problems from morning till night." In 1922, he did come to Cambridge to deliver a series of lectures. Later that year, when Bohr was awarded the Nobel Prize, he replied to Rutherford's congratulations: "I have felt so strongly how much I owe to you not only for your direct influence on my work and your inspiration, but also for your friendship in these twelve years since I had the great fortune of meeting you for the first time in Manchester." In 1923 the Royal Society offered Bohr a research professorship to work with Rutherford in Cambridge, but he confined himself to receiving a Cambridge Doctorate and visiting Rutherford privately, although he suggested regular meetings instead of the professorship. Rutherford officially opened Bohr's institute and visited Copenhagen several times more. Bohr accepted an invitation to give more lectures in Cambridge in 1925, but wrote to Rutherford that what he really wanted were "privit [sic] discussions . . . about our present theoretical troubles". And so the correspondence went on, the mutual visits went on, and students passed from Cambridge to Copenhagen or the other way around in considerable numbers.

Those authors who omit Rutherford from their descriptions of European physics in its heroic age distort the facts by failing to mention this northern connection. Bohr continued to rely on Rutherford and his workers primarily for new experimental results. Thus in 1930 he wrote: "In view of the latest theoretical developments almost every problem has acquired new interest and we are all hoping for new experimental facts." And again in 1932, immediately following the discovery of the neutron and artificial nuclear disintegration, Bohr wrote:

By your kind letter with the information of the wonderful new results arrived at in your laboratory you make me a very great pleasure indeed. Progress in the field of nuclear constitution is at the moment really so rapid that one wonders what the next post will bring, and the enthusiasm of which every line in your letter tells will surely be common to all physicists. One sees a broad new avenue opened and it should soon be possible to predict the behaviour of any nucleus under given circumstances.

Then after again wishing that he was nearer Cambridge he carried on: "I have been exceptionally interested in the foundation of the new mechanics and its limitations . . . the possible failure of energy conservation as regards electrons in nuclei . . . If it should be possible to excite electron emission from nuclei by means of the recently discovered powerful agencies it would perhaps be possible to settle this fundamental question." This sounded remarkably like suggesting experiments for Rutherford to perform in order to provide answers for the theoreticians.

George von Hevesy not only worked with Bohr and visited most European laboratories, but was also one of the first to pick up his relationship with Rutherford across the former battle line. In 1920 he wrote to Rutherford from his university post in Budapest, describing the terrible post-war state of affairs there, with three universities packed into a city that could not support one. Hungary, he wrote, was "a small country with a huge capital, like a man with a large head and a tiny body". Because politics were now so deeply embedded in university life, von Hevesy confided that he proposed to resign and take a post in Copenhagen, but meanwhile he wanted the latest news on "the smashing of nuclei". He wrote: "We are all waiting with extreme keenness the news of further progress in this most fascinating field of research." Two years later von Hevesy wrote to Rutherford about the first news of his own pioneering experiments in the technique of radioactive tracer work, following radioactive lead in plants, and Rutherford continued to help with the English publication of von Hevesy's work right into the 1930s.

The faithful Geiger would not let the war interrupt his relationship with his former Professor. He had written during hostilities, once sending his letter through a neutral friend. In May 1919 he "took the first opportunity to write a few lines". And he made his position quite clear: "I need hardly say that all that has happened here these last four years has had no influence on my personal feelings for you, and I hope, dear Professor Rutherford, that you still take a little interest in your old pupil who keeps his years in Manchester always in pleasant memory." Rutherford, of course, replied cordially and this correspondence also continued until Rutherford's death. Geiger told him all about the development of the radioactivity counter which has made his name a household word in our time; and he named his third son Ernst "in remembrance of all the help and encouragement which I have received from you in those happy days in Manchester and which have been of decisive influence on my life [sic]".

With Hahn the same sequence recurred – the pre-war relationship was quickly reestablished and letters continued to flow, though Hahn was near enough Rutherford's age and fame to demand that Lise Meitner be given full academic and scientific credit for her joint work with Hahn which Rutherford had been inclined to pass over.

Relationships with Dutch scientists had never had to be broken off, and the coming of peace saw much correspondence with Lorentz over the financial and administrative arrangements for re-restarting the activities of the Solvay Foundation. "Of course the Germans cannot come at the moment," wrote Lorentz in 1919, and he proposed that all the original members of the "Institut International de Physique" (including Rutherford), should resign to allow for a complete restructuring. The minutes of the institute meeting held in Brussels on March 30th and 31st, show that Rutherford was there with Marie Curie, Kammerlingh Onnes and Lorentz. The first problem they faced was that the value of money had fallen so much that there were financial difficulties in mounting a big conference. All to the good, Rutherford argued. A small conference was just what was needed; international conferences were usually too big to be truly fruitful and small conferences were what Ernest Solvay had originally visualised. But he pressed for the Scandinavian countries and the USA to be included in the new organisation. He carried this point without difficulty, but then there was an argument about the subject to be discussed. Kammerlingh Onnes said that relativity was too abstract, and Rutherford said they wanted a subject that would interest both the experimentalists and the mathematical physicists so that the two groups could help each other. Once again there was a successful alliance between Rutherford and Marie Curie, and overnight they produced a coherent programme for a conference with four different sections to discuss problems under the title "Electrons–atoms –radiations". Kammerlingh Onnes got his own latest discovery – superconductivity – included at the amendment stage, and Rutherford insisted that the "rapport" from each section must be short enough to be read in one hour. Everybody behaved well and exactly as their popular images suggested. Rutherford was also in touch with Marie Curie over some small grants from the Solvay Research Fund, and, perhaps more important, over the names for the emanations from radium, thorium and actinium. One of Marie Curie's suggestions was "radioneon" and she added, "If your lab and mine would decide the immediate use of names, the best would follow." Very politely Rutherford turned down radioneon and later "radion" and the world has finally settled for radon, thoron and actinon.

Lorentz sent some of his pupils to Manchester, notably Professor A. D. Fokker, and in May 1920 he wrote to congratulate Rutherford on being made a Foreign Member of the Academy of Sciences of Amsterdam, in the same election as Einstein, "you both being the poles of my physical interests," Fokker wrote, "on the one side the most concrete problem of how the atoms may actually be constituted; on the other side the theory of gravitation leading into abstract mechanics." Fokker had recently spoken to Einstein who was "very enthusiastic about the British achievements and results of the eclipse expedition".

Rutherford had, of course, met Einstein at the first Solvay Conference in 1911, which has been named the gathering of "the real revolutionaries of the twentieth century". It was on the eve of the war in 1913 that Einstein had produced his most revolutionary work – the special theory of relativity. It was not until 1919 that the first chance came to test relativity and ironically it was the British who chose to do it, by sending an expedition to Brazil and the South Atlantic at the time of an eclipse to test whether the light from the planet Mercury was actually bent by the presence of the massive sun. Despite bad weather and delays, Sir Arthur Eddington was able to obtain six photographic plates – "one plate that I measured gave a result that agreed with Einstein". The formal measurements were completed back in England and Eddington declared "they gave a final verdict definitely confirming Einstein's value for the deflection". Eddington was predicting that this would be the result at the 1919 meeting of the British Association and one of Lorentz's Leiden men took this gossip back to Holland where Einstein heard it on a private visit.

The formal announcement of the fact that relativity had stood up to its first experimental check, the first time the world at large was let into the secret, was at a special meeting of the Royal Society and the Royal Astronomical Society in London on Thursday, November 6th, 1919. J.J. as President of the Royal Society was in the Chair, Sir James Jeans, Sir Oliver Lodge, and Sir Alfred Whitehead were all present. J.J. described the confirmation of the Theory of Relativity as "one of the greatest achievements in the history of human thought". The universe, he declared, was now different. This theme was taken up by *The Times* in an editorial the following morning on "The Fabric of the Universe": "enough has been done to overthrow the certainty of ages and to require a new philosophy of the universe . . . [to] sweep away nearly all that has hitherto been accepted as the automatic basis of physical thought."

So on the morning of Thursday, November 7th, 1919, Einstein was world famous, but quite unable to receive or accept the plaudits of the world, for he was trapped and isolated in Berlin, a city wrecked and cut off by violent civil disturbances. This isolation meant that Einstein was unable to take part in the vigorous discussion of his ideas which now developed among scientists, many of whom remained sceptical, while others balked at the difficulties of understanding relativity. George Ellery Hale, for instance, wrote from the Mount Wilson Observatory in the USA, "I confess that the complications of the theory of relativity are altogether too much for my comprehension," and he wrote to Rutherford in similar tone, that the new theory seemed "to complicate matters a good deal". Rutherford's doubts about the new theory were not connected with its truthfulness, which he seems to have accepted from the moment that experimental confirmation was offered. He replied to Hale that the problem of relativity was that the interest of the general public was so great – "almost without precedent" – precisely because no one could offer an intelligent explanation to the average man. As a result he feared that it "might tend to draw scientific men away from experiments towards broad metaphysical conceptions".

Rutherford spoke publicly about relativity when he gave his presidential address to the British Association in 1923; he cast absolutely no doubt on the theory itself, but he did rebuke those who had said that Einstein had demolished the physics of Newton; the relationship between Newtonian physics and relativity was that the new work was "a generalisation and broadening of its basis, in fact a typical case of mathematical and physical development". Ten years later, admittedly when lecturing to the Royal Society of Arts, Rutherford was to say, "The theory of relativity by Einstein, quite apart from any question of its validity, cannot but be regarded as a magnificent work of art." Rutherford's verdict was therefore that relativity was true, but of little relevance to working scientists except astronomers. This view was not just his own, not just the bullyboy New Zealand farmer's attitude – it was essentially the attitude of the English school of experimental physicists, who did not regard abstract theorising as the best way forward for their science. For instance J.J.'s attitude, as reported approvingly by the younger Lord Rayleigh, was that "he accepted these results . . . but he never seemed particularly enthusiastic on the subject". And Rayleigh remembers J.J. warning "that he thought too much attention was being concentrated on it by ordinary scientific workers with the neglect of other subjects to which they were more likely to be able to make a useful contribution".

Nevertheless the international scientific reputation of Einstein, in addition to the fact that everyone liked and admired him personally, meant that he had to be invited to the Solvay Conference of 1921, which Rutherford, Lorentz and Marie Curie were organising. Lorentz had to explain the unpopularity of Germans in Brussels to Einstein, though as scientists there could be "no mention of formal exclusion"; however "the door will be held open for you so that in future it will be possible for everybody to work together again". In letters to others, Lorentz explained that Einstein could be invited because of "ill-defined nationality – Swiss, I believe". Rutherford wrote to another outsider: "The only German to be invited is Einstein, who is considered for this purpose to be international." Lorentz told Rutherford that Einstein had accepted the invitation with pleasure and had promised to speak – he also explained that Einstein had made himself unpopular with the Germans in Berlin by his pacifist views.

Einstein was indeed "international" as Rutherford had implied, and correspondingly he was no "man of business", no manager of "affairs" as Rutherford was rapidly becoming. Einstein looked back to that first Solvay Conference of 1911, as a biographer has suggested, when "he was embraced as a full member of the small truly international group whose work was concentrated on the single task of discovering the nature of the physical world. For a short while it was a scientific world without politics, an unusual state of affairs to which Einstein always looked back as though it were the norm rather than the exception which occurred as science prepared to tackle the riddle of the atom." And Einstein did not, in fact, turn up at the 1921 Solvay Conference – he was off in America speaking for the cause of Zionism.

It was probably about this time that Rutherford developed his "slight allergy to Einstein himself". In September 1920 Sir James Jeans wrote to Rutherford from Zermatt in Switzerland, referring to a conversation in which they had discussed the need to build up the mathematical physics side of Cambridge by appointing a first-class brain to attract a school of youngsters. Jeans had this suggestion: "I have been wondering what you would think of Albert Einstein . . . It would seem quite likely that he will be leaving Berlin soon – there has been a good deal of disturbance over him there . . . and he would probably consider an English offer." There was the drawback that Einstein did not speak English, but at the age of forty-two, said Jeans, "I imagine he has still plenty of creative power left". Rutherford of course was the best judge of the practicability of such a suggestion, but "you will never reestablish the school of mathematical physics out of the material at present available in England". There was no offer from Cambridge, perhaps precisely because Einstein was now prepared to speak for Zionism in the USA. His trip there in this cause in 1921 was an undoubted popular success, but it was not well regarded by scientists. Boltwood, for instance, wrote to Rutherford, "Thank heaven Yale did not give Einstein a degree. We escaped that by a narrow margin. If he had been here as a scientist and not as a Zionist it would have been entirely appropriate, but under the circumstances I think it would have been a mistake."

Einstein did, however, receive a firm invitation to visit Oxford, and to exchange papers with the new Professor of Physics there, F. A. Lindemann, later to be well-known as Lord Cherwell, scientific adviser to Winston Churchill during the Second World War.

It is one of the ironies of Rutherford's life that he was personally largely responsible for the election of Lindemann to the Chair of Physics at Oxford, yet Lindemann was one of only two men whom Rutherford was known to detest in his later years (the other was J. D. Bernal). There are but five letters from Lindemann in the vast Rutherford collection of correspondence – the first is dated February 20th, 1919, and in it Lindemann announces that he is going to stand for the Oxford position, which carried with it the Directorship of the Clarendon Laboratory and requested Rutherford's support. The post had been occupied by Clifton for many decades of complete inaction. Rutherford, who had certainly met Lindemann when the younger man was rapporteur to the early Solvay Conferences, immediately promised his support as one of the Electors. Two further letters were exchanged briefly discussing such points of current interest as war reparations and the fate of several German scientists known to both men. In March 1919, Lindemann, who knew the Berlin scientific scene well, having worked in Nernst's laboratory, remarked that "Einstein, of course, always was anti-militarist, and signed a public protest against the violation of Belgium in 1914".

Lindemann was duly elected to the Oxford post, and when Rutherford congratulated him on the result of the election Lindemann responded that "the favourable outcome of which, I feel sure, is largely due to you". He went on to say that he had not congratulated Rutherford on his election to the Cavendish post "because I felt, in common with everyone else, that Cambridge really deserved the congratulations in having persuaded you to take it . . . I am most anxious to work in the closest cooperation with Cambridge and the other schools of physics, both to prevent overlapping and because I believe there is no stimulus to new ideas like personal intercourse with other workers in allied branches."

Lindemann's last letter to Rutherford was on June 9th, 1919. He offered "a possible alternative explanation of your remarkable results in nitrogen [i.e. Rutherford's recently published work on the disintegration of the nucleus]. Could there be alpha-particles which have picked up an electron and whose range therefore should be quadrupled?" He went on to detailed matters:

> . . . the electron is much larger and therefore more likely to be hit than the hydrogen nucleus. If you had not measured the mass of the particles in hydrogen I might even have been inclined to question these . . . I do not of course suggest that your interpretation is not the right one, I only put the possibility before you as I think you will agree that the more epoch-making a discovery is, the more closely should all alternatives be examined. I daresay you will be able to squash this one out of hand.

It appears that Rutherford did not even make the attempt to squash the suggestion; there are simply no more letters between the two men. Lindemann, although a clever scientist, has left no corpus of work to match Rutherford's, though he attained much greater political power. The building of the reputation of the Clarendon – Lindemann's greatest achievement – was based on his successful and generous policy of offering posts to some of the finest scientists expelled from Hitler's Germany, a policy Rutherford did not emulate. Lindemann's style was totally different to Rutherford's, in that "the Prof" openly used social success with the most aristocratic circles, and a declared party-political position, to achieve his ends. But one is left with the feeling that Rutherford resented the younger man's questioning, on such dubious grounds, of his carefully performed work.

A happier episode in this period of rebuilding the war-shattered world of international science came in the resumption of relations with Stefan Meyer at the Radium Institute in Vienna. All through 1914 there had been correspondence between the two men about the proposed Vienna Congress of Radioactivity, which was scheduled for 1915, and of which Rutherford was to be president. Occasional messages passed between Manchester and Vienna even during the war years, and from the start of 1920 letters started flowing again between the two old friends and colleagues. On January 22nd, 1920, Meyer confided, "We felt sure always your feelings towards us, as ours towards you, could not be changed by the psychoses of the surroundings" – obviously Freudian terminology had already infected Vienna. But much of Meyer's letter described the terrible conditions in the former capital of the splintered Austro-Hungarian Empire – it was not just impossible to do any research, the scientists were lucky to be able to live. Rutherford's response was swift – he would start sending reprints and copies of *Nature* immediately so that his colleagues could at least keep up with their professional world. At the start of 1921 Meyer reported that while it was now just possible to keep body and soul together, there seemed no hope of starting scientific work again – the annual income of the Institute consisted largely of an official "donation", which still stood at its pre-war level of 2,000 kroner – but this was now worth just £1. Furthermore, "the academic people who had interests in scientific work is [sic] now reduced to poverty and not able to sustain the Institute".

Rutherford took up the problem on an official level and enquired about the precise state of the Institute. Meyer replied that "we cannot look forward with much hope," but he did suggest that the Cavendish Laboratory might buy the radium which the Vienna Institute had loaned them before the war. The problem was immediately discovered to be that the radium, if it really belonged to the Vienna Institute, was "enemy property". Rutherford found a way round this by saying that the radium was a personal loan to him, intended to be returned when he had finished with it. "Of course it was a personal loan to you," Meyer confirmed on February 24th, 1921. "We never mentioned it to our government and therefore it cannot be officially registered [as enemy property]. We're just proud it's been used in such important work." And so Rutherford was able to persuade the Royal Society to give him a grant of £500 to buy the radium outright and convert it from a personal loan to Cavendish Lab property. On this money the Vienna Institute was able to survive – and the friendship with Stefan Meyer became closer than ever.

But perhaps Rutherford's most significant contribution to the restructuring of science after the war lay not so much in his many practical acts – and he was instrumental in such moves as the restocking of Belgian university libraries that had been razed during the fighting – as in a statement of principle. In 1921 the organisation the universities of the Empire held its second congress; Rutherford was the delegate both for Cambridge University and for New Zealand. The Congress was a pretty ponderous affair and it is difficult to see any clear results from its deliberations on such subjects as "Finance" and "The Exchange of Teachers and Students".

Rutherford chiefly interested himself in the sessions on "The Universities and Research". This, said Rutherford, was "the most important subject" for scientific men, and it was by "the magnitude of their contributions to knowledge" that universities would be judged. All was not as well as the complacent English might think: "This country is judged, not by the size of its exports or its fleet, but by its contributions to knowledge, and as a whole I have found that the opinion in which we are held is not so high as one would like to think." The main practical problem at the moment, he saw, was "the great flood of postgraduate students now in movement", many of whom would normally have gone to German universities, where there had been a very large number of teachers, an enthusiasm, and a welcome for foreign students. The first priority therefore must be to develop postgraduate education and research programmes and to do it immediately. But that in turn would lead to the danger of overwhelming the university staff with teaching duties and not leave them time for their own research.

Postgraduate students come to a university primarily to work under a particular teacher, and so Britain needed to find teachers who would attract students. At that time however the universities were so full of ordinary undergraduate students that there was already too little time for the staff to do their own research. The only way out of the problem was to set up and expand postgraduate schools and "there has been very little of this in either the Arts or the Sciences", he pointed out. "The first need is men," he emphasised; it was impossible to have each department a "one-horse show". It is impossible to claim that this was an entirely original doctrine, invented by Rutherford himself, but he was certainly one of the first to state it as a programme. The significance of the statement is that this doctrine has been the objective of British science policy right up to the present. "Support the man who can get good results, who does good science" and the rest – the flow of bright students, the influx of further supportive funds, the discoveries and inventions that will rejuvenate industry – should follow. That has been the policy, implicit or explicit, of those bodies such as the research councils which have been the channels for official or governmental financing of science since the 1920s, when most of them were founded. It is only in the 1970s and 1980s that this policy has been seriously questioned, when it was found that although the policy of supporting "the man" had undoubtedly made British universities among the world's leaders in providing "major contributions to knowledge" other peoples as well as ourselves were, in fact, judging us on "the size of our exports".

But this Rutherford doctrine was certainly "official" as early as 1924. Then, when arranging for the grant to pay Chadwick's salary as Assistant Director of the Cavendish, Sir Frank Heath, Rutherford's wartime ally and at this time his best "lever" in the corridors of Whitehall as Secretary of the Department of Scientific and Industrial Research, wrote on March 23rd, 1924: "As I think I have told you before, I consider that the money we are trusted to dispose of in the interests of scientific investigation are most profitably expended in enabling men of proved capacity to make discoveries to give the time to making them."

At the start of his Cambridge period Rutherford was indeed making discoveries at a great rate. He arrived at the Cavendish with the publication of his disintegration of the nitrogen atom immediately behind him. He set about proving that this was no fluke, that it really was a "major contribution to knowledge". And while he was so heavily involved in the reorganisation of his war-weary laboratory as part of the larger structure of science, he nevertheless had time to prove that he could do the same to at least six other of the lighter elements. In 1921 and 1922 he published four papers on "The Disintegration of Elements" showing that boron, fluorine, sodium, aluminium and phosphorus, as well as nitrogen, could all be disintegrated by alpha-particle impacts. For all these six elements gave off "H-particles" which could be detected by scintillations at a range far greater than that achieved by "natural" H-particles. Furthermore the fast-moving particles shot out from the target materials could be deflected by a magnetic field to show that they were what we would now call "protons" with a single positive charge – the bare nucleus of a hydrogen atom.

By succeeding in these experiments Rutherford kept up the flow of shattering and exciting results for which he had such a reputation. In finding the disintegrations of these six elements, Rutherford had, however, tested all the other light elements up to potassium – the nineteenth element – excluding only the "rare" gases such as neon. And he had quite definitely shown that similar effects were not observable in the other elements he had tested. This did not mean that he was not able to disintegrate the other elements – it simply meant that he could not observe undeniable protons coming away from the bombarded atoms. It was still possible that protons might be shooting off the other elements but at lesser velocities so that they could not be distinguished from contaminating "natural" hydrogen. He went to great lengths to combat this hydrogen contamination, coating the inside of glass tubes with platinum so that hydrogen atoms would not be knocked out of the glass by alpha-particles, using brominated grease on all taps and stopcocks, coating brass plates with hard pitch. To Boltwood he wrote, "I wish I had a live chemist tied up to this work who could guarantee on his life that substances were free of hydrogen. With this little detail set on one side I believe that I could prove very quickly which of the lighter elements give out hydrogen, but it is very difficult to do so without the chemical certainty as the effect is so small" (August 19th, 1920). Much of Chadwick's contribution to Rutherford's work at this time – and Chadwick was co-author of two of the main "Disintegration" papers – was on the chemical side.

The result of finding that these six particular elements could be disintegrated led Rutherford into the one major blind alley of his scientific career. The metaphor of blind alley is chosen with some care; Rutherford's successes were based on his vision of the microcosm, his feeling for the dimensions and forces at work on the atomic level, which enabled him to "see" more clearly than others what was likely to be wrong or right. In this case he saw something obvious – and the years to come would show that what he "saw" was not the reality. The six elements that he could disintegrate were boron, number 5 in the atomic table, a nucleus which he believed contained five positive charges and which seemed to have an atomic mass of about 10; nitrogen, atomic number 7, mass 14; fluorine, atomic number 9, atomic mass 19; sodium, atomic number 11, atomic mass 23; aluminium, number 13, mass 27; phosphorus, number 15, atomic mass 31. He saw six successive odd-numbered elements, whose atomic masses could be visualised as all being in groups of four plus "odd" units left over (or two odd units in the case of nitrogen). He concluded that nuclei therefore probably consisted of alpha-particles, the groups of mass 4, with extra H-particles as the additional units. From nuclei with patterns of this sort, it was easy to visualise single, unattached H-particles being knocked out. Elements with even numbers must have some other arrangement which did not leave such vulnerable "odd" units to be knocked out.

Rutherford pursued this false trail for five years until it led him into a scientific impasse. He constructed in his own mind a picture of the nucleus which was a version of his own creation, the nuclear atom, but an atom within itself, a nucleus that was like the atom. The nucleus in this vision had a nucleus within itself, around which circulated satellites, probably complete four-unit alpha-particles. The details of the structure of each different type of nucleus, varying from one element to another but with regularly repeated patterns, could be worked out by examining the behaviour of the different nuclei under the impact of different probes. By the alphas and betas and gammas emerging from them, by the different patterns of these undoubtedly nuclear emanations, it should be possible to discover the structure of the nucleus.

This was the nuclear physics into which Rutherford pushed the Cavendish and its staff – though, as we have seen, the direction was chosen not entirely for scientific reasons. Very soon new leaders emerged from among his latest recruits – C. D. Ellis, the army officer who had learned his physics from Chadwick in the prison camp at Ruhleben; P. M. S. Blackett, who had come to Cambridge as one of the Royal Navy officers on a special post-war course, and then chosen Cambridge rather than the navy; Peter Kapitsa, the brilliant new recruit from Russia. All of these men soon became leaders of small teams of two or three youngsters working on their special problems, and these special problems were usually concerned with alpha- or beta- or gamma-rays, coming probably from nuclei. Rutherford had seen a "way to attack" the problem of the nucleus and he hurled his forces into that attack, admitting that earlier he had been too pessimistic: "In a discussion on the structure of the atom ten years ago, in answer to a question on the structure of the nucleus, I was rash enough to say that it was a problem that might well be left to the next generation, for at that time there seemed to be few obvious methods of attack to throw light on its constitution," he confessed in his presidential address to the British Association in Liverpool in 1923. Others, too, felt that he was making remarkable progress; in that same year Jacques Loeb wrote from California:

It is perfectly amazing how rapidly you are settling the problem of atomic and nuclear structure. I still remember one afternoon at Berkeley I heard you give a forecast of the planetary atom such as you have shown it today to be a reality [and here Loeb must have been remembering back to 1906]; but that in these few years you should also have succeeded in solving the problem not only of the electronic structure but also of the nuclear structure, takes one's breath away. I do not think any man in the history of science has been able to make the progress that you have made in a comparatively short number of years.

The theory of the nucleus that Rutherford seemed at first to be revealing – his "view", or model of the nucleus – was finally expressed in a great paper, "Structure of the Radioactive Atom and Origin of the Alpha-Rays", which was published in the *Philosophical Magazine* of September 1927. Rutherford's opening statement reads:

> In the course of the last twenty years a large amount of accurate data has been accumulated on the emission of energy from radioactive atoms in the form of alpha-, beta- and gamma-rays, and in the majority of cases the average life of the individual elements has been determined. This wealth of data on atomic nuclei, which must have an intimate bearing on the structure of these radioactive atoms, has so far not been utilised as our theories of nuclear structure are embryonic. We are not yet able to do more than guess at the structure of even the lighter and presumably less complex atoms . . .

His second paragraph begins: "The simple theory of the origin of alpha-rays which I shall outline in this paper had its inception in the endeavour to reconcile apparently conflicting results on the dimensions of the nucleus obtained by different methods."

After discussing this conflicting evidence at some length Rutherford puts forward his proposed structure:

> On the views put forward in this paper, the nucleus of a heavy atom has certain well defined regions in its structure. At the centre is the controlling charged nucleus of very small dimensions surrounded at a distance by a number of neutral satellites describing quantum orbits controlled by the electric field from the central nucleus. Some years ago I suggested that the central nucleus was closely ordered, semi-crystalline arrangement of helium nuclei and electrons.

These electrons must be orbiting at speeds very close to the velocity of light, the central nucleus must be very small indeed, and there may be particles of no charge at all and mass equivalent to the proton (mass 1), the possible neutron. (Further reference to the possibility of neutrons will be made in later paragraphs of this chapter.) Radioactivity can then be visualised in this way:

> One of the neutral alpha satellites, which circulates in a quantised orbit round the central nucleus for some reason becomes unstable and escapes from the nucleus losing its two electrons when the electric field falls to a critical value. It escapes as a doubly charged helium nucleus

with a speed depending on its quantum orbit and nuclear charge [the familiar alpha-particle]. The two electrons which are liberated from the satellite fall in towards the nucleus, probably circulating with nearly the speed of light close to the central nucleus and inside the region occupied by the neutral satellites. Occasionally one of these electrons is hurled from the system giving rise to a disintegration electron [this is the beta-ray]. The disturbance of the neutral satellite system by the liberation of an alpha-particle or swift electron may lead to its rearrangement, involving the transition of one or more satellites from one quantum orbit to another emitting in the process gamma-rays of frequency determined by quantum relations.

This was the magnificent mental concept of the nucleus of the atom which Rutherford had built up. By a study of the emissions of rays it should be possible, he held, to provide a picture of the detailed structure of different atoms. By this time, working with Chadwick and using more sophisticated methods than he had used by himself in his earliest disintegration experiments, he had shown that virtually all the light atoms could be disintegrated by alpha-rays. There seemed every possibility that he had mapped out the course of nuclear physics for years ahead. But he was wrong – the nucleus of the atom is not at all the ordered structure he visualised. The evidence proved him wrong within a few months of the publication of this paper. Yet good judges, notably his rapidly-developing colleague C. D. Ellis, consider this paper one of Rutherford's finest pieces of pure science – one of his greatest feats of "imagining a possible universe". Rutherford had seen the anomalies in the evidence about the size of the nucleus, he had pursued this to create a nucleus of the nucleus, and he had built a system which was in accord with all the evidence available to him at the time. It was furthermore a model, a hypothesis, which showed the way to further crucial experiments – it was not his fault that these experiments proved him wrong.

The basic anomaly that Rutherford had spotted arose out of the historic conversation in Rutherford's home the evening he announced the discovery of the nucleus, when Charles Darwin immediately raised the question of what happened when the bombarding alpha-particle impacted a nucleus that was only the same size as itself. Rutherford's original nucleus theory had assumed that the forces acting between the invading alpha and the atomic nucleus must be forces acting according to "the inverse square law" – that is forces, probably electrical, which acted in the same way as Newton's force of gravity. Almost continuously since then Rutherford had had one experimental programme or another looking at this problem. In Cambridge in the early 1920s he made it one of his own lines of enquiry, working again with Chadwick. In simple terms these experiments had shown different results for the probable size of the nucleus than had other experiments using heavier atoms as targets, and Rutherford and Chadwick had had to start considering whether the alpha-particle – itself the nucleus of a helium atom – had some particular shape.

But these were precisely the years when the great theoretical physicists of Europe, by applying the quantum principle to atomic structure, were revolutionising our concepts of the nature of matter and the nature of the principles that govern our universe. These were exactly the years when de Broglie and Schrödinger were producing the equations that showed that elementary "particles", such as the electron, must also be regarded as "waves"; when Niels Bohr was producing his principle of complementarity to explain this paradox; and when Max Born was developing the whole basis of what we now call "Quantum Mechanics". The new physics made it clear that such simplicities as Newtonian Inverse Square Laws could not apply to bodies within the atom. Although no one denied the existence of the nucleus, it appeared that the theoretical basis on which Rutherford had proposed the nucleus was wrong. Only in later years, when his Cavendish team had been strengthened by a new generation of mathematical physicists who also performed experiments – Neville Mott, a later Director of the Cavendish, was the man who did the vital work – was it shown that in the case of alpha-particles striking the nuclei of light atoms the new theories produced exactly the same mathematical result, in this particular case, as the original simple theory.

In fact, as Norman Feather points out, in the second half of the decade of the 1920s experimental atomic physics was barred from further progress not only by the lack of any satisfactory theory, but quite as much by the problem that Rutherford and his colleagues had reached the end of the technological power of their experimental apparatus. Further major steps forward could not be taken until the development of new techniques for detecting and counting particles were forthcoming. The invention of methods of artificially accelerating particles – primarily a Cavendish invention – was an even more important step for the advancement of nuclear physics.

Five years later Maurice, Duc de Broglie, contributed an article to *Nature* in a series entitled "Scientific Worthies". The issue was No. 129, dated May 7th, 1932, an historic number. The Duke's article was an appreciation of the work and achievements of Lord Rutherford – he had been ennobled and taken the title of Baron

Rutherford of Nelson in 1931 – and it is in itself an interesting commentary on the coincidences that brought the paths of the members of this distinguished French aristocratic family so often across the path of the son of a New Zealand pioneer. The article was highly laudatory of Rutherford's past work, his "sureness of intuition" in disentangling the knots of radioactivity, his "decisive attack" on the problem of atomic structure. Yet on the whole it wrote him off. The Duke began: "It is often a matter of surprise to note the profound differences between the methods the human mind adopts towards the problems it is seeking to solve, and any narrow classification would necessarily be inexact because of the variety of possible temperaments", and he goes on to distinguish at least the two main types of scientist, the abstract theoretician and the experimenter. The crucial paragraph reads:

> Taking Lord Rutherford's work as a whole one finds that underlying it there is always a concrete picture of the problems based on direct experiment on them; avoiding all mathematical complications these experimental facts are connected up by means of a theory which gives a direct visual picture of the phenomena. Recent theoretical views suggest that such a mechanistic view of nature cannot be pushed beyond a certain point, and that the fundamental laws can only be expressed in abstract terms, defying all attempts at an intelligible description.

In other words, Rutherford's simple methods have been overtaken and finally swallowed up by the greater brilliance of the theoreticians. To sugar the pill, perhaps, the Duke added: "The philosophy of science has always swung between these points of view. The work of the great physicists to whom these lines are dedicated shows, however, to what brilliant discoveries the method followed by Lord Rutherford can lead."

It was as well the Duke included this last sentence, for this same issue of *Nature* contained not only Chadwick's announcement of the discovery of the neutron, predicted twelve years previously by Rutherford on the basis of his simple, concrete view of the nature of the atom, but also the first expansion, by Rutherford himself, of the bald announcement, made only a few weeks earlier by Cockcroft and Walton from Rutherford's laboratory, of the first artificial disintegration of atomic nuclei, with the first nuclear accelerator ever built in the world.

Rutherford's prediction of the existence of the neutron had been made in his Bakerian Lecture (a Royal Society honour) in 1920. His words then are an example of his simple visual "modelling":

It seems very likely that one electron can also bind two hydrogen nuclei and possibly also one hydrogen nucleus. In the one case this entails the possible existence of an atom of mass nearly 2, carrying one charge, which is to be regarded as an isotope of hydrogen. In the other case it involves the idea of a possible existence of an atom of mass 1 which has zero nucleus charge. Such an atomic structure seems by no means impossible . . . Such an atom would have very novel properties. Its external field would be practically zero except very close to the nucleus, and in consequence it should be able to move freely through matter. Its presence would probably be difficult to detect . . . On the other hand it should enter readily the structure of atoms, and may either unite with the nucleus or be disintegrated by its intense field . . . The existence of such atoms seems almost necessary to explain the building up of the nuclei of heavy elements . . .

We do not now regard the neutron as a combination of an electron and a proton, as Rutherford here suggested – quantum electrodynamics has shown us that this is too simple an idea – but we do know that the neutron can disintegrate, producing an electron and a proton (plus a neutrino), and so we can understand why Rutherford should have "viewed" the neutron in this way. But we must accept that the prediction of the existence of the neutron was not just a piece of speculation – Rutherford "saw" why it must exist, to account for the mass of the various atomic nuclei, and he correctly predicted the main features of its behaviour, which were to have such an enormous impact on the scientific theory of the nature of matter and on the practical development of atomic energy. And let it be noted that the "abstract theoreticians", of whom de Broglie approved so strongly, did not predict the existence of the neutron or discover the particle. (Whereas abstract theory did predict the existence of the "positron" or positively charged electron, and helped the experimentalists to discover it.)

The search for the neutron was, for ten years, the steady and frustratingly unsuccessful background to the scientific life, collaboration and rare disputes between Rutherford and Chadwick. It is the essence of Chadwick's complaint against Rutherford's policy for the running of the Cavendish, that by depriving the laboratory of special equipment for special jobs and devoting all the resources in hand to immediate use for research and education, chances of "getting" the neutron were lost. It was over this problem that the two men split up for a short time, and it was, according to Chadwick, the hunt for the neutron that brought them together again. Sometimes Rutherford threw himself into a fresh attack on the problem; often he would seem to lose interest. Rutherford had said in his Bakerian Lecture that he thought possibly neutrons might be found, or formed, in intense electrical fields in hydrogen gas, perhaps thinking that in the simple electron-proton structure of the hydrogen atom there might be electrical forces strong enough to push the electron into close binding with the proton. This approach produced nothing.

Chadwick has explained the way they went on:

> I . . . did a lot of experiments about which I never said anything. Some of them were quite stupid. I suppose I got that habit or impulse or whatever you'd like to call it from Rutherford. He would do some damn silly experiments at times, and we did some together. They were really damn silly. But he never hesitated. At times he would talk in what seemed a rather stupid way. He would say things which, put down on paper, were stupid or would have been stupid. But when one thought about them, you began to see that those words were inadequate to express what was in his mind, that there was something in the back that was worth thinking about. I think the same thing would apply to some of these experiments I have said were silly. There was always just the possibility of something turning up, and one shouldn't neglect doing say a few hours' work or even a few days' work to make quite sure . . . But until 1932 you see, it was in many ways a frustrating time.

And a little later he added, "But I just kept pegging away. I didn't see any other way of building up nuclei. So I did quite a number of quite silly experiments, when it comes to that. I must say the silliest were done by Rutherford."

Rutherford's position was the same as that of Chadwick – he could see no other way of building atomic nuclei except by having particles of the mass of a proton but with no charge. Particularly this was the case for isotopes, where the nuclei of the chemically identical elements differed only in weight – there must be mass without charge somewhere in the nucleus. It is significant that the only major reference he made to the neutron after his prediction of 1920 and before the discovery of the neutron in 1932, was in the 1927 paper quoted above, where he was writing directly on "The Structure of the Atom".

Rutherford's laboratory notebooks tell the story of these ten years of frustrating failure and ever-decreasing productivity. His working methods were the same as before, but since he often had parallel lines of research running at the same time, he often kept them apart in separate books. Thus there are two parallel books marking the start of his laboratory work in Cambridge – Notebook 26, which begins in Manchester in August 1918 with some of his later experiments on the disintegration work continues in the same vein with the simple marking for November 4th, 1919, "Cambridge"; Notebook 27, however, begins "In Cambridge, November 14th, 1919", and starts up fresh work on gamma-rays. The first results on the disintegration work deal with the troublesome problem of hydrogen contamination, and carry the conclusion: "So results indicate no sensible increase at slanting angle which is to be expected if H atoms arise from H as impurity in metal of source" – in other words, another possible source of contamination had been eliminated. What he called "Important experiments" on January 7th and 8th (though again Rutherford forgot to change the year at New Year's Day and marked them 1919), examined "long-range scintillations" from nitrogen and concluded, "Results alpha-rays bent exactly to middle of field of view – so clear all scintillations much more bent than alpha-particle and equal (within error) to H." On March 19th, he is testing "whether boron gives H-atoms", and later in March he is working with "short range N-atoms" and concluding, "Therefore seems clear short-range atoms from nitrogen are not nitrogen atoms but particles more bent than alpha-particles under same conditions but bent like H-atoms." He then discusses with himself whether these could be particles of "Mass 3 and charge 2" or even things of "Mass 2, charge 2".

It is worth noting here that Rutherford thought he had found particles of "Mass 3 and charge 2" which we would now consider to be the rare isotope of helium – helium three. He published his belief in his Bakerian Lecture of 1920, at the same time as his prediction of the neutron. Though there is little doubt that he had not discovered such atoms in 1920, nevertheless he predicted the existence of helium three, and before he died he found it, working with Oliphant; this piece of work is very rarely credited to him, just as his discovery of tritium – the isotope of hydrogen which he would have described as "Mass 3, charge 1" – is almost always ignored by modern (especially American) writers.

The parallel notebooks, Numbers 17 and 28, look distinctly different. Many of the readings are in Chadwick's neat and uninformative hand. Beside them are Rutherford's scribbled calculations and occasional, rather sad, remarks when he has been recording himself – "Eyes not good". Chadwick certainly noted this and devoted much time and effort to working out new and easier drills for scintillation counting and purchased new and more useful wide-aperture eyepieces for the microscope. But there was a typical burst of Rutherfordian energy when his first term was over, and there is almost an entire notebook taken up with work which went right up to December 23rd, 1920. Then an equally typically Rutherfordian gap for more than a month until January 27th.

February 1921 was another period of intense activity, as, with new methods, he examined a whole series of elements for traces of disintegration. Most experiments carry his informal remarks, "Good", "New screen", "Contamination". But in that one month he looked at iron, sulphur, sodium, phosphorus, beryllium, lithium, boron and aluminium. "Not much confidence in boron results," he writes to himself, but "Definite in aluminium source". By March 1922 he had perfected his arrangements for spotting protons flung out at nuclei in directions other than straight forward: "Now testing by both forwards and backwards particles," he notes. By June he had tested calcium, fluorine, nickel, manganese, chlorine and even vanadium.

The next spell of Rutherford excitement does not come until August 1922, when he hurled himself into a series of experiments on the interaction between alpha-particles and hydrogen gas. There are pages of readings, calculations and notes all in his own hand; there follow long discussions with himself about doubly-charged and neutral hydrogen atoms, and the heterogeneity of emerging alpha-rays. He describes the experiments as "Good". Slowly the enthusiasm falls off. In October he works briefly with Blackett, then a couple of weeks later with Ellis. One day, working with Chadwick, he notes: "Actual results this day of little importance and accuracy." From January to April 1923 there is a long pause, and the whole impetus decreases into finding material for two papers he published in 1923 and 1924 on the "Capture and Loss of Electrons by Alpha-Particles". In a parallel notebook for the same period we see Rutherford working on what were referred to as "Shimizu tracks". These are what would now be called "bubble chamber photographs", for a Japanese researcher at the Cavendish had started to automate the Wilson cloud chamber, allowing the expansion of the gas in the chamber to occur just as the elementary particles were about to enter, and photographing the results. This became grist to Rutherford's alpha-particle mill and the solution of his problems about the ionisation caused by alpha-particles. He was puzzled by the behaviour of the particle tracks, noting, "It only seems possible to account for the observed (tracks?) on the assumption that the forces are greater than those calculated on simple theory; in addition the effect appears mainly near the end of the path and more so than is to be expected from probability considerations." And again, "How to account for diverging tracks indicating helium plus helium??"

But there is no doubt, from the ever-increasing gaps of two or even three months in the laboratory notebooks, that his output was becoming smaller as the middle of the decade approached. He

Rutherford, spreading tobacco ash and noise, rules the laboratory as a tribal leader. He talks to Ratcliffe and ignores the delicate electrical counter on the lab bench on the left.

Lord Rutherford in his role as grandfather. Peter Fowler is on his knee.

published only three papers, two of them very short, in 1925, and in 1926 there were none at all. 1927 saw the appearance of his "summing-up" paper on atomic structure, referred to above, with two other papers, but there was nothing again in 1928, two very small pieces in 1929, and but one in 1930. Chadwick was quite frank about it – in the second half of the 1920s Rutherford took less and less interest in his own research and took on more and more work outside the laboratory.

From 1925 to 1930 he was President of the Royal Society, officially Britain's scientific "Number One Man". It was at the end of this period of office that he was ennobled. He was approaching sixty years of age; he had a career behind him containing more "major contributions to knowledge" than any other living scientist. It ought to have been the slow movement towards the dignified close of an eminent academic career.

But underneath the surface there may well have been the feeling that everything had gone sour. There was no progress towards sighting the neutron nor towards deciding the structure of the nucleus. There was not even any clear lead towards a method of "attack" on these problems. The whole Rutherford approach to physics was becoming increasingly less "fashionable" and progress was being made instead by the theoreticians on the European continent. Nor was his personal life very satisfactory. His only child, Eileen, had married Ralph Fowler, the mathematical physicist, who was most closely connected with Rutherford's work, but the girl led an unconventional life – what was called "fast" in the rather stuffy Cambridge society of those days. Rutherford's very few intimate letters refer less and less to her and more and more to his grandchildren. And it was shortly after the birth of his fourth grandchild that Eileen died, tragically young. Meanwhile Lady Rutherford had found the transition to that title from the New Zealand simplicity of Mary Newton increasingly difficult. To cover her basic shyness she became haughty and overbearing, particularly to young research men and their wives, who found the occasional compulsory visits to the Professor's house more and more daunting, for they might well find themselves virtually ordered to help in the gardening which was their hostess' chief hobby.

In his laboratory, too, Rutherford was being forced into policies he did not really like; he was being forced into the introduction of big machines. It started with the persuasiveness of Kapitsa, who obtained Rutherford's approval for the building of a large generator of intense magnetic fields. The pressure for spending more money on equipment continued from Chadwick. The only possible way forward in experimental nuclear physics appeared to lie in producing either very intense electric fields or finding some way of accelerating particles to energies they did not possess in nature. In 1927 the process took a big step when Rutherford invited the young T. Allibone from Metro Vickers in Manchester to help in designing and building a high-tension electricity laboratory. If it must be, it must; that seems to have been Rutherford's attitude, but he never learnt to interest himself in the design and construction of machines. What he wanted, and he was prepared to have machines if they would provide what he wanted, was results, readings, measurements, of the behaviour of atoms and the particles that made them.

His public lectures, in which he always allowed himself more speculation than in his formal scientific publications, show the steady build-up of depression. In his presidential address to the British Association in 1923 he could say:

> I have become more and more impressed by the power of the scientific method of extending our knowledge of nature. Experiment, directed by the disciplined imagination either of an individual, or still better of a group of individuals of varied mental outlook is able to achieve results which far transcend the imagination alone of the greatest natural philosopher. Experiment without imagination, or imagination without recourse to experiment, can accomplish little but, for effective progress, a happy blend of these powers is necessary.

And he could be optimistic: "We may confidently predict an accelerated rate of progress of scientific discovery, beneficial to mankind certainly in a material, but possibly even more so in an intellectual, sense." But the signs of wariness also creep in for the first time – the difference between positive and negative electricity is "an enigma" and "In the present state of our knowledge it does not seem possible to push the enquiry further." Quantum theory is accepted at least as having "proved of great value in several branches of science and is supported by a large mass of direct experimental evidence". And while the theory developed by Bohr and Sommerfeld about the electronic structure of the outside of the atom was a "great advance", it was "too soon to express a final opinion on the accuracy of this theory". He found difficulty in the philosophical problem that "We cannot explain why these orbits are alone permissible under normal conditions or understand the mechanism by which radiation is emitted", and he concludes: "The atomic processes involved may be so fundamental that a complete understanding may be denied us. It is early yet to be pessimistic on this question for we may hope that our difficulties may one day be resolved by further discoveries."

A year later, addressing the Franklin Institute in 1924 – one of the few occasions on which he mentioned the possibility of the neutron in public – he spoke of the nucleus as "a world of its own which is little if at all influenced by the physical or chemical forces at our command". And the problem of the structure of the nucleus was "much more difficult than the corresponding problem of the arrangement and motions of the planetary electrons". The reason was that "the facts known about the nucleus are few in number and the methods of attack to throw light on its structure are limited in scope". Furthermore the cause of radioactive decay remained "a complete enigma". He outlined the theory of rotating satellites in the nucleus but admitted that his experiments with bombarding light nuclei had given a "disappointing result" in this respect. And when he turned to considering how nuclei of heavy atoms might be built up, so as to give some account of the evolution of the elements, he admitted: "It has always been a matter of great difficulty to imagine how the more complex nuclei can be built up by the successive additions of protons and electrons since the proton must be endowed with a very high speed to approach closely to the charged nucleus," though here one may see the germ of the idea of the nuclear accelerator which was to come to fruition seven years later.

In 1927, speaking at the Centennial Congress in memory of Volta – an Italian holiday from which he returned with a rather severe attack of paratyphoid which he believed he contracted from drinking lemonade – he was still outlining his proposals for a nuclear structure with satellites and a very intensely charged core. His caution, however, was even greater – he spoke of "reconciling conflicting evidence". Although there was much evidence from radioactive nuclei he was compelled to admit that "so far our theories of nuclear constitution have been in too embryonic a state to make use of this information". So it is perhaps not surprising that in his Presidential addresses to the Royal Society late in 1927 and 1928 he eschewed these favourite subjects altogether, and spoke instead of his hopes that "recent advances in the production of very high voltages for technical purposes and the application of these voltages to highly exhausted tubes" would enable him to get high-speed electrons and atoms and very high frequency radiation.

But Rutherford, even in his late fifties, was not the man to let old age come upon him quietly. It is probably the most creditable achievement of a life filled with achievement that he could come back to science still young and flexible in mind, after five years of apparent failure, and end his scientific career with further triumphs. And he had filled the five lean years with much work in other fields – indeed he had created an entirely new career for himself in the grey area where government and science meet and interact.

15

Politics and Power

There is one feature of Rutherford's life that appears completely inexplicable: how did this brash, noisy, flamboyant, pure scientist, hailing from the backwoods of New Zealand, mesh in so perfectly with the quiet, unostentatious public-school men at the Athenaeum and on the fringes of Whitehall who controlled the first government support for science and who founded and formed those institutions by which public money is still channelled into research. Rutherford worked in nearly perfect harmony with these men, many of whom were not themselves scientists, and he was himself very much responsible for setting the policy and "tone" of those highly successful institutions.

When we switch attention from the Rutherford seen in large gatherings of young scientists, the Rutherford who was tribal leader of his team of researchers in the laboratory, where the big, bluff, hearty, pushing pioneer of his science ruled – and that is the Rutherford that so many of his scientific memorialists remember – to the man who worked in "the corridors of power", a totally different persona is seen. It is now that those small phrases in the official farewells from faculties and Vice-Chancellors of McGill and Manchester Universities appear significant, those few words of thanks and appreciation for help and advice and for valuable suggestions in the world of Senate discussions and university politics. Rutherford had shown from the start of his university career that he was an expert "man of business", a powerful "man of affairs". He expressed good policies lucidly and forcefully, but without irritating those who did not agree with him. He could sway a meeting by the force of his argument. He was honest and straightforward in his "politics" and, above all, he was truly discreet; he abhorred bringing an argument into public view. He was also far above what is called a "good committee man"; he was what we should now regard as a supreme "fixer". Rutherford believed in the small private group who, over lunch or dinner, in college or club, set the agenda and made the decisions that even the "small sub-committee" inevitably followed. Fortunately we have documentary proof of the fact that a great deal of important

453

business was, in fact, settled at lunches at the Athenaeum.

Rutherford's influence and "business" spread far beyond Britain into the USA, Europe, India, South Africa and his home territory of Australia and New Zealand. This is the Rutherford who could settle the agenda, and the outcome, of an international scientific congress in a quiet talk with Marie Curie and a letter to Lorentz.

Rutherford's entry into the world of government science came from his involvement with the wartime research work of the Board of Invention and Research – and from his disillusionment with the achievements of that Board. It will be remembered that the BIR had been pressing for the setting up of a Naval Engineering Experimental Establishment as early as 1915, and when the work of the BIR was brought to an end by the Sothern Holland Report, that Report made very strong recommendations for the rationalisation of naval research, the formal appointment of what we would nowadays call a chief scientist, and the eventual establishment of a central Naval Research Laboratory. The Royal Navy had done its best to avoid putting these recommendations into action, but had at least been forced to appoint Sir Charles Merz, the distinguished engineer, to the position of Director of Experiments and Research, though without the powers that Sothern Holland proposed.

Rutherford and his immediate colleagues, in this case Sir J. J. Thomson and Sir Richard Threlfall, did not let the matter rest. Early in 1918 they prepared a memorandum advocating "the establishment of a 'Naval Research Laboratory for Physics'", and they had enough political punch to force it first to the Admiralty and then to the War Cabinet. Merz acknowledged its arrival on May 25th, 1918, promising to "discuss it with the First Lord, and let Rutherford know how I will deal with it officially". The timing was right, the worst moments of the war (March 1918, the peak of the final German offensive) were past, and by November 1918, the matter had got to War Cabinet level, after a joint letter from Rutherford and J.J. had advised Merz to ask for £400,000 for the development cost rather than the £300,000 that had originally been estimated. By now the title of the proposed establishment had become the clumsy "Physical Research Institute for the Navy", but the objective was the same; it was "intended primarily for scientific research of a fundamental and pioneer character which might have a bearing on Naval development, and it may safely be assumed that no existing Naval establishment covers this field. In this respect it is necessary to draw special attention to the distinction between research and the experimental development of new appliances and apparatus." Rutherford and J.J. had learned from their wartime work that research and development are not the same thing – a lesson that has had to be relearned many times since in other fields.

Their memorandum to the War Cabinet admitted that there were a number of naval establishments developing new devices, "but during the war it has not been possible to go in for research except to a very limited extent". They emphasised that those scientists who had been employed by the navy during the war had not been doing research; they had been doing development work, because there had been no time and no suitable institutions for doing basic research. "The present war has shown the desirability and urgency of an adequate naval research institution. I do not consider that establishments arranged and equipped merely to deal with the experimental development of apparatus, etc., will have the influence on the progress of the Navy and the scientific efficiency of the naval officer that may reasonably be expected to follow from the establishment of a naval research institution." The existing establishments, such as Parkeston Quay, would continue to have the facilities for doing actual experiments at sea and they would be fed by the new research laboratory with "both ideas and personnel".

More recent history has shown the wisdom of this advice as applied to the development of either sonar or radar, but at that time the Royal Navy was not convinced. Merz had to write back to J.J. and Rutherford that "the chief objection to an all naval research institution, separate from universities and the National Physical Laboratory, is the lack of intercommunication between officers and scientists at all levels". There were, of course, naval officers receiving scientific instruction at the Cavendish Laboratory when Rutherford arrived there in October 1919, and this line of argument could now be dealt with. Captain E. Fullerton, of the Royal Navy Office in Cambridge, wrote to the new Professor, saying that he wanted to put a new scheme to their Lordships of the Admiralty concerning the "scientific training and requirements of naval officers". "There is no doubt that the present facilities available to naval officers for research at Greenwich and Portsmouth, etc., are inadequate and unsatisfactory, and that the resources of the Cavendish are far beyond any or all of them," he wrote, and ended by asking whether naval officers could be trained in Rutherford's laboratory on some permanent basis.

Rutherford drafted a long and careful reply in his own hand, making such a careful job of it that one cannot but wonder whether Captain Fullerton's letter came to him out of the blue. He carefully distinguished between two objectives: firstly, the instruction of selected naval officers in physics by a fairly standard course in the lecture room and laboratory; and secondly, training specially selected officers in research and the direction of research. The Cavendish, he pointed out, was already engaged in the first of these activities, and had forty or fifty officers learning physics, and presumably the future researchers of the navy would be found among their number. The second problem was more complex:

> When I gave evidence before the War Cabinet Committee on the future of science in the navy, I was of the opinion that the navy should have a research laboratory of its own on a fairly large scale, devoted mainly to research in basic physics. It was understood that the personnel of the research department should keep closely in touch with the universities and should be allowed to work at intervals in such institutions . . . As there does not seem much likelihood that this scheme will be carried out, the next best solution . . . would be to get specially interested officers . . . to undertake research in the physical laboratories of our universities.

Rutherford's plan for a naval physics research laboratory never quite materialised in the way he envisaged, but his campaign was very far from a failure. The Admiralty Research Laboratory, which closely approaches his ideal, was set up eventually at Teddington, next door to the National Physical Laboratory, and the government tried to improve research facilities for all three Services by giving the responsibility for a coordinated research programme to the newly-founded Department of Scientific and Industrial Research. Rutherford, along with J.J. and W. H. Bragg, was appointed to the Physics Board which supervised and developed the appropriate part of this programme, and Rutherford continued to interest himself in naval research, visiting such establishments as the RN Mining School near Portsmouth and even being taken out on a destroyer for a demonstration in 1921.

At exactly the same time as he was attempting to persuade the Royal Navy to invest in research in Britain, Rutherford was also pressing exactly similar ideas upon the Americans – and he possessed very powerful leverage behind the scenes in the USA. He had, of course, led the Allied mission to the USA in 1917 to coordinate anti-submarine research but, more important, his personal friends and scientific colleagues were in highly influential positions – H. A. Bumstead, whom Rutherford had known since the McGill days, was Scientific Attaché in London; G. E. Hale was on the Inter-Allied Anti-Submarine Committee; and America's best-known scientists, R. A. Millikan and Irving Langmuir were involved in the same work. Rutherford pressed his two favourite points: the importance of building up good postgraduate research schools in the universities, and the importance of setting up a naval research laboratory which would do basic physics research.

Two letters from his American friends tell much of the story. On May 18th, 1919, Irving Langmuir wrote:

> The information that you brought us and the suggestions and advice you gave me started our work on the submarine problem in the right direction and guided us throughout the work. The problem was certainly an interesting one and, as you know, was beset with many difficulties other than the purely scientific ones. However, at the moment nothing pleases me more than our ability to forget the whole episode and trust that the navies of both countries will continue in the work far enough to make us really secure against any future submarine attacks.

A few weeks earlier R. A. Millikan had written from the University of Chicago: "Hale and I are particularly grateful to you in helping us out in the campaign for getting the Rockefeller Foundation to contribute funds for research in connection with *educational institutions*." The original Rockefeller proposals, he explained, had ignored the universities altogether, but now the Foundation had agreed to give half a million dollars over five years mainly in research fellowships. "I do not know that we could have done it without your strong support for the general type of plan," wrote Millikan, who was no man to flatter.

Rutherford's correspondence with Hale had turned to the problems of long-term naval research rather earlier. As the war ended (November 13th, 1918) he had written to give his support for the idea of building a new naval research lab in the USA on the New Plymouth site which had been developed during the war. He had emphasised "the important effect such a special laboratory would play in giving good research opportunity to the best young people and in advancing the study of chemistry and physics . . . The progress of technical science is dependent mainly on advance in pure science. My little experience in war problems has convinced me that it is the applications of basic ideas of physics and chemistry that are of the first importance." In a later letter Rutherford considered the ideal research organisation, and confided to Hale that probably a "mixed" system would be best; there would be considerable advantages in having a few small laboratories connected with the most progressive universities and a large central naval laboratory, all linked together. The problem was that "a well-endowed central lab with a large personnel would tend to damp out research in the universities . . . There would always be the danger that such a department would not as a rule take kindly to criticisms or rivalry from what they considered less important centres. This peculiarity of human nature is even now displayed by central institutions of various kinds."

This particular letter to Hale has great fascination, for in it Rutherford allows himself to dream – the dream of a perfect physics laboratory, made all the more interesting by the fact that this was the time when he was struggling with the difficulties of rebuilding and developing the Cavendish. His "Ideal Laboratory" would cover not just physics, "but technical physics, physical chemistry, even metallurgy". For "I am confident if a good physicist and crystallographer had been in close touch in the old days, the diffraction of X-rays would have been discovered years ago". His department of technical physics was not there just to do its own work, "but to bring physicists in touch with technical methods and machinery and to help devise apparatus for special purposes. As a rule your pure physicist is very weak on the methods which technical physics can offer." The pay of the head man of this super-lab should not be fixed but should depend on his ability, Rutherford wrote, and continued, "The physical and chemical labs should be in the same building or close by, for the most interesting (and most valuable) thing in science is to know how the other fellow thinks and to respect him for it if you can." The problem was that chemistry "lags in ideas behind physics and they want a broader philosophy of science, while the physicist is woefully ignorant of the reasons why a chemist's mind works in such particular grooves and why he has acquired such an impenetrability to the flying colours of the physicist".

So much for a dream. Rutherford, always practical, ends: "If you are really anxious to advance science in the English-speaking world a similar lab in this small island would be a damn'd good thing, but I don't know whether your people have advanced to that stage of internationalism – our millionaires are pretty difficult to tap for such purposes." The US Navy Research Laboratory, while not attaining Rutherford's standard of the ideal, probably came rather nearer to that ideal than the Royal Navy's efforts. But the Americans introduced the remainder of his suggestion, that of having service laboratories as part of university organisations, and there are many academics nowadays who question the wisdom of this development in so far as it affects academic freedom.

The best example of Rutherford's style in handling a conflict without causing offence or raising a public outcry, while remaining firm is found in the "Affair of the Royal Artillery College". This began, as far as Rutherford was concerned, when his former pupil,

Edward N. da C. Andrade, came to visit him in Cambridge on Saturday, February 6th, 1926. Andrade had had a successful career in the Artillery during the war, and had taken up the post of Professor of Physics at the reorganised Artillery College at Woolwich, an institution which had been reformed after the successful use of scientific methods during the war, notably W. L. Bragg's system for sound-ranging on enemy guns, and which had the objective of giving a higher technical education to gunner officers, especially in connection with the obvious need to mechanise the transport system.

Andrade brought a whole series of serious complaints, and Rutherford must have taken some time to check them out, for it was four weeks after Andrade's visit that he drafted a letter to the Secretary of State for War, Sir Laming Worthington-Evans, and added a letter of support from the Master of Trinity College – that is J.J. He began firmly: "I have heard, both directly and indirectly, that an unsatisfactory state of affairs exists at the Artillery College, Woolwich, and that the scientific staff there are labouring under a sense of grievance." But there was no attempt to make a public scandal; Rutherford simply asked for an interview with the Secretary of State to bring the matter "to your personal attention in an unofficial and private way".

On the other hand there was no intention of "fudging" the issue. "I understand that the position of the civilian staff is being made intolerable by the discourtesy and lack of sympathy of the commandant of the College, Colonel Wilkinson . . . I am surprised to learn that there is nobody at the War Office with any knowledge of higher education to whom these gentlemen can appeal. There seems to be a deliberate policy to treat the professors as civilian subordinates of low grade." It was the problem of the scientist and the navy officer all over again, essentially a social problem, and Rutherford shows his sensitivity to English snobbery by deliberately referring to the professors as "gentlemen", which is a title which would probably not have been given to them by Regular Army officers. And Rutherford was quite prepared to give "chapter and verse" to the complaints – the intention had been to create at Woolwich a service college of university ranking – and universities were the province of gentlemen – but "The atmosphere created by the Commandant and his immediate superiors is deplorable. The Commandant walks into classes without warning or the customary civilities." Syllabuses were altered by officers at the War Office, who had no technical knowledge, without consulting the Professors. And the conditions of service for the Professors and lecturers were unsatisfactory in respect of holidays, leisure for study and research, and security of tenure. Andrade himself had been forced to resign from the National Aeronautical Research Committee "by grave incivility by the Commandant and the War Office".

So far, so bad, but right at the end comes both the stick and the carrot. Rutherford warns that he has tried to "hold off parliamentary action" on the grievances, and asks for this personal intervention by the Secretary of State because "it would be a pity to let the situation become public, as it would be made political capital against the army".

He got his interview with Worthington-Evans on March 25th, and his diary shows how carefully he managed such things; he had a final talk with Andrade and one of his colleagues just five days before the interview, and he made it clear to the War Office that he was speaking, not personally, but formally as President of the Royal Society, as the official spokesman for scientists, but at the same time "unofficially" – a nice distinction, which officialdom understood.

The interview with the Secretary of State was "satisfactory" according to Rutherford's diary, and the War Office was plainly forced to accept negotiations. Rutherford himself returned to the War Office on April 13th to see a Colonel Peck, and this gentleman obviously tried the oldest evasive action in the governmental book, for four days later Rutherford wrote to him saying that the present differences "are not so much a matter of the personality of the staff as you are inclined to think but that the staff suffer a genuine sense of grievance. These differences would, I am sure, vanish if the staff were taken into the confidence of the administration on matters affecting the purely educational side of their work." He then went on to advise Colonel Peck to "have a chat" with F. E. Smith at the Admiralty: "He has had experience both on the administrative as well as the scientific side and I have a good deal of confidence in his judgment on such matters." Frank E. Smith had first come to Rutherford's notice in the latter days of the BIR, and was to become his closest ally as a civil servant who dealt with scientific affairs when he became Sir Frank Smith, Secretary of the DSIR.

None of this is to make any claim that Rutherford's intervention in the affairs of the Artillery College had anything but a short-term effect – that institution never fulfilled its objective of becoming a university-level focus of higher education for serving officers. But the effort was not totally wasted, for institutions of this type have eventually forced their way into the pattern of the defence services, for instance in the shape of the Royal Military College of Science at Shrivenham.

Rutherford's basic attitude – that he did not desire to be seen to win victories and that he deplored public controversy – was undoubtedly one of the chief reasons for his success in quasi-governmental organisations and affairs. This attitude was equally one of the chief reasons why he attained such influence in inter-national scientific circles, and is well illustrated in his dealings with the Radium Institute of Vienna, the scientific home of his life-long friend Stefan Meyer, when dispute threatened. In July 1924, Rutherford received a letter from Hans Petersson, a Swede work-ing in Meyer's institute, asking him to arrange for the publication of a paper in *Nature*. The paper described the latest work in Vienna on atomic disintegration, claiming to have developed a new method for the photometrical measurement of the brightness of scintillations which allowed the scientists to claim that they had disintegrated oxygen and carbon and had found the first examples of helium (alpha-particles) as the byproduct of disintegrations.

Now this was exactly the field in which Rutherford and Chad-wick were working at the time and the two Cambridge men knew – simply knew – that there was something wrong with the Vienna results. Rutherford's reply was to tell Petersson privately of his own latest results – that is to say results which showed that the Vienna readings must be wrong somewhere. Rutherford had also written, privately, to Meyer warning him that "Chadwick and I have definite evidence that many of the results are wrong or wrongly interpreted . . . You know me well enough to appreciate that I would not interfere unless I thought the situation was serious." Petersson visited the Cavendish early in 1927 for dis-cussions, but the final word remained with Chadwick when he went over to Vienna in December 1927.

Chadwick regarded the whole incident as throwing a very vivid light on Rutherford's character, "to explain his attitude towards public controversy. He would never indulge in it. He thought it was wrong. If you disagreed with a man on a scientific topic, have it out with him in private, but don't start quarrelling in public." When he arrived in Vienna Chadwick found that "the observers, the coun-ters of the scintillations, were three youngish women . . . of what Petersson called Slavic descent because he believed (I'm only repeating what he said to me) that . . . Slavs had better eyes, secondly, and mainly, that women would be more reliable than men as counters of scintillations because they wouldn't be thinking while they were observing them."

The problem, Chadwick found, was that the three friendly young women knew what was expected to happen all the time. He arranged two simple experiments in which the women did not know what was supposed to be happening – and the results came out exactly as they had been found in Cambridge! There was no question of any cheating, Chadwick insisted; the "blind trial" had simply shown "they were deluding themselves. They were seeing what they were expected to see." Chadwick advised that the best thing to do was to drop the experiment and say no more about it, and he explained more than thirty years later: "I took it on myself to act in that way because I knew that was what Rutherford would have wanted. I didn't need to consult him. I knew that was what he would do because he would never have done anything that would have caused pain to Stefan Meyer. This was a personal matter, but at the same time he would never have indulged in this kind of public argument to settle a matter of that kind. And so it was left like that." But – and this was the strength of Rutherford's attitude – he and Meyer and Chadwick and Petersson were good friends for the rest of their lives.

Most of the crucial action of the Petersson affair took place while Rutherford was out of touch. From July to December 1925 he toured Australia and made a triumphal return to his home country of New Zealand. He lectured wherever he went, he was entertained by university vice-chancellors in Australia, and had his car triumphally drawn by students through the streets of Christchurch. He met innumerable relatives in New Zealand, and even more people who claimed to have known him as a child when he visited the scenes of his birth and early childhood. The *Sydney Morning Herald* carried no fewer than five articles chronicling his arrival in Australia and giving long coverage to the lectures he gave in Adelaide and Sydney. Rutherford himself wrote an account running to ten foolscap typed pages. This is one of the most monumentally dull pieces of writing that anyone could imagine – indeed it seems almost immature, and might have been written by a rather uninteresting child of fifteen.

Before he left for this long trip – certainly the first time in his career that he had been away from his laboratory for such a time – Rutherford made no mention of being "stale". But we have seen that his scientific progress had become very much slower by 1925, and he was beginning to feel the twinges and minor ailments that become more noticeable as the sixtieth birthday approaches. In some sense he seems to have felt that a change in his life-style and interests had become necessary or inevitable. He returned from his grand tour to start his term as President of the Royal Society, and for the only time in his life he began to keep a diary. The effort lasted little more than a year, covering 1926. Being Rutherford the entries are laconic, but taken together they give an unusual picture of the largely honorific life he was to live for five years.

Throughout this year he travelled up to London at least once every working week, and often twice. He would leave Cambridge one morning, and get to London in time for lunch (often at the Athenaeum), attend meetings and perhaps a formal dinner in the evening, sleep at the Savile Club or the Great Eastern Hotel, and travel back to Cambridge the following day "by the breakfast train". He took up his Presidency of the Royal Society formally on January 11th, his diary informs us, and, after lunch at the Athenaeum, "met many friends". He had a business meeting between the Royal Society and the Research Council, and he had to meet Professor Wood, Secretary of the British Academy, "to discuss arrangements for admission of Central Powers". This business of the formal readmittance of German and Austrian representatives to various bodies recurs a number of times in the year – the same matter was discussed when he went on an official visit to Brussels in June.

In February he totted up the number of invitations to formal dinners that he had already received. He found that he had accepted no fewer than twenty, including the Royal Academy of Arts, the Royal College of Physicians, and the Institution of Electrical Engineers. But he had refused a grand total of sixty similar invitations. Sometimes he enjoyed himself: on an April day he noted, "Dinner, Armourers and Braziers [one of the City of London Livery Companies] – made final speech, good dinner, pleasant evening." And on Thursday, March 4th, after two Royal Society meetings, one for business, one for science, he notes: "Took chair Royal Society Club in evening, Appleton and Lindemann my guests. Told story of porpoises in Red Sea and visit to Sakkara." But it all added up to a punishing schedule – there were meetings of the Radio Board and the Physics Research Board, meetings of the Board of Research Studies, and the National Physical Laboratory Research Committee. One day saw "Lunch with Pope, discussion Mineralogy Chair. Meeting re salaries of staff," but May 21st had "Attack rheumatism in shoulders. Saw Cunney and took salicylate. Passed off after 3 or 4 days."

The social highlight of the year was plainly the meeting of the British Association in Oxford in August. On Wednesday, August 4th, "escorted P of W [Prince of Wales] into Sheldonian" and on the following day "again with Prince, gave paper before Prince" and "sat next to him at luncheon", followed a few days later by "Blenheim, Lunch with Duke of Marlborough. Duchess not present. Queer company."

Golf was the chief relaxation – the word occurs regularly throughout the year's events, though sometimes marred by such comments as "flat tyre". Rutherford survived the pressure mainly by taking a number of total holidays, times when he went away, usually in his car, and had nothing to do with work. He motored round the South Coast in April, he went to Italy in June and in September he took off again for the Yorkshire Moors. This latter holiday marred somewhat by "Sept 5th . . . carburettor blocked near Barnby Moor; delay 1 hour" and "Sept 5th; wife dislocated finger, walked" which did not prevent "Sept 8th; climbed Whernside".

There are no further diaries, so we must presume that life went on in this hectic way right through to 1930. There was very little scientific output in these days, though clearly Rutherford kept up well with what was going on. It is perhaps indicative of these points that there are no Rutherford letters to Niels Bohr from 1926 to nearly the end of 1929 when he writes: "I have heard rumours that you are on the warpath and wanting to upset the Conservation of Energy both microscopically and macroscopically. I will wait and see before expressing an opinion but I always feel 'that there are more things in Heaven and Earth than are dreamt of in our philosophy'." To Stefan Meyer in 1927, at the height of the Petersson difficulty, he complained that he was too busy as President of the Royal Society – he was having to cope with the Annual Meeting of the Greenwich Observatory and the Annual Visitation at the National Physical Laboratory – "it is all amusing enough if it did not take so much of one's time". In September he returned from the Volta Centennial Congress at Lake Como in Northern Italy with what he describes as "a bad cold, several days in bed with a temperature and now a bad digestion upset", which was obviously paratyphoid. The congress was good, but there were too many papers, twenty a day, he groans. "I got very weary of such a mixed mass of papers," he reported in this letter; and in another, also to Meyer and also about the congress, he says it was very interesting "but was very hard work for all concerned and was in no sense a holiday". From this date onwards his feeling against large international meetings became a positive prejudice.

The Rutherford of the first fifty years of his life almost disappeared in this welter of honorific ceremonies. In 1928 he was awarded the Albert Medal of the Royal Society of Arts and made an Honorary Fellow of the Royal College of Physicians. He opened the new physics laboratory at Bristol University, where his grandson, Professor Peter Fowler, now has one of the Chairs, and he presided over a big meeting on nuclear structure at the Royal

Society, pointing out that just fifteen years earlier they had held an historic discussion on the structure of the atom. In the summer of 1929 he went with his family – his wife, daughter, son-in-law, and first grandchildren – to South Africa for a lengthy holiday mixed with the business of the annual meeting of the British Association. The disturbing feature of his personal life came with occasional news of the deaths of some of those he loved – Bertram Boltwood died by his own hand in a fit of depression in 1927; in Holland his old ally, Lorentz, died; and in 1928 his own father, aged eighty-seven, reached the end of his days.

The records of the Royal Society contain little but formal notice of Rutherford's work as President. But that eminent body was going through a rather difficult period, as is evidenced by a major revolt against its "establishment" led by Soddy in the years immediately following Rutherford's period of office – a revolt which was not connected with Rutherford in any way, though he did not respond actively to Soddy's requests for support. There were, however, rumblings of discontent in Rutherford's time – in 1926 H. E. Armstrong wrote a couple of times complaining about the Society's "indifferentism". In one of these letters Armstrong added a postscript: "Your attitude in the Chair is delightful; to have a President asking questions and promoting discussion is an astounding departure. You may restore a dead body to life, if you persevere and get a little human feeling into the show." And three years later Armstrong wrote in a similar vein:

> You were very clear and most interesting on Thursday (at the R.S.) and Aston was good – but what muffs the others as speakers. You should have a private rehearsal of these discussions so as to coach people at least to speak up, speak out and address the house, not the screen. You might with advantage add to your officials a teacher of elocution. Money spent in that way would give a far better return than much of that spent on research, so called!

Armstrong, of course, was very "conservative" and Rutherford may not have approved of his advice.

But there was a considerable amount of public business to do behind the scenes of the Royal Society, notably in the reform of the National Physical Laboratory. Rival schemes were put forward by Sir Frank Heath, well-known to Rutherford as the senior civil servant of the BIR, by Henry Tizard, the rising new star of the DSIR and always an ally of Rutherford's, and the Director, Sir Joseph Petavel, known to Rutherford from his Manchester days. Rutherford, according to Eve, used plenty of oil to get the resulting compromise salad.

It was while he was President of the Royal Society, and probably because he held this post, that Rutherford became involved in a new interest – or perhaps it would be more accurate to say returned to an old interest – radio and broadcasting. In 1928 he was appointed to an advisory position with the British Broadcasting Corporation, and he became steadily more involved in this work for the remaining ten years of his life, ending up as one of the chief motivators for the expansion of the BBC's Empire Service into something more like its present Overseas Services.

When Rutherford was President of the British Association in 1923 and gave his presidential address at Liverpool he was involved in a pioneering experiment. "A Miracle of Broadcasting – the BBC's Biggest Experiment" was the headline in the *Radio Times* of September 28th, 1923. The article started:

> An historic milestone in the History of Wireless was reached the other night by the broadcasting of the Presidential Address of the world famous scientist Sir Ernest Rutherford . . . It was the first occasion in this or any other country on which the voice of a public man had been transmitted simultaneously through six wireless stations hundreds of miles apart and also made to operate loud-speakers at overflow meetings . . . Perhaps the most amazing result of the experiment was that the sound of the speaker's voice was heard in the North of Scotland before it reached those who were sitting at the back of the hall in which he was actually speaking.

The explanation given for this was that it took a fifth of a second for his words to reach the back row of the hall, whereas it took only a fiftieth of a second for his broadcast sound to reach Scotland. The *Radio Times* article gives a detailed explanation of the amplification and lines-links that were so new then, but are so familiar now, and it ends with a paragraph headed "Shoals of Congratulations", saying: "Probably nothing has ever happened that has not had some critics and there have been a few who have criticised the broadcasting of a single speech on a scientific subject lasting for an hour and a half; but the BBC has been overwhelmed with congratulations." The ratio in favour of the experiment was fifty-five to one, and the BBC had decided, on the strength of this performance and reaction, to change its rule of never broadcasting speeches of more than ten or twelve minutes' length.

Certainly five years later, in 1928, the BBC was considering starting an annual prestigious lecture which it called the "National Lecture", predecessor to the Reith Lectures of today, except that the National Lectures were single performances rather than a series. Rutherford was invited to join the Panel of Advisers for

Broadcast National Lectures, charged with finding the men to give
the lectures. Lord Reith considered that "nomination to a Lec-
tureship [should] be regarded as a definite distinction" and was
prepared to offer a fee of £100 for a forty-five minute broadcast.
The panel first met on Wednesday, October 31st, 1928, and
Rutherford found himself in the company not only of old friends,
such as Sir W. H. Bragg, Sir Oliver Lodge and Sir James Jeans, but
also of men such as H. A. L. Fisher, the historian, Professor T. F.
Tout and Sir Israel Gollancz. Later the panel was joined by the
poet Walter de la Mare, aristocrats such as the Marquesses of
Crewe and Zetland, and the poet, Sir Henry Newbolt, invited as
"another representative of literature".

There was not much work involved, as the panel only met once a
year, but plainly Rutherford impressed the stately masters of the
BBC. In 1930 he became involved in an even more dramatic
broadcasting experiment. His friend A. S. Eve was that year's
President of the Royal Society of Canada, and he invited Rutherford, as President of the original Royal Society to give a broadcast
fifteen-minute speech to a meeting in McGill. The Royal Society of
Canada meeting was to be at nine p.m. on May 21st, which was two
a.m. on May 22nd in London. The broadcast, Eve instructed him,
would have to be made from an hotel room in London, and would
be transmitted by the joint efforts of the GPO, the Canadian
Marconi Company, and the Canadian branch of the Bell Telephone Company. It would therefore be what Eve called "the all
British route"; Canadians had heard the King broadcasting to the
Peace Conference shortly before, but that had been relayed to
Canada via the USA so Rutherford's words would be heard by "an
all-British effort of some importance". Eve may have been patriotic but he was not foolish, and he added: "We hear discussion of the
freedom of the seas, but certainly freedom of the Heaviside layer
[the ionised layer of the upper atmosphere which reflects radio-
waves back to Earth] can never be disturbed by army, navy or air
forces of any country – it may be the last time the all-British route is
used, as well as the first, because it may not be viable commercially."

There was a slightly comic last-minute panic over this broadcast,
when Eve suddenly realised that "Summer time" was likely to be
ruling in both England and Canada when the speech was made,
and all arrangements had been made on the basis of Greenwich
time. There was a flourish of transatlantic cables before everything
was sorted out and the broadcast was successfully made.

When Rutherford's term on the BBC Panel of Advisers came to
an end he was asked to give one of the National Lectures himself,

and in 1933 duly broadcast a forty-five minute dissertation on the "Transmutation of the Atom". This was, of course, after the great discoveries of the neutron and the Cockcroft–Walton accelerator smashing atoms at the Cavendish, and Rutherford was back again at the top of a booming branch of science. His National Lecture showed what had been done and was bubbling with optimism for the future about what would soon be done, as the new machines allowed for more experiments. It is full of the wonders of Cambridge's particle accelerator; of Lawrence's big cyclotrons in California; and of the new Van dem Graaff machines.

The lecture was also full of what was, at that time, sound common sense. Describing the energies involved in Cockcroft and Walton's transmutation (or smashing) of lithium atoms by fast protons he said:

> Considering the individual process, the output of energy in the transmutation is more than 500 times greater than the energy carried by the proton. There is thus a great gain of energy in the single transmutation. But we must not forget that on an average more than a thousand million protons of equal energy must be fired into the lithium before one happens to enter the lithium nucleus. It is clear in this case that on the whole the energy derived from transmutation of the atom is small compared with the energy of the bombarding particles. There thus seems to be little prospect that we can hope to obtain a new source of power by these processes. It has sometimes been suggested, from analogy with ordinary explosives, that the transmutation of one atom might cause the transmutation of a neighbouring nucleus so that the explosion would spread throughout all the material [this is what we now know as the chain-reaction]. If this were true we should long ago have had a gigantic explosion in our laboratories with no one remaining to tell the tale. The absence of these accidents indicates, as we should expect, that the explosion is confined to the individual nucleus and does not spread to the neighbouring nuclei, which may be regarded as relatively far removed from the centre of the explosion.

This is the viewpoint for which Rutherford is, perhaps, best known – the first man to split the atom and release its energy yet the man who insisted that the practical use of atomic energy was "moonshine". Like nearly all popular labels, this one is wrong. The opinions expressed in his National Lecture in 1933 were entirely sound, good common sense, on the evidence he had available at that date. When fresh evidence became available he changed his position in accordance with it, as we shall see.

It is only fair to look at the final paragraph of his lecture as well:

As one whose scientific life has been largely devoted to investigations on the structure and transformation of the atom, I watch with much interest and enthusiasm the development of these beautiful experiments to add to our knowledge of the constitution of nuclei. No one can be certain what strange particles or unexpected phenomena may not appear. I know of no more enthralling adventure of the human mind than this voyage of discovery into the almost unexplored world of the atomic nucleus.

Apart from the obvious qualification of his pessimism about the use of atomic energy, and apart from the intellectual humility, observe the exceptionally lucid and human style. This was Rutherford at his best – how any man could write so well about science, yet so badly about a trip to his homeland in New Zealand, remains a mystery.

But it was the style that impressed the BBC. Shortly after this lecture Charles Siepmann, the Director of Talks, wrote an internal memo on the question of getting the BBC's experts to give advice to speakers – what we should now call "production". Siepmann wrote of one lecturer "who eminently failed to realise our expectations of him and of his subject" largely because he "was not very responsive to suggestions made to him by us". He compared this man's attitude to "Lord Rutherford who, on a difficult subject, succeeded, I think, in interesting a very wide circle of listeners, [and] proved the virtue of liaison with the Corporation by being quite unusually friendly and open to suggestion".

Whether or not the success of his National Lecture was the reason, Rutherford's name was among the many being lobbied among the senior men of the BBC throughout 1934 as they went about forming a "General Advisory Council", a body of people distinguished in various fields, which still exists. Right from the start, although the members were not intended to have their own specialities as "constituencies", Rutherford's name was down for "science". In a short series of memos between Lord Reith and the Director of Administration, Charles Nicholls, there is a list, a very private list, showing the perceived political positions of the people on this Advisory Council. Sir William Beveridge, Sir Walter Citrine, Mr Lloyd George, the Marquess of Lothian and Dame Sybil Thorndike are grouped under "Left". The group under "Right" includes J. J. Astor, MP, and L. S. Amery, MP. Rutherford comes in a very small group marked "Central or Unknown", and with him is C. B. Cochran, the film-producer who subsequently refused the invitation to sit on the council.

The General Advisory Council represented a much greater involvement with the development of broadcasting. It met three times a year, had long discussions including matters of programme policy, and involved considerable correspondence for those who were really interested in its work. This number included Rutherford who was a most regular attender at the meetings and who spoke fairly frequently. He found among his fellow members the Archbishop of York, the composer Walford Davies, the author John Buchan (Lord Tweedsmuir), and the comedian George Robey. And on one occasion shortly after the first meeting of the Council in 1935, the Minutes show Rutherford allying closely with George Bernard Shaw in questioning "whether a series of twelve talks about Galsworthy's plays was not excessive". A more significant intervention came when the very delicate subject of broadcasts about birth-control was raised. Even the cold Minutes make it clear that Rutherford made a fairly powerful plea for what he called "the broad approach" – urging the liberal view. He pointed out that the economic situation of families, the different attitudes of different nations and the decline of the population, were all arguments against refusing to make any mention of this difficult subject.

But Rutherford's biggest impact on the BBC came in the year of his death, 1937. From the start of the year he had been pressing for a discussion of "Empire Broadcasting" to be put on the agenda as the major topic of a meeting of the General Advisory Council. The BBC Archives show a curious reaction to this by BBC officials; apparently for the first time the Chairman was fully briefed with prepared notes on the subjects likely to be raised: the technical problems of overseas broadcasting, the history of the Empire Service, which, after experiments in 1927, the BBC had started in 1932, and the "propaganda efforts of other countries which had overseas services". A BBC memo speaks of "the preparations for, and the stage-management of, the forthcoming GAC discussion of Empire Broadcasting", and with this came a whole file of papers, the Empire Service Booklet, memos on "Broadcasting and the Colonial Empire", the use of foreign languages, the use of recordings, the problems of time differences.

Rutherford himself led the discussion at the meeting of the GAC on June 14th, 1937. There seems to be no record of what was said, but four days later Lord Reith wrote to Rutherford, thanking him for bringing the matter up and for raising such an "interesting discussion". C. A. Cliffe, the Acting Director of the Empire Service, seems to have been a little irritated by Rutherford, for he wrote another internal memo on June 16th, in which he "proposed that Lord Rutherford, who had initiated the whole discussion on the Empire Service, should be got out of bed in the small hours in order to give a talk".

The crucial question that had been raised was the question of propaganda. Some countries, notably Germany, had already opened short-wave services that were open vehicles of propaganda, not in the best sense of the word. The BBC, true to its Reithian principles, would have nothing to do with this sort of work. There was also the problem of money – Rutherford's GAC meeting had been told that the Empire Service was paid for out of the ordinary Englishman's ten-shilling-a-year licence fee, and that the fees paid to speakers for the Empire Service were clearly less than those paid for the Home Service. How then could one expect famous men and women to broadcast for the Empire Service?

It is difficult to tell to what extent Rutherford had acted independently in raising the question of Empire Services and their possible expansion into Overseas Services using foreign languages. It is possible that there was "stage management" in the raising of the whole subject. Certainly BBC memos flowed rapidly throughout the summer of 1937, and in October that year the British government formally requested the BBC to start foreign language broadcasts, and of course eventually paid for this extension of the broadcasting service. But Rutherford did not attend the October meeting of the GAC, which was scheduled to continue the discussion he had started. In that month he died.

Rutherford's involvement in the development of broadcasting services shows the extraordinary width of his contacts and his growing reputation for flexibility and "usefulness" outside his own area of science. His main impact on the formation of national institutions and even national policy came through the DSIR, the Department of Scientific and Industrial Research. The creation of this totally new government organ in 1915, in the middle of the war, was the largest practical result of the new influence of scientists on government thinking, that same influence which appeared more obviously but less effectively in the creation and activities of the BIR. In ministerial circles Balfour and Arthur Henderson were again the chief motivators, but they were, of course, pressed, or supported, by the influence of the Royal Society under J.J. and Sir Arthur Schuster. "The primary objective is to encourage scientific research which may ultimately benefit the industries, but there is no intention of interpreting this very strictly," Schuster wrote to Rutherford, and advised him to apply for some sort of grant immediately; even if he could not use the grant at the moment it would give him a prior claim for the future and it would encourage people in government circles who were supporters of the idea of official funding for science, especially if they could claim knowledge of important research work that was held up for lack of funds.

The White Paper (Cd 8005) which proposed the new machinery as a Department of Scientific and Industrial Research in the summer of 1915 spoke of the "strong consensus of opinion among persons engaged in both science and industry" for the need for some additional State assistance for science, not least because of the wartime shortage of certain articles and materials which had largely been manufactured abroad "particularly in Germany because science has there been more thoroughly and effectively applied to the solution of scientific problems bearing on trade and industry and to the elaboration of economical and improved processes of manufacture".

When the war came to an end, and the necessity arose for setting up peacetime machinery to manage those changes which had evolved under the pressure of war, the role of the DSIR was greatly expanded. It soon became the chief channel through which all government funding for scientific and technological work was to flow. The National Physical Laboratory, the Fuel Station, the Food Investigation Board, were all brought under the DSIR umbrella. Furthermore, it was decided to bring all the research in basic science needed by the armed services into a single stable and to get the DSIR to coordinate it all. To this end four Research Boards were set up by the services – for physics, chemistry, engineering and radio – and many of the scientists who had worked in military research during the war were appointed to these boards. It was thus, through membership of the Physics Research Board with J.J. and W. H. Bragg, that Rutherford came into close working touch with the DSIR. The idea of bringing all service research together was, of course, soon foiled by the Admiralty which, as we have seen, set up its own scientific service which it eventually made independent of DSIR, with the effect that there remained a strong aeronautical flavour in that body.

Rutherford first started tapping the resources of the public purse, through DSIR in 1924. Plainly he had come to realise that Cambridge University finances would not support the expansion of the Cavendish that he wanted; and we have seen that he would not tap private sources. In 1924 DSIR offered him Chadwick's salary as Deputy Director of the Laboratory, in order to free Rutherford for creative work. Rutherford, for his part also needed support for an entirely different project – the creation of intense magnetic fields proposed by Peter Kapitsa, which required machinery, heavy electrical machinery, of a size and sort never seen before in the Cavendish. To get this sort of grant for a special research project in a university laboratory is nowadays a standard procedure. (That is not to say that the grant is always given whatever the request, and the DSIR has now been replaced by research councils covering different fields of scientific effort.) But obviously it was something new in 1924, and Rutherford, Sir Richard Threlfall, on the Science Advisory Council, and the two chief officers of the DSIR, Frank Heath and Henry Tizard, had to work out how to go about it. Rutherford wrote to Threlfall about the basic proposal on March 25th, and Threlfall replied, three days later, that he would propose to his Council that the grant be given. Then he had to see both Heath and Tizard. Tizard thought that Rutherford would need an outright grant of some thousands of pounds immediately, Threlfall thought that £4,000 over four years would be better. Would Rutherford please prepare a formal document, wrote Threlfall. "I am not quite sure how we stand financially at the moment . . . but it is certain to get sympathetic consideration."

A week later, on April 1st, 1924, Threlfall wrote again: "We will see what can be done – but I expect it is a thing for which Heath would have to get Treasury sanction. It is all right, really. The Treasury people are quite decent and reasonable and appreciate a straight fellow like Heath." Rutherford would have to put up a formal, costed proposal, and then:

> When your proposal is formulated, Heath will find a way, i.e. if he is convinced that he ought to do so. In these matters I always act on Machiavelli's advice – I show clearly the worst that can happen at the start. That saves coming round later for "just a little more", and is a course of action in unison with what I take to be your nature . . .
>
> P.S. I see you can't make a close estimate, but you can't possibly be too frank and detached in what you put forward. We all know the difficulties.

Rutherford, however, was a quick learner in this new game. He applied for a grant of £8,000 over four years for Kapitsa's work on "intense magnetic fields" to cover the purchase of large accumulators, a special dynamo and transformer for sudden discharge and special cameras for taking X-ray photos. Kapitsa's salary was also increased from £450 to £600 a year, apparently on the extra grounds that "he has a large Russian family of relatives whom he keeps and feeds". But Rutherford was also able to apply a form of pressure which is familiar to the present day. Early in May Frank Smith wrote to Tizard: "I saw Rutherford on Friday and he is not quite certain whether he will be able to retain Kapitsa. Kapitsa has had many good offers and it is not unlikely he may return to Russia." Then Rutherford had to make his formal plea, as Director of the Laboratory, and wrote:

I have great confidence in the ability of Dr Kapitsa both on the theoretical and practical side, and he is the only man I know capable to carrying out such a difficult research to a successful conclusion . . . While I cannot guarantee results from investigations in such an entirely new field of research . . . the work will be followed with much interest by physicists and electrical engineers everywhere. I feel confident however that, with good fortune, such an investigation cannot fail to yield results of great scientific interest and incidentally of possible industrial importance.

By July, when Rutherford was at a meeting of the British Association in Toronto, the whole matter had been fixed, with Tizard and Frank Smith (for the work was claimed to be of potential interest to the Admiralty) completing the administrative work. Thus was Rutherford launched on to the road towards big science. It was to be a few years yet before he became convinced that his own line of work needed big machines, so it was his affection and professional respect for Kapitsa that started him down the road that all physics laboratories have subsequently been forced to take.

Rutherford was not, however, only a taker from the public purse. His common sense and wise advice, as well as his scientific knowledge, began to be used by government at this time. In 1921 there were rumours of a German development of a "death ray"; such rumours arose with great regularity, and the story has often been repeated of how one such rumour, fifteen years later, led by chance, and by the right man asking the right questions, to the development of radar in Britain. In 1921, however, the situation both politically and scientifically was quite different. Rutherford, Sir William Bragg and Sir Henry Jackson, looked into the matter of the "death ray" on the request of Tizard, who was then Assistant Secretary of DSIR. They all decided the proposals were "unworthy of serious consideration". J.J. and his physics board were a little more cautious and ordered experiments to be carried out to see whether explosives could be detonated by the reported equipment. The results were a complete negative. Again in 1924, Rutherford was consulted privately by the then Secretary of State for Air as to the advisability of the Air Ministry appointing a Director of Scientific Research as a counterpart to Sir Frank Smith's job as DSR for the navy. Rutherford favoured the idea, and favoured Tizard for the job, and though the proposal did not come to fruition at that time it did blossom some six years later into the setting up of that scientific advisory committee under Tizard's chairmanship which found, promoted and developed the war-winning radar weapon.

Despite the desire for independence by the Royal Navy and the

newly-founded RAF, the DSIR became more and more the prin-
cipal interface between government and professional scientists.
And it is in the development of government scientific effort that
Rutherford, working with the DSIR, made his greatest impact. In
a sense it was a coincidence of timing that brought about this
development. In 1930 Rutherford reached the end of his term as
President of the Royal Society, and in March of that year Sir
William S. McCormick, the Chairman of the Science Advisory
Council of the DSIR, died suddenly. McCormick had been what
was then called an "administrative chairman" – that is he had
performed some executive duties. Rutherford, primarily employed in running the Cavendish Laboratory, could not do this
work, but he accepted a redefined post as Chairman of the Council,
a position in which he had no office duties but was "more than a
figurehead". He accepted the offer in May 1930 and started
officially on October 1st of that year with his first formal meeting of
the Council on October 8th.

At this first meeting the Council discussed such problems as the
Lenses in Railway Signals, Research on Dry Rot and the Death-
watch Beetle, Atmospheric Pollution and the setting up of a
Box-Testing plant for Forest Products Research. But there was
also a special meeting later in the day for the formal acceptance of
the Annual Report and it had been Rutherford's first job to begin
the drafting of this report. His corrections in pencil of the first draft
are still extant, and they show that he inserted his own view of the
job of DSIR: "At the end of the last century, and even later, Great
Britain held a pre-eminent position among the industrial nations of
the world. This position has been largely lost . . . [because of many
factors but particularly because of] increased use of scientific
knowledge and scientific method by our competitors abroad. It is
when times are bad that research is most important."

On the one hand, Rutherford pointed to the lack of research on
coal, one of the country's few natural resources, and on the other
he pointed to the growth of radiotelephony as "a great new
industry" which had arisen out of research. He admitted that there
was always the problem of foreseeing the results of research and
what might be its effects on industry because "many years frequently elapse before it can become of benefit to mankind".
Nevertheless:

> We can play our part in the making of such future discoveries only by
> doing our best to ensure that really competent scientific workers are
> enabled to pursue lines of work which display promise of making
> important additions to existing knowledge . . . But making new dis-

coveries which will form the basis of new industries is not the only way science can help. Perhaps even more important is the day-to-day application of scientific knowledge in every phase of our industrial life.

After this rather serious start Rutherford found a quieter period. In mid-December the Secretary wrote to him: "The next meeting will practically be limited to discussion of the Million Fund . . . we shall not want a great deal of time to discuss business beforehand. I do not think you need trouble to come in on Wednesday morning . . .

"P.S. We can meet at the Athenaeum."

Rutherford inherited at least one major initiative, the proposal to set up a Locomotive Experimental Station and this was much discussed through 1931, though it eventually came to naught. He himself was much more responsible for pushing forward a big programme to invest £33,000 on coal research to include the development of both high- and low-temperature carbonisation processes, and in particular to work "to the production from coal of oil fuel and motor spirit". This was the subject of one of Rutherford's two important speeches in the House of Lords. He had become a Baron at the start of 1931, a distinction marked even in the Minutes of the DSIR when he changed his signature from "E. Rutherford" to "Rutherford".

Early in 1931 Rutherford's Council was considering the Effluent from Sugar Beet Factories and the Restoration and Preservation of Oil Paintings, but from this time onwards the effects of the Great Depression began to grip the DSIR. Late in 1931 Rutherford had to set up a special sub-committee to consider cuts in finance. A proposal to save £1,000 that year does not seem very stringent but it was reported that there was a "serious fall in receipts at the National Physical Laboratory due to the industrial depression", and the plans for expansion in 1932 had to be limited to £28,000. For the next few years Rutherford's chief business at the DSIR was to nurse the ailing research associations through the depression and try to get them on to a sounder financial footing. Research associations were new scientific establishments devoted to solving the problems of particular industries and funded mostly by the industry itself, often through a levy on manufacturers. The official history of the DSIR says: "The first half of Lord Rutherford's seven year tenure of office coincided with a period of grave economic crisis and financial retrenchment and the second half with growing international certainty and anxiety. It is not surprising therefore that no very marked changes in DSIR activities occurred during that period. Nevertheless there were some." It goes on to record that it was during this difficult time that a new basis for helping the research associations was established – something that would have been a major feat of administrative construction at any time; something which in fact was one of Rutherford's major achievements, for the research associations are still flourishing today, fifty years after he preserved them. The history records – surely too coolly for the effort that was demanded from Rutherford – that the new foundation for the RAs "involved an increase in the volume of government as well as industrial support, and this too at a time when the Million Fund [the original sum of one million pounds for setting up the RAs which had been raised in the early 1920s] having been exhausted, Parliament was called upon for the first time to vote annual sums for the support of the RAs. Indeed, in this period of economic crisis, began the real rise, after long hesitation, of the Research Association movement."

Throughout 1932 and 1933 Rutherford turned up at every one of the Council's monthly meetings except for two. In October 1932, having reported on the "parlous state" of the research associations, he led a deputation to a meeting with bankers headed by Montagu Norman, but this only led to a promise of £70,000 a year. Some of the research associations – such as the Refractories RA were too small to be viable with an income of only £4,000 a year – "no basis for a research organisation . . . which might have a profound effect on important basic industries". Some small RAs had to be cut, but Rutherford insisted on saving the research association of the British Rubber Manufacturers when the Dunlop Company and some others withdrew their support: his council decided this RA must continue, even if it was only getting £4,000 from its few remaining industrial members, and the DSIR would back it with a further £5,000. The successor to this RA still exists. Yet this minor struggle within the major battle revealed a further problem, for the point over which the big rubber companies such as Dunlop refused their cooperation was the assignment of patents arising from work at the research association laboratories. The DSIR had to work out its patents policy and this was no easy task where the patent might arise from work that was partly funded by money from industry and partly funded by state finance.

Nevertheless new research associations were founded in the first five years of Rutherford's work at DSIR for the boot and shoe industry, for the car industry, for the cotton industry, and for iron and steel. Cast iron, leather manufacturers, the linen industry, the jam-makers, the launderers, and the printing industry were also successful in setting up RAs once Rutherford had established the new basis for securing grants.

The records of the DSIR make it clear that Rutherford himself took part in the negotiations for the setting up of individual research associations and with the government for instituting a new system of support for the RAs. The new arrangement was that the Treasury would give some agreed block grant for at least the first five years of the existence of an RA, provided the industrial members guaranteed some sum above a minimum required to make the organisation viable – so a small new RA might be offered £7,000 a year for five years if the industry guaranteed £10,000. If the industry raised more than the agreed minimum the government would match it £100 for £100 – so the RA might get £12,000 from the state if its members contributed £15,000. This system lasted for almost exactly thirty years until it was reorganised in 1962.

Parallel negotiations were also being carried on with the Treasury over the relationship between the DSIR and the Empire Marketing Board. In the 1920s this Board had been contributing £50,000 a year to research carried out by DSIR in subjects of interest to the food industry – strictly into "The preservation and transport of food and its scientific basis to benefit both producers of food [i.e. countries of the Empire] and home consumers". The Empire Marketing Board cut this contribution under the impact of the depression and Lord Rutherford and the Secretary were authorised to negotiate "with the Board demanding a stable grant of £30,000 a year for three years". These negotiations had a "satisfactory outcome", but Rutherford had to point out to the Treasury that this still amounted to a cut of forty per cent in this source of income, and he took the opportunity to point out that food research was not essentially an Empire problem but a national problem: "The primary needs of man – his food, fuel, building, his use of timber and the protection of his drinking water from pollution, this is the research work initiated by the DSIR."

In the same field Rutherford made a special report on food research to the Lord President of the Council – the minister formally responsible for DSIR – in which he pointed out that apparently there were to be no representatives of science or research at the forthcoming Ottawa Conference (an attempt to weather the storm of the world slump by setting up Imperial Preference in world markets). He pointed out that food markets were of vital interest to most countries of the Empire and he warned that the Dominions might very well bring up the problem of the Empire Marketing Board and the research done with its funds. A little later, in July 1933, Rutherford was empowered to negotiate for DSIR with the Import Advisory Committee.

It is quite clear that Rutherford personally set great store by the Empire, and believed that by Imperial cooperation, if not necessarily by Imperial Preference, ways could be found out of the worst of the depression. His correspondence with Ernest Marsden is particularly revealing here, for Marsden, whose work with Geiger at Manchester had provided the evidence for the existence of the nucleus, had now given up research, after a spell as Professor of Physics at Wellington, and was Secretary of the New Zealand DSIR. In January 1934 Rutherford had received a report from Marsden about the wool industry in New Zealand, and was plainly in close touch over a number of similar problems for he wrote that "something definite will have to be done to get a more standard product [of wool] which will command a better price. It is quite clear to me that a Dominion like New Zealand will have to be prepared to spend a good deal of money on research for her own benefit in order to hold or extend her market." In England there was clearly an intention to produce more food at home, but this "need not necessarily mean a very serious reduction in the imports of butter, etc., from New Zealand. The real trouble is that so many countries are trying to get a strong foothold in the English market. The Danes in particular are very disturbed by the bacon quota." Strange to read these words from 1934, words which could be written again forty years later as the argument about Britain and the EEC goes on.

In later letters to Marsden that same year Rutherford reported first that he had prised £40,000 out of the government for his RAs, then the sum was up to £70,000, and he had high hopes that it would rise to £200,000 the following year. But while Whitehall was susceptible to his pressure, industry was more difficult, as both men found on opposite sides of the world. In May Rutherford heard that the New Zealand wool producers had refused to pay the toll of threepence a bale for Marsden's new research organisation and he commented, "It seems to me they are a very short-sighted generation and have the mentality that is characteristic of those who live by the land." But he had similar difficulties himself. Using the DSIR contribution "as carrot to the animal" he had got a really good response to the new RA scheme from some big industries, and their RAs would soon "be big enough to act as real centres for the industries concerned". But in smaller industries, with more old-fashioned managements, there was much trouble. "Sometimes one has half-a-dozen men round the table dickering over the question of £500 when each of them individually makes so much money that they could almost run the RA on their own. It leaves me with a tired feeling, but I have no doubt time is an important factor in eliminating the survivals of the Victorian regime . . . It is obvious there is a lot to do to get the whole of our industries scientifically minded."

The few men who knew Rutherford well enough to know his politics say that he considered himself a Liberal, but would undoubtedly have voted with the Conservative party on most major issues. His work at DSIR shows that he was completely uninterested in any sort of party politics – as the BBC noted, he was "Centre or unknown" – he was simply interested in social engineering to achieve greater efficiency, not as any form of reactionary, but as a technocrat who wants to improve the standard of life.

The welter of DSIR business varied from persuading the insurance companies to help finance a fire research station, to considering the vitamin C content of apples, and having a special report commissioned on trees and timber, which was always a Rutherford favourite subject because of what he had seen of forest destruction in New Zealand. In addition Rutherford was appointed by DSIR as its representative on the governing body of an organisation for research into massive radiation by radium. This episode, though of no great importance in the history of medicine or nuclear physics, gives the best of all illustrations of Rutherford at work, because one of his wide circle of official contacts, the Secretary of the Medical Research Council, Walter Morley Fletcher, had a habit of writing "aides-mémoire" for himself which have been preserved in the MRC archives.

Rutherford had first come across the Medical Research Council in 1920 over the subject of radium. J.J. approached Walter Morley Fletcher (later Sir Walter Fletcher) pointing out that various medical authorities in Britain controlled at least five grams of radium which had come from the government's Surplus Property Disposal Board. Could they release about one gram of this radium for physical research, since even an establishment like the National Physical Laboratory had virtually no radium at all? Rutherford and W. H. Bragg, working through the infant DSIR, pressed for this release of medical radium all through 1920 and 1921. Eventually, in March 1921, Rutherford by direct letters to Fletcher managed to get nearly a half gram transferred to him from Professor Sidney Russ at the Middlesex Hospital, but only by offering to come and pick it up himself, and promising that it would be returned at any time, given six months' notice. It was not that the MRC were unwilling to share out the radium, it was just that it was difficult to get any single medical institution to give up any of its supply. Finally Tizard, Bragg, Fletcher, Russ and Rutherford set up a small, informal conference to discuss the whole supply of radium including such needs as gun-sights and mechanical research. The half gram which Rutherford had obtained from Russ in 1921 was officially passed over to DSIR in 1928, after some of it had been transferred to the Cambridge biologist, Dr Strangeways.

But it was in 1933 that we really see Rutherford at work in this field. From the Belgian firm, Radium-Belge, a subsidiary of the international Union Minière de Haut-Katanga, had come an unofficial offer of a loan of ten grams of radium to be used as two five-gram "bombs" to test out high-intensity irradiation as a treatment for cancer. Smith and Fletcher of the DSIR and MRC got together on February 7th and drew up a scheme under which their two organisations would be joint trustees and the two "bombs" would be placed at the London Hospital, as the clinical research centre, and at the Mount Vernon Hospital, in North-West London, where the Radium Institute was already situated as the experimental centre for research into medical physics. The next day an "informal discussion" took place "at the Athenaeum" between these two administrators, Professor Lennan, a Mr Souttar, the doctor through whom the Belgian offer had come, and Rutherford. Fletcher's notes show that his memorandum of agreement met with general approval "but Lord Rutherford was strongly of opinion that the minimum period of two years was essential for the investigation suggested". A meeting with the Belgians, also at the Athenaeum, was fixed for the following week.

However it was not until February 22nd that all the parties could get together; then they met for "Lunch at the Athenaeum . . . Rutherford acted as Chairman". Fletcher again preserved his notes which show that the original approach to the Belgians had come from the medical profession through a conference organised largely by Dawson – later Lord Dawson. The Belgian representative, Monsieur G. L. Lechien was quite frank; there had been an experiment with a smaller bomb at the Westminster Hospital and that had been a failure. This failure had "greatly discouraged the sale of radium on that scale, and he admitted that if this free loan made now should lead to better use of big masses in this country, it would ultimately benefit the commercial suppliers". Nevertheless everyone agreed that the free loan offer was "very generous", with Rutherford, however, again insisting that it would have to be for at least two years to make worthwhile experiments. Furthermore everyone agreed that the radium would have to be used for research purposes with the medical men, in particular, free to choose cases from which they could learn, rather than for routine treatment. Smith, Rutherford and Fletcher rapidly reached agreement on the arrangements for joint MRC and DSIR holding of the radium and payment for its keeping. It was all so "obvious" in

Fletcher's words. Dawson, however, raised one point after another – all the interests of the medical profession would have to be looked after; the Radium Commission might be annoyed; the Post-graduate Scheme would have to be consulted – "Heaven knows why!" Fletcher comments. The Rutherford–Smith–Fletcher axis was quite satisfied – it was their job to organise research and they were doing it both for clinical medicine and for the physical side. Dawson was sent off with promises of full consultation and the suggestion that he should either organise a new medical committee to supervise the programme or should reform the existing committee which had first started the ball rolling but which was very large.

Two weeks later there is another of Fletcher's "aides-mémoire" to himself:

> April 5th – Casual talk with Rutherford, Smith, etc., at Athenaeum. General dissatisfaction that Dawson seems to have made no progress towards his formation of a supervisory body and is obviously not going to "pull the trigger" within any reasonable time. Rutherford is lunching with D tomorrow and is going to propose to him that Mr Baldwin (Stanley Baldwin, then Lord President of the Council) should be asked to appoint the Body. I warned them that D would not like that because he hates any interference by what he calls the government.

But Fletcher had no objection himself to a Board appointed by Baldwin if Dawson accepted it – "the great thing was to get busy quick". The following week Fletcher called on Dawson at his home and another interesting conversation ensued: "Dawson said he and Rutherford got on well but he did not like Mr Baldwin being brought into it . . . to meet Rutherford he has agreed to the DSIR being one of the bodies represented on his Body if Rutherford would represent it. I gather D rather funks Smith and thinks Rutherford more amenable." After overcoming medical hesitation, Fletcher, Smith and Rutherford paid a joint visit to Mr Baldwin, and eventually the "Body" was set up, with Rutherford representing DSIR on it. It started work in July 1933, but essentially the whole matter had been organised by Frank Smith, Fletcher and Rutherford at that first informal meeting at the Athenaeum, and all that had held up progress had been the need for others to fall in with their plans.

Behind this work there was a real conviction about the relationship between science and human welfare which was to be expressed in the help science could give to industry. Rutherford did not regard himself as a "fixer", he regarded himself as an evangelist, spreading the rising gospel of science to the largely indifferent ruling class of British politics and industry. His attitude is clearly expressed in a short speech he made in the House of Lords, again in 1933, when the Rubber Industries Bill was being discussed, a Bill which made compulsory for five years the contribution of the industry to its own research association. The nub of this speech was:

> From some points of view the question before us is a small one. From another point of view it is one of real importance. It is a test of the attitude of the forces that govern this country towards scientific research in its application to industry. As a man of science who has had some opportunity of seeing the fruits of scientific inquiry into industrial problems I am convinced that research is a potent weapon for combatting the evils of waste and inefficiency in industrial production. I hope that the House, in passing this measure, will give an unmistakable declaration of its faith in the application of scientific methods and knowledge as a means of keeping this country in the forefront of progress.

But at the very time that Rutherford was making this statement, he was hurling himself into another much more public battle on a very different front. In April 1933 he was planning to visit George von Hevesy, who had obtained a post as Professor at Freiburg University in Germany. Rutherford, like everyone else, had read of the recent ascent to power of the National Socialist party, and found Hitler's anti-semitic policy hard to understand. On April 3rd, in a letter largely devoted to his travelling plans, and the latest achievements of the Cavendish Lab, Rutherford included this paragraph "We have of course all been very interested in following the progress of the new government in Germany and in particular the anti-semitic troubles. I hope you are in no way affected by this strange effervescence. I see that Einstein has resigned his Berlin post, but I presume he is well fixed financially in the U.S.A. due to the special endowment there." The visit to Freiburg never came about, for in little over a month Hevesy had written that he was, as a Jew, having to give up his post. Rutherford replied offering all his sympathy. "It is a tragedy," he said and added that Hevesy should "not hesitate to write to me about your plans for the future". Rutherford told him "privately" that there were big plans for an appeal in England for support for displaced scholars.

William Beveridge, later Lord Beveridge, one of the principal authors of Britain's welfare state, had heard this sort of news a little earlier than Rutherford. He had in fact been enjoying himself with friends in a Vienna café when he read of the first dismissals of Jewish professors in Germany in an evening paper in the last weekend in March. He immediately decided to do something and on his return to the London School of Economics he started trying to raise funds to help such cases. Shortly afterwards he went to stay with George and Janet Trevelyan in Cambridge for the weekend May 6th to 8th.

> I talked about nothing else. In that weekend the Academic Assistance Council of the future was conceived, largely in discussion with George Trevelyan, Frederick Gowland Hopkins and Lord Rutherford. Most important of all I persuaded Rutherford, after a first refusal on the grounds that he was up to the eyes in other work, and against strong opposition by Lady Rutherford, to become President of the Council. I was helped in this by the long friendship between his family and my Janet (now Lady Beveridge) who came with me to our second talk. But in the end it was our cause rather than our friendship that brought him over. As we talked he exploded with wrath at Hitler's treatment of scientific colleagues whom he knew intimately and valued. He would have been miserable not to be with us if we went ahead. He did everything and more to make our going ahead possible.

Rutherford worked in this new cause both on a personal basis with those he knew, and on an official and organisational campaign. On the more formal side it was at first simply a question of raising money to help refugee scholars, and attempting to do so without arousing too much trouble. Rutherford confided to W. H. Bragg as early as May 1933 that he was trying to raise assistance to help scholars and scientists from Germany, but "My one anxiety is whether or not we can really help. It is possible I suppose to do more harm than good, by angering the people in power in Germany. It is a matter of balancing possibilities of good against those of evil."

At this early stage Rutherford wanted to proceed quietly, avoiding publicity, he told Bragg. But soon the enthusiasm of the movement swept such caution aside and the effort to help refugees from Hitler reached its first great climax on October 3rd, 1933, with a monster public meeting in the Albert Hall in London. Rutherford was chairman of this meeting and both opened and closed the proceedings. The chief speakers were Einstein and Sir Austen Chamberlain.

It was Einstein who moved the public: "It cannot be my task to act as judge of the conduct of a nation which for many years has considered me as her own; perhaps it is an idle task to judge in times when action counts . . . How can we save mankind and its spiritual acquisitions of which we are the heirs? How can we save Europe from a new disaster? It is only men who are free who create the inventions and intellectual works which, to us moderns, make life worthwhile . . ." And the meeting was counted a great success.

But behind it there was both disorganisation and doubts. The Academic Assistance Council, with Rutherford at its head, had been created on May 24th. Through his good offices, two secretaries were obtained, Walter Adams and Professor C. S. Gibson, and offices were made available by the Royal Society. But the Royal Society would play no official role, and none of the forty signatories of the public announcement of the objectives of the AAC signed as members of the Royal Society. Furthermore the Society advised that care must be taken to see that secretaries of the AAC should not be of Jewish origin. The AAC made it clear from the start that it was not solely concerned with helping Jewish refugees; any scholar of any nationality would be helped; and "our action implies no unfriendly feelings to the people of any country; it implies no judgment on forms of government or on any political issues between countries. Our only aims are the relief of suffering and the defence of learning and science." To most of the post-war generation the caution – or "appeasement" – of the middle 1930s seems slightly ludicrous. A couple of days before the Albert Hall meeting Walter Adams wrote to Rutherford: "Sir Austen Chamberlain is particularly anxious that there should be no implication in the speeches of hostility to Germany, and would prefer that the word Germany should not occur. As this is precisely what you wanted there need be no fear that the meeting will have a political colour."

But after the initial enthusiasm and the great public meeting there was much hard slog. Rutherford himself wrote to Lord Halifax, Chancellor of Oxford, and Stanley Baldwin, Chancellor of Cambridge, in an attempt to get a letter signed by all the Chancellors of Britain appealing for more funds and especially "to provide a type of Fellowship or Professorship for ten or twenty of the more distinguished exiles whom we would like to keep in this country". In typical fashion Rutherford makes it clear that he does not want the support of the Chancellors in their official capacities, but as "distinguished representatives of the university world", and he adds, "The proposed letter is (of course) entirely non-political." More publicly Rutherford wrote an article for *The Times*, which appeared on May 3rd, 1934, entitled "Wandering Scholars", and later that year he wrote another public appeal for funds which also was given space in *The Times*. These were drafted by Walter Adams, but corrected and redrafted in detail by Rutherford – who as always performed with his scrubby pencil. He involved himself, too, with individual cases – in 1935 for instance he interviewed Miss M. Hertz, a zoologist, only "a quarter or one-eighth Jewish", and suggested that a wireless firm might raise money for her since it was her father who had discovered radio-waves. His feelings showed in a letter to Adams: "When we consider how much the prestige of Germany in the old days was heightened by the work of Hertz, it is a tragedy that his daughter feels it necessary to leave her country." Some of the cases were indeed tragic: F. G. Houtermans, an assistant professor at the Technical High School in Berlin, obtained a job in an industrial laboratory in Russia – "on recommendation by Lord Rutherford" – but later got caught in Russian persecution of foreigners, was handed back to the Gestapo in German-occupied Poland, and only after world scientific opinion had been mobilised again did he end as Professor in Berne.

There was a long correspondence, too, with Imperial Chemical Industries, its officials and even its Chairman, Sir Harry McGowan, about fourteen German scientists who had been given jobs by ICI for three years and allowed to do whatever research they chose. But in 1936 the company obviously felt its generosity had gone far enough – or perhaps the accountants had asserted control. At one stage of this correspondence Rutherford had to be fairly sharp – though, always cautious, he made a draft of his proposed letter in pencil – and he wrote: "The suggestion that these men could return home is not a practical one . . . The Rumanian is in fact a Hungarian, and a Jew, so he cannot go home . . . I am convinced it is only a matter of time before these scientists will be absorbed once more into the normal work of organised research."

The most famous, certainly the most articulate, of those he helped personally was the Nobel Prize-winner Max Born, of Göttingen. Born recorded that Rutherford was "certainly the greatest experimental physicist of our period and in addition a wonderful man and character". The two men had first met in a doorway at the Volta Centennial Congress at Como in 1927. Born found one of the lectures dull and the air in the hall intolerable; Rutherford was quite frank – "I cannot stand this any longer – nor can you, I assume. Let's go for a drive round the lake." And thus they spent the time, driving up the lakeside to the Swiss border, crossing the lake by ferry, having a good lunch, and talking all the way – politics, physics, and most of all the problem of language in science, on which Rutherford held strong views: "[He] rejected all artificial languages and declared that English was already in international use and ought to become more so." Born remembered this day very distinctly – "a day spent in this way brings people together" – and a few years later, in 1932, he gave the official address, in both English and German, when Rutherford was awarded an honorary Doctorate of Göttingen. Our best record of Rutherford's voice and lecturing style comes from this occasion because the whole proceedings were secretly recorded by keen physicists and later put on permanent record by Rutherford's admirers.

When Born became a refugee, Rutherford exerted all his kindness both in practical matters, such as helping to find accommodation in Cambridge, and in smaller matters. At the first tea-party given for the Borns at Rutherford's home the refugees were worried about their family pet dog, who had been quarantined. A few days later Rutherford arranged to take them, in his car, to the quarantine kennels, and even brought books for himself and his wife to read, so that the Borns need not feel hurried in visiting their dog. On the more practical level it was arranged that Born could work temporarily at the Cavendish, until eventually he was successful in obtaining a post – the Chair of Physics at Edinburgh.

The problem turned out to be that of getting the refugees new jobs – raising money to tide them over a period was not too difficult. Rutherford and Walter Adams were only too well aware that in Britain, where unemployment was still the major social problem, there could easily be antagonism towards foreigners who got positions. In 1936 Adams was writing to Rutherford that "hostility could be aroused by competition for university or general labour market posts". Beveridge records official support for this sentiment: "HM government supported this jealousy; in May 1936 for instance the Home Office announced that no refugee doctor or dentist would be allowed to set up in practice in Great Britain even if he were admitted by the General Medical Council." Recent research has shown that anti-semitism was fairly strong among doctors in the 1930s. There was also opposition from the Fascist party – a gang of their toughs threatened to break up a reception for refugee scientists at the Royal Society's headquarters in Burlington House – but fortunately the hosts included a fair number of rugger players, led by the Secretary, Griffith Davies, and the Fascists retreated before blows were struck.

By 1936 the Academic Assistance Council had given help to some 1,300 Germans and to a few from other countries, such as Russia and Portugal. Rutherford, in the final report of the AAC, was able to say that they had re-established 363 scholars who had left Germany in permanent posts around the world, and were still helping to maintain 324, who had secured temporary posts in universities and learned institutions. At this stage the AAC was wound up; the flow of learned refugees had dried up and the nations of Europe were drifting into more warlike postures. To replace the AAC, a "Society for the Protection of Science and Learning" was set up, and again Rutherford took the lead. This society expanded its field and set up an international body for the same purpose, with Niels Bohr as chairman, in an effort to coordinate the activities of similar bodies in various different countries, though open hostilities soon brought this work to an end. The thanks Rutherford received for this work was almost entirely private. Lionel de Rothschild wrote to him personally on behalf of British Jewry, and his old friend Chaim Weizmann sent his "admiration and profound gratitude" after one of Rutherford's articles appeared in *The Times*. But Weizmann was as cautious as Rutherford: "Jews outside Germany have avoided expressing their feelings of indignation in case it upsets those who, for the sake of peace, are trying to negotiate with the Germans." And so he felt: "Your voice raised yesterday with so much dignity and restraint brought, I am sure, comfort and solace to many sufferers, and you have placed us under a debt of deep gratitude."

Some were not so restrained. A. V. Hill wrote an angry letter to *Nature* in February 1934, protesting about the state of affairs in German universities. Rutherford got the backlash in the shape of a letter from Professor J. Stark, the Berlin physicist who had done original work on atomic spectra which had much interested Rutherford and Bohr when they were working on the first development of the nuclear atom. Stark protested that Hill's letter contained "gross falsehood of a political and cultural kind . . . against Germany", and he complained of such political matters being published in a scientific magazine like *Nature*. German scientists wanted to remain friendly with their English colleagues and even understood English research men offering to help Jews "who have left Germany". "Therefore in the name of the majority of German scientists, I appeal to you, as the leading representative of English science, to help in putting an end to the mischievous propaganda against Germany in English scientific circles, and in avoiding an estrangement between English and German scientists."

Rutherford replied at considerable length on March 14th, 1934, but only after taking a fair amount of advice; he had been to see the editor of *Nature*, Sir Richard Gregory, and had taken "an opportunity of consulting with a few friends". He made it clear that he, personally, had felt rather critical of some of the material that had appeared and he quite agreed that it is "in general undesirable to mix science with politics". However, *Nature* was a privately published magazine belonging to an independent company, and it had no connection with any scientific society or committee; furthermore *Nature* felt free, and "quite properly", to criticise our own government in its attitude to science or scientific men. But eventually, after the cautious approach, Rutherford was quite firm:

> This country has always viewed with jealousy any interference with its intellectual freedom, whether with regard to science or learning in general. It believes that science should be international in its outlook and should have no regard to political opinion or creed or race . . . We all sincerely hope that this break with the traditions of intellectual freedom in your country is only a passing phase and does not indicate any permanent change of attitude towards the freedom of science and learning.

This belief in intellectual freedom, in the necessary independence of science from political systems, was a genuinely "Liberal" view that was the core of Rutherford's public attitude to political problems. He was equally firm, though more privately, when P. M. S. Blackett, who had by this time left Cambridge for London, asked him to support a Russian Exhibition in 1936. Rutherford made it quite clear that, after his struggles with Moscow over Kapitsa, he had no intention whatsoever of being seen to support Russia either.

But as the war-clouds gathered over Europe, Rutherford's mind turned more and more towards the relationship between science and the defence of the country. In this area Rutherford's influence was exerted with and through Sir Henry Tizard at the climax of an alliance which had lasted more than twenty years. This friendship began by chance when Tizard was offered, at the last minute, a place aboard the liner *Euripides* sailing to Australia in July 1914 with many berths allocated to scientists who were going out to attend the annual meeting of the British Association: "To Tizard as to so many others Rutherford gave the friendship and encouragement which was always remembered with gratitude," according to Tizard's biographer.[1] And he quotes Tizard himself on his memories of Rutherford on that voyage: "I partnered him at deck-tennis; he used to stand at the back of the court where he was worth a good many points to the side by keeping up a running commentary on the looks and behaviour of the opponents." Together they won the doubles, enabling Tizard always to claim that he had shared in one of Rutherford's triumphs. Tizard also recalls that Rutherford lectured on radium and emphasised its rarity and value and the danger of keeping it close to one's skin;

[1] Ronald W. Clark, *Tizard*, Methuen, London, 1965.

then he passed round a small tube of what he declared to be radium bromide, but which Tizard afterwards found was a mixture of common salt and sand.

The alliance was resumed after the war when Tizard became Assistant Secretary, and later Secretary, to the DSIR. Tizard had, like Rutherford, that quality of commanding the respect of "men of affairs", of being understood and trusted by senior civil servants and politicians, service chiefs and industrialists. The encouragement he gave to the invention and early development of radar was not the least of his contributions in this field; his own experience in aeronautical research in the First World War enabled him to perform the even greater function of leading RAF officers into the construction of an operational system which linked radar with the defending fighter squadrons. But Tizard's work in this vital area was certainly not helped by the irruption of Lindemann, backed by Churchill, on to the scene. This terrible story of the two former friends, Lindemann and Tizard, becoming unyielding enemies as they both strove to move Britain into a posture of self-defence in face of the Nazi threat, has been often told and argued over. Lindemann triumphed politically when Churchill came to power, and he had won many battles before his final victory. What has rarely been noticed is that Tizard's "power base" was Rutherford, just as much as Lindemann's lay in Winston Churchill. Lindemann's victories began only with Rutherford's death. Tizard had successfully got Lindemann off his committee – and off the battlefield – in 1936, but two years later, with Rutherford dead, Lindemann overpowered Tizard. Tizard was quite frank about it in a letter to A. V. Hill on November 24th, 1938, explaining that the Churchillians were "pressing Lindemann's claims to be the brains on defence. It is rather clever the way they have waited until Rutherford was well dead, so to speak, before pressing this again." All that Tizard could fight for was to get Blackett or R. H. Fowler – both staunch Rutherfordians – on to the new committee which Lindemann was to dominate.

Rutherford took a much more positive role in preparing Britain's defences than the traditional story of the Lindemann–Tizard battle tells. And possibly Tizard may have helped to undermine his own position by rejecting Rutherford's final attempt to improve the provision of scientific help for the services. Tizard won the first skirmish of the war against Lindemann and Churchill when he reformed his committee on scientific air defence without Lindemann in 1936. It was but a temporary advantage, and Rutherford may have sensed this, for almost the last act of his "official" life was to try to organise an overall defence science committee which

490

would coordinate the entire work done for all three armed services and bring in the expertise of university and industrial scientists as well. He was, in this last year of his life, 1937, quite powerful enough to force such an idea upon the attention of the most senior government ministers and Cabinet officers. And he was not alone in his efforts; he had obtained the backing of a number of eminent Cambridge colleagues. The whole move was a repeat of the attempt by senior scientists, led by J.J. and the Royal Society in the early months of the First World War, to impress on the government that science had something real to offer in the case of war. But this time it was done by Rutherford who had much clearer lines of communication through the machinery of Whitehall.

The whole proposal is revealed in the letters of Sir Charles Merz to Rutherford in March 1937. The first mention of the scheme shows that it had been put up, in Rutherford's usual way "unofficially", and had already run into trouble. The first letter from Merz (dated March 19th) reveals that the minister having been approached a second time "asked that the whole matter of the Committee should remain in abeyance for the moment!!". The minister was Sir Thomas Inskip, then recently appointed Minister for Coordination of Defence, who has been described as "honest but uncomprehending". Merz then admitted that the navy was also opposed to the scheme – the First Sea Lord had replied to Inskip "that the organisation which we started during the war was working so well that he did not wish to add to it". Nothing had been heard from the War Office, so Merz ended by merely asking Rutherford to "let our friends at Cambridge . . . know the position".

A few days later Merz sent Rutherford the final draft – coded "Secret" – of the "Memorandum on the proposal for the Formation of a Research Council" which had been sent to the chiefs of the three services. It shows in the first paragraph that Rutherford was officially in the position of consultant to Tizard's committee for scientific study of air defence, and that this organisation "corresponds to the BIR which was set up by the Admiralty during the war". Air attack was now seen as the "major menace of modern war" as submarine attack had been in the earlier conflict. There was no suggestion of interfering with the Air Ministry's successful structure but

There is a mass of research work of all kinds in progress in government departments. Often the departments have the assistance of eminent "outside" scientists . . . It has been suggested however that full use is not always made of the labours of the several committees and that it ought to be someone's duty to see that the recommendations or conclusions are not merely pigeonholed in a department. A great deal of the work is necessarily secret and, unless there is in existence some machinery of coordination, researches may lose half their value on account of the veil of secrecy which prevents other departments from hearing of them and putting the results into practical effect.

This is a plain echo of the criticisms the scientists of BIR had made against the navy twenty years before. But the scientists had learned their lessons. They now knew the difference between research and the development of new devices and the memorandum went on to express the "hope . . . that by the different approaches of the pure scientists engaged or interested in research, and of the industrialist concerned with the application of research to commercial purposes, a more rapid solution of the problems involved in our defence measures would be possible". Further there was "the need, in connection with defence, for keeping abreast with scientific and industrial developments and for taking up the latest ideas, and in particular with the difficulty of decisions as to when scientific processes should be tested or developed on a manufacturing scale". The simple coordination committee, was, therefore, also to be a consultative committee, and in putting forward the idea Sir Thomas Inskip claimed the support of Rutherford, Sir Frank Smith and the chemists Sir J. Bancroft and Sir Harold Hartley, "in the light of their knowledge of existing government machinery, its methods of contact with outside organisations and the potentialities of the country's scientific resources".

The machinery of Whitehall and the centrifugal independence of the three armed services managed to abort the attempt to set up what was derisively described as another clumsy committee, ignoring the positive developments proposed, in their mutual determination to have no one overlooking their work. Tizard was among the leaders of opposition to the new consultative committee:

Lord Rutherford told me about these proposals at the time . . . I disagreed with them and told him so. It seemed to me that the kind of central committee he had in mind under the chairmanship of Mr Merz was liable to grow into a somewhat clumsy administrative machine and to do more harm than good, the harm being done by the unnecessary calls on the time of hard-worked executive officers in the service departments, and duplication of the work of other committees. I fancy that Lord Rutherford was shaken by my criticisms, at any rate the proposal fell through and so far as I know he did not press it.

Thus Tizard wrote to Inskip two years later when Tizard may well have been having some regrets, for by that time he was losing the battle against Lindemann, and was himself proposing something very similar to Rutherford's idea, though confined to aeronautics.

It was to take another twenty-five years and the Second World War before the three services were formally integrated into one Ministry of Defence which had one chief scientist, and few people would claim, even now, that research (or anything else) has been finally coordinated between the needs of the different services. But Rutherford had undoubtedly played an important role in pressing the necessity for coordination and, perhaps more important, in clearing the minds of both scientists and military men as to the differences between research and development. Furthermore, he undoubtedly prepared his younger colleagues at the Cavendish for the role they would have to play in wartime. He talked occasionally about the problems of defence in the euphoric, ivory-tower peace of pre-war Cambridge, and would bluffly repel defeatist talk about "the bomber always getting through". Snow recalled vivid memories of an occasion when Rutherford was ambling down Trumpington Street booming out, "We've got something up our sleeves that will give bombers a big surprise."

Following the trend of Rutherford's thought and public pronouncements on the possibilities of atomic energy, it becomes clear that from about 1935 his opinions began to change. Whereas in the early 1930s he had been trumpeting loudly that the practical application of atomic energy was "moonshine", and this analysis was based firmly on the evidence then available, he later became very interested in the experiments of Enrico Fermi in Rome which showed that slow neutrons – quite literally neutrons whose flight had been slowed down – had greater powers of penetrating and disturbing the nuclei of any atoms they crashed into. By the time of the Watt Lecture, which he delivered in 1936 in Glasgow, his public statements were at least changing their tone. It is now known that in the early years of the decade he had given a private warning to Sir Maurice Hankey, the Secretary of the Committee of Imperial Defence. In private conversation with Ronald Clark shortly before his death, Hankey told how he

had been drawn aside at a Royal Society banquet by Lord Rutherford . . . There was something, Rutherford explained, which he felt he should say to the young man. The experiments on nuclear transformation which he was supervising at Cambridge . . . might one day turn out to be of great importance to the defence of the country. He did not quite know in what way this would be so . . . but some inner sense, for

493

which he apparently saw no scientific justification, correctly warned him that someone should, as he put it, "keep an eye on the matter".

There is, quite typically, no mention of this in Rutherford's correspondence; it was the way he worked in public affairs, to keep as much as possible private and to confide in the men who were at the centre of the working machinery of government. Rutherford was always a very fine judge of men – the young civil servant he had warned about atomic energy in the early 1930s was Lord Hankey by 1941, the man at the centre of the scientific defence effort, Chairman of the Scientific Advisory Committee. So when the "Maud" Committee reported that they believed the atomic bomb – a "uranium bomb", as it was then called – was going to be feasible, they found that Hankey's interest and knowledge stretched further back than they suspected – back to Rutherford himself.

It has to be said that Tizard tended to oppose the use of scarce resources on the atom bomb project, which he felt could not be brought to fruition in time to affect the war against Germany, while Lindemann pushed atomic research. The Cavendish, Rutherford's legacy to the nation as far as manpower went, proved vital in both fields – radar and nuclear energy – for while most of the younger men went straight into radar, by arrangement, in the crucial early months of the war, Cockcroft, Chadwick, Massey, Appleton, Egon Bretscher and Fowler were all involved in the nuclear programme.

Rutherford has now emerged as a powerful figure, commanding respect and political leverage, at the interface between government and science. His judgment of men and matters was admired, his determination to avoid public controversy was more than welcome, his common sense was valued, and his enthusiasm was appreciated. But he had more than all these things – he commanded love from many, and one must seek a reason why. Tizard tried to capture the reason in a great memorial lecture delivered to Rutherford's old enemies, the chemists, and he compared him closely with the great Dr Samuel Johnson: "Like Johnson he never considered whether he should be a grave man or a merry man, but just let inclination for the time take its course. He had a boisterous sense of fun and a loud laugh. By precept and by example he helped keep our minds free from cant. He hated pomposity and artificiality. He loved simple people and simple ways, and lived a simple life. He brushed little annoyances aside." But more than this, and unlike Johnson, Tizard found Rutherford an aboundingly healthy man who was usually at his best at breakfast; correspondingly he was remarkably little subject to mental depression. Continuing the comparison Tizard said, "Johnson lived on the whole a

494

life of laziness interrupted by periods of feverish activity to which he was driven by lack of money. Rutherford . . . lived a life of feverish intellectual activity relieved by short periods of magnificent idleness. When he took a holiday it really was a holiday."

But Tizard adds another aspect of attractiveness which very few of Rutherford's scientific colleagues noted – the man was extraordinarily widely read and well informed. Every memorialist records that Rutherford was a voracious reader consuming vast quantities of novels in his younger days. Tizard saw more: "Later on in life he gave up the reading of novels or any form of imaginative literature and confined himself to history and biography, to books that dealt with facts. I think he must have surprised many people from time to time by his knowledge of ancient history." Other close friends have remembered his astounding knowledge of military history, and his letters show that he was quite capable of initiating a correspondence with authors such as Bertrand Russell and Lord Simon when he had been interested by something he had read in one of their recently published books. Quite simply, Rutherford was an excellent companion, a great and cheerful and knowledgeable talker, and most people delighted in his company. One of his favourite phrases was "Well, it's a great life," and this was no empty cliché – he meant it and people loved him for it.

16

Kapitsa

Peter Leonidovich Kapitsa came to the Cavendish Laboratory in 1921. For the next fifteen years he was the most important single figure in Rutherford's life. His own observation of Rutherford, from the standpoint of a background and character quite different from anyone who had previously entered the scene, reveals a person markedly different from the revered Professor, or "Papa" seen by the young scientists of Rutherford's "tribe". Kapitsa's actions and talents pushed Rutherford into fields and positions he might not otherwise have entered. And Kapitsa's presence and influence in the Cavendish Laboratory changed the balance of personal relationships, and probably, indirectly, of political opinions, too.

In 1921 Kapitsa's Professor in Leningrad – A. Ioffe – led a small team to Western Europe to negotiate the purchase of scientific equipment to rebuild Russian laboratories after the devastation of the Revolution. Because of a problem with exit visas, Kapitsa was delayed at Riga and did not catch up with the party until the very end of the tour at Cambridge. He immediately decided he wanted to stay there to work, and, largely on Chadwick's advice, Rutherford accepted the twenty-seven-year-old Russian as a research worker.

The newcomer was a phenomenon of a sort that Rutherford had not met before. He was very clever, but there had been clever Russians in his laboratory teams in the past. The two crucial differences were that Kapitsa was an extraordinary combination of engineer and scientist, and that he was deeply interested in his personal relationship with Rutherford. He seems to have loved and admired Rutherford from the very first and to have set out, quite unsentimentally, to understand and charm the Professor. Probably because he was not dependent on Rutherford, Britain or the Cavendish, Kapitsa was able to build up a relationship of mutual appreciation, but as between independent individuals, that no one else could achieve. On Rutherford's side there seems to have been a need for a surrogate son; his only daughter, Eileen, was now grown up, leading an independent life, and, although she was soon to marry his colleague Ralph Fowler, she was not close to her father in spirit or scientific interest. Chadwick might have filled the son role, but he was naturally dour and unexpansive, and he, too, married in the mid-1920s and drifted away from the leader on matters of laboratory policy. After Kapitsa both Oliphant and Norman Feather filled the position for short periods.

But over and above all this Kapitsa was (and still is) a man of enormous charm, generosity and attractiveness. He talked far more, and far more adventurously, than was acceptable from an Englishman of his time; he got his own way with Rutherford and university society far more often than his Cambridge contemporaries. Although his success, especially in getting money and machinery with Rutherford's support, aroused some jealousy in Chadwick and others, he fitted with consummate ease into the essentially élitist life of Cambridge University; what, elsewhere, would certainly have been regarded as "bad form", was admired in Cambridge; for instance the Kapitsa Club, which met on Tuesday evenings in his rooms at Trinity, and which was deliberately limited to about thirty members, "apparently because Kapitsa wanted to irritate people doing physical subjects he disapproved of". Several of the biggest physical discoveries of the 1930s, including the dramatic discovery of the neutron by Chadwick, were first spoken of privately to the club and the confidences were never broken.

Of Kapitsa C. P. Snow wrote:

> He had a touch of genius; also in those days before life sobered him, he had a touch of the inspired Russian clown. He loved his own country, but he distinctly enjoyed backing both horses, working in Cambridge and taking his holidays in the Caucasus. He once asked a friend of mine if a foreigner could become an English peer; we strongly suspected that his ideal career would see him established simultaneously in the Soviet Academy of Sciences and as Rutherford's successor in the House of Lords.

But Snow also describes the man (correctly) as strong, brave and good. Rutherford came as much under Kapitsa's spell as anyone – it was with Rutherford's backing that he was elected a Fellow of Trinity, and it was Rutherford who pushed his election to a Fellowship of the Royal Society, despite background murmurs from the traditionalists that, since Kapitsa remained a Russian national he was only eligible for foreign membership. But Cambridge had taken Kapitsa so much to its heart that such minor difficulties were overridden. There is a story still told in the university that when Kapitsa returned to Britain in the 1960s, laden with honours after an absence of thirty years, there was a momentary embarrassment when he came to dine at Trinity and it was discovered he had no gown, but the problem was immediately solved by the college butler, who, after enquiring, "Are you Dr Kapitsa, sir?" immediately produced the visitor's original gown left behind during the dramatic events of 1934.

Underlying the stories and the personalities, the true link between Rutherford and Kapitsa was an identity of views on the subject that was most important to both of them – their science. An eighteenth-century Ukrainian philosopher, Skovoroda, wrote, "We must be grateful to God that He created the world in such a way that everything simple is true, and everything complicated is untrue." Apart from the attribution to God, Rutherford would have agreed to that wholeheartedly, but the significance of the quotation is that it was Kapitsa who unearthed and used the thought. And it is Kapitsa himself who has written, "Good work is never done with someone else's hands. The separation of theory from experience, from experimental work, and from practice above all, harms theory itself."

When the time came for obituaries and panegyrics, Kapitsa described Rutherford as "essentially an experimentalist endowed with an extraordinary intuition. His intuition led him to the experiments which enabled him to find simple and clear solutions for the most difficult and fundamental problems of science." But in a more comprehensive memoir the Russian qualified and expanded the simple concept of "intuitions":

Many admire Rutherford's extraordinary intuition, which, figuratively speaking, told him each time how to set up the experiment and what to look for. Intuition is usually defined as an instinctive process of the mind, something inexplicable which subconsciously leads to the correct solution. In my view this may be partly true, but at any rate it is strongly exaggerated. The ordinary reader is simply unaware of the colossal amount of work done by scientists. He only hears of those specific works which have produced results. Anyone who has closely observed Rutherford can testify to the enormous amount of work he did. Rutherford worked incessantly, always in search of something new. He reported or published only those works which had a positive result; these, however, constituted barely a few per cent of the whole mass of work he did; the rest remained unpublished and unknown even to his students. Sometimes in conversations with him one could hear a hint to the effect that he had tried something but unsuccessfully. He avoided speaking about his future projects and would rather tell about works that had yielded positive results.

Many of those unsuccessful, unmentioned experiments are presumably the ones that Chadwick described as "silly", and it becomes easier to understand why Rutherford preferred the sympathetic Kapitsa to the discouraging English colleague of longer standing.

Rutherford and Kapitsa also shared other attitudes which bound them together under strains which would have broken lesser friendships. They were both internationalists in science, and simple patriots in politics. Kapitsa was a Russian – if his country chose to be Communist, then he accepted this, and what is more relevant, so did Rutherford, whose essential loyalty was to the British Empire. Yet both men believed that science should overlap national boundaries.

These, however, are the conclusions at the end of a long relationship. When Kapitsa first came to Cambridge, in July 1921, he was sent to work in the attic room of the Cavendish where Chadwick ran a regular course in basic radioactivity techniques for beginners. Kapitsa finished in two weeks a course which often took six months, and by early August he had come to his first tricky moment, the selection of the subject for his first independent research. On August 12th he wrote to his mother: "Yesterday for the first time I had a talk about scientific themes with Professor Rutherford. He was very kind, took me to his room and showed me his apparatus. About this man there is definitely something fascinating, although at times he is gruff. So my life here flows like a river without whirlpools and without waterfalls . . ." This latter was a favourite Kapitsa metaphor, which will recur in stormier circumstances.

By September he was writing to his mother that he was worried that he was beginning to be too bold in his scientific conceptions . . . "And then this Rutherford is a puzzle to me. Shall I be able to unravel it?" Kapitsa was by now established between an American and a Japanese in a room near Rutherford's, working with alpha-particles and plainly beginning to feel on a sounder basis with his new Professor. By October 12th he was telling his mother: "Rutherford is more and more kind to me, he greets me and asks how my work is progressing. But I am slightly afraid of him. The room in which I work is almost next to his study. This is bad as one has to be very careful with smoking; if he ever sees me with my pipe in my mouth I shall get into trouble. But thank God, his steps are heavy and I can distinguish them from the others . . ." And a couple of weeks later he confides that he calls Rutherford "Crocodile", an appellation which has continued to baffle Englishmen, as Kapitsa might explain that the word stands for father in

Russian, which is the truth but perhaps not the whole truth, or he might say more expansively that in Russian folklore, the crocodile never turns back and "for that reason it can symbolise Rutherford's penetrating mind and his thrusting movement forward. In Russia one looks upon the crocodile with a mixture of horror and admiration." Certainly Kapitsa got Eric Gill to carve a crocodile over the main door of the Mond Laboratory, next to the Cavendish, when it was built to house his work in the 1930s, and its meaning had to be explained to Rutherford.

By November it was clear that Kapitsa's first research was working out well, and that he would become a valuable member of the Cavendish team. He wrote back to Russia:

> Rutherford is satisfied, as his assistant told me. This can be seen from his attitude to me. Whenever we meet he greets me with kind words. On Sunday he invited me to tea at his house and I observed him at home. He is very homely and kind . . . But . . . when he is displeased, then hold on. He will so go for you that you will not miss the point. But he has a marvellous mind! His is a quite specific intelligence; a colossal feeling and intuition. I could never imagine such a thing previously. I attend his course of lectures and addresses. He explains a subject very clearly. He is an absolutely exceptional physicist and a very original man . . .

What Kapitsa did not mention in his letters to his mother, but the obvious reason why he was so concerned with Rutherford's reaction to his work, he only revealed more than forty years later, in 1966, when he gave the Royal Society his "Recollections of Lord Rutherford". Then he said:

> On the first day I started work in the Cavendish I was surprised to hear him saying to me that in no circumstances would he tolerate my making Communist propaganda in his laboratory. At this time this remark came quite unexpectedly. It not only surprised me but also shocked me and to a certain extent even offended me. Undoubtedly it was a consequence of the current atmosphere . . . Later on when my first experimental research was published I presented Rutherford with a reprint and I made an inscription on it that this work was proof that I had come to his laboratory to do scientific work and not to make Communist propaganda. He got extremely angry with this inscription, swore and gave me the reprint back. I had foreseen this and had another reprint in reserve with an extremely appropriate inscription with which I immediately presented him. Obviously Rutherford appreciated this foresight and the incident closed. Rutherford had a characteristically hot temper but cooled down just as quickly. Eventually we had many conversations about political questions; we were

especially concerned about the growth of Fascism in Europe. Rutherford was an optimist and thought that all would soon be over . . . Rutherford, like most scientists who work in the exact sciences, had progressive political views.

On an entirely different note Kapitsa described to his mother one of the Cavendish dinners of the 1920s:

At the dinner there are present about thirty to thirty-five people . . . They sat at a three-sided table while the chair was taken by one of the young physicists. I cannot say that they drank much, but English people get tipsy very soon, and this is soon seen in their faces. They become mobile and lively and lose their rigidity . . . Whenever possible the toasts were of a comical character. The English people are very fond of cracking jokes and being witty . . . between the toasts they sang songs. In general one could do what one liked at the table: to squeal and shout, etc. The whole scene had a rather wild appearance although very original. After the toasts all stood on the chairs and held each other's crossed hands and sang a song in which they remembered all their friends [this, presumably, was Kapitsa's version of "Auld Lang Syne"] . . . It was very amusing to see such universal celebrities as J. J. Thomson and Rutherford, who stood on their chairs and sang at the tops of their voices.

Kapitsa added that he had to spend the early hours of the morning helping home those who were less sober than himself: "My Russian stomach is apparently more adjusted to alcohol than the English one."

Obviously Kapitsa managed to avoid Rutherford's customary ban on Christmas holiday working, for on December 22nd he told his mother that the success of his experiments was virtually assured: "Then I shall be able to solve the question which since 1911 neither the Crocodile himself, nor the good physicist Geiger had been able to solve." By constructing a form of radiometer, and by working fourteen hours a day, Kapitsa had managed to measure the rate at which alpha-particles lost their energy while travelling through matter. Rutherford, however, had spotted that he had been overworking and finally insisted that he take a holiday early in January. This resulted in some motorcycle riding in the frosty, flat, Cambridgeshire countryside, and on one occasion Kapitsa took Chadwick with him. "I was foolish enough to let him drive, as a result of which he overturned the machine at high speed and we both fell off."

In that spring of 1922 Kapitsa became involved in another tremendous bout of work – a characteristic similar to Rutherford's methods – and obtained "special permission" to work late in the laboratory against the rule that everything must close promptly at six o'clock. To his mother he reported:

> Then I used to go home and calculate the results until four or five o'clock in the morning so that I could, on the following day, start my work in the morning . . . During this time I had three long conversations with the Crocodile (an hour each). It seems to me that he is now well disposed towards me. But I feel somewhat frightened – somehow he compliments me excessively . . . He is a person of great and unbridled temperament. But with such people there are bound to be ups-and-downs. But his mind, dear mother, is indeed exceptionally wonderful. He is devoid of any scepticism, is bold and is passionately enthusiastic. It is not surprising that he is capable of making thirty people work. You should see him when he flares up . . .

Kapitsa goes on to quote some of Rutherford's remarks, the sort he regularly made on his tours of the laboratories: "When on earth will you get results?" "Will you potter about for a long time to no purpose?" "I want you to give me results, results, not your chatter."

These gruff remarks, which awed the young researchers, were eventually discovered by his intimates to be meaningless, semi-automatic phrases which were almost greetings. On one occasion, when urgent modifications to the laboratory equipment were needed which involved cutting a hole in the outside wall of the building, it turned out that a strike of building workers made it almost impossible to get the work done. At length a small, independent bricklayer was discovered and persuaded to do the job, but shortly after he had started work he walked off the site. When eventually he was retraced he stated firmly that he was not going to work in the laboratory again because he was not going to put up with rude and gruff old gentlemen who wandered round and asked him why he was not getting results.

Kapitsa reported to his mother on the care with which Rutherford helped him prepare his first scientific publication; he found the whole affair of his first appearance in print "frightening and terrifying". But now he felt really accepted into the Cavendish school: "I feel myself to be at the centre of this school of young physicists. This is undoubtedly the most advanced school in the world and Rutherford is the most outstanding physicist in the world and the most prominent organiser . . ." This might have been only a young man's enthusiasm and loyalty to the new "tribe", but Kapitsa goes on to explain, unconsciously, the secret of Rutherford's success: "Only now do I feel my powers within myself. Success lends me wings and the work carries me away . . ." he wrote, revealing the enthusiasm and sense of power and purpose that Rutherford managed to generate. Shortly afterwards the Russian analysed the process more carefully:

> With us in Russia everything was cut according to the German pattern, there was little contact with the English scientific world . . . But England provided the most outstanding physicists and I now begin to understand why: the English school develops individuality extremely widely and provides infinite room for the manifestation of the personality. Rutherford does not press one and is not so demanding with respect to the exactness and polish of the results . . . Here they often do work which is so incredible in its conception that it would be simply ridiculed in Russia. When I asked why they had been carried out it emerged they were simply ideas of young people, but the Crocodile values so highly that a person should express himself that he not only allows them to work on their themes but also encourages them and tries to put sense into these sometimes futile plans. The absence of criticism, which undoubtedly kills individuality, and of which [Ioffe] has too much, is one of the characteristic phenomena of the school of the Crocodile. The second factor is the urge to achieve results. Rutherford is very afraid that a man may work without results, as he knows that this could kill in one the desire to work. For that reason he does not like to set a difficult task. And if he sets a difficult task then it simply means that he wants to be rid of a man. In this laboratory it could never happen that I would spend three years struggling on one work, struggling with excessive difficulties.

But Kapitsa also penetrated more deeply into the interior of the man. He spotted that Rutherford had a very sound grasp and a subtle understanding of the psychology of the people he met. He loved, of course, people with strong personalities, and, strangely, it is only Kapitsa who records how much Rutherford was impressed by Winston Churchill: "His description of Churchill was, like all his descriptions, short and clear, and in due course I found out it was quite correct. I well remember that Churchill in those days already regarded Hitler as a real danger to peace and called him 'a man riding a tiger'. Possibly this conversation somewhat altered Rutherford's optimistic view of the future." And he went on to comment: "Of course Rutherford's correct evaluation of people and his understanding of them was due to the fact that he was a subtle psychologist. People interested him and he had a faculty for understanding them. His assessments of people were always very outspoken and direct. As in his scientific work, his description of a man was always brief and very accurate . . . Possibly his approach to people was also a subconscious process and could be called intuitive." But against this bluntness and outspokenness, "Rutherford's interest in understanding human psychology and his kindness to others was undoubtedly felt by them. This explains why Rutherford's excessively direct way of speaking, which was sometimes not very tactful, was completely compensated for by his kindness and cordiality."

And precisely because he achieved a greater intimacy with Rutherford's mind and personality, Kapitsa was able to throw more light on the sources of this psychological power. He saw that Rutherford had in his maturity become a very widely-read man, and he records how Rutherford went to see a production of Chekov's *Uncle Vanya* in what was then Cambridge's most progressive small theatre. Rutherford was enormously interested in the play, gave an accurate and comprehensive account of the psychological problems of the intellectual characters, which provide the motivation of the plot, and was completely on Uncle Vanya's side when that character takes a pot-shot at the pedantic, meretriciously famous Professor. Similarly the Russian was one of the very few who knew how Rutherford helped Ehrenfest, the Austrian theoretical physicist who had taken over Lorentz's chair at Leyden. Paul Ehrenfest was a superb critic, much admired by Bohr and Einstein in the great revolution in physical thought which marked the 1930s. But probably because of his critical acuity, he failed to produce much original work himself, and this frustration led him to a positively suicidal depression in which he desired at the very least to retire to some small university in Canada. Kapitsa sought Rutherford's help, and Rutherford, although he barely knew Ehrenfest, took over the whole problem, not by finding him a place in Canada, but by writing him long letters explaining how valuable he was in the world of international physics and how much his colleagues needed his particular talents.

Kapitsa's perception of these different aspects of Rutherford's personality make comprehensible what has always appeared as one of his weaknesses – his slowness to accept that the age of big machines had arrived in the physics laboratory.

The peculiar character of Rutherford's thinking could easily be followed when talking to him on scientific topics [Kapitsa recalled]. He liked being told about new experiments but you could easily and immediately see by his expression whether he was listening with interest or whether he was bored. You had to talk only about fundamental facts and ideas without going into the technical details in which Rutherford took no interest. I remember, when I had to bring him for approval my drawings of the impulse generator for strong magnetic fields, for politeness sake he would put them on the table before him, without noticing that they were lying upside down and he would say to me, "These blueprints don't interest me. Please state simply the principle upon which this machine works." He grasped the basic idea of the experiment extremely quickly, in half a word. This struck me very much, especially during my first years in Cambridge when my knowledge of English was poor, and I spoke it so badly that I could only vaguely explain my ideas, yet in spite of this Rutherford caught on very quickly, and always expressed very interesting opinions. Rutherford also liked talking about his own experiments. When he was explaining something he usually made drawings. For this purpose he usually kept small bits of pencil in his waistcoat pocket. He held them in a peculiar way – it always seemed to me a very inconvenient one – with the tips of his fingers and thumb. He drew with a slightly shaky hand, his drawings were always simple and consisted of a few thickly drawn lines, made by pressing hard on the pencil. More often than not the point of the pencil broke and then he would take another bit from his pocket.

On at least one occasion Kapitsa gave Rutherford a present of a special propelling pencil, but it was soon discarded.

Remembering the Cambridge of the 1930s, C. P. Snow wrote that Kapitsa flattered Rutherford outrageously, that both of them knew he was doing it, and both enjoyed it. The relationship was, however, slightly more subtle than brazen flattery; there was a barb in most of Kapitsa's remarks; there was as much teasing of the great man as flattery, and it was this element that Rutherford loved, for no one else stood up to him in quite this way. For instance, Kapitsa remembered a conversation at Trinity High Table, sparked off by a discussion of a book called *Genius and Madness* by Lombroso. He held forth to his neighbour that every great scientist must be to some extent a madman. Rutherford overheard this and boomed out his challenge: "In your opinion Kapitsa, am I mad too?" Kapitsa took on the battle immediately and offered to prove his point that Rutherford was mad, too. "Maybe you remember a few days ago you mentioned to me that you had a letter from the USA, from a big American company [Kapitsa thinks it may have been General Electric]. In this letter they offered to build you a colossal laboratory in America and to pay you a fabulous salary. You only laughed at the offer and refused to take it seriously. I think you will agree with me, that, from the point of view of the ordinary man you acted like a madman."

Nevertheless the relationship between the two men was based on Kapitsa's scientific success. Little more than a year after starting at the Cavendish, the Russian had achieved another feat by obtaining what we should now call cloud-chamber photographs of the flight of alpha-particles deviated by an intense magnetic field. By September 1922, Rutherford was saying to him, "I would be very glad if I had the opportunity to create for you in my establishment a special laboratory so that you could work in it with your pupils." And by this time, as Kapitsa told his mother, he had two rooms at his disposal in the crowded Cavendish and had two young researchers working with him. Kapitsa seems to have felt he was living on a knife-edge, for he had not yet learnt the full and final techniques for handling Rutherford. He was "horrified and scared" by his own success both with his science and with his new master – he saw that Rutherford was "a man of colossal temperament who can go very far in one direction, as also sweep in the opposite one". Sometimes Rutherford would suddenly demand economy, and Kapitsa was aware that he was getting a very large share of the available resources; one experiment cost £150, he admitted. And yet . . . "I feel I am a member of the collective headed by the Crocodile. I feel that I really turn one of the wheels of European science."

And indeed he was doing so. Even larger scale experiments in 1923 were, Kapitsa told his mother, "giddily successful", but involved an enormous amount of hard work. "The one thing which makes my work easier is such care of me by the Crocodile that it could easily be compared with that of one's own father." Rutherford lent Kapitsa the money needed for the university fees for the Doctorate of Philosophy, and he encouraged the Russian to apply for the Maxwell Scholarship. Kapitsa was probably the wiser of the two on this subject:

> I replied that that which I already receive I consider quite sufficient, and think that as a foreign guest I should be modest . . . He told me that my foreign origin did not in the least prevent my receipt of the scholarship . . . For me, like for a passing bird, of course this [honour] makes no difference. But apparently the Crocodile could not understand my psychology and we parted rather drily. My refusal of course somewhat puzzled him and hurt . . . Despite that I feel that I was right. But I still have the feeling that I had hurt the Crocodile, who is so infinitely kind to me . . .

Rutherford, however, persisted, and Kapitsa submitted his application for the three-year scholarship which he received in August 1923. By this time he was starting to conceive his plan for producing the most intense magnetic fields that physics had known up to then, which involved building a large dynamo and accompanying heavy electrical equipment. Again Kapitsa got Rutherford's full support, but the finances of the Cavendish would not begin to approach the large sums of money necessary, and Rutherford, as we have seen, moved into entirely new territory by negotiating and obtaining support from the DSIR. It was, therefore, Kapitsa who pushed Rutherford into the area which we now term "Big Science", where large machines are necessary for the performance of laboratory experiments, and where, of necessity, the corresponding large finance has to be sought from the public purse.

One of the points made by Rutherford in urging the provision of the grant of £8,000 over four years for research into "intense magnetic fields" was that Kapitsa had received many offers from other establishments and it was not impossible that he might return to Russia. This was taken seriously by both Rutherford and DSIR (notably by Tizard, then the Secretary). For instance in the negotiations at the beginning of 1926 for the annual renewal of the grant, Tizard officially minuted a qualification which read: "It being understood that if any unforeseen circumstances, such as the total disappearance of the staff engaged, interfered with the progress of the work, the monthly payments . . ." and the formal agreement in March 1926 included a sentence: "If for any unforeseen circumstances it be not found possible to provide fully the assistance and facilities contemplated, or if for any reason these researches are delayed or stopped . . ." and there is a special marginal mark against this sentence, possibly put there by Rutherford when such terms became significant. As a result it was agreed that the apparatus purchased under the grant would eventually become the property of the Cavendish Laboratory, which was to prove enormously important ten years later.

The financial supervision exercised by DSIR was nevertheless extremely close. Although "one special generator" was purchased for £2,420, items such as "one blowpipe @ £1.5s.0d" were listed alongside, and at a later date the amount allowed for one experiment performed by John Cockcroft was reduced from £8.13s.4d to £8.12s.9d.

Kapitsa necessarily soon became involved in correspondence with DSIR. His early handwritten letters show that his claim to have suffered from lack of command of the English language were quite justified, for the civil servants were doubtless surprised to receive letters in which the "wrighting" was unclear, which asked questions about "payement" and which requested "let me know the contain of this letter". At one stage Kapitsa requested an allowance of, say, £100 a year to pay a young assistant, who also helped him with the typing of letters and keeping of accounts. The administrators in London cannot, of course, be blamed for failing to recognise in this "Mr Colcroft" or "Mr Corkcroft", as they variously described him, the future Sir John Cockcroft. They felt that an "honorarium of £1 a week for assisting Dr Kapitsa in clerical or other work" might be more suitable. One of them, L. C. Bromley, seems to have been more generous: "I am not sure whether Mr Corkcroft can type. If he can, perhaps an alternative arrangement would be to pay him rather more, say £1.10s.0d a week and arrange for him to take over the whole of the clerical work." At this Kapitsa called on the aid of the Crocodile, who snapped at DSIR, pointing out that the potential clerical assistant was the possessor of a First-Class Honours degree in mathematics at Cambridge and a First-Class Honours degree in electrical engineering at Manchester and was a research student supported by Metropolitan Vickers. A more satisfactory arrangement was then made securing Cockcroft's salary as Kapitsa's assistant.

It was in this year, 1926, that Kapitsa paid his first return visit to Russia for a summer holiday, and with the security of a promise, signed by no less a personage than Trotsky, guaranteeing that he would be allowed to return to his scientific work in Cambridge. The threat to keep him in Russia, or the desire of the Soviet authorities to regain his services, was very real, and was the subject of considerable unofficial negotiation between Kapitsa and the Russian leaders. Thereafter he visited Russia almost every summer, often by invitation, and from the start of 1929 the pressure to return to his homeland on a permanent basis became very strong. In January 1929, L. B. Kamenev, writing on behalf of "The High Council of People's Economy, Board of Science and Technology", informed Kapitsa that a Physico-Technical Institute, similar to that run by Ioffe in Leningrad, was being set up in Kharkov; Kapitsa was formally invited to take a post as consultant in the organisation of the Institute, which would involve him spending two to three months of every year in Russia at a salary of 2,000 roubles yearly. Although this is the first document on the subject, it was obviously a compromise solution offered for a problem which had been simmering for some length of time; Kamenev wrote somewhat threateningly: "Acquainting myself with your position, I was surprised that, at this time there were no official negotiations with you about the transfer of your work, as a Soviet scientist, to the USSR." He promised however, that there would be further discussions when Kapitsa arrived and added, "I fully guarantee to you that no pressure will be put on you for your immediate transfer here and the whole question will be decided in the best and most satisfactory way not to interrupt your work."

Kapitsa's reply was a masterpiece of evasiveness; he appreciated the offer which would "give me an opportunity of helping my own country in its scientific development which is now developing very rapidly"; he highly approved of the idea of building institutes outside Moscow and Leningrad: "spreading over Russia will make easier the very important task of picking up young research workers and will improve the standard of scientific education in provincial universities". This however led him on to criticise some of the decisions: "I do not think sufficient care is taken to prepare the young generation who will fill these laboratories and do research work" and he pointed out that both Britain and Germany were spending more on undergraduate teaching than Russia. He suggested that he should be paid an annual retainer of £200 for his consultancy plus £150 travelling expenses, but he could not possibly take on the obligation of coming to Russia at certain definite times every year because of the responsibilities which fell on him at unexpected moments in his own laboratory at Cambridge. As to the full-time "possibility of my returning to Russia, that could not be contemplated for some time" because "the next few years will be so full that it will be impossible for me to leave this laboratory which was specially built for me".

Kamenev combined firmness with diplomacy in return. On April 15th he wrote, "I am looking to your visit as a first step of your definite removal to USSR," but in the meanwhile he thanked Kapitsa for agreeing to take on the Kharkov Institute consultancy, started paying him his retainer and simply hoped that he "will later work in one of the new institutes". Perhaps Kapitsa managed to take his problems higher up the hierarchy that summer in Russia, for there is no trace of further correspondence from Kamenev, and the invitations to visit Russia in the summers of 1930, 1932 and 1933 were signed by Bukharin.

Certainly Kapitsa had no intention of leaving England if he could help it. In April 1930 he addressed to Rutherford, as "Chairman of the Committee of Magnetic Research" formal proposals for the construction and financing of an entirely new form of academic institute which would be somewhere between the traditional university laboratory and the basically practical – or "applied" – objectives of the National Physical Laboratory. In the preamble he pointed out that the magnetic work at the Cavendish had started by producing fields of 100,000 gauss (gauss are the units for measuring the strength of a magnetic field). But over the eight years since the work started they had now trebled that figure until they could produce fields of 300,000 gauss – "without exaggeration our magnetic laboratory possesses quite unique facilities for further investigations of a new region of modern physics".

This new field of research was so wide that it "would take several men's lifetimes to cover" Kapitsa declared, and he pointed out that it would also be necessary to develop low-temperature techniques, for low temperatures showed most magnetic interest. (Low-temperature work eventually proved to be Kapitsa's greatest interest and he showed his acute feeling for the likely progress of physical science in suggesting entry into this field as early as 1930, for some of the most remarkable discoveries of the late 1930s were made at the Cavendish, and also at Oxford, in this speciality.)

His proposals were clear and magnificent. Such an institution would need more trained mechanics and technical assistants than ordinary university departments; it would also need very much larger facilities. He asked for £15,000 for construction, £10,000 for equipment and an annual provision of £5,500 for salaries. He wanted the laboratory to be built in Cambridge, and he wanted to work in it and run it himself. But while he offered the carrot of further scientific success, "still I cannot neglect certain offers which have been made to me to develop such an institute somewhere else." And he concluded with what seems an extraordinary proviso: "In case of the possibility of my leaving, I am providing for the complete reimbursement of all the original expenses incurred in the equipment of the laboratory which I think will amount to £15,000 to £20,000. This will also simplify my position in a new place as I will be able to transfer the equipment." This proposal was dated April 16th, 1930, and this arrangement about the equipment was precisely what finally occurred five years later, so Kapitsa was clearly referring to the possibility of a forced removal to Russia, rather than threatening (as his words seem at first reading to imply) to accept the lures of some rival Western university.

Clearly Kapitsa and Rutherford had discussed the whole matter in detail before this formal proposal was written, for Rutherford had no very great problem in raising at least the majority of the money required. Some came from DSIR, but the vast bulk came from the Royal Society, which had at its disposal a munificent bequest from the Mond family – ICI, the industrial giant of present-day Britain traces its ancestry back to the heavy-chemical business founded by the Mond family. And thus the Mond Laboratory was built, complete with sculptured crocodile over the door, on land immediately adjacent to the Cavendish Laboratory, and Kapitsa was installed as its Director with a Royal Society Professorship and an official position in the hierarchy of Cambridge University.

During this period, while the Mond Laboratory was being built and starting work, Kapitsa regularly visited Russia. According to his wife, he several times discussed the question of his ultimate return to Russia, and he freely talked about this with high Soviet officials. His attitude was "that technical and other conditions were not suitable" for doing his work in Russia and that he felt that by remaining a Soviet citizen and yet continuing to work in Cambridge he was being of most usefulness to his country.

It has never been established what finally caused the Russian authorities to act. Three different explanations have been put forward, and Rutherford himself finally came to believe that Kapitsa, in his ebullient conversational mode, had spoken rather boastfully that his work on low temperatures and high magnetic fields would revolutionise the production of electricity in the near future. It must have been technically possible in the state of knowledge at that time for Kapitsa to have foreseen the possibility of constructing a superconducting electric generator, and he may well have made a misjudgment about the time span necessary for converting this possibility into a practicality – it is certainly known to be possible now, though no one has yet put the idea into large-scale practice. It is also possible that one of the other Russians then working at the Cavendish could have reported his conversations back to the Russian authorities. It must be remembered that the Russians themselves emphasised the scientific nature of the Marxist doctrine and revolution; that the modernisation and reconstruction of Russia was officially based on electrification and that the Russians had already begun the construction of various enormous hydroelectric generation schemes. Certainly none of the principals in the ensuing controversy ever denied that the Russians had the complete right to demand the services of Kapitsa in his own country.

The blow fell at the beginning of the 1934–5 academic year. At the start of October 1934 Kapitsa did not return from his summer holiday in Russia. Instead Mrs Kapitsa turned up in Cambridge alone, on October 10th, and went immediately to inform Rutherford that her husband had been detained in Leningrad and was not to be allowed to return to Cambridge. On October 12th Rutherford wrote to Ivan Maisky, the Soviet Ambassador in London, asking him to use his good offices "to hasten the return of Professor Kapitsa". "I cannot but believe that his detention is due to some misunderstanding that will soon be cleared up," he continued, and, in his capacity as Chairman of the Committee of Management of the Royal Society Mond Laboratory, he stressed the importance of Kapitsa's return to Cambridge as soon as possible in order to run the laboratory and supervise his students. Furthermore, scientific men all over the world were interested in Kapitsa's research into the magnetic properties of matter, so "If it should become public that Professor Kapitsa is not allowed to return to this country it would have a most unfortunate reaction on the scientific relations between our two countries. This I would greatly deplore."

A few days later a scribbled postcard from Kapitsa arrived for Cockcroft, his chief assistant at the Mond. It contained simply a request for the laboratory mechanic to construct another helium liquefier of a pattern that Kapitsa had pioneered, and asked about the possibility of selling the machines to other universities. It was addressed from Leningrad and dated October 14th. Then there came another postcard, this time addressed to Rutherford, also from Leningrad. Again it contained some small instructions about running the laboratory, and the first admission that something serious had happened couched in the careful phrase, "I am gradually recovering from the shock". Any doubt that Kapitsa's detention was just some temporary misunderstanding must have been brought to an end by a very firm reply from Ambassador Maisky, dated October 30th. The Soviet government, Maisky explained, planned not only the economy of the country, but also the distribution of labour, including even scientific manpower. Now, as a result of the extraordinary development of the national economy of the USSR in its Five-Year Plans, the availability of scientific workers had simply become insufficient. "Now it needs all scientific workers who have hitherto been working abroad," the Ambassador concluded.

During November Rutherford seems to have accepted the idea that Kapitsa was certainly not going to return soon, perhaps not going to return at all. The question now was what action could be taken, what policy should be pursued that might get Kapitsa back on some terms. It was time for the heavy artillery to be deployed, but not necessarily for it to be used. Rumours of Kapitsa's detention were already widespread among the scientific community by the end of November and at about that time Rutherford must have communicated officially with the Vice-Chancellor of Cambridge University, Dr J. Cameron. Certainly Cameron wrote to Rutherford on December 4th. Their mutual problem was that neither of them knew who was officially in a position to give Kapitsa leave of absence, which would have to be done to keep the situation formally under control. But the Vice-Chancellor agreed "as to the desirability of keeping the information from the newspapers so long as that may be possible". Replying a week later, Rutherford re-emphasised this point; it was important to avoid publicity "as anti-Soviet papers would make political capital" out of the affair and this would make the Russians harden their attitude. He recalled the "Metro-Vick trouble", when British engineers from the Metropolitan-Vickers company, engaged on a contract in Russia, had been detained and accused of spying. The initial publicity had made the Russians "obstinate in refusing to listen to our requests", in Rutherford's opinion. He must still have regarded the whole affair as some misunderstanding, for he commented, "To save face seems to me to be one of the chief occupations of the Soviet government. I think them quite as bad as the Chinese in this respect." The Vice-Chancellor however was now beginning to see things differently. He had been talking to Sir Robert Vansittart, the Permanent Under-Secretary at the Foreign Office, and he told Rutherford on December 12th, "He says that publicity would do no harm and that it might do a great deal of good." Vansittart's idea was to write to Maisky telling him that if Kapitsa was not released there would, in fact, be publicity, but he was unwilling to do this unless he could be certain that the newspapers would take up the story. Since Rutherford was working in entirely the opposite direction, the Vice-Chancellor suggested a meeting to discuss a plan of action.

Rutherford was trying to mobilise international scientific opinion. He believed – just as he acted in internal politics – that the best way of achieving things was to keep publicity at bay and to seek to sway the opinions of powerful individuals. On December 6th he had written to Niels Bohr, telling him of the problem and of his own approaches to Maisky. Various "informal representations had already been made to the Russians", Rutherford continued, and Langevin had approached the Russian Ambassador in Paris, but there was "no indication of a change of view" in Moscow. Rutherford had had an appeal for Kapitsa's freedom drawn up and had had it translated into French, because he thought it more hopeful to work through the French, who were on better terms with Russia officially than was the English government. He continued, "It is believed the pivotal man is Litvinoff. It is known he was absent from Russia when the decision to detain Kapitsa was taken by the government. He is the member of the government most likely to realise the danger of international repercussions." Therefore he wanted, "in view of the recent Franco-Russian entente", to get some Frenchman who was in touch with Litvinoff to press the appeal: "Essential that the English source of the memorial should not be mentioned."

Bohr already had a shrewd idea of what was going on – "Although through Gamow I had heard rumours about Kapitsa's difficulties, I was quite shocked" – but his basic idea of how to treat the matter was similar to Rutherford's: "I agree with you in the greatest caution to avoiding premature publicity in a delicate case like this." Bohr was gloomy about the prospects, for if the Ambassador had replied to Rutherford that Kapitsa's detention was a matter of national planning and needs, then there was not much leverage that the Russian Academy of Sciences could bring to bear. However Bohr wrote immediately to Bukharin, whom he had met through his contacts with Ioffe. On December 19th Rutherford sent Bohr a copy of the memorial already translated into French, and confided that he had little hope of getting anywhere through the official channels of the Foreign Office; there was "not much sympathy between our FO and that of the Russians". Furthermore, he admitted, "Of course I have always had at the back of my mind the probability that Kapitsa would return to Russia eventually . . . but it is exceedingly awkward for him and for us to leave his laboratory and his students in the air in this way."

This tone, of rather sorrowful disappointment that better arrangements had not been made, permeates the first paragraphs of the memorial Rutherford had drawn up for foreign scientists to sign – an appeal which started by admitting the legal right of the Russians in the case.

It is only in the second half of the document that we find stronger views about the international nature of science, views expressed by Rutherford in 1934 which are exactly paralleled by modern scientists writing protests about Sakharov or others forty-five years later. A scientist cannot work best when he is alone, the memorial says; he needs the constant exchange of scientific ideas, the stimulus of contact with workers in other fields, the sharing in a common scientific atmosphere, to give of his best. "To deny a scientist the means of pursuing his investigations in the milieu most favourable to them is one of the most grievous injuries he can suffer, and through him the whole scientific world suffers, too." And the concluding paragraph declares: "Science has no national frontiers . . . Scientists throughout the world . . . cannot do effective work in isolation but are members of a single team pushing forward the frontiers of knowledge."

Niels Bohr, however, was doubtful about the likely effect of all this; in the first week of 1935 he wrote that it "might harden the case prematurely and make it a matter of prestige of the Russian government to keep to its original decision". He was also more aware of the pressures that the confrontation between Fascism and Communism were bringing upon scientists on the Continent; he warned that the "deplorable political development in several other

514

European countries" might make it difficult for several of their friends and natural allies to sign. Nevertheless many scientists did do their best to apply pressure to help Kapitsa – Debye, in Liège, certainly wrote to Ioffe in the cause, and P. M. Dirac, nearer at home in Cambridge, asked what he could do and was told by Rutherford that the cause of the trouble was a false notion that Kapitsa had been doing some war-work for the British government. However Kapitsa was now at least able to write to his wife, "but of course he has to be very careful in wording his letters as everything is delayed for several weeks by the OGPU"; additionally two detectives watched him at all times and a Soviet lady was present at all discussions.

The Royal Society, or at least its Council, met to consider the problem of Kapitsa, who was one of its Research Professors and Director of its Mond Laboratory, on December 13th. Rutherford, whose tenure as President had ended nearly five years previously, was not present, but Gowland Hopkins, Professor at the Dunn Biochemistry Laboratory in Cambridge, put the case and read a letter urging secrecy. He reported to Rutherford that the members were much interested and concerned. The official decision of the council was to grant Kapitsa leave of absence, which at least meant that his salary would continue to be paid and his wife and children would therefore have an income. But further it was decided to send a deputation to the Prime Minister, Stanley Baldwin, to urge either that Sir John Simon should approach the Russian representative at the League of Nations in Geneva, or that the British Ambassador should raise the matter in Moscow. All this was to be done in greatest secrecy: "In consequence there is no record even in the Minutes," Gowland Hopkins wrote.

However Gowland Hopkins' letter also contained less pleasant news. He asked Rutherford, "Have you heard the unpleasant rumour that is current here just now – the story that Kapitsa has himself engineered the whole incident with an ulterior motive. The scandal is perhaps unworthy of notice, but it is evidently widespread." Possibly as a result of the deputation to the Prime Minister, the Royal Society was in contact with the Foreign Office over the Kapitsa affair. Sir Henry Dale, for the Society, was told that if they formally approached the Foreign Office, then representations would be made in Moscow. Whether it was apprehension that this might let the secret out, or whether it was simply unwillingness to get involved with such obviously unhelpful partners is not clear, but the Royal Society never pushed their protests to an official level.

Rutherford's own campaign to enlist the support of foreign

scientists also petered out in a Europe which was rapidly becoming polarised between Fascism and Communism. It was the "hawk" of the Foreign Office, Sir Robert Vansittart, who proved to have come up with the right answer. In March 1935 the story of Kapitsa's detention "broke", to use the journalistic term. On March 9th a small-circulation anti-Soviet newsletter, which was produced by White Russian refugees in Paris, put the story into print. Anna Kapitsa knew about the article immediately, and on March 11th reported to Rutherford that the dam of silence was starting to break. Her hurried translation of what *Last News* reported was that Kapitsa was detained and "taken by Bolshevists as a hostage for Professor Gamow". The crucial paragraph read:

> Not so long ago went abroad also for scientific work the other young physicist, Gamow, for whom, as is reported, Prof Ioffe and Prof Kapitsa stood warranties. But once in America Gamow refused to return to Soviet Russia. Prof Kapitsa, who did not know of Gamow's not returning went for a short time to Moscow. Bolshevists refused him permission to go back to England and stopped him as a hostage for Gamow . . . English academical organisations have protested through Soviet Embassy against the Middle Age system of hostages.

It seems unlikely that it was pure coincidence that at exactly this time the attitude of the Soviet authorities should have changed. But Rutherford had, in fact, just received a letter from Ambassador Maisky, dated March 1st, which cautiously proposed the opening of some form of negotiation and asked Rutherford to receive a visit from S. B. Cahan, a Counsellor at the Russian Embassy in London. The meeting was fixed for March 12th, and three days later Rutherford wrote a formal letter to Cahan on behalf of the University of Cambridge, who "did not feel they were able to accept your proposal". That proposal, the letter makes clear, was nothing less than that the Russian government wished to buy all the special equipment and apparatus that had been installed for Kapitsa in the Mond Laboratory. While admitting that the university had "always envisaged . . . the possibility that Professor Kapitsa might ultimately decide to return to work in his own country", they reserved to themselves the right to continue the lab and "they deeply regret that circumstances over which they have no control . . . prevent Professor Kapitsa from carrying out those researches in Cambridge for which special provision was made and at a time when the prospects of gaining new knowledge seemed so promising".

Rutherford, however, was not slamming doors; he was prepared to use his bargaining powers to help his friend. "At the same time under the present unfortunate circumstances I am prepared to help in any way I can to facilitate the starting up of the work of Professor Kapitsa in Russia. As I told you, this is only possible if Professor Kapitsa is allowed to communicate with me freely," he wrote, and concluded with a similar mixture of plea and threat: "I would also like to stress the plea I made to you for according full trust and liberty to Kapitsa, for it is only under such conditions that any scientific man can be expected to do scientific work of an original character."

Rutherford knew a great deal about the conditions under which Kapitsa had to live, for the Russians did allow him to write to his wife, and Anna Kapitsa provided Rutherford with translations of all his letters, many of them very long. It is in these letters that we can see the justification for Snow's description of Kapitsa as "Dostoyevskian" – they are full of philosophical gloom followed by numb acceptance of the fate that has befallen him. "Life is like a stream", one of his favourite metaphors throughout his life, "and we are no more than bits of solid matter on its surface." From the very start the authorities made it quite clear what they wanted from Kapitsa: he was to build his own Institute in Moscow, in exact parallel to the Mond Laboratory, and there he would revolutionise the production of electricity. It was openly admitted in a letter written before the end of 1934 (letter 29 of the series) that Kapitsa had just seen Krajanovsky and they had talked of the "initiation of electrification of the Union" and "He knew of my work, but unfortunately from somebody in England, or even in Cambridge, who told of it with many mistakes. All this is very stupid and very bad. You know one often misrepresents scientific work and some people even enjoy it, as if they did it on purpose."

But isolated there in Russia, Kapitsa was not only having to cope with his forced detention, he was also having to deal with the envy and opposition of his fellow scientists: "The scientists here [in Moscow] are definitely opposed to my transfer here: 'There he enjoyed himself while here we experienced all sorts of privations, and now he comes and wants to be a boss.' This is why such cold attitude of the Academy and complete unwillingness to underline my transfer here." And a little later in the same letter of April 12th, 1935 we hear: "Of course, once my work is understood, and after finding out that all is purely scientific they do not see any use in me, but even if now they are not pleased themselves that they started all the trouble there is the prestige and for this reason it is all formally supported but I do not see . . ." At this moment someone came in and Kapitsa broke off writing, continuing later that he feels himself "a useless and foreign element" and so will try to find a new and

517

modest field in which to start again so that he can convince his colleagues that he really is "just a scientist who does not want to be a boss and who does not want material wealth".

Taken as a whole, these letters of Kapitsa make a wonderful commentary on the problems of integrating science into the social- ist state – a process which the author of the letters fully supported as a loyal Russian. The problem he reports again and again, is that the formal machinery for doing this – the Academy of Sciences – has come to be dominated by men who are second-rate as scientists but who are expert at playing the new system of the Communist state. The result is that science suffers as science, and is also not seen correctly for what it can, and cannot, do by the non-scientific politicians. It is a fascinating discussion but not part of Ruther- ford's story; on the other hand these letters from Kapitsa to his wife placed Rutherford in an extremely strong bargaining position with the Russian Embassy in London, for he probably knew more about what was going on in Moscow than they did.

Kapitsa's first decision was to switch to bio-physics, the quiet field in which, he felt, he might convince his colleagues that he really was a pure scientist. More than that, he doubted his ability to build a new institute, isolated and in Russian conditions: "The question of the transfer of my laboratory and more than that the question of the building of the new one for work in low tempera- tures and magnetism, these questions are over," he wrote in November 1934, thus proving that the Soviet authorities' only original desire had been to transfer all his work to Moscow. He continued:

> You remember that in Cambridge I could not decide for a long time to build a laboratory, and the year it was being built and I was still working in the old one, was not the happiest of my life. Rutherford is quite right, I am not made for the administrative work; it is torture for me and I only accept it as a sad necessity. In Russian conditions the building of the laboratory like mine will mean colossal administrative difficulties and will spoil my life.

He goes on about the complicated bureaucratic processes he would have to deal with and reminds his wife how cross he used to get, even in Cambridge, over problems with their car ". . . and the building and administration of a laboratory would be a continuous cursing and 99% chance that I should end in the lunatic asylum. I have no collaborators here and that is all the tragedy."

Little by little, however, as time passed and as the physical conditions of his life became easier, he settled down to the idea that accepting what the authorities wanted was, in fact, the only way to continue his work. At first anything he wanted from his home in England was charged such import duty that it was equivalent to two months of his salary. First, this problem was eased, then he was given a better flat, and the use of a car. In May 1935 he reported cautiously, "The guardian angels went back to heaven so I do not see them any more", referring to the withdrawal of his OGPU guards. By this time, too, he had successfully pressed his way up the Soviet hierarchy, through Valerii Ivanovich Mezhlauk, the Chairman of the State Planning Committee, Gosplan, until he achieved an interview with Molotov himself in May 1935. By this time Kapitsa had come to accept the situation and the Russians were negotiating with Rutherford in England. Kapitsa's formal statement to Molotov started by submitting the letter he proposed to write to Rutherford and asking if any changes in it were required: "My opinion is that it honestly portrays the present position and my 'credo'." But, he went on, "I want to tell you once more that Rutherford is quite the most remarkable and unique man, and that I respect and love him very much and will never do anything that can hurt him or be unpleasant to him." Kapitsa then asks for a visit from his wife with a guarantee from Maisky that it would be quite safe for her to come. "You must not be offended about this but .you know yourself your ways with people make them nervous sometimes. If I am now and again threatened, I do not mind it; I am only sad, but I do not get frightened. But there is no need for my wife to live through it all . . ." – brave words from a brave man at the time of the great purges. He goes on to ask why the Russian authorities feel they need him: "You thought that I was useful to English science and so could be useful to you." Now it has been found that he is not of much practical use – and, in a foretaste of the style that Khrushchev made internationally famous, Kapitsa points out that what is good in one country is not necessarily so in another; the English drink Epsom salts but the Russians manage without them. Finally comes the most important point, the problem of how to manage science in Soviet Russia, in a one-party state. Kapitsa continued to be bold: "You said to me that you have got plenty of Kapitsas among your youth. I am certain that you have got not only Kapitsas but even super-Kapitsas, but with your methods you will never 'fish' them out of the 160 million . . . I never will admit that the attitude shown to me is the right one. I am not offended for myself. I am afraid for the other Kapitsas. This I see very clearly and cannot pass it in silence for the sake of the Union." Finally this extraordinary letter ends with a plea for his own freedom from restrictions and then he will work "for the glory of the USSR and for the use of all the people".

By this time his story was fully known in other countries. It was not until April 24th that the London newspaper, the *News Chronicle* picked up the story that was circulating among most scientists and among the refugee Russian papers. "Cambridge has shock from the Soviet – Famous scientist recalled" screamed the headline, and the article began, "Great Britain has lost and Soviet Russia has gained one of the greatest physicists of the day", and then it gave the gist of the story. Although a great new scientific research lab was being built in Moscow for Kapitsa, the *News Chronicle* insisted that he still wanted to return to Cambridge at least to finish his experiments. However the Soviet Union needed the services of its scientists because of the "necessity for technical and industrial advance". Then the newspaper described Cambridge's attempts to get Kapitsa back. The Chancellor (Balfour) had interested himself in the case; Prime Minister Baldwin had been approached and Anthony Eden, the junior minister at the Foreign Office had been told all about it before his recent visit to Moscow.

Rutherford had given a statement outlining the course of events, and then commented rather carefully on the sudden action which he described as "in many ways unfortunate . . . Even under the best conditions it will take a considerable time to collect in Russia the unique equipment provided in Cambridge . : . I have enquired at the Russian Embassy whether he will be allowed to come back but am informed that he is needed in Russia." Ambassador Maisky got the last paragraph, but scored most heavily: "Cambridge would no doubt like to have all the world's greatest scientists in its laboratories, in much the same way as the Soviet Union would like to have Lord Rutherford and others of your great physicists in her laboratories."

The *News Chronicle* report rather overstated the strength of the university approach to the government. It seems to have been the breaking of the news story that brought a government approach to Cambridge, for on April 29th Rutherford wrote to Stanley Baldwin, the Prime Minister, obviously in reply to his enquiries, "I have some reason for believing Kapitsa was commandeered as the Soviet authorities thought he was able to give important help to the electrical industry in Russia and that they have now found out they were misinformed. It is not unlikely they might wish to get out of the tangle if they could do so without losing face." Rutherford told the Prime Minister of Kapitsa's letters and how he seemed to be in a very excitable state and added, "I'm afraid if we can't set him free he will in the end be of no use to anyone." Both the university and the Royal Society had so far refrained from bringing the matter officially to Foreign Office notice, Rutherford continued, but now that Anthony Eden's visit to Moscow had apparently brought about a better spirit, perhaps "a little informal pressure brought on the Russian authorities might be helpful at this stage". Baldwin's reply went no further than promising to speak to Eden about the affair, but the Foreign Office, through an official, P. Leigh Smith, had been in touch with Rutherford in a series of letters ever since the story appeared in the papers. The Foreign Office was courteous but not very helpful, although they did put the matter officially to the Moscow Embassy. Rutherford reported this to Baldwin on May 10th simply commenting, "It seems to me that unless the Soviets alter their mind during the next month the matter will have to be considered as closed from my point of view. I have done everything I could and we must only hope for the best."

Rutherford was by this time at his lowest ebb in the whole affair. He was being pushed by Anna Kapitsa but he could see no way through. As early as February he had been in touch with W. L. Webster, a friend of Harold Laski, who was preparing for a trip to Russia. Webster had been asked, probably by Mrs Kapitsa, to try to get in touch with her husband but Laski "disapproved profoundly" because he thought such contact would simply aggravate the situation and lead to rougher treatment for Kapitsa personally. However, Laski had been talking at great length to Maisky and had made several suggestions for possible compromise solutions that Rutherford and Maisky could discuss. Rutherford's only card seemed to be a suggestion that Western scientists should boycott the forthcoming Physiology Conference in Russia, and here was the germ of an idea that eventually bore fruit.

It was undoubtedly Mrs Kapitsa who got the whole affair made public, for on the day the first story appeared in the *News Chronicle* Frank Smith wrote from the DSIR to Rutherford. Despite a last-minute attempt to suggest an appeal to the newspaper not to publish, which Smith had deprecated because it was clear that "the reporter had been informed of certain facts", he remarked, "When Mrs Kapitsa saw me just before Easter I concluded that she thought publicity was the only channel left open which might result in the Soviet authorities permitting K to return to Cambridge." Smith himself doubted whether there would be such an effect. But there were others who were equally worried about the lack of progress and there were other sources of information that confirmed Anna Kapitsa's worries about her husband. On April 14th, for instance, Cockcroft wrote to Rutherford that he had seen Leipunski, another Russian who worked at the Cavendish, and heard that Kapitsa was in a "very bad state" and had "almost lost control of himself". The same source reported that Kapitsa had agreed to draw up a plan for the new Moscow laboratory, and agreement to start building had been given; however Kapitsa was quite incapable of designing apparatus in his present state; "Kapitsa would not write to you to ask for anything"; but the only way is to get his apparatus from Cambridge or have duplicates made. The following day Cahan of the Soviet Embassy wrote emphasising "the second suggestion I put to you in conversation", which was that Cambridge University should donate the equipment of the Mond Laboratory "to the new Institute which is being organised in Moscow for the purpose of enabling Professor Kapitsa to continue his researches". In return Cahan offered that the Soviet government would make a grant to Cambridge equal to the value of the equipment. A face-saving formula was thus offered, and with it came something else, for Cahan wrote, "Your readiness to help in any way to facilitate the starting of the work of Professor Kapitsa in the USSR is greatly appreciated and there is nothing which can hinder Professor Kapitsa communicating with you on all matters concerning the scientific work on which he is embarking in the USSR."

Furthermore Rutherford rapidly discovered that public, or at least scientific opinion was not solidly behind him – possibly envy of Kapitsa had done its work. On April 25th, the day after the first *News Chronicle* story appeared, although that paper followed up with another, shorter, story including an interview with Anna Kapitsa, and other popular papers carried stories, from the editor of *The Times* came a less propitious response. The editor understood from his Cambridge correspondent that Rutherford was preparing a letter to *The Times* and would welcome an editorial comment on the subject. Fuller information about Kapitsa's arrival in Britain and the circumstances of his last visit to Russia was requested, and the interrogatory ended: "Is there any ground for the statement that is being made by young university people of 'advanced' ideas that the Soviet government had been in treaty with Dr Kapitsa for some time and had offered him excellent conditions of work and had prepared a large laboratory for him and that all they have done in fact is 'to make up Dr Kapitsa's mind for him'?"

Rutherford's reply was, as usual, measured and cautious. The letter for publication was "non-controversial", really nothing more than "an appeal *ad misericordium* to the Soviet authorities to do the right thing and release Kapitsa before his health is hopelessly impaired . . . I have reason to believe that the Soviet feel they have handled this matter in the wrong way and are anxious to find a way out of the tangle without losing face." He admitted that the question of Kapitsa's ultimate return to Russia had been raised often both in Russia and in Cambridge, and he mentioned that both Kamenev and Bukharin had been involved in the discussions, concluding, "I, myself, of course, realised that it was likely that Kapitsa, when he had brought some of his work to a conclusion might feel it necessary some day to return to work in his own country." The problem lay in the suddenness of the action, which left university, laboratory, and the Royal Society in grave difficulties. As to the young persons of advanced ideas, Rutherford felt that their explanation of the crisis "has been probably disseminated through the official channels of the USSR. I have heard it many times before. Many of our young Communists are very disturbed by the effect of Kapitsa's retention on the political side and naturally they seize hold of any explanation they can." Rutherford also sent a copy of his letter to *The Times* to his ally, Sir Richard Gregory, Editor of *Nature*.

The letter was published in *The Times* on May 1st. Again Rutherford made it clear that he did not dispute the Russians' legal claim in Kapitsa who was a Soviet citizen and a loyal one. But the way the Russians had acted, the suddenness of the decision, had placed British organisations in a difficult position. "Science is international and every scientist hopes it may remain so" – that was the essential sentence in the letter and it was the central theme of Rutherford's own "credo". So it did not matter in principle where Kapitsa did his work, but in practice it was essential for a creative scientist to have the right equipment and the right atmosphere of tranquillity. He ended with an appeal to the Soviet government to grant Kapitsa freedom of movement. His letter was supported on the same day by an appeal from Gowland Hopkins, putting the view of the Royal Society, which had, he pointed out, spent the money on Kapitsa's laboratory. He, too, accepted the Russians' legal rights, "but ventures to believe that with full understanding of the position they would wish [Kapitsa's] services to be rendered wherever they can best promote the interests of pure and disinterested science as an international pursuit".

Both Rutherford's and Gowland Hopkins' letters contained passing reference to the forthcoming International Physiology Conference to be held in Russia. It would be pleasant if the Soviet government would show their belief in "the internationalism of science" at this conference. And so a bargaining counter was moved quietly on to the board.

But the position was weakened the next week when *The Times* carried a letter from H. E. Armstrong, the chemist, claiming to be the senior Fellow of the Royal Society and also a member of the Russian Academy of Science, yet pleading that he had been urged by many younger men to put forward a different point of view. He denied there was any feeling of "shock in the scientific world", rather there was a feeling of "relief at least among the younger men". Why should we import foreign labour? Surely Englishmen could have constructed a magnet of great power? "Surely Cambridge has placed too high a value on Kapitsa? Some of us fail to see that proof has been given of his genius . . . Is not relatively too much importance being attached to the doings of the atom-smashing brigade led by Lord Rutherford?" Armstrong voiced the opinion that the Royal Society was too academic and that the use of the Mond bequest for Kapitsa's laboratory had been unjustifiable. Which was more important, studying atom-smashing or trying to do something about foot-and-mouth disease? Furthermore we should look at the political situation – Germany was now virtually persecuting those who would pursue pure knowledge, so the Russians were right in considering that Kapitsa would be better employed there than in "leading a lotus life in Cambridge".

Others however took a more sinister view of the whole affair. "What a plot" was the headline of an article in *The Aeroplane* on May 1st. This explained that the probable reason for the Soviet government wanting Kapitsa to itself lay in the fact that he had been developing, according to Rutherford's own statement, a helium liquefier: "The one thing that is wanted more than anything else by any belligerent nation in these days is the secret of how to get, how to store, and how to transport helium cheaply." The reason for this was that helium-filled airships would be invulnerable to all known anti-aircraft defences and to intercepting fighters. The paper threatened airship raids by helium-filled Russian airships all over civilised Europe. "Whether Kapitsa has joined the USSR, taking with him all the knowledge which he has gained at our expense, or whether he has been 'captured' and will be forced to work under duress, does not affect the issue," *The Aeroplane* concluded.

Throughout May and June Rutherford kept open his lines of communication with the Russian Embassy in London. As a result of the revelation of the story in the newspapers a slow, but genuine, pressure built up among scientists, and both Europeans and Americans who had Russian contacts. In Moscow Kapitsa's negotiations had reached the level of Molotov. Preparations for the big International Physiological Conference were reaching their final stages. On June 14th Rutherford again wrote to Cahan, the Russian

Counsellor in London, complaining that neither he nor any member of the Royal Society had received any personal word from Kapitsa, though information received through Anna Kapitsa made them "very concerned" about his bodily and mental health. "I have heard nothing definite about his scientific work or attitude with regard to it or whether he wishes us to help him make a fresh start in Russia," Rutherford continued, and then proposed that perhaps two of the British delegates to the Physiology Conference be "granted free facilities to talk over matters with Kapitsa". He suggested a meeting on June 10th to discuss this proposal and asked to bring Sir Frank Smith, who was now Secretary of both the DSIR and the Royal Society, and who therefore represented the principal owners of the equipment Cahan wanted sent to Russia.

At almost exactly this time Rutherford received a letter from Kapitsa, the letter which had been approved by Molotov. The essential point here was that Kapitsa, in what he was perfectly prepared to have issued as a public statement, accepted that he would henceforward have to do his work in Moscow: "I am and always was in sympathy with the work of the Soviet government on the reconstruction of Russia on the principles of socialism, and I am prepared to do scientific work here. Secondly the Soviet government does its best to build me a laboratory." Although this was an acceptance of the unavoidable facts, it was not surrender. Kapitsa went on to explain that there were still a number of points on which he and the Soviet government were in disagreement – major subjects such as the relationship between pure and applied science and the most efficient way of managing scientists. Furthermore he was personally very miserable; he missed Rutherford, he missed his laboratory, and especially he missed his work; above all "the stupidity of the created position is that it is based on a complete misunderstanding as everyone concerned really acts with the best intentions".

From this time onward events moved fairly smoothly to their conclusion. Rutherford obviously accepted that Kapitsa was prepared to stay in Russia; Kapitsa accepted the same; the Russian government was prepared to deal on this basis, too. Rutherford and Sir Frank Smith had a meeting with Astakov, who replaced Cahan, and serious negotiations about the "transfer" of Kapitsa's equipment to Moscow, with its corresponding "grant" from the Russians to Cambridge, began in earnest. In Moscow Kapitsa was pleased to hear of Webster's forthcoming visit and had been in touch with the Moscow correspondent of the *New York Times* and, at some remove, with Ambassador Bullitt of the USA. He managed all such talks carefully, usually having them in the presence of fairly high Soviet officials. Ambassador Maisky wrote to Rutherford (July 27th), giving assurances that delegates to the Physiology Congress could certainly talk freely with Kapitsa. Anna Kapitsa's proposed visit to Moscow was also the subject of correspondence with Maisky. Meanwhile Rutherford asked E. D. Adrian, the distinguished physiologist (later Lord Adrian, Master of Trinity College) to act as ambassador to Kapitsa. Referring to Maisky's bland assurances that any delegate to the Congress was perfectly free to talk to Kapitsa, Rutherford commented, "It is a humorous document, when I recall that the Counsellor of the Embassy told us that he was not sure it could be allowed and in any case he must get the permission of the Russian government, and it took them five weeks to make a decision . . . It looks to me as if Kapitsa has been able to make some more reasonable arrangement for his work in Russia with the Soviet authorities."

Adrian found Kapitsa in a highly excited state, half exuberant, half conspiratorial, pouring out words, full of ideas about what could be done. Reading what Kapitsa in Moscow and his wife in Cambridge and London said and wrote over the months of August, September and October, gives an atmosphere of high drama, although in fact the major decisions had already been taken. Adrian gave Rutherford a long written report when he returned from Moscow in August, nearly all consisting of a digest of what Kapitsa had told him. There were, for instance, three reasons for Kapitsa's detention: firstly unfounded reports from a well informed Cambridge source that Kapitsa was doing war work for England; secondly that Gamow, once he was outside Russia, had written to Molotov demanding the same standing as Kapitsa had in Cambridge, and making such standing a condition of his return to Russia (and presumably therefore leading the Russians to action against Kapitsa); and thirdly the belief that Kapitsa's abilities would be useful to Russia in any forthcoming war.

Kapitsa therefore felt, Adrian reported, that he had three alternatives: he could rely on international scientific pressure to get himself returned to Cambridge, though this had the disadvantage that it would strain relations between his native country and the rest of the world to such an extent that it was not justified to regain the freedom of one individual; secondly he could get his apparatus from Cambridge and work in Moscow, accepting restricted freedom; or thirdly he could give up physics and take to physiology. He felt that the right course, in view of his moral obligations to English scientists, must be the second, or middle, way, which could be achieved by getting his government to pay all the costs of the last fourteen years, some £30,000–£50,000, and by Rutherford arranging that he retained his Royal Society Professorship which would enable him to come to England to lecture every year at a salary of, say, £400. He would also ask for his assistants, Laurman and Pearson, to be seconded to his work in Moscow, with guarantee of their jobs and their pensions if they returned to Cambridge in three or four years. In legal terms Kapitsa considered that he was under the protection of international law. He had been detained by *"Force majeure* and *la volontée de prince"*; since he had made no voluntary application to be released from his duties in England, the status quo remained there and he hoped this would strengthen Rutherford's position. He repeatedly expressed his gratitude to Rutherford and his other friends and put forward another bargaining counter: "The best guarantee that he will have complete control of the apparatus will be that he will be able to visit England to report on his work. He expects that the government will agree to give a written statement that he shall remain in control of the apparatus as long as he wishes," Adrian reported. But he emphasised that speed had now become essential and everything must be settled in the next two months.

To this semi-official report Adrian added his own notes, which plainly arose from some conversation he had been able to have with Kapitsa when they were certain they could not be overheard. Kapitsa had confided that the Russian government would probably make it a condition of any agreement that he should sign a public statement of satisfaction with his working conditions. He had asked Adrian to be non-committal when anyone enquired about how he was getting on, and he requested information about any approaches by the Russians during the last year for the purchase of the laboratory apparatus, as this information would help him in his own negotiations with the Soviet authorities. Kapitsa also explained that he had written to no one but his wife because he thought it would weaken his legal position if he did so. Finally there was the question of keeping open the line of communication that Adrian had established – Adrian had not felt there would be any personal danger to Kapitsa if letters were exchanged in the discussion of the plans which Kapitsa had put forward. But an alternative might be to use the services of P. A. M. Dirac, the Cambridge theoretical physicist (later a Nobel Prize-winner) who was spending the whole summer in Moscow.

The straightforward prospect of business/diplomatic negotiations, now that both Rutherford and Kapitsa had accepted that there could be no change in the basic situation, was, however, interrupted by Anna Kapitsa's desire to go to her husband in Moscow, by Kapitsa's natural desire to have her there, and by

Rutherford's worries that she would never be allowed out again. The request for a written guarantee that Anna would be free to return to England had led to a violent scene at the Russian Embassy in London in the first weeks of September 1935; plainly official Russian pride had been wounded by this suggestion and by the fact that Anna had insisted on taking a witness with her for an interview.

Dirac's reports from Moscow on the strain Kapitsa was suffering and Anna's determination to get to her husband, added to the fact that she could take Rutherford's reply to Kapitsa's proposals, defeated Rutherford's caution. She made all arrangements to travel on September 27th, and she did in fact receive a "return" visa just before she left. But before this vital document arrived she wrote an impassioned letter to Rutherford; she must have half-believed that he was right and she would never be allowed to return, and this was a provisional farewell. "There are many people who are only too glad to see the back of Kapitsa but I should not mind this if they would not at the same time spread all sorts of damaging rumours about him. Of the many things that are said about Kapitsa one is that for a sufficient amount of money he will be willing to do anything," she wrote, and denounced this as "the greatest slander and lie".

She accepted (though her husband did not) that he was often outspoken and offended people, "but never anyone could blame him that he did wrong to another man to benefit himself". Then she went on to tell Rutherford what Kapitsa's attitude to him meant; it was not just one of overwhelming gratitude for all that Rutherford had done for him, much more "if he did not commit suicide in Russia during the last year it was not for love of me or the children, it was only for love of you, not to let you down after all you did for him, after all the trust you put in him, he would do anything now to compensate you ever so slightly". His decision to keep plugging away, his rejection of his own weakness in wanting to retire to physiology, was for Rutherford: "Why does he bargain with our government how many thousands of pounds he would be allowed? He does it all with one thought, that your reputation should not suffer. He feels guilty and responsible for the stupidity of our people, he feels that he did not succeed to explain what a scientist is."

There was also Anna's own gratitude: "I am absolutely certain that I owe to you the life of Kapitsa, without his love and gratitude to you, without your invaluable help all through this year, he would not be alive. I still want one more thing from you, never to believe anything wrong about Kapitsa. He will sooner go to prison or be shot than do anything against you. Whatever you hear, however authentic the report may sound, always remember that Kapitsa will remain true to you."

In the stuffy, gentlemanly, "stiff-upper-lip" Cambridge of the 1930s – an atmosphere faithfully mirrored in the understated flatness of most of Rutherford's correspondence – letters from the Kapitsas stand out like glowing tropical orchids. This one, however, ends with a most businesslike postscript. Kapitsa had just telephoned from Moscow to say that a man called Rabinovitch was coming to England. Rabinovitch was a member of the Communist party, but seemed genuinely interested in Kapitsa's affair and would ask to see Rutherford when "he will try to find out if you really feel what you wrote in the letter to *The Times*". Anna added, "I think there is a real chance of impressing someone who feels interested in what an unfortunate business it all was. I think K does rather think this one is better than many others."

Anna soon announced her arrival in Moscow with a letter from the Metropole Hotel, dated October 4th, reporting cautiously, but favourably, on Kapitsa's mental state. To Rutherford her arrival in Moscow meant that Kapitsa was now in possession of the long letter containing the Cambridge proposals for settling the affair.

Firstly in the field of science, Cambridge (and the Royal Society) would continue to use the Mond Laboratory and would concentrate the work in the field of cryogenics – that is, very low temperatures – which Kapitsa had just started before his detention, but they would, at least for a time, keep out of the field of very high-intensity magnetism until they could at least see whether Kapitsa was making progress in Moscow. Kapitsa's assistants, Laurman and Pearson, would be needed for a short while to start up the Mond equipment, but after that they would be free to go to Moscow if they wanted – Rutherford refused to advise them either way but promised to look after them. Kapitsa's proposals for retaining an English professorship were not, however, feasible.

On the crucial question of the equipment in the Mond Laboratory and the money invested in it, Rutherford recalled that the Soviet government had offered to buy it all six months before, but neither Cambridge University, nor the Royal Society, nor Rutherford, had, at that time, been willing. Rutherford had, however, offered personally to try to get duplicates made "on condition that there was a free and direct consultation with you as to your wishes – I was not prepared to help anyone but yourself, as my sense of gratitude to the USSR is not particularly strong". Furthermore there was no necessity for Kapitsa to try to persuade the Russians to pay the full cost of his entire stay in England – neither the university nor the Royal Society was asking for compensation, and Kapitsa had been free at all times, like any other university worker, to take another post anywhere he wished. "After all, the help was given unconditionally on broad lines to promote investigations in what seemed a promising line of enquiry, and apart from the sudden and enforced interruption of your work we have no grievance of any kind," Rutherford wrote, sticking firmly and fairly to his own position on the internationalism and freedom of scientific workers, and with at least one eye on the probability that his letter would be read by others as well as Kapitsa. He concluded his statement of position by adding, "You did the best you could with the facilities given you and – apart from personal feeling – there is no obligation on you or on the USSR for compensation."

As for the detail of sending Kapitsa the equipment from the Mond, Rutherford considered it might be possible to send the big generator and its accessories straight away, but he did not want to send the hydrogen and helium liquefiers until he had had duplicates made, and for this he insisted "the whole cost is financed by the USSR". And on a personal basis Rutherford was prepared to go much further: "We could send over any old apparatus and personal mementoes of special interest to you. I know that Cockcroft will be very pleased to get any special things you require from this country to equip your laboratory."

Things now started to move very fast. At the beginning of October Philip Rabinovitch, as promised by Kapitsa, came to see Rutherford, and it seems likely that Rutherford gave him a copy of his letter to Kapitsa, for on October 3rd, Cahan wrote from the Soviet Embassy in London that Rabinovitch had reported the conversation and shown the text of "the letter". Cahan formally handed over negotiations, saying, "I beg to confirm that Mr Rabinovitch is officially authorised by the Soviet authorities to negotiate with you on this subject and the proposals he makes in his letter are also presented by him on behalf of the authorities of the USSR." Rabinovitch's letter, written on the same day, confirms the conversation and the full discussion he had had with Rutherford. It states that the building of a special laboratory for Kapitsa had been started in May and should be finished by November, and that Kapitsa was taking a great interest in this, "in accordance with his own desires". By personal request from Kapitsa, and with full authority from the Soviet government, Rabinovitch had proposed "the whole of the equipment should be transferred to Moscow in order to enable him [Kapitsa] to equip his new laboratory there as quickly as possible" and that "all the costs that have been incurred in the construction of the said equipment will be reimbursed to Cambridge University".

Rutherford replied to this immediately by calling a special meeting of the Mond Laboratory Management Committee to put the proposals in a constitutionally acceptable form; there were the Royal Society and the DSIR as well as the university to be considered. He wrote to Rabinovitch about this on October 5th, saying he was "uncertain how the university will regard the matter and there are certain obvious difficulties, but I will do my best to see that we can make some arrangement convenient to all parties concerned . . . I am naturally very anxious to give a helping hand to Kapitsa to start up his work again in Russia."

It was at this point that Rutherford's enormous power at the interface between government and official bodies and scientific research could be exerted to full effect. In less than a week he had got the whole thing settled. On October 8th he wrote again to Rabinovitch, reporting that the Mond Committee had advised the university that "the transfer of the apparatus should be favourably considered at a total cost of £30,000". This would be put to the Council of the University within a week and "it is not anticipated that any serious difficulty will arise". The manoeuvre would be accomplished by the university buying up the share of the property which formally belonged to DSIR. Furthermore, "no pressure will be brought on Pearson and Laurman", Kapitsa's technical assistants, if they desired to go to Russia. But there was a *quid pro quo*: "The university will expect a written statement that Professor Kapitsa will remain in control of the apparatus transferred for as long as he wishes. The transfer is only made to help Professor Kapitsa to continue his work in Russia." And Rutherford was using all his power and influence to get things moving in the practical sense: "I am prepared at once at my personal expense to arrange that work should start on the duplication of the helium liquefier." He concluded by asking Rabinovitch to come up to Cambridge to look over the Mond Laboratory in the next couple of days.

At this stage Rutherford was being disingenuous, to say the least. It would be perfectly reasonable to say that he was being dishonest with the Russians, but it seems more likely that he was taking his revenge; he was reverting to his farming-pioneer mentality, determined not to be done down by the other man. Attached to his copy of this letter to Rabinovitch is an unsigned schedule of the true figures of the case. It shows clearly that the true cost of the apparatus in the Mond was £12,000, split up into £5,300 for the cost of the apparatus not requiring replacement, and £6,700 for the apparatus that would need to be duplicated. The famous generator and its auxiliaries were the largest single item at £4,100. Allowing for the expertise involved the note concludes "correct value at least £15,000". Yet Rutherford had agreed with the Russians a repayment of £30,000. In fact he *was* claiming full compensation, for there are further figures showing that DSIR grants from 1925 to 1935, plus the special grant for liquefiers, plus university grants totalled £32,707.

But more than this, Rutherford believed that Kapitsa's method of generating very intense magnetic fields was not the best way forward for science. In a letter to Sir Frank Smith, dated October 5th, giving the Secretary of the DSIR the essential features of the letter brought from Kapitsa by Adrian and the main outlines of the reply Rutherford had sent via Anna, he announces that he and Cockcroft are agreed that they must keep out of the field altogether and that the generator was "a white elephant". What Rutherford wanted to replace the Kapitsa generator was a large electromagnet – "probably by far the biggest in the country" – to enable them to work "probing the very lowest temperatures". For while the Kapitsa affair had been dragging on, Cambridge, and Cockcroft, the electrical engineer, had got to know of developments in France, by Professor Cotton, of the construction of permanent electromagnets of a power previously unattainable. Rutherford was, therefore, proposing to let the Russians buy something that was already obsolete, so that he could use the money for bringing his laboratory into the forefront of the new field of cryogenics, and one can well imagine him believing that thus he was getting his revenge on the USSR.

Surprisingly, however, this was not concealed. The *Cambridge University Reporter*, the official gazette of the University, noted, on October 11th, the vital meeting of the Mond Management Committee – apart from the Vice-Chancellor its members were all close friends and colleagues of Rutherford who was himself a member – Ralph Fowler, John Cockcroft and P. A. M. Dirac, now back from his weeks in Russia and officially reporting his discussions with Kapitsa. The Committee's reasons for recommending that the university transfer the equipment to Moscow start in the most gentlemanly fashion: "Since the whole of Dr Kapitsa's personal work was, in a sense, preliminary to these experiments, Lord Rutherford and Dr Cockcroft would feel under a strong sense of obligation not to take up the experiments using the large generator if Dr Kapitsa wishes to resume work at once on these problems . . ." But the argument goes on: "The situation has also changed since the installation of the machine. Methods have recently been developed for the production of temperatures within one-thousandth of a degree of absolute zero by the use of large electromagnets. This field of work is of fundamental importance and is of particular interest to Cambridge since it offers possibilities of study of the magnetic properties of the atomic nucleus." Furthermore a large magnet would enable the laboratory to produce particles of extremely high energy for nuclear research. And so they proposed, officially, to replace the Kapitsa generator with a large electromagnet.

This left Rutherford with a great deal of wire-pulling to do. On October 26th he wrote to Sir Frank Smith, asking him to tell the Foreign Office, "informally", of the negotiations that were proceeding satisfactorily. In November he had further meetings with Cahan and Rabinovitch, and was put in touch with a man named Sokolov who was in charge of the Russian trading organisation Arcos, with an office in London. There was much discussion about the problem of getting Pearson and Laurman to Russia, yet safeguarding their positions in England. Early in November Rutherford received the first handwritten letter from Kapitsa himself, one of his most Russian and Dostoyevskian outbursts:

> Life is an incomprehensible thing. We have difficulties in clearing up a single physical phenomenon so I suppose humanity will never disentangle the fate of a human being, specially as complicated as my own. After all we are only small particles of floating matter in a stream which we call fate. All that we can manage is to deflect slightly our track and keep afloat. The stream governs us. The stream carrying a Russian is fresh, vigorous, even fascinating, and consequently rough. It is wonderfully suited for a reconstruction economist but is it suited for a scientist like me. The future will show it. All is new here and the position of science has to be newly determined. We must not be too harsh judges . . . I have no ill-feeling.

But it ends up, "I miss you very much, more than anybody else." To this Rutherford replied with bourgeois common sense. "A little advice" was in order, he wrote. "I think it will be important for you to get down to work on the installation of your laboratory as soon as possible . . . I think you will find that many of your troubles will fall from you when you are hard at work again," adding that his relations with the authorities would then also improve. In response to Russian eloquence came English clichés: "Your chances of happiness in the future depend on your keeping your nose down to the grindstone in the laboratory – too much introspection is bad for anybody."

Ambassador Maisky returned to the stage in November, officially confirming the result of the negotiations, and in the middle of that month the Russians promptly paid up the first instalment of £5,000. From Moscow, however, Anna Kapitsa wrote that the material arrangements were not all that mattered: "The question of freedom and confidence is much more serious for him than people here understand . . ." Additionally Kapitsa was depressed and upset by all the bureaucracy he had to deal with, and his assistant director was less than helpful, "a thorough scoundrel, he makes a terrific noise of himself but little work . . . Peter has to bear him as he is appointed from outside." But Ambassador Maisky provided Rutherford with a written letter of confirmation of the deal which could be sent to Kapitsa. And, on November 23rd, Kapitsa wrote and sent his formal resignation from the Directorship of the Mond Laboratory, asking Rutherford to communicate it to the Vice-Chancellor. Personally, Kapitsa added, "This is a terrific pain for me especially as it happened in this abrupt and unexpected way . . . [but he hoped his case] will not affect . . . the broadminded attitudes towards the internationalism of science which the Cambridge University has shown so splendidly towards me and which ought to be an example for all."

Rutherford's disentangling of the administrative complications of what amounted to selling most of the equipment of a university laboratory to a foreign country, was brief and masterly. He represented the university through the Mond Management Committee; Sir Frank Smith, who was both Secretary of the DSIR and Secretary of the Royal Society, represented the other two interested parties. On one day Rutherford wrote three letters to Smith, one personal, and one each to him in his two different posts. All three letters suggested the same administrative arrangements for transfer of ownership of the apparatus to the university, which would then transfer it to Russia. And that was all that was necessary, leaving Smith to do the paperwork at which he was a past master. The Management Committee of the Royal Society Mond Laboratory held a meeting on June 8th of the following year, 1936, at which the finances were finally cleared up. They had disposed of equipment worth £12,610.6s.3d., which sum included buying duplicate helium and hydrogen liquefiers, and also included £593 expended on buying a larger lathe and a vacuum pump compressor for re-equipping their own lab. So "from the sum of £30,000 to be paid by the government of the USSR it seems probable that a sum of the order of £17,000 will be available after all the expenditure on the transfer has been met". With this clear profit of £17,000, in a laboratory whose annual expenditure was of the order of £2,500, they proposed to buy a big electromagnet which would cost them about £3,500, and they decided to set up an endowment fund of £9,000 and a research fund of £5,000.

The burden of transferring the equipment from Cambridge to Moscow fell upon John Cockcroft. At times Kapitsa was intolerably demanding and impatient – he even asked for the Yale locks and insisted on receiving the special telephone equipment and internal exchange system which had been designed by Wynn-Williams. He got all these things, and he apologised for the difficulties he was causing.

On occasion he went too far in his demands. Early in 1936 Rutherford wrote a very stern letter: "I feel that these drains on the laboratory should come to an end . . . I feel you are a little ungrateful and inconsiderate in the way you write to Cockcroft about your wishes." Cockcroft, he insisted, was working like a horse to fulfil Kapitsa's demands and he concluded: "I hope in future you will discontinue complaining letters of this type. Otherwise we shall begin to consider you a nuisance of the first order. You know from old experience that I am not slow to express my frank views to you and I trust you will, as of old, give them serious consideration." But this was frankness between friends and all was soon smooth again as both sides understood each other's position. Rutherford was soon replying that he had read "with some amusement" Kapitsa's explanation of why everything should be done for him, and he explained that at one time he had expected "demands for the paint off the walls".

In the summer of 1936 Rutherford repeated his advice: "Get down to some research even though it may not be epoch making as soon as you can and you will feel happier. The harder the work the less time you will have for other troubles. As you know 'a reasonable number of fleas is good for a dog', but I expect you feel you have more than the average number." In other letters Rutherford passed on all the gossip – Kapitsa was one of the first to learn of Lord Austin's colossal gift of a quarter of a million pounds to the Cavendish in May 1936, and Rutherford told this tale: "He told me that on the way down he [Austin] had the narrowest squeak of a motor accident he has ever had in his life . . . I may mention that he had the securities (the whole magnificent donation) in his pocket at that time, so one could make a good story of it."

Some of the problems had their comic side, as when Anna Kapitsa had to write to Cockcroft begging him not to wrap up the smaller items of equipment in old Western newspapers, which were of course forbidden in Russia: "The Custom people had to get every scrap of it and were very upset." But Cockcroft and his wife were able to visit the Kapitsas in Moscow in the summer of

1936; the Kapitsa children had by then been allowed to rejoin their parents. Gifts such as tobacco were welcome, and Anna was able to explain that she had been having trouble with the bureaucracy who couldn't understand that "the Cavendish is not a trading organisation and the usual rules don't apply".

Kapitsa, too, settled down just as Rutherford had advised – "I agree with you by 100% I must keep my nose to the grindstone" and things did get better – he got a new assistant director, a lady engineer whom he found "excellent", but he was still asking for many small items to be sent to him, and he wrote: "I decern [discern] in your letter a certain conflict between the fatherly attitude towards me and desiring to help, and the Director of the Cavendish Laboratory trying to keep the business end as high as possible. Let the father win!" Kapitsa must have had a very shrewd idea of how Rutherford had scored in the final deal with the Russians for he went on to estimate that Rutherford would have a five-figure reserve of capital when everything was finally settled – and he was right.

The letters became more and more frank. In August 1936 Rutherford confided that he was "alarmed by the potentialities of the situation in Spain. It looks a very bad business. Re-armament is making the UK flourish." Kapitsa for his part admitted, "Our bosses who manage perfectly well masses of people are, from my point of view not accustomed to deal with scientists . . ." In January of 1937 Rutherford described how Lindemann was anxious to force his way into public life "under the aegis of his friend Churchill", and how Lindemann had written a letter to *The Times* in which "he made it clear that he alone could save the country in these difficult times of air warfare". He added a few weeks later that he and Fowler and Appleton "have been entangled in advising on defence problems and there is great activity in connection with matters of air and naval defence". But at the same time Rutherford was boasting, "I still can do a day's work without serious effort and I may tell you in confidence that my temper improves with age – at any rate my wife says so!" One reason for this feeling of good health was that at long last his knee was better: "This is the result of the ministrations of a local pork butcher, Mr Walker, who is an unlicensed expert on such matters and a most interesting and amusing fellow." Rutherford's last letter to Kapitsa, and one of the last he wrote to anyone, is dated October 9th, 1937 and repeats that he is physically "pretty fit", looks forward to a visit by Niels Bohr and hopes "it will not be too long before you are able to come over and see us all again". He adds that he has "had to help a little on various problems connected with anti-aircraft and submarine defence". None of this carries any implication that Kapitsa had ever been involved in defence work in Britain, and this increases the irony in the fact that one of the two positions at the Mond Laboratory made available by using Kapitsa's salary was offered to a young refugee from Germany, the physicist Rudolf Peierls, one of the two men who first showed that an atomic bomb was possible.

Kapitsa was devastated by Rutherford's death. He wrote to Cockcroft as soon as he received the news (November 1st, 1937): "It is difficult to believe that there is no more Rutherford. We all had the feeling that Rutherford is immortal, not only by his work, but as a human being, he was so strong and so full of life." Kapitsa has never ceased to praise Rutherford's memory in lectures, books and articles, and it must surely be due to his influence that Rutherford is in some ways more honoured in Russia than in Britain.

17

Final Triumphs

Rutherford's life ended, scientifically, as triumphantly as it had begun. He went out with a bang, not a whimper, and after the decline of the late 1920s, and despite his growing absorption in public affairs, he finished on a high note of achievement, making discoveries until the very end. But the triumph of his final years was above all the triumph of the Cavendish Laboratory, Rutherford's Cavendish.

In 1930 the Cavendish was, beyond dispute, the greatest experimental physics centre of the world. In Europe Lise Meitner, working with Hahn, and Marie Curie's daughter Irene, working with her husband Frédéric Joliot, ran the only major units in the same field of nuclear physics, while all the other important figures were theoreticians. In America the major move of personnel and resources was only just getting under way: Ernest Lawrence was building his first cyclotron; in Boston Van dem Graaff was constructing his first large accelerators; Oppenheimer had left Cambridge in a state of nervous breakdown, had been happier in Göttingen and was just starting the first American school of theoretical physics in California. The only large flow of experimental results came from the Cavendish, and every physicist of note visited Cambridge and lectured there, or spoke at the Kapitsa Club.

The outward appearance of the Cavendish belied this world leadership. The laboratory was situated down the tiny Free School Lane, its main entrance looking across the ten-foot-wide street to the almost blank rear wall of Corpus Christi College. Newcomers were often shocked by the dingy passages with uncarpeted floors of bare board, the varnished pine doors and stained plaster on the walls. Waiting in such a corridor "indifferently lit by a skylight with dirty glass" and shepherded by the "formidable" Mr Hayles, who acted as secretary to Rutherford, though his real position was lecture assistant, newcomers such as Mark Oliphant were not impressed; but then:

> I entered a small office littered with books and papers, the desk cluttered in a manner which I had been taught at school indicated an

untidy and inefficient mind . . . It was raining and drops of water ran reluctantly down the grime-covered glass of the uncurtained window . . . I was received genially by a large, rather florid, man with thinning fair hair and a large moustache, who reminded me forcibly of the keeper of the general store and post office in a little village in the hills behind Adelaide . . . Rutherford made me feel welcome and at ease at once. He spluttered a little as he talked, from time to time holding a match to a pipe which produced smoke and ash like a volcano.

Nor was Rutherford's domestic life any more distinguished in style or appearance. It was the style of a Cambridge don's life, rather than that of the leader of one of the world's most powerful research groups. There was nothing of the "jet-setter" in the comfortable round of a daily life that involved walking from his home at Newnham Cottage across the river to his laboratory, no more than a ten-minute stroll. Regular evening dining at Trinity College varied his return route, but not by much.

Cambridge life was safe and provincial, little stirred by the Great Depression and the rise of political fanaticism in Europe – indeed Rutherford's official life, dealing with Jewish refugees and the problems of Kapitsa, was much nearer the real world than the social backwater that was Cambridge society. People who still remember those times seem to feel slightly guilty that they lived such unadventurous, unworldly lives: the whole social setting was similar to a vicarage tea-party, a mixture of languorous summer afternoons on lawns, with conventions and conversations remote from the world outside. Tea-parties on lawns were indeed a major feature of this life, where the young and unconventional sat on the ground or on rugs rather than on garden chairs. Rutherford's daughter, Eileen, led this sort of life, which seemed so daring, until her early death in 1930. The Fowlers shared a house with another family, the Cooks. Lady Rutherford disapproved, and disapproved, too, of Ralph Fowler her son-in-law, regarding him as hopelessly impractical, bullying him into domestic activities and issuing streams of contradictory instructions as she supervised whatever operation she had compelled him to perform. Likewise she issued a stream of contradictory instructions to young research men when she dragooned them into helping in the interminable reconstructions of her garden.

Several friends have noted that there was never any outward sign of affection between Rutherford and his wife, though it was clear that there were still deep bonds between them. Lady Rutherford never hesitated to rebuke her husband for his personal defects. Often the only remark at the breakfast table would be her sharp

"Ern, you're dribbling again", and the great man would have to remove the marmalade that had fallen on to his ample waistcoat. She could be equally stern in public, making no secret of her teetotal attitudes and shouting at him across a room of distinguished people if she saw him drinking a glass of wine. Perhaps to avoid this plaguing, Rutherford played more and more golf, – perhaps it would be more accurate to say frequented the golf course more and more. Nearly every Sunday morning there would be a noisy foursome, playing with three balls on the course on the Gog Magog hills. Aston, G. I. Taylor, Fowler and Frederick Mann, a Cambridge chemist, were regular members, with several others joining from time to time. The standard of golf was, apparently, very far from good, the brash "joshing" style of humour and laughter obviously being more important to all the participants. In later years this golfing party took to spending a few days each spring on the milder courses of Sussex and Kent.

The noisiness of the party and possibly, too, the fact that they did not take their game very seriously, tended to shock and upset other members of the Cambridgeshire club and members of the southern clubs they visited. Rutherford and his friends seem rather to have enjoyed the disturbance they caused, joking about the comparative weakness of others against the strength of a "Trinity College foursome".

But this was all very innocent, as was the remainder of Rutherford's domestic life. Many of his summer holidays were spent on the beaches of Dorset with large family parties of his grandchildren and their friends. After Eileen's death the grandchildren were largely cared for by Mrs Cook, whose house was shared with the Fowlers. Rutherford loved his grandchildren and got great pleasure out of their company. He knew his own limitations and theirs, and he liked to see them often but only for short periods; sixty-year-olds and six-year-olds can mix well, but can find too much of each other's company wearisome. In return the grandchildren loved him very dearly. Grandmother, however, was more difficult – it is significant that the children always called her "Lady Rutherford" or "Lady R", rather than "grandma" or any similar term. She was the sort of elderly lady who was always "expecting", or demanding certain types of behaviour and formal demonstrations of affection and, as a result, failing to inspire exactly that affection she hoped for.

One unusual and brighter spot in the unexciting home life of the Rutherfords in the early 1930s was the presence of Bay de Renzi. She was a distant relative of Lady Rutherford. Her mother was a nurse in London and she herself was "a mercurial mixture of Irish and Latin ancestry", according to Oliphant, who added, "Whereas Lady Rutherford was cool, competent and never demonstrative, Miss de Renzi's great affection for Rutherford was shown in countless ways, and he obviously enjoyed being fussed over at times . . . The Rutherfords and her many friends were upset when suddenly Bay left Cambridge, for all had grown very fond of her."

In our time one immediately wonders whether this woman, in her thirties, was Rutherford's mistress, since she certainly looked after his house during a long holiday that Lady Rutherford spent in New Zealand. But this is to misunderstand the social tone of the Cambridge of those times as well as to misinterpret Rutherford's character and his clear sexual diffidence. There is not a shred of evidence that Eileen de Renzi (her proper name) was anything but a somewhat impoverished, distant relative who helped in the household management, in a manner common in the earlier years of the century.

In addition there has recently emerged one piece of evidence from that same solicitor's file which included the itemisation of Rutherford's personal belongings. It is an undated letter from Bay to Rutherford, written at the Red Lion Hotel near Buntingford. She addresses him as "Grandpapa" and regrets that she has to ask him for cash – she wants a cheque for £6 by return of post because her cheque from "Harman" in New Zealand has not arrived. She says she has been ill and has not wanted to ask his help again. Someone called "Nurse Amy Chidgey" at Londonderry House is going to send her £4 a month until she is quite strong again, but the nurse has not said who is making the allowance of £4. "I do realise that I have in the past been careless about money and allowed little things to run on and mount up . . . The bills for my illness are appalling; it would have been much cheaper and better if the Doctor had let me die in the first few weeks of illness." There is no trace of whether Rutherford sent the money or of any other response, though since the letter, quite unlike anything else in his vast files of correspondence, emerged from a solicitor's office he presumably dealt with the matter at arm's length. It is certainly not like the letter of an ex-mistress, but rather a begging letter from an embarrassed former dependant, affectionate but with no real claim upon him.

Even in his sixties Rutherford continued to show the work-pattern of his earlier days – bursts of intense and concentrated work interspersed with spells of total relaxation. His only concession to advancing years was that the holiday periods became more frequent. Throughout the 1930s he had a country cottage to which he retired for the occasional week, or long weekend. Foreign travel was almost completely abandoned.

The first cottage was at Celyn in North Wales, near the foot of Snowdon. The dwelling was fairly isolated, a whitewashed single-storey building probably made from the amalgamation of two workers' cottages. It was a place of great charm with marvellous views when the weather was fine; but the facilities were primitive, the village shop carried only the smallest of stocks, and in the shadow of Snowdon the weather is not always fine. The Rutherfords would often take a couple of guests with them on their trips to Celyn, perhaps the Chadwicks or the Oliphants, but there were few local "entertainments" unless one was willing to enjoy the magnificent walking and hill scrambling in the area. Rutherford, however, was not fond of walking; he would soon get tired and irritable, especially if the weather was warm, and then he would sit on a rock and smoke his pipe, leaving the remainder of the party to go on ahead. On the other hand he found great fascination in the conversation of the local people, and he became very interested in the remains of the local narrow-gauge railways which had once served the many slate quarries of the neighbourhood.

The journey from Cambridge to Celyn, was, however, long and mostly uninteresting. It was not made any easier by Lady Rutherford's passion for fresh air, so that the touring car they used would have its folding hood down, and all the passengers would have to wear heavy coats, scarves and hats, as well as wrapping themselves up in rugs. Even the driver was provided with a hot-waterbottle.

Celyn, therefore, proved too Spartan as the years crept on, and after some searching the Rutherfords bought a plot of land near the village of Chute in Hampshire, on a farm which, by an odd coincidence, had the historic name of New Zealand. On this plot they built a modern semi-bungalow country cottage, which, although it was heated by wood fires and lit by paraffin lamps, was much more warm and comfortable. Chantry Cottage, so named because there were ruins nearby which almost certainly were not those of a medieval chapel, became Rutherford's haven for the few remaining years of his life. There was good accommodation for younger friends to stay; there was little walking to be done, but plenty of local places of interest, such as Salisbury and Stonehenge, which could be visited by car. And there was even more local history to be discovered in an area of downs and ancient trackways and early British hilltop "forts". Even the exercise of clearing the scrubby trees and brambles from the overgrown patch they made into a garden was suited to Rutherford's taste, temperament and early training. Though he made many and very audible grunting and wheezing and groaning noises, he plainly enjoyed himself with axe and billhook. Lady Rutherford, demonstrating wide botanical knowledge in the shape of correct and properly pronounced Latin names for all the plants she tried to nurture, showed an almost hilarious lack of botanical wisdom as she tried to persuade rhododendrons to grow where the chalk lay only a few inches below the surface.

After the exercise of woodcutting Rutherford might walk up the hill to chat with the farmer over the economics of running a dairy herd, and in the afternoon, following a post-lunch snooze, he might wander down to the nearest cottage to have a word with a retired Royal Navy petty officer and examine the little electric light plant he had installed in a shed. The evenings were spent talking with whichever friends might be staying, reading books or discussing world affairs. Lady Rutherford would knit, or sew or play patience, keeping an ear on the conversation, correcting her husband on some point of fact when he got too carried away in his description of an event of pomp and ceremony which he so much loved, or in gossiping about the entanglement of the heir to the throne with Mrs Simpson. Oliphant, who played the role of surrogate son in Rutherford's later years, records the delight Rutherford showed when they attended a military tattoo on Salisbury Plain; the great man stamped his feet and clapped noisily when an item of military drill appealed to him and revelled with childish delight in the military band music, the only sort of music for which he had any appreciation. Afterwards he teased Oliphant, "That expressed the true spirit of England far better than the nonsense you read in the *New Statesman*."

The political atmosphere of Cambridge in the 1930s has come to have great importance, for it was the seedbed in which the quartet of spies, Philby, Burgess, Maclean and Blunt were recruited to Russian service, through the élitist Apostles' Club. The atmosphere has been described in *The Climate of Treason* by Andrew Boyle with its mixture of secret Communism, homosexuality and social and intellectual snobbery in which the spies were moulded. This was all very close to Rutherford's life in physical terms, for a number of the men involved were from his college, Trinity, and there were certainly Communists at the Cavendish, and one could just as easily convict the physicists of intellectual élitism. It is true, too, that one Cavendish man, Allan Nunn May, was one of the quite separate group of so-called "atom-spies".

Why then was the Cavendish, at the time considered to be the most left-wing of all Cambridge institutions and which provided dozens of men for all the most secret of wartime programmes on radar and atomic energy, not another source of spies? The answer lies in the totally different social make-up of the Cavendish and in its much more robust and open left-wing proclivities.

An entirely different approach to the Cambridge political history of the 1930s is provided in *The Visible College* by Gary Wersky; it is a study of the lives of four notable scientists who were of strongly socialist persuasion. Wersky describes the typical Cambridge scientist as a man from an aristocratic, bourgeois or professional family, often with a private income and educated at Cambridge for both his first and second degree. He had therefore spent six years in a town and university setting "notorious even by English standards for its ritualised observance of class distinctions". The man was a member of "an intellectual élite, entry to which was largely reserved for one class, one sex (and one race) only", his society was stratified according to sex, male-dominated, and virtually without contact with the working class. Whether or not this is true of Cambridge scientists in general, it was certainly not true of the Cavendish. Rutherford, Chadwick, Cockcroft and Appleton, were all men of little wealth who came to Cambridge from North Country, mostly Manchester, backgrounds. Aston was probably the only senior man who was believed to have private means, and this was itself a talking point. Of the next generation both Ratcliffe and Allibone were from the North of England. Grammar school boys were more common than those with public school educations. Men such as Oliphant and Harrie Massey came from Australia. There were Chinese, Japanese, Indian research workers, alongside Americans, Canadians, Danes, Russians and Swiss. There are one or two women in every annual photograph of the Cavendish research team, sometimes four or five, and it was quite impossible to avoid contact with the "working class" in the shape of laboratory technicians. The Cavendish was, in fact, a powerful machine for social mobility.

Furthermore, there was a robust set of left-wing, sometimes openly Communist, men among the scientific leaders in the 1930s. Blackett, Bernal, and Wooster were among the most "advanced", but younger men such as Oliphant and C. P. Snow held views which were certainly socialist in the British Labour party sense. Very many of the young research men joined the Communist party or left-wing groups such as "Scientists against War". Rutherford himself chaired a meeting of an organisation called the Democratic Front, and also signed a public appeal for support for "Tory Democracy" programmes – a left-wing conservative movement associated with the brand of policies favoured by Harold Macmillan.

There were, however, significant differences of experience between the "left-wingers" of the Cavendish and their more secretive contemporaries in "The Apostles". The Cavendish men knew a great deal about Russia at first hand, and were less likely to confuse the intellectual appeal of Marxism with the actual policies of the Russian state. Kapitsa was but one of three or four Russians who worked at the Cavendish between 1925 and the outbreak of the Second World War. Kapitsa's misfortunes were known to everyone in the Cavendish, and many who worked there had first-hand experience of Russia. Allibone, Cockcroft, Dirac and Shoenberg all spent weeks or months working in Russia on fellowships or exchanges of one sort or another.

Furthermore the example of their own Professor was sufficient to show the young physicists that liberal or even anti-Fascist activities were not the sole prerogative of Communists or socialists – Rutherford's activities for the Academic Assistance Council made this quite clear, and the presence of refugee Jewish scientists among them kept this fact in front of their eyes. Wersky, while deploring the weakness and inactivity of the left-wing movements in Cambridge, nevertheless admits these political points:

> If the radical tide was not overwhelming, it was at least tolerated by the laboratories' Directors, Lord Rutherford and Sir Frederick Gowland Hopkins. Though politically a conservative, Rutherford was himself disturbed by the signs of industrial stagnation in Britain and Fascist repression in Germany . . . He could therefore at least sympathise with those of his "boys" who also wanted to do something about those problems. As long as their "Communism" did not get in the way of their experiments, Rutherford could lump their politics . . . Knowing what we now do of the low support that the left received from Cambridge's two most radical laboratories we must also recognise how critical were the stances adopted by Rutherford and Hopkins. For what would have become of the scientific left had the Cavendish been headed by a reactionary physicist like Oxford's F. A. Lindemann? . . .

The same author also emphasises the élitist nature of the Cavendish and its practice of High Science – "research which academic scientists hold in the highest esteem", fashionable, "hot", science which specialised in "an aversion to speculative theorising which runs far ahead of empirical investigations". And he writes of Cambridge as the "locus" for such "first-rate science", saying, "No other university had such an abundance of human and physical resources to devote to disinterested investigations . . . By sitting on important government committees and controlling the major laboratories in and out of Cambridge, they were able to determine the style, the ethos, and, not least, the direction of pure science in Britain." Rutherford and Chadwick would not have recognised themselves or the Cavendish in this description; they felt they had not enough money or resources even to make the Cavendish what they thought it should be. But Wersky nevertheless is perceptive in spotting the values that Rutherford and the Cavendish demanded of those who were to be considered of the élite, those values which made "a first-class man". They were indeed values which included loyalty to your laboratory, extreme devotion to hard experimental work, a good intellect, and a strong aversion to speculation beyond that justified by experimental results.

Rutherford accepted, perhaps even created, the idea of the "first-class man", and those were the men he wanted working in his laboratory. But he almost invariably left the man he did not consider "first class" to find out for himself, and the discovery was usually a piece of self-discovery. It was then that the young man would find Rutherford at his best, for those who knew him agree that he would go to endless trouble to find jobs and places in teaching, in industrial laboratories, or in government service for those men he had taken on and who elected for themselves to give up. For those who showed that they were "first class" and who elected to stay at the Cavendish there were very few official "jobs" to go to. There might be a college fellowship for the very lucky, or a studentship for a few short years, which could be eked out by tutoring a few undergraduates. For another few there might be grants from the DSIR where there was a salary for an assistant included as part of the finance of some piece of research. A few more positions were financed when the Mond Laboratory was opened with the backing of the Royal Society. But the whole structure was far from clear, even to those working in it: "Cambridge was like that in those days. If you hung around long enough people thought you were part of the scenery and they started to pay you," according to Professor J. Allen.

Superimposed on this background of comfortable, essentially kindly, totally non-commercial, English middle-class academic life, are a series of rapid, exciting discoveries. Experimental physics, led by the Cavendish, produced in the decade of the 1930s the most astonishing ten years of development that science has ever seen. These discoveries brought nuclear physics from the most recondite of "pure" laboratory sciences into the atomic era, the age of nuclear power engineering and nuclear weapons.

The first and most important of these discoveries was Chadwick's finding of the neutron – for, just as the existence of the neutron brought into focus a whole mass of findings which elucidated the essential theories of atomic structure, so also did the use of the neutron, first as a probe, and then as a projectile, open the way to the use of the enormous energies locked up in the structure of atoms. Historically, too, the discovery of the neutron was a Rutherfordian piece of work, perhaps the last of the old "simple" experiments before the new era of heavy machinery came to the forefront. It arose out of spotting an anomaly, an inexplicable small phenomenon; it was essentially the proving of the existence of something which was intuitively "known", or in more pragmatic terms it was the confirmation of the reality of an intellectual "model". Both Rutherford and Chadwick "knew" that the neutron had to exist, it was the only way of "building" the atomic nuclei whose existence and size they had experimentally proved. And the method that Chadwick used to demonstrate the existence of the neutron was the method of elimination which Rutherford had used in so much of his work.

With the benefit of hindsight, Chadwick himself was the first to point out a number of experimental results which could or should have led to the earlier discovery of the neutron. Rutherford had clearly predicted the neutron in his Bakerian Lecture of 1920, and, though he had rarely mentioned it publicly since that time, he had again and again started off hunts for the neutron inside the Cavendish – many of them "silly" experiments. At least one researcher, J. K. Roberts, who was set to search for the energy due to the production of hypothetical neutrons in electrical discharges through hydrogen gas, complained that he had "wasted a lot of time" on "a crazy idea of the Prof or Chadwick". J. L. Glasson, and H. C. Webster, an Australian, were both set tasks at different times which were aimed at finding evidence of the neutron, and Chadwick himself, working with two students, Constable and Pollard, "for a short but exciting time . . . thought we had found some evidence of the neutron. But somehow the evidence faded away."

Sheer chance seems, as so often, to have played some part in achieving final success. Chadwick had noted the value of polonium as a source of experimental alpha-particles when he had gone to Vienna to sort out the controversy between the Cavendish and Petersson. He had made himself a similar source when he returned to Cambridge, but it was extremely small and weak – an example of the way Rutherford would not provide satisfactory materials and resources for following up leads. But in 1930 Norman Feather was spending a year away from the Cavendish in Baltimore, where, at the Kelly Hospital, he found some old radon tubes, small glass tubes which had once contained radium emanation for (unsuccessful) experiments in cancer treatment. With the passage of time and the steady radioactive decay of the various members of the radium family, these tubes now contained comparatively large quantities of "radium-D" which is the element polonium, originally discovered by Marie Curie. Feather had these tubes shipped back to the Cavendish – presumably there were no international regulations covering the "Transport of Radioactive Materials" – and from their polonium Chadwick was able to make a much more powerful source of alpha-particles. With this source H. C. Webster was able to show that, when he bombarded beryllium with alpha-rays, the resulting radiation emitted in the forward direction from the beryllium was much more penetrating than what was emitted backwards towards the polonium source. This experiment was performed in the first half of 1931 and the results excited Chadwick: ". . . It could only be explained if the radiation consisted of particles, and, from its penetrating power, of neutral particles." However no amount of examination of this radiation in expansion chambers (Wilson cloud chambers) showed any trace of the tracks of particles.

Beryllium, a toxic, difficult-to-handle, metal is the fourth lightest of the elements, with an atomic mass of nine. It had attracted some interest during Rutherford and Chadwick's earlier disintegration experiments, because it firmly refused to disintegrate.

Since it did not emit any protons to scintillate on the screen, Rutherford and Chadwick wondered (but not in public) whether the nucleus might be splitting into two alpha-particles (each of mass four) and one neutron, to make up the beryllium mass of nine units. They, quite wrongly, felt that this view was supported by the fact that the mineral beryl (from which beryllium, the chief constituent, had taken its name) contained a high proportion of helium (alpha-particles).

On the other hand beryllium, when bombarded by alpha-particles from polonium did emit intense gamma radiation, and there were a number of suggestions "floating around" that this might be due to the excitement of the nucleus in some way which might give a clue as to the structure of that nucleus. In Germany W. Bothe and H. Becker were examining this problem and publishing their results, and in Paris, too, the matter excited interest enough for Irene Joliot-Curie and her husband Frédéric to start a series of experiments. But the basic problem was that there was no theory of nuclear structure to point the way to the right experiments, while the wave-particle theories and the developments of quantum mechanics by the continental theoreticians, great achievements of the intellect though they were, offered no guidance either. In Chadwick's own words, he "was groping in the dark". Feather and

Massey, who were both working in the same field at the Cavendish at this time, when they came to write Chadwick's memorial, point out that there were no awards for physics by the Nobel Prize Committee in 1931 or 1934, which was a "significant comment on the uneasy state of fundamental physics at the relevant time". Chadwick gained the prize in 1935 for his discovery of the neutron and it has been given every year since then. Chadwick's memorial continues, ". . . From about 1927 onwards progress in the subject was temporarily at a low ebb – in nuclear physics, in particular, difficulties of interpretation were mounting more rapidly than they were being resolved by the assimilation of new knowledge derived from experiment. There was a growing need for some reorientation of outlook."

Then in January 1932, the Joliot-Curies published the first results of their work on beryllium bombarded by alpha-particles from polonium. This confirmed that no protons were released by the beryllium, but there was intense gamma radiation. However when they tried to block out these gamma-rays by putting blocks of paraffin wax, or similar substances rich in hydrogen, in their path, numbers of swiftly moving protons were shot out of the paraffin wax. The Joliot-Curies had been trying to measure the ionisation produced by the gamma-rays in an ionisation (cloud) chamber, but when they put paraffin wax between the bombarded beryllium and the ionisation chamber they actually found increased ionisation typical of the passage of protons through the chamber. They suggested that the gamma-rays must contain particularly powerful quanta of energy which were transferred to the protons (hydrogen nuclei) in the wax by some variation of what is called "the Compton effect".

The arrival of this report, in the French scientific publication *Comptes Rendues*, had an electrifying effect on the Cavendish. Chadwick read his own copy as soon as he arrived at work and found it "most startling". A few minutes later Feather rushed in, equally "astonished" at the report. And when Chadwick showed it to Rutherford at their regular eleven o'clock meeting, the great man roared, "I don't believe it."

On this particular occasion Chadwick was ready and the Cavendish could provide the equipment – there was Webster's experimental arrangement and Chadwick's new and beautiful polonium source of alpha-particles. It was really comparatively easy, Chadwick found:

> I started with an open mind, though naturally my thoughts were on the
> neutron. I was reasonably sure that the Joliot-Curie observations could

not be ascribed to a kind of Compton effect, for I had looked for this more than once. I was convinced that there was something quite new as well as strange. A few days of strenuous work were sufficient to show that these strange effects were due to a neutral particle and to enable me to measure its mass.

It has been suggested, probably because only a few days' work was involved, that Chadwick's great achievement was a piece of opportunism. The record shows this must be rejected; there were many, many days of fruitless work in the past which had eliminated other possibilities and there was the fact that both Chadwick and the Cavendish were right up in the front of the field, experimenting with polonium and beryllium with the possibility already in mind that this work might lead to a better understanding of nuclear structure.

Chadwick posted his letter to *Nature* by February 17th – it was printed on February 27th, 1932 and entitled "Possible existence of a neutron". Chadwick had placed hydrogen, helium, lithium, beryllium, carbon, air and argon in the path of the radiation from the bombarded beryllium. From the hydrogen he found that protons were shot forward, from all the other targets whole atoms of the target substance were expelled. He measured the speeds of the particles and their energies in terms of the numbers of ions they produced, and he showed that, unless the basic physical laws of the conservation of energy and momentum were regularly violated, it was impossible for gamma radiation to have imparted such energies. Furthermore any protons ejected from hydrogen targets in a backwards direction were much less energetic than those ejected in the forwards direction. He proposed instead that the radiation from beryllium under bombardment included "particles of Mass 1 and charge 0, or neutrons" and he showed that such a view provided a "simple explanation" of all the phenomena observed. Further than this he did not go for the moment, concluding his letter with: "Up to the present all the evidence is in favour of the neutron, while the quantum hypothesis can only be upheld if the conservation of energy and momentum be relinquished at some point."

This, he admitted in a letter to Bohr at the time, was perhaps excessively cautious; he really "knew" already that it was neutrons from beryllium that were smashing into the target atoms and driving them forward. "It might appear at first sight that here was a piece of opportunism and little more. That would be altogether too simple a verdict. The real merit which he displayed at the time was that of critical judgment, part logical and part intuitive – and that degree of conservatism which in a novel situation disdains the easy belief that accepted laws may be flouted with impunity," wrote Massey and Feather. If Chadwick "knew in his bones" that he had found the neutron, Rutherford must take much credit, for it was he who had predicted the neutron and it was he who had again and again inspired the search for it; it is significant that Irene Joliot-Curie admitted to Chadwick that she had never read Rutherford's prediction but that, if she had done so, she would surely have realised just what she and her husband were observing.

In a very few weeks Chadwick had got the whole business neatly tied up. On May 10th he sent off to the Royal Society the definitive paper, "The Existence of a Neutron". It was, of course, much fuller than the letter, yet essentially the argument is the same: no quantum hypothesis can explain the effects of the radiation from the bombarded beryllium, only the hypothesis of a neutral particle will fit the facts, and there are considerably more facts than in the original Joliot-Curie observations. The French couple had noted that similar effects arose from bombarded boron, and Chadwick confirmed and extended this, so that he could show that neutrons were also produced by boron, a different element. The whole structure was buttressed by separate, shorter, papers from Feather and Philip Dee in which they recorded their observations of what must be neutrons striking nitrogen nuclei and electrons in cloud chambers. There is a first analysis of the spectrum of radiation from beryllium, suggesting that there are both neutrons and gamma-rays of different energies produced by different nuclear reactions when the alpha-particles strike the beryllium nuclei. But, most important, there is the establishment of the mass of the neutron (to a remarkably good accuracy considering that it was the first such measurement) and the demonstration that such a mass accurately fits into the calculations and observations of the energetics involved. There is the first demonstration of the properties of the neutron, but there is also one interesting misjudgment: Chadwick plainly states a preference for regarding the neutron as a proton and an electron in a closely bound state rather than as being an elementary particle in its own right, as we now regard it.

The Cavendish Laboratory as a whole revelled in this great discovery. For them the moment of triumph came at a meeting of the Kapitsa Club, where Chadwick, for once mellowed by food, wine and applause, gave a brilliant and lucid "pre-publication" lecture, which for many of those present was the first clear knowledge of the importance of the discovery that had been so much gossiped about.

Rutherford himself publicised the neutron and its importance shortly after Chadwick had formally staked his claim to the discovery by his first letter to *Nature*. A series of lectures at the Royal Institution in London was a regular feature of Rutherford's life for many years – hardly a chore for he always enjoyed talking to people about his work. Preparing the demonstrations he gave was, however, something of a chore for his chief technician, George Crowe, and sometimes he got junior colleagues to do some "devilling" on the details of the lecture. His talk for March 18th, 1932 was billed as "The Origin of the Gamma-Rays", but the title was changed at the last moment to "Recent Researches on the Gamma-Rays", which enabled him to bring in the discovery of the neutron in a natural way. The lecture started by giving the latest opinions on gamma-rays and their origin, however – they arise "from the nucleus of the radioactive atom and represent in a sense some of the characteristic modes of vibration of the nuclear structure". There is mention of the wavelengths and quantum energies in the complicated gamma-ray spectrum, but "it has been difficult to determine with certainty the origin of this radiation" and "very little is known about the structure of the nucleus". The final minutes of the lecture were devoted to "The radiation from beryllium and the neutron", and it is characteristic of Rutherford, in his energy and buoyancy, that he should tell the lay audience of the Royal Institution of research work that had been done only three or four weeks previously. The story is told briefly, simply and clearly and everyone is given their proper due, including some of those former Cavendish men who had carried out unsuccessful experiments in search of the neutron. There is proper (and no more) credit given to "Rutherford" – the original "I" is crossed out in his own hand on the typescript – who "discussed the properties of such a neutron" in 1920.

But obviously what excited Rutherford that night were the photographs of neutron impacts taken by Feather and Dee. Feather had found, in addition to the straight tracks of nitrogen nuclei struck by neutrons, a number of cases where there were branching tracks:

> which indicate that the nitrogen nucleus has disintegrated in a novel way. These branch tracks are believed to be produced by the recoiling nucleus and by an alpha-particle which is ejected from the struck nucleus. It will take time to analyse the results obtained and to examine the effects produced in other gases. The peculiar properties of the neutron allow it to approach closely, or even to enter, nuclei of high atomic number, and it will be of great interest to study the effects of such collisions. It is however evident that this new radiation has surprising properties and there is every promise that it may prove an

effective agent in extending our knowledge of the artificial disintegration of elements. It will, for example, be of much interest to decide whether the neutron is captured in such disintegrating collisions, or whether it merely passes through the nucleus on which it has such a catastrophic effect.

It would be an overstatement to say that the discovery of the neutron alone revivified Rutherford after the dismal years of the late 1920s, but here plainly is the old Rutherford back on the warpath again, leading the next "attack" on the problem, seeing correctly which effects should be studied most urgently, seeing immediately what will be the next steps forward, and how the new knowledge can be most effectively applied to produce yet more knowledge. He was at this moment riding the crest of the wave, one of the many waves which he, quite correctly, knew he had created himself. The next great discovery of the Cavendish Laboratory, the second of the "annus mirabilis" of 1932 was on its way. This was Cockcroft and Walton's first artificial disintegration and transmutation of an element. And this, too, had its origins in the search for the neutron.

Five years previously, in 1927, Rutherford had said, "This sounds interesting" when a young man, T. E. Allibone, came to him with a proposal to start his research by trying to accelerate electrons by supplying a very high voltage. The young man came from an unusual background; he was an electrical engineer at the Metropolitan-Vickers engineering company in Manchester who wished to switch into academic research. His earlier education had been at grammar school and university in Sheffield, the traditional north-country background from which so many Cavendish men had sprung. Allibone's decision to come to Cambridge arose largely from meeting Chadwick at a lecture and being advised by him that there were scholarships, such as the Woolaston Scholarship at Caius College, which could provide for men who had not been Cambridge undergraduates. But working in the laboratory at Metro-Vick, where Sir Arthur Fleming was Director, had made the young man familiar with the latest ideas and techniques on providing voltages, potential differences, of as much as half a million volts. In America, at this time, Coolidge was able to make X-ray tubes working on a voltage of 300,000, and both Chadwick himself and American researchers had considered using Tesla transformers to provide large voltages to accelerate electrons. Allibone got much support from Sir Arthur Fleming when he decided to go into academic work, the promise of £100 a year for living expenses, and the prospect of getting the special transformers, with their huge cores, made by Metro-Vick without labour charges, so that the cost to the Cavendish would be very small.

John Cockcroft had preceded Allibone from Metro-Vick to the Cavendish, but that had been at the very start of the decade and Cockcroft's work had been largely in helping Kapitsa with his very large electrical machines. Allibone's was the first proposal for bringing heavy electrical equipment into the mainstream of Cavendish nuclear physics, and it was the start of a close cooperation between Metro-Vick and the Cambridge scientists. Later the company was to manufacture the large electromagnet which replaced Kapitsa's equipment in the Mond, and there were many orders placed with them for the high-voltage and high-vacuum equipment which was needed as the concept of modern nuclear accelerators and cyclotrons slowly took shape in the hands of the scientists.

Eventually five men followed the same route from the Metro-Vick laboratories, through the Cavendish to end up as Fellows of the Royal Society, and the links between the Manchester engineers and the Cambridge scientists were much firmer and more common than any of Rutherford's scientific biographers seems to have cared to mention. Rutherford, for instance, performed the formal opening of the new Metro-Vick High Voltage Research Laboratory in 1930, when its director was Brian L. Goodlett, a man born in Russia who had shot his way out of the Revolution, become a naturalised Englishman, and was deeply involved in making the equipment that Cockcroft and Walton designed for their first nuclear accelerator. When Goodlett himself decided to turn to academic research at Cambridge, Allibone returned to Manchester, at the end of 1930, to direct the industrial laboratory.

Rutherford took a great interest in Allibone's project, and personally showed him round the Cavendish when he arrived in order to find out if the room allocated to him would be high enough to contain his proposed high-voltage transformer. Allibone agreed that it would do for at least the provision of half a million volts. The actual height of the rooms in the old laboratory now became a matter of real significance. It was only when a large and high lecture room was vacated by the engineering department and became available to the physicists that Cockcroft and Walton were able to contemplate building a machine which might generate a potential difference of one million volts, the machine that was to lead them into successful acceleration of nuclear particles.

Certainly in 1927 Rutherford believed that the generation of very high voltages for the acceleration of particle-projectiles was a possible line of advance; as has been seen earlier, he devoted his annual address as President of the Royal Society that year to this subject, a complete break from the general line of his public speeches at the time.

Probably the greatest single criticism which can be levelled at Rutherford concerns his apparent reluctance to invest in big machines for the Cavendish, a failure which led, some have averred, to the loss of the British lead in nuclear physics to the American universities during and after the Second World War. It is not unreasonable to argue that Rutherford refused to be dragged into, or to lead the way into, the world of Big Science. Some critics – mostly Americans – have even written in such terms as that he "had no interest whatever in technology and technical problems. It even seemed that he nursed a prejudice against them . . ." The record, however, shows that Rutherford encouraged and supported Kapitsa's development of heavy electrical engineering to produce intense magnetic fields, and later the replacement of his equipment by the largest electromagnets that could be constructed at that time. He supported the first provisions of high-voltage equipment in any British university laboratory, and he encouraged and supported Cockcroft and Walton in the building of the world's first nuclear accelerator. He had Oliphant build a smaller nuclear accelerator on which he worked himself. The machines he had built in the Cavendish from 1927 to 1932 are probably the original inspiration for the pictorial fantasies of the film of H. G. Wells' *The Shape of Things to Come*.

Even more important in establishing Rutherford's attitude towards technology as such were two further developments that he nurtured in the Cavendish at exactly this time. The first was the development of the electrical counter. Although based on an invention by Greinacher this was developed in the Cavendish by a team led by the ebullient Welshman, C. E. Wynn-Williams, with W. Bennet Lewis, who later made a distinguished name for himself in the Canadian nuclear physics field. Every modern physics laboratory nowadays contains innumerable counters; the steady flicker of their illuminated numbers as they record events occurring perhaps every millionth of a second is such a normal background that we scarcely realise that they are there – and they seem to indicate little more than that the main machinery is "on" and working.

The devices developed in the Cavendish around 1930 used what now seem incredibly clumsy, yet delicate, thyratrons and similar valves, but they were the first and true ancestors of the modern counters, and they were developed under Rutherford's auspices. Who, better than he, knew how much they improved the physicist's lot from the wearisome and tiring strain of trying to count minute scintillations on a flickering screen in a light-proof box? There is a famous snapshot of Rutherford, cigarette in mouth, talking to Ratcliffe in one of the rooms of the Cavendish in 1932. Directly above his head is a home-made illuminated sign reading, "Talk SOFTLY Please". Wynn-Williams himself took the shot, switching the warning sign on especially to make the picture. The point of the composition was not, as many have assumed, to show a great scientist in action; it was a joke against Rutherford, for just as everyone in the Cavendish knew that the earliest counters were so oversensitive that they would provide false recordings if they were upset by loud extraneous noises or thumps, so everyone knew that it was quite impossible for Rutherford to talk softly at any time; he always boomed. The job of the electrical counters was simple: they counted and added up the arrivals of particles in a detection chamber. Chadwick's papers on the discovery of the neutron mention clearly that he used electrical counters in his work, and that is almost certainly the first instance of the use of counters in any major discovery in physics. This proves the Cavendish was right at·the forefront of technological development in laboratory techniques.

The second development was that of automating the "Wilson cloud chamber". This was done finally by Blackett in the Cavendish, building on earlier work done in the same laboratory by the Japanese, Shimizu. "Wilson's expansion chamber is to the atomic physicist what the telescope is to the astronomer," according to W. B. Lewis. Blackett's success came through linking the newly-developed Geiger counter, which signalled the arrival of an ionising particle, to a rapid expansion of the gas inside the chamber, so that the particle would reveal its presence, line of flight, energy, and nature, by ionising the expanded gas. The development was completely a Cavendish piece of work, and it is, as we have noted earlier, the ancestor of all the modern "bubble chambers" used all over the world to examine the results produced by our huge nuclear accelerators.

Therefore we must accept Cockcroft's statement that "Rutherford would never have objected to trying something new in the way of innovations". On the other side of the coin, we have to accept the evidence of almost everyone who worked with him that Rutherford was notoriously impatient with machinery and intolerant of all delays caused by machinery. More than that, those who knew him well, such as Kapitsa, realised that he was totally uninterested in machinery as such. What Rutherford wanted were results; results that contributed towards knowledge of the behaviour of atoms and the structure of nuclei; figures and measurements that could be fed into that model, concrete and quantitative, that existed in his mind of the micro-universe of the atom.

This attitude is crucially important in understanding the most important evidence against Rutherford – the evidence about his reluctance to build a cyclotron in Cambridge. There is no doubt that Rutherford had serious disagreements with his closest collaborators and colleagues, Chadwick, Cockcroft and Oliphant, over this subject, but the matter will be treated more fully later in this chapter.

More relevant to his attitude towards heavy investment in machinery was his attitude to money. Three strands of evidence have bearing on this: the evidence of Chadwick and Oliphant that Rutherford was disturbed by the criticism which went on right into the 1930s that he was educating young physicists in the wrong way; the evidence of Chadwick about Rutherford's complete unwillingness to beg for money or to accept the responsibility for its use; and the evidence of C. P. Snow, of the deep streak of diffidence and unease within the man. It was not a prejudice against technology which made Rutherford slow to develop the Big Science possibilities of the Cavendish, it was the lack of self-confidence in his own ability to manage this type of development. When the resources were made available in 1935 by the Austin donation, Rutherford showed a distinct disinclination to work on the planning of the new Cavendish himself and delegated the task largely to Cockcroft and Oliphant.

By 1930, although Allibone had got his high-voltage machinery working and had produced potential differences well in excess of the half a million volts he had promised, it had become clear that accelerating electrons through hydrogen was not producing the evidence for the neutron that was one of its objectives. Technically, too, they had discovered that there was as much difficulty to be encountered in maintaining high vacua in the accelerator tubes as in achieving high voltages. Two other important developments also took place on the technical side. E. T. S. Walton joined the Cavendish from Dublin with a project for accelerating electrons in a different way, by circulating them in a changing magnetic field. In other hands this later became the successful device called the betatron, but in 1929 and 1930 Walton failed because there was neither the theory nor the technology available for focusing the beam of electrons. Secondly, Oliphant, who had started by working on the effects of ions striking surfaces, had discovered how to accelerate positive ions with discharge tubes.

The final thread which led to the totally artificial transmutation of atoms was provided by the Russian theoretician, Gamow, the same man who was later believed to be an indirect cause of Kapitsa's being detained in Russia. In 1929 Gamow was a true wandering scholar, going from Göttingen to Niels Bohr in Copenhagen, and being sent by him to Cambridge. Gamow has told how he literally did not have the price of a good night's accommodation when he first called on Bohr. He has also stated clearly that it was an otherwise undistinguished paper of Rutherford's which he came across in Göttingen that gave him the first of his important ideas. As he described it himself:

I was sitting in Göttingen . . . at this time everybody was applying wave mechanics to atoms and molecules and getting more and more complicated and I hate such things . . . and I was . . . trying to think how the alpha-particle comes out and maybe spirals and so on and why it takes such a long time. And then in the library at Göttingen I found the *Philosophical Magazine* with the article of Rutherford in which he was shooting the fast alpha-particles . . . on uranium and hoping to observe the abnormal scattering and there was none. It was inverse square scattering which was a contradiction. And Rutherford in this article proposed a theory which was quite amusing. You see Rutherford had thought about neutrons . . . before neutrons were really discovered and the Cambridge people were doing experiments, never published, which tried to kick neutrons out of the nucleus and failed.

Rutherford's theory, in this paper,[1] tried to deal with the problem of how the alpha-particle managed to get out of the nucleus despite the fact that the energy equations seemed to show that it would need to have more energy to break out than it possessed when it emerged; connected with this is the problem of where and how the alpha-particle collects the two electrons which make it into a helium atom. The answer, Rutherford suggested, was that the alpha-particle emerges as a unit of four neutrons (though he did not use this word). Four neutral particles would need very much less energy to break out of the electrical fields of the nucleus. He furthermore regarded each of the neutral particles as a close combination of the positive proton and the negative electron. At some stage early in its emergence, Rutherford proposed, two of the electrons were dropped off and fell back into the nucleus, leaving the alpha-particle with its mass of four units and its electrical charge of two positive units that Rutherford himself had identified so many years before.

Gamow could not understand where or how the two electrons

[1] "The Structure of the Radioactive Atom and Origin of the Alpha-rays", *Philosophical Magazine*, September 1927.

got accelerated back into the nucleus: "At this point when I read this, I thought 'My gosh it isn't'. It is exponential solution of the equation of Schrödinger." (He refers here to the famous Schrödinger equations, published only two years earlier, which implied that matter could be regarded as a probability wave-function.) And Gamow claims that in twenty-four hours he had written his famous paper which shows that, in order to leave a nucleus, a particle does not necessarily have to possess enough energy to "leap over" the full energy barrier that the electrical forces of the nucleus provide. It is perfectly possible for a particle of lower energy to "tunnel through" the energy barrier to escape from the nucleus. In fact two Americans, Gurney and Condon published similar ideas in *Nature* in September 1928. But Gamow himself came to Cambridge, bearing his new idea, in the first month of 1929. He spoke of his theory at a meeting of the Kapitsa Club and also at a Royal Society meeting. (Later in 1929 he started a full year at the Cavendish supported by a Rockefeller grant which Bohr had arranged for him.)

It was Cockcroft who made the crucial leap of the imagination, by putting Gamow's theory into reverse: if a particle could tunnel out of a nucleus then a particle must also be able to tunnel into a nucleus. This meant that a proton with an energy of only a few hundred thousand electron volts should be able to enter a nucleus by tunnelling, whereas Rutherford and all the others had been thinking that particles must be accelerated to many million volts to achieve more disintegrations of nuclei than natural alpha-particles, which themselves had energies of a few million volts. But because only some particles could do the tunnelling – it was only a statistical probability that such events would occur – it would need a beam of many thousands of millions of protons to produce an observable number of disintegrations. Rutherford was immediately impressed by Cockcroft's rough calculations and gave him permission to go ahead. Walton, who had started work on accelerating caesium ions in a discharge tube after the failure of his electron acceleration plans, joined Cockcroft in what was to develop into a three-year slog. "With a miscellaneous collection of bits and pieces they assembled a voltage-doubling rectifying circuit, which they connected with an evacuated accelerating tube of the type developed by Allibone," Oliphant remembers, and with this apparatus "they obtained erratic beams of a few microamperes and bombarded elements from all parts of the periodic table". They were primarily looking for gamma-rays, which theory predicted should emerge from nuclei which had captured an incoming proton. Walton did the searching using a gold-leaf electroscope, an instrument that was considered fairly primitive in the days when Rutherford had first studied radioactivity, but Walton insists that he "had had considerable experience with such an electroscope and it was a simple experiment to try". On their best days they managed to accelerate protons to 280 thousand electron volts but, says Walton, they "did not get around to looking for scintillations produced by particles emitted in a nuclear reaction. We would have had to lie on the floor and peer into a microscope – not a relaxed position for reliable counts."

Everyone who remembers those days agrees that a turning point was reached by the good fortune of being able to move into the disused lecture room with a higher ceiling. Cockcroft had developed a new voltage-multiplying circuit which could quadruple the voltage. There were fifteen-inch glass cylinders stacked on top of each other carrying the rectifiers and steel tubing for electrodes. New oil diffusion pumps provided the vacuum system, and sheets of galvanised iron separated some of the components. One of the crucial features of the apparatus was a new form of plasticine which could make airtight joints at low vapour pressure. "There were frequent vacuum troubles due to heat softening the plasticine or to puncturing one of the glass cylinders by a spark. Cockcroft and Walton spent a large part of their time perched on ladders, locating and repairing such leaks or just rubbing over every plasticine joint with their fingers in the hope that they would eventually make the system vacuum tight again," observed Oliphant. Each time there was a breakdown they had to go back to square one and build up the voltage slowly again. The move into the lecture room had been made in the middle of 1931, but it was not until the very end of that year that the apparatus could be persuaded to give steady voltages of five or six hundred thousand volts. The first months of 1932 were spent in making the apparatus more reliable and doing the vital first measurements on the range and velocity of the protons they were successful in accelerating.

By the end of March and the first week in April, Rutherford was beginning to get impatient with Cockcroft and Walton and their unstable machinery – it was the old complaint; he wanted results, he wanted something bombarded. It seems certain that it was Rutherford who suggested that the first target should be the light element lithium. Dee recalls that Rutherford was "itching" to shoot protons at a target and that he "rather bullied" Cockcroft and Walton into putting in the lithium target, which "rather annoyed" them because they wanted to complete their measurements on the protons themselves. Oliphant seems to have similar memories: "Rutherford was impatient while these [proton] experiments were being done. He was not interested in the technique but wanted to know whether the beam could produce any nuclear effects." Walton's memories are slightly different: "My recollection is that we did not spend a long time on this [the proton measurements], and my notebook records only two days on it. Rutherford told us we should go straight to looking for scintillations. I don't agree that he 'bullied' us – a word which suggests resistance on our part. There was no such resistance."

It is beyond dispute however that Thursday, April 13th, 1932, was the crucial day. The lithium target was already in place when Walton gave the machine its morning conditioning and started it up – Cockcroft was working with Kapitsa that morning. Walton recalled:

> When the voltage and the current of protons reached a reasonably high value, I decided to have a look for scintillations. So I left the control table while the apparatus was running and I crawled over to the hut under the accelerating tube. Immediately I saw scintillations on the screen. I then went back to the control table and switched off the power to the proton source. On returning to the hut no scintillations could be seen. After a few more repetitions of this kind of thing, I became convinced that the effect was genuine. Incidentally these were the first alpha-particle scintillations I had ever seen and they fitted in with what I had read about them. I then phoned Cockcroft who came immediately. He had a look at the scintillations and after repeating my observations he also was convinced of their genuine character. He then rang up Rutherford who arrived shortly afterwards. With some difficulty we manoeuvred him into the rather small hut and he had a look at the scintillations. He shouted out instructions such as "switch off the proton current"; "increase the accelerator voltage", etc. but he said little or nothing about what he saw. He ultimately came out of the hut, sat down on a stool and said something like this: "Those scintillations look mighty like alpha-particle ones. I should know an alpha-particle scintillation when I see one for I was in at the birth of the alpha-particle and I've been observing them ever since . . ." He did not stay very long but came back the next day and did some scintillation counting. He did about a dozen counts of which the details are in my notebook.

During this second visit on April 14th Rutherford did a most unusual thing. He swore Cockcroft and Walton to strict secrecy, an act which was quite opposed to the practice and ethos of the Cavendish. Other rules were also broken – late evening working was permitted to the two men, against the strict discipline of the six o'clock closure of the lab which Rutherford normally enforced. Dee, the master of cloud chamber photographs, was drafted in to get pictures of the particles produced by the accelerator bombarding the lithium target, to prove that alpha-particles were indeed being produced. The phenomenon that they were observing was that lithium nuclei of mass 7 were catching the accelerated protons of mass 1, and disintegrating into two alpha-particles, each of mass 4. Not only were the simple masses in good accord, but the energies of the particles, which they were able to measure, fitted well with the equations generated by applying Einstein's now famous $E = mc^2$.

The secrecy Rutherford imposed worked well by preventing interruptions from visitors, and that weekend, late in the evening, Cockcroft and Walton were able to go round to Rutherford's house and draw up a letter to *Nature* announcing their results and their great discovery. The work, however, speeded up rather than slowed down in the weeks immediately following. Cockcroft and Walton's first major paper – communicated by Lord Rutherford – reached the Royal Society on June 15th, and by that time they had already tested targets of fifteen other elements, varying from the lightest, beryllium, to the heaviest of all elements, uranium. All produced some alpha-particles when bombarded by protons, but the most spectacular results, certainly disintegrations, came from lithium, boron and fluorine. There was, however, a significantly large amount of alpha production from a uranium target as well. The atom could now be split at will.

There was a third important achievement by the Cavendish before the year 1932 was out. In Pasadena, C. D. Anderson had been taking cloud chamber photographs of the particles produced in the atmosphere by the impact of cosmic rays from outer space. He produced photographs of tracks which were best explained by proposing they were caused by a new particle, the positively charged electron, or positron. Such a particle had been predicted by the Cambridge theoretician P. A. M. Dirac, as a result of his work on bringing together the wave theories and the quantum mechanics which had been developed in European centres. Furthermore a number of mysterious cloud chamber photographs had been in circulation among the fraternity of physicists for three or four years: a Russian named Skobelyzin had shown such tracks at an international conference in Cambridge in 1928, tracks which puzzled everyone including Chadwick, for they seemed to show an electron "going backwards the wrong way". There was an implication in such theories that mere radiation, such as gamma-rays, could turn into "matter" in the shape of creating a pair of particles, an electron and a positron. There was also the implication, especially from Dirac's work, that a positron represented "anti-matter". It was at this point that the automated cloud chamber developed by Blackett and Occhialini in the Cavendish came into its own. Before 1932 was out they had shown that positrons did exist; they had fully confirmed Anderson's results; they had demonstrated that positrons could be created and identified in the laboratory and they had established the main outlines of their properties.

According to C. P. Snow:

The year 1932 was the most spectacular year in the history of science. Living in Cambridge one could not help picking up the human, as well as the intellectual, excitement in the air. Sir James Chadwick, grey-faced after a fortnight of work with three hours' sleep a night, telling the Kapitsa Club how he had discovered the neutron; P. M. S. Blackett, the most handsome of men, not quite so authoritative as usual, because it seemed too good to be true, showing plates which demonstrated the existence of the positive electron; Sir John Cockcroft, normally about as given to emotional display as the Duke of Wellington, skimming down King's Parade and saying to anyone whose face he recognised "We've split the atom! We've split the atom!". It meant an intellectual climate different in kind from anything else in England at the time. The tone of science was the tone of Rutherford: magniloquently boastful – boastful because the major discoveries were being made – creatively confident, generous, argumentative, lavish and full of hope. The tone differed from the tone of literary England as much as Rutherford's personality differed from that of T. S. Eliot or F. R. Leavis. During the 'twenties and 'thirties Cambridge was the metropolis of physics for the entire world.

Certainly in scientific terms 1932 put Cambridge clearly in the forefront. Rutherford and his "boys" and their results were the main attractions at the Solvay Conference of 1933 and the International Physics Conference in London in 1934, where London acted as a sort of anteroom for the most distinguished visitors to be taken up to Cambridge and shown round the Cavendish. In more practical terms the Joliot-Curies drove along their old lines with even greater determination and discovered artificial radioactivity, the fact that some nuclei could be transformed into radioactive isotopes rather than disintegrating when bombarded. Chadwick has recorded his chagrin at "missing" this when it was perfectly well within the powers of the Cavendish to have added this to their list of triumphant "firsts". Perhaps more important, Cockcroft and Walton's results spurred the American teams of Ernest Lawrence and Van dem Graaff to use their accelerators to bombard nuclei and look for disintegrations, and then to go on to build even more powerful accelerators. The shift of leadership towards the bigger resources and less cramped spaces of the USA began at this time, and within a couple of years Cockcroft would find himself touring American universities to learn their techniques for achieving greater accelerations.

It was at this period, too, that Rutherford began to show a return to something like his former rather arrogant attitude towards other branches of science; on occasions he could drop out of his role as the leader of science in general in British society, and revert to the narrower view of himself as leader of physics, nuclear physics in particular and the Cavendish "tribe" most of all. Lord Snow tells of the day when he was the only competitor against the Cavendish's Philip Dee for a studentship and Rutherford was on the appointment board. "Rutherford took out his pipe and turned on to me an eye which was blue, cold and bored. He was the most spontaneous of men; when he felt bored, he showed it. This afternoon he was distinctly bored. Wasn't his man, and a very good man, in for this job? . . . What could spectroscopy tell us anyway? Wasn't it just 'putting things in boxes?'." Snow's story continues with a description of how he defended spectroscopy vigorously, how Dee (much the finer scientist, Snow admits) got the award, but how Rutherford trumpeted about Snow's "spirit" and supported him for the next available studentship. Here, however, the point of the story is Rutherford's disregard for allied branches of science. Sir Frederick Dainton has a very similar story to tell of a meeting at about the same time. The young Dainton was starting his research career in the Physical Chemistry Department, in a building adjacent to, but not part of, the Cavendish Laboratory. He was sent to Rutherford to obtain the use of a device which would enable him to control the temperature of a reaction with great accuracy, but found that he had also walked into a storm, the great man declaring that Britain did not even have a "best" school of chemistry, for Oxford chemistry was no good, and Cambridge chemistry was "as dead as a dodo", and for physical chemistry and the related colloid science, Rutherford had only a profound contempt. A fairly vigorous defence of chemistry earned the young man the right to borrow and use the temperature control machine, however, though Dainton doubts whether it was worth expending the courage required to retort to Rutherford, for the machine was not much use in his experiments when he got it.

More serious than these small personal incidents was the extension of Rutherford's dislike of chemistry and spectroscopy to cover the whole work of Bernal and the school of X-ray crystallography which he developed. There can be no doubt but that Rutherford let his scientific judgment be distorted by his personal dislike of

Bernal, for the science of X-ray crystallography had been founded by his friends, the Braggs, and was to be the route by which the Cavendish again became one of the world's most important laboratories after the Second World War.

Those who knew Rutherford well recall that J. D. Bernal was one of the two men whom Rutherford loathed – Lindemann being the other. Bernal's open Communism and espousal of left-wing causes did not endear him to Rutherford, and his personal life probably did not recommend him to the sexually diffident Rutherford either. But it was Bernal's scientific method that was at the core of the problem; he was a man who "sprayed out ideas", but left it to others to pick them up and work them out. "Rutherford loathed this impressionistic mode of research. He demanded that the author of the bright idea should prove it by elegant experiment and irrefutable logic," according to the official history of the Cavendish Laboratory (although it must be admitted that the author of this history, J. G. Crowther, was a former ally of Bernal). Nevertheless, the history goes on, "Rutherford could not abide the untidy prophet of the permissive society and he transferred his distaste to the sciences that happened to be associated with him. Bernal's laboratory was starved of resources . . ." A recent biography of Bernal adds to this picture: "Bernal had to be circumspect. Rutherford, when in control of the Cavendish Laboratory, dominated an atmosphere in which speculation was frowned upon. Not only did he dislike Bernal's politics, he warned with great vigour, 'Don't let me catch anyone talking about the universe in my laboratory'." Crowther recorded being present when Rutherford called Bernal into his office and rebuked him severely because something was not as clear as he wanted it: "Bernal stood by, hanging his head but still unable to forget the complications. It was a clash of temperaments and ways of looking at things."

It was Bernal's inspiration that showed the way to the X-ray analysis of large molecules, such as proteins, of which living matter is made. Molecular biology and the whole of the wider field of bio-physics can look to him as one of its chief founders; and his ideas, spread through bodies such as the Theoretical Biology Club, propagated such notions, of which Rutherford might well have approved, that physics is at the basis of biology and that biology can and must be studied by physical methods. Rutherford, who was capable of telling one of the founders of X-ray analysis, at the Trinity High Table, that his science was "not fundamental physics", opposed the suggestion that a Chair of Crystallography should be set up, and at one time even seems to have wanted to throw Bernal and his small team out of the Cavendish altogether. Although much of the evidence for the disagreements between the two men comes from the one source, Crowther, there is the single, most telling, letter from W. L. Bragg, written in 1931 from a holiday address in Munich, although his official position was as Rutherford's successor in the Physics Chair at Manchester.

An official report to the University Council had proposed the down-grading of the crystallography unit, which, although part of the Cavendish Laboratory, had a certain constitutional independence. Bernal "seems utterly discouraged by the report", wrote Bragg and he appealed to Rutherford for support for both Bernal and his unit, and for giving Bernal a more independent position. European laboratories and universities gave great support and fine facilities to the whole subject of crystallography and X-ray analysis, Bragg continued and added, "I do think the work has a tremendous future and all the more so because it lies on the borderline between physics and chemistry and so is often not directly backed by either." Pleading for Bernal himself, Bragg, who seems to have been aware of Rutherford's personal antipathy to the man, wrote that he was "extraordinarily good, though he talks rather much sometimes". It has to be said in Rutherford's favour that he did keep the crystallographers going, though in poor physical accommodation. Although Bernal left for Birkbeck College in London in the late 1930s, there was enough left for Bragg to rebuild immediately he himself took over in Cambridge.

Rutherford's influence on physics, and science in general, was by this time literally world-wide. It was achieved through the implications and extensions of his own work, and also through the vast network of his correspondence. There can have been very few major physicists anywhere in the world who were not in touch with Rutherford in one way or another. His correspondence in a couple of typical years at this stage of his life, 1933 and 1934, shows him in communication with Ernest Lawrence, and Urey and Lewis and Gamow in different places in the USA stretching from Boston to California, while he is also in touch with Fermi in Italy, just starting his experiments with slow neutrons, and with Born in Germany threatened by anti-semitism; he is consoling his wartime colleague, R. W. Boyle, over the financial difficulties besetting the newly founded National Research Laboratories in Canada, and receiving complaints from a former pupil, R. R. Nimmo, who is in danger of being sacked from the University of Western Australia where "the university as a whole is permeated with the idea of doing work of direct economic value to the state and hence instead of trying to foster in a few students a love of learning for its own sake we teach them subjects that are likely to help them to obtain a post after-wards". From even further away his faithful New Zealand protégé, C. C. Farr, is sending him appeals for support for a magnetic observatory at Apia in Samoa. Also in these years he returned to his old intimacy with Niels Bohr, for now that the structure of the nucleus brought together the experimental physicist and the theoretician at the frontier of knowledge, Rutherford visited Copenhagen and lectured there and Bohr visited Cambridge almost every other year.

In that famous Bakerian Lecture given by Rutherford in 1920 in which he predicted the existence of the neutron, he also predicted the existence, or at least suggested the possibility, of "heavy hydrogen" – an atom of mass 2 and atomic charge 1, or an atom of hydrogen which had a neutron as well as a proton in its nucleus. The existence of such an atom in nature was proved by Harold Urey in 1931 by spectroscopic evidence – and Rutherford did not grumble at this piece of spectroscopy. At Berkeley G. N. Lewis set about obtaining specimens of the new type of atom and succeeded by using a process of electrolysis – the basic process which is still used commercially for separating "heavy water" from normal water. Samples of heavy hydrogen atoms were soon made avail-able to Ernest Lawrence for trial in his cyclotron, and then Lewis presented Rutherford with a sample. In his covering letter he wrote that this piece of generosity was "in case someone in the Cavendish Laboratory would like to experiment on the new pro-jectile". Lawrence, he added, had "already found that it is going to be a remarkable aid in the study of the nucleus".

In his letter of May 15th, 1933, Lewis also said that they had decided to call the new particle a "deuton". A friendly controversy followed, for Rutherford objected that the name sounded like "neutron" spoken by a man with a cold in his head. He favoured "diplon" with "heavy hydrogen" called "diplogen". The Amer-icans won the little argument, more by having the words in regular usage than by right of discovery, though they gave a little to Rutherford's objection and we now call the particle a "deuteron" and the substance "deuterium".

Rutherford had been impressed by Gamow's argument about the tunnelling power of comparatively low-energy particles, and, of course, by Cockcroft's success in putting the theory to work with such dramatic results. The same calculations, now based on ex-perimental facts, showed that even lower-energy particles should be able to penetrate into atomic nuclei, so Rutherford persuaded Oliphant to build a new particle accelerator, less powerful than the Cockcroft and Walton machine, but capable of producing many more particles, even if these projectiles were travelling with less energy. It even proved possible to build this machine in the existing low-ceilinged rooms of the old laboratory. As Lewis had suggested, the deuterons proved most valuable projectiles, and the last three years of Rutherford's scientific life were occupied almost entirely in experiments with deuterons accelerated in his low energy machine. A simple count of the scientific papers he published shows this final acceleration of his output quite clearly. Seven papers on this subject were written in his last three years, all of them producing new and significant results, ending in his final paper with his last triumph, the discovery of the "Isotopes of Hydrogen and Helium with mass 3" – which we now know as "tritium", the material of the hydrogen bomb and of nuclear fusion, and helium three, the happy hunting ground, still, of those scientists who specialise in low-temperature studies near absolute zero.

Shortly after Lewis had sent him the first samples of "heavy hydrogen", and when he had only just started his own experiments with it, Rutherford was giving his usual series of lectures at the Royal Institution, a series entitled "Transmutation of Matter". His lecture of March 23rd, 1934 was originally billed as "Heavy Hydrogen" but Rutherford changed the title (in his own hand) to the easier "The New Hydrogen". He seems to have understood instinctively that for a lay audience, in an era when matters nuclear were totally unfamiliar, there was grave difficulty in arousing interest in the idea of "a new isotope of hydrogen consisting of a nucleus which contained both a proton and a neutron". He chose instead to lead his audience into the new domain through the use of a medium which they all understood – water.

"For more than a century scientific man believed with confidence that pure water was a well-defined chemical substance, H_2O, of molecular weight 18," he began, and explained that water had been so well understood that it was even made the basis of all standard weights and measures – the kilogram, being defined as the weight of 1000 cc of water. More accurate measurements showed that this was slightly in error and the world standard was changed to a lump of platinum-iridium metal, but no one thought this was of any fundamental importance. "It was only about four years ago that this confidence was slightly disturbed," he continued, and this disturbance came when it was discovered that all oxygen atoms were not identical, that in addition to the normal oxygen atoms of mass 16, there were naturally occurring isotopes with masses 17 and 18 in very small proportions. This was still a matter of only theoretical interest and of no great practical importance; but in the last two years there had come "a revolutionary change in our ideas of the constancy of the constitution of water". This, of course, was the discovery of an isotope of hydrogen with twice the normal weight – only about one in six-thousand hydrogen atoms was of the new type, but the difference between these and normal hydrogen was so great that the new atoms could easily be concentrated (Lewis's electrolysis process) and we could obtain "heavy water", water that was more dense than ordinary water, and which boiled and froze at different temperatures from ordinary water.

The "almost romantic history of this rapid advance in knowledge" was primarily due to the "recognition of the importance of small differences observed in accurate measurements of density" – and he drew the clear parallel with the discovery of argon. Since this sort of science was not Rutherford's style or interest, the tribute was generous and shows that he was, at bottom, perfectly willing to accept the contributions of other forms of questioning nature. He went on to give even more generous praise to Urey and Lewis, congratulating "our American colleagues for the masterly way they have opened up and so rapidly developed this new field of knowledge" and thanking Lewis publicly for sending samples.

Rutherford emphasised the biological importance of deuterium first – "its effect on animal and plant life" – in which his sureness of touch now seems lacking, but he was on much firmer ground when he spoke of the deuteron in his own, nuclear-physics, terms. It could be made from all sorts of combinations of electrons, positrons, protons and neutrons, he allowed cautiously, but "if we assume, as seems not unlikely, that the D-nucleus consists of a close combination of a proton with a neutron" it should produce some very interesting results when used in accelerators either as projectile or as target. So far, he admitted, experiments were "inconclusive" but deuterium had already been shown to be "a new form of projectile which has proved markedly efficient in disintegrating a number of light elements in novel ways".

By this time Rutherford, well into his sixties, had changed his working style. He was, quite literally, a "thundering nuisance" in the laboratory, though his enthusiasm still drove him and his team onwards and his inspiration and intuition led to real results. His hands, by now, shook so badly that he had great trouble in fixing the little sheets of mica in front of the window of the accelerator to act as absorbers. This could often lead to his shouting at Crowe, the technical assistant whose skilled hands had made the mica sheets. Afterwards, of course, there would come the apologies, for Crowe's hands, too, were losing their skill, though in his case this was due to radiation burns which were slowly consuming his finger ends. Eventually Crowe had to wear leather gloves throughout the day, and a long series of operations and grafts failed to clear up the lesions. For a short time Rutherford and Oliphant used the old scintillation methods for counting the particles produced by nuclear disintegrations in their accelerator, but soon they switched to the new electronic counters, which recorded the passage of particles on long strips of film, which was developed and fixed in the laboratory.

Oliphant has recorded:

> Rutherford's participation in the experiments was limited to discussion about what to do next and to deep interest in the results. He gave us a completely free hand in the design of experiments and running of the equipment but he kept us on our toes all the time. Like all Cavendish equipment at the time, ours was hastily assembled from whatever bits and pieces were available so that it often gave trouble. Rutherford was very irritated by delays of this kind but was singularly uninterested in finding the money to buy more reliable components. However he was extremely pleased when things went well, giving us a triumphant feeling of something accomplished.

On normal days Rutherford would visit his machine twice, at the opening of proceedings and finally just before "closing time" at six o'clock. He might well turn up at other times if an exciting stage was being reached and then there was often a catastrophe: "On two occasions Rutherford himself, whose hands tended to shake, pushed something through the mica window . . . letting air rush into the apparatus and creating panic till we had the oil pumps cooled down and everything shut off. He was humbly apologetic, but disappeared for hours, or even days, while we cleaned up the mess and got going again."

With the photographic recordings Rutherford was even more dangerous, if they promised to hold anything of interest. He would not wait for the strips to be properly fixed and dried, and he dribbled pipe-ash all over the photograph, and fixing solution all over his clothes. His stubby pencil messed up the still-damp emulsion and sometimes he even got the long strips tangled in his feet and trod them into the ground until they were completely ruined. On occasion his research team would deliberately try to hide the recordings from him. Oliphant recalled:

> Once, at the end of a particularly heavy day, when the experiments had gone well, we decided to postpone development till next morning when we were fresh and we could handle the long strip in new developer and fixer without damage. Just as we were leaving Rutherford came in. He

became extremely angry when he heard what we had decided, and insisted that we develop the film at once. "I can't understand it," he thundered. "Here you have exciting results and you are too damned lazy to look at them tonight." We did our best, but the developer was almost exhausted and the fixing bath yellow with use. The result was a messy record which even Rutherford could not interpret. In the end he went off muttering to himself that he did not know why he was blessed with such incompetent colleagues.

The apologies came by phone that evening and the following morning the great man was even more contrite, for his wife had pointed out that he had also ruined his suit with the wet chemicals.

The other side of this coin was Rutherford's contribution to understanding what was going on. Among the showers of particles that were produced from the reactions of deuterium on various targets there were various groups of different ranges and different energies. Some could be seen to be neutrons; others protons; others mysteries. It was here that Rutherford's "intuition" came into play, though Oliphant believes it was a profound and quantitative understanding of the way atomic nuclei might be expected to behave. When they believed they were on the track of tritium, the hydrogen isotope of mass 3 with a nucleus which consisted of two neutrons bound to the normal proton, "it was impressive to experience Rutherford's enthusiasm and the extraordinary process whereby he calculated by approximate arithmetic, the range-energy relationship of tritium nuclei from the known range-energy curves for alpha-particles and protons."

Later they prepared an even thinner mica window which would enable them to observe particles of very short range, and they duly found particles with a double charge, just like alpha-particles, but all mixed up with protons and tritium nuclei, where no alpha-particles should have been. This caused "consternation among us".

Rutherford produced hypothesis after hypothesis, going back to the records again and again and doing abortive arithmetic throughout the afternoon. Finally we gave up and went home to think about it. I went all over the afternoon work again, telephoned Cockcroft who had no new ideas to offer, and went to bed tired out. At three o'clock the telephone rang . . . my wife . . . came back to tell me that "the Professor" wanted to speak to me. Still drugged with sleep I heard an apologetic voice express sorrow for waking me, then excitedly say "I've got it. Those short-range particles are helium of mass 3." Shocked into attention, I asked on what possible grounds could he conclude that this was so, as no possible combination of twice two [deuterons hitting deuterons] could give two particles of mass 3 and one of mass unity. Rutherford roared: "Reasons! Reasons! I feel it in my water!"

Once again he was right and it needed only one more morning's work for the group to prove that he was so.

This story is a version of the archetypal scientific success story – a version of Archimedes leaping out of the bath and shouting Eureka. It is important that it should not be seen in so simple a light – it was much more a case of the elimination of arithmetical, quantitative impossibilities. The reaction of deuteron on deuteron, we now know, could produce either a proton and a tritium nucleus or a neutron and a helium 3 nucleus with the same probability which was scientifically unexpected, but arithmetically possible, and quite easy to "see" for someone who liked "concrete" models of atomic behaviour.

Although he believed that publicity rarely helped in a controversial situation, Rutherford had no inherent dislike of journalists and communication with the public on the subject of his great discoveries. Lord Ritchie Calder was one of the first journalists to write about scientific matters for the popular press and he visited Rutherford and the Cavendish on several occasions, finding that "one of the things about which he was embarrassingly blunt was the misdemeanours of those who professed to make science popular". Rutherford made no attempt to avoid such writers, but kept a rough file of cuttings, some of which were even in his pocket, and which he produced for Calder, describing them as "rot" and "drivel". It was useless trying to explain nuclear physics to the public he boomed, journalists would always want to write about the atom blowing everything up or driving engines. Then he pulled out of his desk drawer a draft of his latest scientific paper "full of signs and formulas", thrust it at the journalist and said, "Try and translate that to the public". As always Rutherford reacted well to "spirit", and he roared with laughter and shouted "It's a deal" when Calder responded by thrusting his shorthand notebook back at the scientist and challenged him to translate it. Then he settled down to try to explain what was being done, and he produced a wonderful image for the neutron, which left no tracks in the cloud chamber, but "you could still follow its movement through; it's as though Wells' Invisible Man were playing in a football match – you couldn't see him but you could follow his passage up the field by the players staggering out of his way as he charged them."

Yet Rutherford was also the man who could say on another occasion that if a piece of physics could not be explained to a barmaid it was not a very good piece of physics. It was in this mood that he said to Calder, "Would you like to see how an atom is split?" and took him to see Cockcroft and Walton and "their toy". Calder was most impressed by the primitiveness of the famous machine which appeared to involve the use of "plasticine, biscuit tins and, I suspect, sugar crates". Rutherford explained that although the splitting of a single atom released sixteen million volts, the accelerator had to apply an energy of one million volts to each projectile and only one proton-projectile in ten million disintegrated a target atom. "It's like trying to shoot a gnat in the Albert Hall at night and using ten million rounds on the off chance of getting it. That should convince you that the atom will always be a sink of energy and never a reservoir of energy." Calder noted that Cockcroft challenged Rutherford's view on this matter but that the Professor remained obstinate.

We have seen in an earlier chapter that Rutherford tended to make some distinction between his public views on atomic energy and the advice he gave privately to those who might one day be concerned. Towards the end of his life the scientific evidence that was slowly becoming available began to change his views. Sir Frederick Dainton remembers overhearing, just after Niels Bohr had given a lecture on the "liquid-drop" model of the atom which he and Wheeler had produced in the last year of Rutherford's life, in the post-lecture informal discussion, the Rutherford boom announcing, "Well Niels, if in a nuclear reaction mass disappears, energy will appear, and ultimately, whatever its initial form, be degraded to heat. It might be used." And, in the Sidgwick Memorial Lecture in 1936 Rutherford said publicly, "Athough the overall efficiency of the transformation process increases with the bombarding energy of the proton there is little hope of gaining useful energy from atoms by such a process. The extraordinary efficiency of slow neutrons in causing transformations, with large evolution of energy seems promising, but neutrons themselves can only be supplied by a very inefficient process of transformation."

Rutherford was fully aware of the important experiments being done with slow neutrons by Enrico Fermi in Rome, for the simple reason that Fermi, in his anxiety to obtain a reputation for his new "school", had personally sent pre-prints of his papers, unasked, to forty of the most influential scientists in the world, including the Director of the Cavendish Laboratory. Fermi was essentially following up the work of the Joliot-Curie team's success in producing artificial radioactivity, but he borrowed also from their earlier experiments. He produced neutrons by bombarding beryllium with alpha-rays from polonium, and he used the neutrons to bombard various elements to see if radioactivity would be induced, implying that he had transmuted the element into a radioactive isotope. Rutherford responded warmly to Fermi's new work, congratulating him on migrating from theoretical to experimental physics, and, more practically, by reading his paper to the Royal Society and inviting two of his colleagues, Amaldi and Segre to Cambridge for several months in the summer of 1934. He was even more impressed when Fermi discovered that slow neutrons caused more disintegrations than normally fast neutrons – perhaps because the discovery had been made in a Rutherfordian manner through the pursuit of an anomaly. The anomaly had been discovered, ironically, by Bruno Pontecorvo, who subsequently went over to the Russians with his knowledge of the atomic bomb; he had found that the radioactivity produced in a sample of silver varied according to the position of the sample inside its lead casing, even according to whether it was standing on a marble or a wooden bench. It was Fermi, however, who guessed what the anomaly meant, and quickly proved his point by placing a screen of paraffin wax between the neutron source and the target, showing that neutrons slowed down by collisions with the hydrogen atoms in the paraffin caused many more transmutations in the silver.

The fact that Rutherford could change his mind about the possible usefulness of atomic energy, indeed that he allowed the evidence to change his mind for him, shows what was undoubtedly the finest scientific quality of his old age, his flexibility. He confided to Ritchie Calder that, in his experience, the truly brilliant leaps of the imagination were most likely to be made by minds that were in their first years of research; it was the task of the older man to keep his tinder dry so that he could accept and encourage the sparks that flew from younger minds.

During the last few years of his life Rutherford gave innumerable public lectures – four or six every year at the Royal Institution, Watt Lectures, Sidgwick Memorial Lectures, the Boyle Lecture at Oxford, speeches at the opening of the Metro-Vick Research Laboratory and the annual meeting of the radiologists, are only a few. The subjects could vary from considering the physiological effects of radiation to an address on the nature and properties and usefulness of helium; sometimes he turned his attention to the application of science to the problems of industry in the Depression, although most often the subject matter was his own work of nuclear physics, made more palatable for a general audience by concentrating on "The Transmutation of Matter" which was his favourite title for the last five years of his life. He also "authored" two more books – the neologism is used because he wrote only one of them in any real sense.

Just as in the earlier periods of his life, we can trace the flow of his thought through these lectures and books more comprehensively, more humanly, than through his published papers. In 1930 the Cambridge University Press published *Radiations from Radioactive Substances* by Sir Ernest Rutherford, James Chadwick and C. D. Ellis. Ellis contributed a few chapters on his specialised subjects; Chadwick did most of the work, and since at that particular time he was waiting for the completion of his new house and was forced to live in temporary lodgings which were very badly heated, he had to sit up to the small hours of the morning, wrapped in overcoat and blankets, with his feet on a hot-waterbottle, toiling at the manuscript. In essence the book is an updating of Rutherford's great early works on radioactivity, and the writing must mostly have been a rather laborious checking of the latest published papers. But the preface bears all the marks of Rutherford's personal style, and can be taken to be his view of nuclear physics before the discovery of the neutron and the positron, before the regular transmutation of elements by artificially-accelerated particles. And he provides the grand overview:

> With the exception of a few outstanding problems, the wonderful series of radioactive transformations of uranium, thorium and actinium are now well understood, and attention today tends to be more and more concentrated on the study of the alpha-, beta- and gamma-rays which accompany the transformations and of the effect produced by these radiations in their passage through matter. This is a subject not only of great scientific interest but of fundamental importance. A detailed study of the nature and modes of emission of the radiations spontaneously appearing during the disintegration of atoms promises to give us information of great value on the structure of the nucleus of the atom and of the energy changes involved in its transformation. In addition, the bombardment of matter by swift alpha-particles has placed in our hands a powerful method for studying the artificial transformation of the nuclei of a number of the ordinary elements. Evidence is accumulating that by this method we are not only able to cause a disintegration of a nucleus with loss of mass, but also in some cases to build up a nucleus of greater mass by the capture of the alpha-particle.

This amounts, in some respects, to a defence of the choice of title for the book since the emphasis has shifted now from radioactivity itself to the radiations emitted.

It has been often noted that the book contained no use of the word neutron. It is typical of Rutherford that in formal written work he would never venture one iota beyond the limit provided by the experimental evidence at that moment. His speculations, therefore, are inevitably wrapped up in negatively-turned phrases. The evidence of what comes out of the nucleus allowed him to assume that alpha-particles and electrons were components of the nucleus. Protons never emerge of their own accord, but can be knocked out by bombardment. Hence "it seems clear that while the proton can exist as an individual unit in the structure of some of the lighter atoms, the alpha-particle is to be regarded as an important secondary unit in the building up of the heavier nuclei and probably of nuclei in general. The helium nucleus itself is believed to consist of a close combination of four protons and two electrons." Excepting the statement about protons, this is not acceptable to modern physicists and Rutherford goes on virtually to predict why it will not be considered satisfactory and why a neutron must exist:

It is very unlikely that the component electrons and protons can exist as separate independent units in a complex nucleus. They no doubt tend to form aggregates like the alpha-particle and other combinations of a simple type . . . On modern views it appears difficult to account for the equilibrium of free electrons in a nucleus containing massive charged particles. It may be that the electrons in a nucleus are always bound to positively charged particles and thus have no independent existence in the nucleus. Unfortunately it is very difficult to obtain convincing evidence on any of these points, for we have little if any definite information on the internal structure of the nucleus to guide us.

The final sentence here seems almost a *cri de coeur*, for the previous sentences have put forward conflicting views and speak eloquently of the existence of a state of considerable confusion.

There is a similar confusion, virtually admitted, even when Rutherford (or Chadwick) writes about the apparently well-covered subject of radioactive nuclei – Chapter Nine in the book. "It is however only within very recent times that a picture, even of the most general type, has been given which will explain satisfactorily the spontaneous disintegration of a nucleus with the emission of an alpha-particle, and no application of this has yet been made to the emission of the beta- and gamma-rays."

But Rutherford shows his complete open-mindedness towards the new theoretical achievements of the continental thinkers in the "Application of the Quantum Mechanics". "A simple explanation of the radioactive disintegration and a solution of the apparent conflict of evidence between the radioactive data and the scattering results was offered simultaneously by Gamow and by Gurney and Condon on the basis of the new quantum mechanics." Thus in 1930 it seemed to Rutherford that the "tunnelling" effect, by which the alpha-particle's escape from the nucleus could be explained, was more important than its corollary, the possibility of disintegrating the nucleus with comparatively low-energy particles, a possibility on which Cockcroft and Walton had just started work. But he shows himself equally willing to appeal to "the wave mechanics" when they provide satisfactory elucidation of experimental problems, and when he comes to "Energy Considerations", in a later chapter of that title, he quotes the famous Einstein equation $E = mc^2$, and uses it to great effect.

Just two years later, in the closing months of 1932, with the neutron discovered and the Cambridge accelerator smashing nuclei of all kinds, the note of loss and confusion was gone when he gave the Institution of Mechanical Engineers their annual Thomas Hawksley Lecture. Entitled "Atomic projectiles and their applications" this lecture showed Rutherford back in a youthful mood of rather masculine ebullience. "This age will for ever be memorable for the development of new and swift methods of transport over land, sea and in the air. The spectacular advances in the speed of the motor car, the motor boat, or still more of the flying machine . . . are triumphs of which the engineer may well be proud"; and indeed this was the period of Sir Malcolm Campbell and the "Bluebird" setting new land and water speed records. Furthermore "much higher speeds may be ultimately obtained by the development of a rocket type apparatus". However, the maximum speed that could be communicated to matter in bulk, as far as Rutherford could see, was about two miles per second; and this was only about the speed of a molecule rushing about in any ordinary gas. Even in the interior of a hot star the smallest atoms, hydrogen atoms, only attained a speed of 1,000 miles a second, he pointed out. Now if one looked at atomic particles one found that a single electron accelerated by a potential difference of just one volt would reach a speed of 590 kilometres a second. So the speeds involved in the particles of radioactive experiments were enormous, for potential differences of ten million electron volts were attainable and "we have direct evidence that ultrapenetrating radiations in our atmosphere give rise to particles of energy reaching as high as 1,000 million electron volts". Here Rutherford was referring to cosmic rays which became an important field of study in the 1930s.

To achieve very high velocities, or energies, in the laboratory so that the particles could be studied, required discharge tubes, and brought problems of a type which Rutherford was now having to deal with as an administrator of science and a director of research. "There is no insuperable difficulty in devising discharge tubes to withstand an accelerating voltage of several million volts, but the size of apparatus and the size of laboratory to contain it, not to mention the cost, rise rapidly with the voltage," he declared and he described this approach as a "rather brutal method" of obtaining laboratory measurements. At least in a public lecture he seemed committed to exploring other means, and went on to describe Lawrence's cyclotron before saying, "Devices of this or similar kinds will have to be utilised in the laboratory if we wish to obtain projectiles for experimental purposes much more energetic than can be produced by one million volts applied directly to a vacuum tube. The development of these new methods is thus of great interest and importance for the future."

In his final triumphant years Rutherford could also trumpet the success of theoretical physics, as well as the measurements of the experimentalists. At the Mendeleev Centenary in April 1934, he addressed the meeting at the Chemical Society on "The Periodic Law and its Interpretation". Naturally he concentrated at the start on the successes of Moseley and his own school in showing that the Periodic Table had a real meaning, and he praised even more loudly the daring and acumen of Bohr in applying quantum mechanics to provide a physical explanation in terms of electron orbits for the differences between the elements. "I remember that in the early days, when I reviewed the enormous complexity of the spectra of the elements and the apparent lack of any relation between them, it seemed to me that a solution of such a formidable problem would be long delayed," he recalled, but he gloried in the triumph of all the contributing scientists that such a solution had, in fact, been found in twenty years' work. Bohr's idea of electrons in planetary orbits "served and continues to serve a very useful purpose in assisting us to visualise the arrangement of electrons", but out of these ideas there had grown the wave mechanics of de Broglie, Heisenberg, Schrödinger and Dirac and "the new theory enables accurate calculations to be made of atomic properties with, so far as we know, complete success". Furthermore, "I am informed on good authority that the success of the new wave mechanics is so complete in its own domain that the Periodic Law of the elements, as we know it today, could be entirely reconstructed from first principles – given a competent mathematician, and he would require to be competent, the laws of quantum mechanics and electrons, plus Pauli's principle [the exclusion principle], he should be able to do it."

Nonetheless we must note that the original idea of planetary electrons "continues to be useful" in "visualising". And later in the same lecture Rutherford enlarged on his position: "The earlier idea of precise orbits must be abandoned and replaced for the present by electron distributions much less definite . . . Nevertheless the old orbits remain a good first approximation to many features of the true distribution and we shall for that reason continue to use them for descriptive purposes." The interesting feature of this remark is that it was not a "last line of defence" position, not the great scientist of former years refusing to leave his entrenched positions; it is still the practical position to this day, the working position adopted by most scientists in dealing with the material world.

At the very end of this lecture Rutherford considered the nucleus rather than the outer electrons, and made the obvious suggestion that nuclei, too, might show some periodicity in their structure in step with increasing nuclear charge. But he was quite prepared to admit that "it may be more complicated" because there was still "little definite information on the detailed composition of nuclei". In the long series of lectures and addresses of his last years, as well as in his correspondence with Niels Bohr, Rutherford gradually gave up his belief in the possibility of finding a structure in nuclei and seemed more and more ready to accept Bohr's new view that the nucleus has to be regarded as a "mush", an unstructured soup of particles, which suddenly presented an event to the outside world from which it was not possible to deduce the existence of the emerging particles before that event. This is the view still accepted by modern physicists, who have been able to impose only a few constraints by consideration of the energy limitations that seem likely to prevail.

It is entirely to Rutherford's credit that he should accept the evidence that his long pursuit of a structure for the nucleus was likely to prove fruitless. He retained both his sense of wonder and his demand for a concrete visual model to the end of his days, and he was not to be moved from this by fashions or abstract theories. He faced up to the dilemma this posed without fear. One of his Royal Institution Lectures of 1935 was entitled "The Propagation of Waves through Space" and he declared: "In the days of my youth it was usual to suppose that the whole of space was filled with an imponderable elastic medium called the ether, whose function was to transmit radiation with the velocity of light." Now, however, the scientist had to admit that there is no evidence whatsoever of the existence of this ether and there is no reasonable concept of its detailed structure, so the word itself has become "taboo". But,

Rutherford continued, "I must confess, however, that I am old-fashioned and still believe the ether to be a useful and necessary conception to understand the propagation of the electromagnetic waves. Indeed the whole idea of waves and frequency becomes almost meaningless unless we can suppose these waves do travel through an elastic medium." It was quite true, and he was quite prepared to admit it, and to use the results, that for X-rays and gamma-rays and light-rays the quantum mechanical equations of radiation were accurate and all important, "but this does not necessarily mean that we can discard the idea of an ether, for otherwise it is very difficult to understand the conception of frequency of radiation which is so intimately connected with the quantum theory". All but fifty years since Rutherford spoke these words, the concept of "ether" is even more taboo, but the basic dilemma, the failure to find an alternative, is still present and is ignored, or skirted round, by the physics of the second half of the twentieth century, which simply denies the necessity of providing a concrete model but which continues to use a concrete model by implication.

The other problem, which continued to perplex Rutherford and which he frequently mentioned, in passing, during his last lectures, although it never appeared as any formal statement in his published work, was the scarcity of positive electricity. He had mentioned this in his early days, when he gave a Benjamin Franklin Memorial Lecture in Philadelphia. The discovery of the existence of the positron by Anderson in 1932, and the confirmation of the particle in his own Cavendish, seems to have brought it to the forefront of his mind again. It is a far from trivial problem: in so far as the positron, the positive electron, represents anti-matter it is the problem of the uneven distribution of matter and anti-matter throughout the universe, or possibly even the predominance of matter against anti-matter. This problem is still unsolved and plays a prominent part in cosmological theorising at the present time. In his Royal Institution Lectures in the last winter of his life (on December 23rd, 1936) Rutherford said: "Positive electrons are rare visitors to our world and only make their appearance under exceptional conditions. They have a very short life outside the atom and rapidly vanish with the emission of penetrating radiation. We live in an electrically unsymmetrical world of atoms where negative electrons play a predominant role."

Yet the subject on which Rutherford spoke most often in these last years of his life was the transmutation of atoms – the new alchemy that was now possible through bombardment by protons, neutrons or deuterons. "Neutrons and Radioactive Transformations", "Recent Researches on Transmutation", "The Transmutation of the Heavy Elements", are typical titles of lectures given in 1935, 1936 and 1937. They are summed up in the last book Rutherford wrote, *The Newer Alchemy*, an expanded version of his Henry Sidgwick Memorial Lecture, given at Newnham College in November 1936. It was very different from anything he had published before, just a tiny volume of fewer than seventy pages, expressed in his lucid, popular style, and well illustrated with many examples of the latest photographs of particle tracks in cloud chambers. But it also includes a "smartened-up" version of his hyperbola diagram of the passage of alpha-particles near nuclei.

Rutherford had a very real feel for history; he read enormous quantities of history in his later years, and he saw very clearly how his science had grown on the foundations laid by his predecessors. *The Newer Alchemy* makes no direct reference to the fact that alchemy and transmutation were in his mind, and even more in Soddy's, when he wrought his first scientific revolution, yet he places his own work on radioactive transformations, correctly, right at the start of his story, and shows no lack of sympathy with the old alchemists in their search for the philosopher's stone which would transmute base metal into gold. It is only the modern swindler who is attacked:

> From time to time there arose a succession of men who claimed to have discovered the great secret, but we have the best reasons for believing that not a scintilla of gold was produced. When we look back from the standpoint of our knowledge today we see that the transmutation was a hopeless quest with the very limited facilities then at the disposal of the experimenters. With the development of experimental science and the steady growth of chemical knowledge, the ideas of transmutation were gradually discarded and ceased to influence the main advance of knowledge. At the same time these old alchemistic ideas have persisted in the public mind and even to this day impostors or deluded men appear who claim to have a recipe for producing gold in quantity by transmutation. These charlatans are often so convincing in their scientific jargon that they disturb for a time the sleep of even our most hardheaded financiers. We shall see that it is now possible by modern methods to produce exceedingly minute quantities of gold, but only by the transmutation of an even more costly element, platinum.

He does not say directly that it was his own work which started the process that has led to this newer alchemy, he simply remarks that the "idea of the permanency of atoms received a rude shock in 1902", but the consistency of his thinking over so many years is shown when he refers to the methods of proving that transmutations have taken place; he can be just as certain now that an atom has been changed as he was in 1902, although the quantities of the substance involved are too small to be weighed or analysed chemically:

> The amount of transformation produced is usually on a minute scale and only rarely is the quantity of matter produced either visible or weighable. Our methods of detection and recognition of the flying particles produced in a transformation are, however, so extraordinarily sensitive that even a minute amount of transformation gives very large effects in our measuring apparatus. The certainty of our methods of detection and analysis of the fast particles is in many cases greater than if the element were transformed in weighable quantities which could be analysed by ordinary chemical methods.

This same book is the best statement, also, of the modification of Rutherford's views, and their soundness on the subject of atomic energy. He explains that the work of Fermi in Italy and Feather in the Cavendish has shown that "neutrons, and particularly slow neutrons, are extraordinarily effective agents in transforming atoms . . . owing to the absence of charge a slow neutron may freely enter a heavy nucleus". This leads him, after an interval, to say:

> We have already discussed the large evolution of energy from the spontaneous transformation of atoms of the natural radioactive bodies. In several cases of artificial transformation by protons and deuterons the energy emitted per disintegrating atom is even greater than from radioactive atoms . . . [he explains that the transformation of lithium under bombardment by deuterons evolves twenty-two million volts per transformation] . . . it is clear that there is a large gain of energy in the individual process. On the other hand only about one deuteron in a hundred million is effective so that on the whole more energy is supplied than is emitted as a result of the transformation. Even allowing for the fact that the overall efficiency of the process rises with increase of bombarding energy there is little hope of gaining useful energy from the atoms by such a process. At first sight the extraordinary efficiency of slow neutrons in causing transformations in certain elements with large evolution of energy seems promising in this respect, but we must bear in mind that neutrons themselves can only be supplied as the result of very inefficient processes of transformation. The outlook for gaining useful energy from the atoms by artificial processes of transformation does not look promising.

This is a very different story from the rough dismissal of atomic energy as "moonshine" a few years earlier. And furthermore it is still correct by our present standards of knowledge. It was only when the natural disintegration of the fissile isotope of uranium, with the release of more than two neutrons per disintegration, was discovered, that the possibility of a chain reaction, and hence of the evolution of worthwhile quantities of energy, became a scientific possibility.

The discovery of natural uranium fission was achieved by Hahn and Meitner, although it was only the interpretation of the results by Meitner and her nephew Otto Frisch which showed what was really happening. What is much less commonly mentioned is that the reason why Hahn and Meitner did not at first realise the meaning of their results was that they had for some time been on an entirely false track. And Rutherford, who had known and trusted Hahn since the German had been his first important pupil in the McGill days, accepted the Hahn–Meitner misunderstandings at the time of his death. Hahn and Meitner had followed up Fermi's work and started bombarding uranium with slow neutrons, using Hahn's preeminent skill as a chemist to analyse the results of the bombardment. They were, of course, looking for the transformation of uranium atoms by the neutrons and therefore expected to find in the resultant product elements with a little more or a little less the atomic mass of uranium, the heaviest of the elements. It seems never to have crossed their minds that the uranium nucleus might split into two roughly equal-sized nuclei of totally different atoms. When, therefore, they found clear chemical evidence of elements such as osmium and platinum in the residues they guessed they had made chemicals identical in behaviour to these known substances, but with nuclei heavier than uranium because of the additions of neutrons to the uranium nucleus. They named the elements they had discovered "eka-osmium", "eka-platinum", "eka-rhenium", and so on. The picture of radioactivity produced by these experiments was also complicated and difficult to understand. Rutherford seems not to have questioned the results or their meaning, although we now know that it was ordinary rhenium, osmium, and so on (or radioactive isotopes of them) that were produced from the splitting of the uranium atom. Certainly in March 1937, only a few months before his death, Rutherford gave a long public description of their work, claiming that "radioactive elements of higher atomic number than uranium" had been discovered, that "nine new and distinctive radioactive bodies have been observed", and that "It does not appear that the bombardment of uranium by neutrons has any effect in accelerating the natural disintegration of this element."

We must conclude, therefore, that at the time of his death

Rutherford had no idea, indeed had very much the wrong ideas, about the vital area from which the practical application of nuclear power would arise.

There is undoubtedly an irony in this, for it was Rutherford's work more than that of any other man which offered mankind the doubtful choice of unlimited energy and unlimited destruction. The irony is heightened when we look at what he said in his Norman Lockyer Lecture in November 1936.

> It is of course true that some of the advances of science may occasionally be used for ignoble ends, but this is not the fault of the scientific man, but rather of the community which fails to control this prostitution of science. It seems to me that scientific men have shown themselves unduly sensitive to these criticisms . . . It is sometimes suggested that scientific men should be more active in controlling the wrong use of their discoveries. I am doubtful however whether even the most imaginative scientific man, except in rare cases, is able to foresee the ultimate effect of any discovery.

Would Rutherford have changed his belief if he had lived five years longer and seen the possibility of the nuclear bomb? His successors stick manfully, on the whole, to his doctrine about the impossibility of foretelling the effects of their results, though most of them, unwillingly, now believe the scientist ought to be more active in community decisions concerning his work.

To Rutherford, at the time, the problems of science in its relations to the community were problems of industry and the world depression, a situation which was easing in 1936 though only under the threatening drive of "Rearmament". The criticism of science was basically a Luddite attitude – that science did people out of jobs. Perhaps the scientist could help society not merely by doing the research to create new industries, but also by giving warning of what the extension of research from the laboratory might do. And so he suggested the

> formation of a prevision committee of an advisory character, composed of representatives of business, industry and science who could form an estimate of the trend of industry as a whole, and the probable effect on our main industries of new ideas and inventions as they arose, and to advise whether any form of control was likely to prove necessary in the public interest. A competent committee of this kind could, no doubt, have foreseen the coming competition between motor and railway transports which had such serious effects upon the latter, and have advised the government on the need of adjustment of competing claims before the difficulties become acute.

This suggestion seemed extremely interesting to Rutherford's contemporaries; Eve for instance makes much of it in his biography. Nearly fifty years after Rutherford spoke these words the competing claims of road and rail are still with us in the shape of demands for finance for motorways and high-speed trains, and they are complicated by the demands of air transport. No one has got anywhere near a prevision committee despite all the chief scientists, and Offices of Technology Assessment, and Think-Tanks. Indeed, if we have learned anything it is that Rutherford was hopelessly optimistic in thinking that a collection of our wisest men could foresee even over a short term what developments were likely to occur.

Almost all Rutherford's problems were now, in a sense, on the industrial scale. In Cambridge, at the Cavendish, the crucial decision was whether to build a cyclotron, whether to push the laboratory further into the world of Big Science – and if so, where were the resources to come from. Both Rutherford and Chadwick had met Ernest Lawrence, the inventor of the cyclotron, at the Solvay Conference in 1933. The American was so enthusiastic, "bubbling over with enthusiasm", according to Chadwick, that both men took to him immediately; Rutherford said afterwards to Chadwick, "Lawrence reminds me of my young days. He's just like I was when I was his age." There was no question of the Cavendish staying out of the field through the old-fashioned courtesy that Rutherford had shown to the Braggs and others during his career – on the contrary Lawrence would have been delighted if the Cavendish had decided to build a cyclotron, and, according to Chadwick, would have supplied all the necessary information. (He did in fact supply all that was asked when Rutherford did finally decide to build such a machine.)

But Rutherford in 1933 would not go into building a cyclotron. Chadwick was quite clear about it:

It was becoming very difficult to push on without some new equipment. I couldn't get any further with what I had. I was at an end with the equipment which I had or could see myself getting, and it was quite clear to me, as it was no doubt to others, that we needed a means of accelerating protons or other particles, particularly protons, at high energies. But that meant more space, particularly more money and particularly engineering. It meant complicated equipment, and Rutherford had a horror of complicated equipment. Quite naturally. He'd done everything with very simple means and he didn't want big equipment in his laboratory. That was the chief reason why I left the Cavendish. It got to that point. I couldn't get any further. I knew

Rutherford's attitude. I understood it. I was not prepared to quarrel with him. It might have come to that. I wasn't going to do it.

Chadwick's version has been accepted by all who have written about this matter, even the official history of the Cavendish by Crowther tells the same story – how Chadwick had told Rutherford the Cavendish must change. There was a fearsome row when Rutherford criticised Chadwick for encouraging Cockcroft to build a cyclotron while Chadwick was paying him a visit from his new post at Liverpool where he was building his own cyclotron.

Reconsidering what was actually happening in the Cavendish we can see that Chadwick's motives for leaving may have been more personal and less cyclotronic than he believed. Complicated apparatus had in fact been installed in the Cavendish, whatever Rutherford's feelings about it. There was the High Tension Laboratory that Allibone had built, the big Cockcroft–Walton accelerator, and the smaller accelerator that Oliphant had built for Rutherford's own line of experiments. It was undoubtedly true that Rutherford felt, and wrongly, that he could go further with low-energy protons than has subsequently been shown to be the case. But again, by the use of hindsight, it has not been the case that a cyclotron was necessarily the only or the right machine to open up larger fields of nuclear physics.

Crowther wrote of this period:

At times all of Rutherford's binding force was needed in order to hold this highly excited and almost spontaneously explosive nucleus of discoverers together. Very able men were competing for a portion of the modest resources of the laboratory and some seemed, in the opinion of their colleagues, to get far more than their share. Nor did it appear that those who had discovered most always received the largest grants for apparatus. Tempers were not improved when Rutherford replied, in answer to expostulation, that expensive apparatus was not necessary in order to make discoveries. He had a tendency to give excessive consideration to some research workers.

Crowther goes on to suggest that the psychological need for a son influenced Rutherford in these judgments, and while it is possible this may be true, it is nevertheless the fact that Cockcroft, who was never in the "surrogate son" relationship with Rutherford, needed and got more money than Oliphant, who probably was.

Cockcroft stated that they decided that they would have to build a cyclotron after he had visited the United States, and Lawrence's Berkeley establishment in 1933, but Rutherford was "not agreeable to our making a start on the cyclotron at that time". The reason was that Rutherford and Oliphant believed they could get all that was needed at low energies, said Cockcroft, and he specifically denied that there was any question of the cyclotron being out of the style and tradition of Cavendish physics: "Rutherford would never have objected to trying something new in the way of innovation." And by the time Rutherford and Oliphant found that they could only transmute the lighter elements with their low-energy accelerator, and could not get results with the heavier elements, the Cavendish had received a colossal injection of finance and he agreed to the building of a cyclotron.

Rutherford certainly began to feel that the cracks in his kingdom were showing. Blackett had gone to London, Ellis had gone and Chadwick went to Liverpool taking Feather with him. When Oliphant was offered the opportunity of going to Birmingham, although Rutherford himself had suggested that the authorities of that university should approach him, the situation became very emotional. Rutherford had told Oliphant that he intended to retire at the age of seventy – in four years' time – and Oliphant asked Chadwick for his advice. Chadwick made it quite clear that the answer must lie in Oliphant's own view of his future – if he wanted a Chair and laboratory of his own he should take the Birmingham offer, but otherwise it would be a mistake. When Oliphant told Rutherford he was very tempted to go to Birmingham, Rutherford exploded, going red in the face, shouting that he "was fated to be surrounded by ungrateful colleagues" and ending with a roar, "Go and be damned to you." Within the hour he was apologising completely, discussing what help he could give to Oliphant in his new post and how his Cambridge work could be carried on. Although Chadwick was by then at Liverpool, Rutherford was also very cross with him, and is believed to have written an angry letter. Although the matter was smoothed over in correspondence, Chadwick was left distressed that Rutherford's sudden death meant that they had not been able to have a last meeting as friends.

Whether Rutherford was comforted in these stresses by an even older colleague, von Hevesy, writing "old assistants, like old elephants, might easily become rather nasty", there is no telling. But there was still the Cavendish to keep going. By 1933 it was clear that more money was going to be needed quite apart from the problem of cyclotrons. The official history of the Cavendish says that the university asked Rutherford to raise money for the laboratory, and that Rutherford did it with "characteristic ability". He certainly asked Sir Arthur Eddington to write a pamphlet about the Cavendish for this campaign, for whatever Rutherford thought about Eddington's science (fairly often there were scathing remarks), he could recognise a better writer than himself. Strangely there is no copy of this pamphlet in Rutherford's files, and hardly any mention of such a campaign in his correspondence. Rutherford's letters to his many correspondents contain plenty of mentions of the solution to his problem, but no mention at all of how it was achieved. There is one comment from Chadwick in a letter of May 4th, 1936: "Begging, like swindling, is only respectable on a big scale." He knew his Rutherford well, and he may have understood that Rutherford was willing to accept the final gift because it was made to the university and not to himself.

Cockcroft, many years later, described the campaign very differently. "We went to the Prime Minister [Stanley Baldwin] because he was Chancellor of the University, to get him to find a benefactor or to sign an appeal for money. And Mr Baldwin was so lazy that he decided he'd do it in one go by going to Lord Austin, who in turn got a peerage for the benefaction. So he produced £250,000 and in return he was made a Lord." Cockcroft does not say what evidence he had for this statement, but the history of the Cavendish appears to support him by implication: "In 1936 the motor manufacturer Sir Herbert Austin gave a quarter of a million pounds for the laboratory's development. Shortly afterwards he was raised to the House of Lords."

The colossal munificence of the gift must be recognised, however. In 1930 the university contributed just over £15,000 towards running the Cavendish, and much of this money came to the university from students' fees. In addition the DSIR contributed just over £2,600 in grants, mostly towards Kapitsa's High Magnetic Field Lab. At almost exactly this time Cockcroft and Walton were building their first accelerator which cost, in round figures £1,000. In 1935, the year before Austin's benefaction, the university bill for the Cavendish was £16,200, and the sum spent by Rutherford on new apparatus in that year was slightly over £1,000. Compared with these sums, £250,000 seems almost out of proportion.

Rutherford himself took little pleasure in the gift or in the spending of it. He knew that he was building for his successor, and he would say so loudly, jingling the loose change in his pocket as he did so. In fact the money from Lord Austin was, in a sense, spent before it was given. Cockcroft, we have seen, had visited Lawrence and his cyclotron in 1933, and he was to visit again in 1937 as Lawrence offered "detailed drawings and specifications" with an invitation to let him actually run the American machine for a few days so that he would know what the problems were. Cockcroft and Oliphant had visited the Philips laboratories in Holland in 1934 where they saw a one million volt discharge tube which Oliphant found "an eye-opener to me and I think that when the Cavendish is rebuilt their design must be borne in mind . . . in the opinion of Cockcroft far superior to any in America". But English companies such as Metro-Vickers were also interested in getting Cavendish contracts for big prestige machines, and soon financial competition started, though quite unofficially. In August 1935 Oliphant wrote to Rutherford at his shock at receiving an estimate of £15,000 for their proposed new High Voltage Laboratory; it ". . . staggered me as I imagine it would you". Cockcroft had estimated £6,000 and "allowing for his well-known optimism, perhaps £9,000".

Serious planning for the new lab was by then well under way, for Oliphant continued, "The plan has been made too elaborate for a laboratory which must of necessity be prepared for changes to follow fresh requirements of future work." Too much attention had been paid to providing a pleasant exterior, when what was needed was something more like an airship-shed: "It would then be a real workshop and recognised as a place liable to change and alteration." This same letter illustrates exactly the pressures that led to the search for money which in turn led to the Austin benefaction; Oliphant wrote: "I can see the new lab receding into the distance if we are not careful. Of course, if we had the money or you thought you could raise it, all would be well." All the usual problems of building by committee had appeared, there were too many changes being made by individuals, many of which Oliphant had never heard about; the Building Committee "have not served us well". "It is a thing which we need urgently and not in some distant future when all the cream has been scooped off by folks whose results we dare not trust too deeply," he wrote of one particular proposal. It was probably Rutherford's rather detached attitude to the exciting prospect of spending a quarter of a million pounds that finally sealed his unwarranted reputation for being totally uninterested in engineering and technology.

In any case his own line of research had dried up and there was also a major theoretical problem facing him – Bohr's idea that the nucleus had no structure at all, but must be considered as a "mush". It must be emphasised that Bohr and Rutherford continued to work closely together right up to the end of Rutherford's life. Bohr still came to Cambridge every year or two, and, quite apart from old friendship, Rutherford was impressed by Bohr's arguments. This is made clear in his letters to Chadwick where he reverts again and again to the "mush" theory in the last year of his life, and can find almost nothing to say against the theory. Furthermore Rutherford also tended to be more mellow on the subject of theoreticians towards the end of his life. Under the influence of

Fowler and Dirac theoretical physics began to be taught formally at Cambridge, although technically outside the Cavendish, and some of his brightest young men, such as Neville Mott who would succeed to the Directorship of the Cavendish forty years later, were now theoreticians. Yet if the nucleus had to be regarded as a mush the search for the structure of the nucleus, Rutherford's objective for nearly twenty years, was pointless.

Despite this, his power and influence and the élan of the Cavendish did not slacken. "Rutherford the individual and the Cavendish, the institution became one and together radiated a brilliance rarely matched . . . The Cavendish was Rutherford's domain, his sphere of influence. But one never felt this stemmed from any formal title, his influence . . . seemed a wholly natural phenomenon. Benevolent guidance, leadership and intellectual authority flowed from him and loyalties were returned." These are the memories of Professor "Sam" Devons, in 1937 one of the youngest Cavendish researchers, but who had experience in physics laboratories on both sides of the Atlantic before he wrote for an informal pamphlet publication which celebrated the centenary of the Cavendish in 1974.

In his last years Rutherford lectured three times a week, at noon on Monday, Wednesday and Friday in the old Maxwell lecture theatre, the largest in the Cavendish. The lectures were primarily for students of physics, but, while they were not compulsory, anyone was allowed to attend. Normally there would be about forty people in the steeply-raked semi-circular rows of wooden seats, only half filling the room to hear Lord Rutherford on "The Constitution of Matter". The lecturer would appear with a few loose pages of notes which he took out of an inside pocket of his jacket, but which he rarely consulted, and the lecture was accompanied by table-top experiments, slides, diagrams and photos. Devons recalled:

> Although formal in the Cambridge pattern in style and subject matter, they were highly personal; a quasi-historical or rather biographical account of the development of atomic physics of the past few decades. It was no easy matter to separate the development of physics from Rutherford's own life work and Rutherford did not try . . . There was no doubt we were listening to a great man relating an epic story, rather like the story of some great scientific exploration as told by its leader. We were being told not so much what Rutherford (or anybody else) thought about this or that, but rather how nature did its work and how this had been discovered.

The lecturer plainly felt no need to consider the normal social demands for modesty "and in Rutherford's case, as has been so often remarked, he had so little to be modest about".

Devons' view of these lectures compares most fascinatingly with the views of the more senior Oliphant, who was often called upon to substitute for Rutherford at the last moment when the great man was called up to London on business.

> On these occasions he would call me into his office to sit beside him while he explained what I should talk about with the aid of an extraordinary collection of notes written on odd scraps of paper and pinned together in the right order. He had used these notes for many years so that there were copious amendments and additions, always in pencil, some of which had become almost too faint to read. We went through these carefully, sometimes for an hour or more, adding, crossing out and getting the papers into a hopeless disorder . . . an experience of great interest for I saw what he regarded as important and what could well be omitted, decisions not in accord with most textbooks. When I argued a point or asked for clarification, his physical insight into the behaviour of atoms and their nuclei became apparent. It was at these times that I grew to appreciate his profound sense of history. He understood how knowledge grew and how spasmodically new ideas were born.

To the younger men and women he presented a problem in what we should now call "credibility". Devons remembers:

> Rutherford in his sixties appeared as a patriarchal figure, somewhat archaic, vaguely Victorian in dress and manner . . . There was a paradox in this combination of an elderly conservative gentleman of the old school and the proponent, nay the discoverer, of the latest word in this most modern field of knowledge: atomic and sub-atomic physics. I remember puzzling over this paradox with fellow students but we made little headway resolving it.

Rutherford very often took the chair at the weekly "special lectures" which were given "in one of the smaller, dimly-lit, recesses of the Cavendish" and which were delivered to "small, often rapidly diminishing audiences". Also he led the weekly discussions, on Wednesdays, of the Cavendish Physical Society where the lecturer might well be a distinguished foreign visitor. On these occasions Rutherford could introduce Debye as "a chemist, but still a very good fellow" and he could thank Heisenberg for "a lot of interesting nonsense". Although he often appeared to doze during the discourse, at the end he would spring to life and give appraisal and criticism of what had been said; he would ask questions and try to stir up reaction from the senior men who occupied the front rows, men such as Fowler, Dirac, G. I. Taylor and Sir Arthur Eddington. Undergraduates and young research men were allowed in at the back of the lecture hall and Devons remembers: "At these meetings, as in his own lectures, Rutherford's attitude to physical problems was unambiguously expressed. There was always the demand for the 'objective', and, if possible, simple reality. Almost invariably there was the question 'What are the facts?' Facts were to be respected and treated quite differently from theory, which was, in a sense, 'opinion'." Rutherford made it quite clear that he respected the achievements of the theoreticians, especially their logical reasoning, but in addition to this "there was an extraordinary transparent honesty and a deceptive simplicity about the clear distinction between fact and theory (opinion). He was impatiently hostile to any attempt to obscure, or to conceal, or to complicate unnecessarily. And no matter how prestigious its proponent, Rutherford would denounce sophistry point blank if he felt it were deliberate." This often made the occasion exciting for students, if some great scientific figure were publicly "debunked" by "our" Rutherford. It is interesting that, although many of the survivors of the Rutherford school tell the same story about their master's ability to attack "sophistry", none of them can, or will, identify any individual so attacked.

Rutherford also taught most successfully during his tours of the laboratory. At the end of his life he only visited each student once or twice a year, but, remembers Sam Devons, for the student these "more or less unannounced visits" were a major occasion at which Rutherford

> would briefly examine the apparatus and then would seat himself on a lab stool and put one through a quite searching examination. What precisely are you doing? How? Why? And of course this rapidly led to the request, "Now let's see what the results are." There is no doubt . . . that Rutherford's real interest was in the results of the experiment and not in the method or technique or difficulties themselves. However on the few occasions when, as a very junior research student, I was cross-examined I recall being very much impressed by the questions and criticisms regarding both the aims and methods of my work. In a few thrusts Rutherford's questions penetrated right through the limit of my thinking and stimulated me to do some more. He was clearly acting in the role of teacher, not critic, and the results of his uncompromising, not hostile, questioning were undoubtedly salutary. This was all very much in keeping with his attitude towards teaching. A student should think for himself, should ask himself questions; a teacher should not so much supply answers as encourage the student to pose questions to himself.

In accordance with the Rutherford tradition, most students went first to the training course in radioactive methods, the nursery, up in the attic of the Cavendish. And just as in the Manchester days, most beginners worked with radon sources in small glass capsules "precariously sealed off". Devons recalls that he "was strongly advised not to get the stuff on my skin or in my lungs". But he was even more strongly warned against contaminating the laboratory and ruining other people's experiments, and it seems that the precaution of wearing rubber gloves, washing hands and changing coats was aimed more against contamination of the laboratory than of the person: "Inside the radium sanctuary itself (the tower room at the top of the lab) the residual activity was so high from contamination everywhere and from the residues of innumerable sources of the past that it was difficult to charge up the gold-leaf electroscope for long enough to measure, even roughly, the strength of a newly prepared source of some 100 millicuries."

It throws our present-day concern with the slightest threat of radioactivity into some perspective to read that Devons regularly walked about with radioactive sources contained in no more than a glass tube with a rubber bung in his pocket, and he adds, "Nor was I myself unduly alarmed when, shortly after a visit to the Tower (where I spent a couple of hours each day) I found that by simply blowing on a Geiger counter its register would rattle furiously or completely choke in an attempt to record the activity. After a day or two of radioactive abstinence my breath always returned to normal."

There was however an underlying worry among physicists as the 1930s drew on – it was of all unlikely things a philosophical, or even metaphysical worry about the meaning or existence of causality. Virtually all logical thought has proceeded on the assumption that any effect must have a cause. Rutherford's own inability to explain why an atom of a radioactive species should disintegrate at some particular moment, the development of quantum mechanics, so that a wave and a particle were but complementary aspects of one reality, Heisenberg's Uncertainty Principle, and eventually Bohr's view that the nucleus might be an unstructured mush of particles, unknowable until some event occurred, all seemed to threaten the existence of the normal human assumption of causality. The question "Why did this happen?" seemed meaningless or unanswerable. The problem did not remain with the physicists, because Heisenberg's Uncertainty Principle was seized upon by all who disliked the determinism of the exact sciences, and particularly by many churchmen, as an uncertainty in physical processes which would allow the concept of "free will" a place even in a fully scientific account of the universe and its history.

A totally unexpected view of the man comes from the knowledge that, through their mutual contact in the Athenaeum, the Archbishop of Canterbury asked Rutherford to explain to him the Uncertainty Principle, since so many churchmen were interested in the question. There is no record of what the scientist told the Archbishop but presumably it was a less orotund version of what Rutherford said to the Royal Academy of Arts in April 1932. Having described Einstein's Theory of Relativity, "quite apart from any question of its validity . . . as a magnificent work of art", and having stated that the Greek potters had left valuable data for science in that their pots contained a record of the earth's direction of magnetisation 2,000 years ago, he went on:

> Quite recently there has been much interest taken by the cultivated public in the metaphysical aspects of science, especially those of theoretical physics. Some of our publicists have boldly claimed that the old ideas that have served science so well in the past must be abandoned for an ideal world where the law of causality fails, and the principle of uncertainty, so valuable in the proper domain of atomic physics, is pushed to extremes. The great army in its march into the unknown discusses with interest. and sometimes amusement, these fine-spun disputations of what is reality and what is truth. But it still goes marching on, calling out to the metaphysicians, "There are more things in heaven and earth than are dreamt of in your philosophy".

The style here may not be to most modern tastes, but it is important to see the soundness of Rutherford's judgment on a matter which might seem to be outside his normal range of interests. Whereas Albert Einstein's emotional denial of indeterminancy was "God does not play dice", Rutherford replied more reasonably that one could not extrapolate rules from the microcosm of the atom to the macrocosm. The speech to the Royal Academy of Arts was not the first time he had had to turn his mind to the problem. In the previous year he had written to Herbert Samuel in appreciation of an article that Samuel had written on the "New Doctrine of Eddington and Jeans" in the magazine *Contemporary*. Samuel followed up Rutherford's approach and wrote to him when he was preparing his address as newly-elected President of the British Institute of Philosophy. He was anxious to show that the views of Eddington and Jeans (on relativity in general) were "far from being the accepted opinion of physicists in England". Samuel had already obtained a letter from Einstein, he went on, in which the proponent of relativity declared "that the advocates of indeterminancy are on the wrong lines". Samuel also proposed to quote from Planck's published work to the same effect, and now he wanted Rutherford to say at least that "the new doctrine has no countenance from you – and if something even harder so much the better". Samuel explained that he felt justified in approaching Rutherford because "the consequences of the new doctrine continuing to receive an almost unchallenged acceptance among the general public would be so grave in the development of thought in this country that I make little apology for troubling you in this way". Rutherford's statement to the Royal Academy of Arts amounts to a public declaration of his sharing Samuel's view.

The discussion of causality was not, however, mere metaphysics; it has been argued that attitudes towards causality represented in fact a political attitude among scientists, who are not as unaware of the general trends of thinking in their society as Rutherford believed. A modern scholar, Paul Forman, claims that anti-causality was popular among German scientists immediately after their country's defeat in the First World War precisely because they were operating in a climate that was hostile to their intellectual enterprise. He claims to have found

> overwhelming evidence that in the years after the end of the First World War, but before the development of an acausal quantum mechanics, under the influence of "currents of thoughts", large numbers of German physicists, for reasons only incidentally related to developments in their own discipline, distanced themselves from or explicitly repudiated causality in physics . . . Extrinsic influences led physicists to ardently hope for, actually search for, and willingly embrace an acausal quantum mechanics, is here demonstrated – for, but only for, the German cultural sphere.

Later the same author declares, "I am convinced . . . that the movement to dispense with causality in physics . . . was primarily an effort by German physicists to adapt the content of their science to the values of their intellectual environment." And this intellectual environment in post-war Germany had a strong existentialist mood, with a craving for crisis and a readiness to adapt its ideology. It was also an anti-intellectual environment in which "the mathematical physicist, the personification of analytical rationality, was often singled out as the prime exemplar of a despicable way of grasping the world".

If this analysis can be accepted we may take Rutherford's defence of causality, at least in the macrocosm, as a reflection of his view that he lived in a conservative society, in which his scientific work was accorded high prestige. Immediately after the Second

World War, in which it was believed that science and scientists had
played a major part in Britain's victory, Rutherfordian doctrines of
the importance of publicly supported basic research, in which the
brilliant man should be supported to make free enquiry of nature in
his own laboratory, reigned supreme. Nor was there any questioning of the political and scientific leadership which took Britain into
the fields of nuclear weapons and nuclear power.

Just as Rutherford in his "official" life on the fringes of government became more and more involved in defence preparations in
the last year or two of his life, so his world-wide correspondence
reflects the growing international tensions and the personal problems that the Fascist regimes of Europe were posing for scientists.
In 1936 Rutherford was appointed to the newly reformed Papal
Academy of Sciences, a strange distinction for a man who was
known to be completely indifferent to religion of any type. But he
accepted the position without demur because, as he told his Dutch
colleague Professor Zeeman, he thought it might do something to
ease the strain in "international relations". In November of that
same year he told von Hevesy that he was frankly puzzled by the
international scene because "the present outburst of ill-will between the Fascist and Communistic states seems to me very
unnecessary and very harmful especially in vie·v of the trouble in
Spain". Even from distant, peaceful California came similar
sounds of worry, when Ernest Lawrence wrote about the problems
of arranging visits, purely scientific visits, to German laboratories
under the Nazi regime.

India and Indian science were the final subjects in Rutherford's
mind because the British Association for the Advancement of
Science was planning to hold its 1937 meeting, for the first time, in
India. The meeting was planned to coincide with the Jubilee
meeting of the Indian Science Congress at the very end of the year
and Rutherford had accepted a request that he should be joint
president of both bodies for the occasion and deliver the most
important of the speeches.

For the early summer of 1937, therefore, Rutherford turned his
mind to writing this keynote address. He sought advice from a man
who had worked long in India, had been a leading figure in the
Indian Science Congress, and had spent years urging a visit of the
British Association to India, Professor J. L. Simonsen. Rutherford
asked for details of the history and past activity of the Congress, as
well as advice on travelling problems and dress. In his letter to
Simonsen of June 18th, Rutherford offered the strategic plan of his
speech – he would outline the organisation of research in Britain
and other Dominions, and refer to the history of scientific efforts in

India. Wisely, "I shall not, except indirectly, refer to what might be done in the future in India, but I shall leave it to be inferred from my remarks on the organisation of research in this country and on the activities of our Dominions on similar lines."

He did not live to deliver this speech, but it had been completely written out and Sir James Jeans read it posthumously. It was largely through Rutherford's influence, as one of the Commissioners of the 1851 Exhibition Scholarships, that one of these awards was now assigned to India, and this was one of the points he made in the speech along with tributes to pioneers of science in India, both British, such as Everest and Ross, and Indian, including Raman. From all of which it followed that "The Indian student has shown his capacity as an original investigator in many fields of science and in consequence India is now taking an honourable part and an ever increasing share in the advance of knowledge in pure science."

But Rutherford's prescription for the future direction of Indian science was a good deal more practical than might have been expected.

> It is imperative that the universities of India should be in a position not only to give sound theoretical and practical instruction in the various branches of science, but, what is more difficult, to select from the main body of scientific students those who are to be trained in the methods of research. It is from this relatively small group that we may expect to obtain the future leaders of research both for the universities and for the general research organisation . . . This is a case where quality is more important than quantity, for experience has shown that the progress of science depends in no small degree on the emergence of men of outstanding capacity for scientific investigation and for stimulating and directing the work of others along fruitful lines. Leaders of this type are rare, but are essential to the success of research organisation. With inefficient leadership it is easy to waste money in research as in other branches of human activity.

So Rutherford transferred directly to India, without making any allowance for cultural differences, the problems that he had encountered when he faced the reconstruction of the Cavendish, and of British science in general, after the First World War. This problem of finding the balance between pure science in the universities and applied science in the "research organisation" has continued to worry Britain, America and other Western countries since the Second World War, and it is of course exactly the problem that Kapitsa pointed out to the Russian authorities when he debated the role of science in the one-party state with them.

Rutherford seems to have had no doubts that the leaders of scientific research must emerge from university research, but he was fully aware of the necessity of applied research especially in a country such as India, for he went on: "It is clear that any system of organised research must have regard to the economic structure of the country. One essential feature at once stands out: India is mainly an agricultural country, for more than three-quarters of her people gain their living from the land, while not more than three per cent are supported by any single industry . . . Research on food stuffs has a primary claim on India's attention." He also stressed the importance for India of research on radio-communications to span her vast distances and he concluded this section with: "I am, however, not unmindful of the pressing need of India to alleviate the suffering of the people from the attacks of malaria and other tropical diseases. I know that India herself is giving much thought to these vital problems in which science can give her vital help." What Rutherford could not see or foresee was that the brightest and best scientists were likely to be attracted into the most "glamorous" sciences, and that India would politically demand industrialisation when she achieved independence – the development of a major nuclear research programme, the ability to build nuclear power stations and explode nuclear devices, the construction of giant steel-making plants, and the establishment of a small but significant space programme, can hardly have fitted into his "Prevision".

But long before this speech was due to be given, just as he came back to Cambridge from his Wiltshire cottage to face the start of another academic year, he died – suddenly, after a life-time without any serious illness, he was cut down by an almost humiliatingly banal misfortune. He had suffered from a small umbilical hernia for a number of years – it never appeared to worry him and he wore the usual truss to control it. A number of his friends and colleagues, Sir James Jeans, Sir Frank Smith, Sir Richard Gregory, Eve, all remember that at a meeting at the Athenaeum to consider the future of radioactive treatment for cancer at the end of September he seemed in fine form. On Thursday, October 14th in Cambridge he complained of indigestion and vomiting. The following day his doctor was called, and that evening, after Professor Ryle had been called into consultation, Sir Thomas Dunhill was summoned from London to operate for a strangulation of the hernia. At first all seemed to be well and there was even talk of his being able to make the Indian trip. But he rapidly deteriorated and died of "intestinal paralysis" on Thursday, October 19th, with the doctors apparently powerless to do anything except relieve him of uncomfortable symptoms.

It seems a strange end, quite out of keeping with the driving success of his life. Cambridge society was critical of the medical attention he received largely, perhaps, because of Lady Rutherford's well-known unconventional views of health matters. She may have felt some implicit criticism, for writing to Oliphant in the middle of the crisis, she added a P.S. "Dunhill was Ryle's choice, an old friend. Great opinion of his diagnosis as well as his surgery." Professor Ryle himself was the Regius Professor of Medicine at the university. In the scientific world Rutherford's death was a great shock, and the announcement of it at least was in appropriate Rutherford tradition. Lady Rutherford had telegraphed to Cockcroft and Oliphant, who were attending the congress in Italy honouring the two-hundredth anniversary of Luigi Galvani's birth. The news was therefore announced by Niels Bohr, from the congress platform with tears in his eyes. Rutherford's body was cremated two days after his death, and his ashes were later buried in Westminster Abbey near the tomb of Isaac Newton.

Death always poses hypothetical questions to the survivors – but it is only in a few cases that the non-existent answers to these questions could be of interest to the world at large, to history. If Rutherford had lived to the not unreasonable age of seventy-five, instead of dying at sixty-six, he could have been active throughout the Second World War, and would surely have influenced the development and use of the first nuclear weapons. In parochial British terms his survival would surely have influenced the battle for power between Lindemann and Tizard over the direction of wartime scientific research.

In a broader field it is difficult to see how the development of the atomic bomb could have followed the same pattern had Rutherford, the acknowledged leader of nuclear experimental science, been alive. What would have happened had Rutherford come face to face with General Groves, the autocratic master of the Manhattan project? More important, would Rutherford – or could Rutherford – have influenced Roosevelt and Churchill over the decision to use the bomb more than did the unworldly Einstein and Bohr? Cockcroft, at least, believed Rutherford would have liked civilian nuclear power; he "would have been thrilled by the powerful atomic research piles generating energy quietly and safely", he remarked in a BBC radio broadcast in December 1950.

It seems likely that the explanation for the oblivion into which Rutherford seems to have fallen is twofold. Firstly he died at the wrong time. To the British public the first widely known scientific heroes of any sort were those who had "helped to win the war" – the Second World War – men such as Whittle, the inventor of the jet engine, Fleming, the discoverer of penicillin, Watson-Watt, the pioneer of radar. Parallel with these men were the "atom-pioneers", which meant the men who had worked on the atomic projects of the war years. Rutherford was ten years dead when these men's names became public property.

In the USA the process was different. There the image of the physicist is the theoretician. Einstein, the most famous of all, largely because relativity is so difficult to grasp yet so exciting, was joined by many other European refugees, Fermi, Szilard, Wigner, Gamow. And Oppenheimer, the bomb-maker, the epitome of the scientist's moral dilemma, was essentially a theoretician too, a man of the Göttingen tradition, for there he had been personally happy, while he had been disturbed and miserable during his short stay at the Cavendish. When the Second World War ended, Americans found they were one of the two super-powers economically, industrially and militarily. America was also a super-power scientifically, and most of all in the field of nuclear physics. Nuclear science laboratories were almost confined to the USA, and other countries had to devote enormous efforts to get back into the field with its promise of nuclear weapons and nuclear power. The Cavendish by the end of the war was old and dirty and run down; very few of Rutherford's men were even in Cambridge – I remember being an unenthusiastic and disenchanted physics student at the time. And so it has come about that the American view of the history of nuclear physics has prevailed.

Epilogue

In the last years of the seventeenth century the parson of the little village of Milton in Buckinghamshire was a Dr William Wotton, a man said to be "remarkable for his learning as a boy, and for no extraordinary wisdom as a man". He decided to write a memorial of Robert Boyle, second only to Isaac Newton among the British scientists of the century, and he asked his friends for details of the man. Among those who replied was John Evelyn, diarist, gardener, and one of the first Fellows of the Royal Society. He had known Boyle for forty years and he wrote:

> But by no man have the territories of the most useful philosophy been enlarged, than by our hero, to whom there are many trophies due. And accordingly his fame was quickly spread, not only among us here in England, but through all the learned world beside. It must be confessed that he had a marvellous sagacity in finding out many useful and noble experiments. Never did stubborn matter come under his inquisition but he extorted a confession of all that lay in her most intimate recesses; and what he discovered he as faithfully registered, and frankly communicated . . .
>
> Neither did his severer studies yet sour his conversation in the least. He was the furtherest from it in the world, and I question whether ever any man has produced more experiments to establish his opinions without dogmatising. He was a Corpuscularian without Epicurus; a great and happy analyser, addicted to no particular sect, but, as became a generous and free philosopher, preferring truth above all; in a word a person of that singular candour and worth, that to draw a just character of him one must run through all the virtues as well as through all the sciences. And though he took the greatest care imaginable to conceal the most illustrious of them, his charities and the many good works he continually did, could not be hid . . .
>
> In the meantime he was the most facetious and agreeable conversation in the world among the ladies, whenever he happened to be so engaged; and yet so very serious, composed and contemplative at all other times; though far from moroseness, for indeed he was affable and civil rather to excess, yet without formality.

In the 1930s this letter of 1696 was read by Sir Henry Tizard. Recognising the extraordinary similarity to Ernest Rutherford's

601

achievements, views and behaviour, Tizard copied it out with his own hand and sent it to Rutherford as a tribute. It was not intended as an epitaph, but it will stand for one.

Notes and Sources

1. *New Zealand Education*   pages 13–49

Both Eve and Feather deal rather briefly with Rutherford's career in New Zealand. There is a good deal more material in the collection made by Eve shortly after Rutherford's death which is kept with all the rest of Rutherford's papers in Cambridge University Library under (CUL) PA 312–PA 318 which has been used here. Undoubtedly the best source is E. Marsden, from whom Feather seems to have obtained much information, for Marsden was a pupil of Rutherford's at Manchester who subsequently made his scientific career in New Zealand and devoted much time to collecting material about Rutherford's early life. Marsden delivered the Fourth Rutherford Memorial Lecture on December 14th, 1949, which is printed in *Rutherford by those who knew him*. He also wrote an obituary printed in the *Transactions of the Royal Society of New Zealand*, Vol. 68. 1938, 4–16. There is additional material collected by him but never publicised in CUL PA 312–PA 317 as above.

One letter from Sir William Marris I have taken from Eve, p. 10. All other letters quoted are from CUL. CUL PA 293 is an interesting loose scrap of paper with examination marks in an unknown hand and PA 295 contains the official testimonials from Canterbury College for Rutherford's 1851 Exhibition application.

In addition, I have been provided with much local material by Mr Frank Allan, of Nelson, a former pupil and teacher at Nelson College. *The Bulletin of Nelson College Old Boys' Association Inc.*, July 1971, No. 34, contains an interesting article on Rutherford and there is also a locally printed pamphlet *The Rutherfords in Nelson*. *The Nelson College Old Boys' Register*, Nelson 1956, throws interesting light on the college history. For general information about the province I have to thank the Library of the High Commissioner of New Zealand in London.

Dr John Campbell, Lecturer in Physics at Canterbury College, who has been responsible for refurbishing "The Den" and other Rutherford historical items, has provided much local information. *The History of Canterbury College* by James Hight and Alice Candy, Whitcomb and Tombs, Christchurch, 1927, is a mine of solid information about people and conditions in Canterbury in the 1890s and before. Sir Henry Dale, in 1949, gave a lecture at Canterbury, "Some Personal Memories of Lord Rutherford of Nelson", printed by R. W. Stiles, Nelson, 1950.

The suggestion that there was an incident over the College Science Society and talks about "evolution", or that Rutherford himself lectured on "The Evolution of the Elements", comes in different forms in Feather and in Russian works on Rutherford. But I can find no original source for the story, and Dr Campbell finds no mention of it in the surviving records of the Science Society, nor any programme giving Rutherford's lecture on this subject.

Rutherford's own story about Littlejohn and cricket is CUL PA 305. For general views of the history and origins of New Zealand society there is still nothing to beat William Pember Reeves' *The Long White Cloud* – I have used the 4th Edition – George Allen and Unwin, London, 1950. And for the development of science and scientific education in the university system of New Zealand, S. H. Jenkinson, *New Zealanders and Science*, New Zealand Department of Internal Affairs, Wellington, 1940.

2. First Research   pages 50–62

Rutherford's first two research papers are, of course, the first two papers in *CPR*. The short introduction to them by Sir Edward Appleton is useful, but I have to thank Dr John Campbell, of Canterbury College, for the first glimmering of the idea that the time order of the work should be reversed.

Eve is uninformative about this period of Rutherford's life and Feather provides more detail. Brown provides most colour about the teaching episode. Once again Marsden is obviously the best informed and his unpublished material provides plenty of useful detail (CUL PA 312/7). Notes made by Mr Evelyn Shaw, formerly secretary to the 1851 Exhibition Commissioners, for Eve (CUL PA 312/8), give this fuller version of Rutherford's winning of the scholarship.

Sir Henry Dale in his 1950 Canterbury lecture (see Chapter 1 Notes p. 603) adds some detail, and the enlightening comments by Lord Bowden of Chesterfield come from a lecture he also delivered there in March 1979 which has not been published at the time of writing. The Chadwick/Eve obituary of Rutherford for the Royal Society acknowledges a debt to Professor C. C. Farr, who worked for many years at Canterbury and was a devoted admirer of Rutherford – some of his letters in CUL add small details.

The working notebook referred to is CUL NB 1.

CUL PA 312/18 contains a pamphlet sent to Eve by Charles M. Focker entitled, "A tribute to New Zealand's greatest scientist". It contains a few extra details of Rutherford at Canterbury. Focker was another who devoted much time to collecting references to Rutherford and drew up the first list of his published work.

The programme for the Science Society for 1894 is also in CUL, but Dr Campbell provided me with my own copy. Hight's *History of Canterbury College* is invaluable for conditions prevailing in the 1890s and for checking on other persons mentioned in passing. Letters – e.g. from Stevenson and Erskine – are all in CUL under authors' names.

3. The Wide, Wide World   pages 63–86

It is for this period of Rutherford's life that Eve is the most valuable, indeed often the only source. The letters written by Rutherford to Mary

Newton, his fiancée, and to his mother have now disappeared. A number of letters to his mother were published in the *Taranaki Herald* in 1935, after her death but during Rutherford's life. He was extremely annoyed and forbade any further publication. There are a number of different and dramatic stories as to the fate of his letters to his mother, letters written at least fortnightly throughout his life, but the only fact seems to be that they have disappeared.

Sir Henry Dale's Canterbury lecture of 1950 provides extra detail and confirmation of the points drawn from Rutherford's own letters.

On the founding and history of the Cavendish Laboratory the fullest work is *The Cavendish Laboratory, 1874–1974* by J. G. Crowther, Macmillan, London, 1974, though there are a number of present-day Cavendish workers who dislike the Marxist/sociological treatment by this author. *The Cavendish Laboratory* by Alexander Wood, Cambridge University Press, 1946, is felt by them, and by me, to give a better "feel" of the place where Wood worked for so many years, although it is very much more modest work. But a book *A History of the Cavendish Laboratory, 1871–1910*, Cambridge University Press, 1910, now rare and long out of print is much the best source for this chapter (and for Chapter 4 following), since it has sections written by Rutherford, by J. J. Thomson, and by several others working in the laboratory in 1910. It thus provides almost contemporary views of the chief participants. *Recollections and Reflections*, J. J. Thomson's autobiography (Bell, London, 1936), a strange, rambling collection of memories, studded with fascinating thoughts and pictures of past academic life, has also been extremely useful.

Finally, through the courtesy of the Librarian of Trinity College, Cambridge, I have been able to use the obituary of Rutherford written by J. J. Thomson for the *Cambridge Review* of November 5th, 1937, and the obituary of J.J. himself from *Country Life* of September 14th, 1940.

4. *Science in Cambridge*   pages 87–129

The best source for Rutherford's scientific work in this period is *A History of the Cavendish Laboratory, 1871–1910*, mentioned above. It contains two sections by J. J. Thomson and Chapter VI (p. 159 et seq) by Rutherford himself on the period 1895–98. Both men must have remembered the time quite clearly since they wrote little more than ten years after the event, yet they would have had time to obtain a perspective on what was likely to be of lasting importance. I have quoted from this book several times in this chapter. J. J. Thomson's autobiographical sketches, *Recollections and Reflections*, G. Bell, London, 1936, gives some additional material.

The fine series of letters to his mother and fiancée in New Zealand quoted by Eve provides the next main source of material, especially of a personal nature. These letters are no longer available and all my quotations come from Eve.

Rutherford's six scientific papers from this period are all printed in the

Collected Papers where they form a formidable block from p. 80 to p. 215. All the European work which opened up his subject, such as Röntgen's and Becquerel's papers are cited there. But the Becquerel papers in particular are more conveniently available in *The Discovery of Radioactivity and Transmutation*, by Alfred Romer, Dover, New York, 1964, and I have also quoted from Romer's valuable introductory material (p. 1).

Both Feather and E. N. da C. Andrade's *Rutherford and the Nature of the Atom*, Anchor Books, Doubleday, New York, 1964, are useful in giving a scientist's view of these early developments, and I have quoted from Andrade pp. 35, 36.

Sir Henry Dale seems the best contemporary witness to have left us with a personal view of Rutherford and Cambridge society. Sir Henry started collecting material for a life of Rutherford in the early 1950s and received considerable assistance from Marsden in New Zealand – indeed it seems the book was intended to be a joint effort. The Royal Society Library kindly drew my attention to this material which is kept among Sir Henry Dale's papers in a separate box. Early drafts of the proposed first three chapters of the book are among these papers along with reminiscences (not very valuable) from James Rutherford and one or two interesting articles by New Zealand scientists. The essence of this research was put by Sir Henry into his Canterbury Lecture of 1950, given at the Cawthron Institute (and mentioned in the Notes to Chapter 3). But his memories of the Coutts Trotter Scholarship and also of Rutherford's attitude to commercial development are in his draft chapters. I owe thanks to Lady Todd for permission to use her father's papers.

Rutherford's own notes in CUL, chiefly NB 4 and NB 5 are not particularly illuminating – in fact they are rather messy, mixing up notes from what seem to be J.J.'s lectures with lab jottings and a considerable amount of mathematical theorising and algebraic working.

Rutherford's memories of Elliot Grafton Smith are to be found in PA 308, CUL. The quotation from Sir Peter Medawar comes from *Induction and Intuition in Scientific Thought*, p. 56, Methuen, London, 1969.

The article by Chadwick and Eve which makes the claim that Rutherford was first to observe gamma-rays is "Obituary Notices of the Royal Society of London" 2 (1938), 395–423. But standard reference works, e.g. *Dictionary of Scientific Biography*, still claim that "Villard was the first to observe a penetrating radiation which he named gamma-radiation". He was working with radium at the Ecole Normale in Paris in 1900 and confirmed his work with a more active radium source provided by the Curies. He referred to this radiation as a type of X-ray which was more penetrating than the charged radiations discovered by Rutherford. Paul Villard, born Lyons 1860 – *Comptes Rendues* 130, 1900, (1010–12, 1178–79, 1614–16).

In CLM 2, CUL we can find Rutherford working up his uranium paper. There is a firm date, February 24th, 1898, but much of the basic work seems to have been done beforehand, for many of the carefully written phrases here recur in the final version of the paper. The preceding notebook, CLM 1, is hardly dated at all. Under "Oct 1897" there is work on the velocity of the carbon dioxide ion and mention of ultraviolet light. Under a section that may be dated November (1897?) there are uranium experiments, comparisons of effects with uranium radiation and Röntgen radiation. Then comes "Lent Term" – possibly 1898 – and the experiments are nearly all on uranium with occasional mentions of ultraviolet light. It is under the possible November work that we first find one of his comments in the notebooks, "Very accurate experiment".

5. *Radioactivity*   pages 130–65

The Canadian period of Rutherford's life is the only one that has attracted major academic studies. The Rutherford–Soddy partnership is the subject of Dr Thaddeus J. Trenn's *The Self-Splitting Atom* (Taylor and Francis, London, 1977). I have followed Dr Trenn's interpretation at all major points because I have found that his coverage of the evidence is sound, thorough and accurate. If, in his effort to reassert Soddy's claim to a full half share of the partnership, Dr Trenn has perhaps overemphasised Soddy's work, he has, I suspect, faced the same difficulty that I have met, namely that Soddy writes so much more appealingly about himself than Rutherford did.

On only one major point do I disagree with Dr Trenn and that is that I think he has underestimated the importance of Soddy's lecture on alchemy and chemistry. This is clearly dated 1900 by both Alton and the Bodleian Library, but it surely must belong to 1901 where Trenn places it (without discussing the matter). A 1900 date would imply that Soddy was considering alchemy and by implication transmutation seriously before he became Rutherford's collaborator. But even given a 1901 date, I feel that Soddy's statements in it, taken with other evidence, imply that Rutherford and Soddy knew "in their bones" where they wanted to get to long before they could formally prove their points. I therefore do not share Dr Trenn's rather serial and logical view of their progress towards the disintegration theory.

Dr Trenn has also produced two important papers, "Rutherford and Recoil Atoms" in *Historical Studies in Physical Sciences*, Vol. 6, 1975, and "Rutherford in the McGill Physical Laboratory". The latter in particular, giving a detailed account of Rutherford's 1902 alpha-ray experiment and showing how Rutherford reached the right conclusion despite two calculational errors, is most valuable, though again I feel that it displays the man knowing where he had to get to, rather than finding his way there; but at the same time he always accepted experimental evidence even when it surprised him, as did the charge on the alphas. This paper of Dr Trenn's was delivered as part of the Rutherford Symposium at McGill in 1977, the proceedings of which are published as *Rutherford and Physics at the Turn of the Century*, Dawson and Science History Publications, New York, 1979. From Professor L. Badash's paper at the same symposium I have taken the material about the popular impact of radium.

All the quotations from scientific papers are taken from CPR. All the quotations from Soddy are taken from his collection of papers, deposited at the Bodleian Library, Oxford, by his literary executor, Mrs Howarth, with the one exception of the final quotation of the chapter, taken from an article in *McGill University News* which I found only in Sir Henry Dale's collection of Rutherford material.

Dr Robert Michel, Archivist of McGill University, and Professor F. R. Terroux have most kindly provided lists of the material in their possession, including a brochure of their collection of Rutherford apparatus, and also photocopies of the minutes of the McGill Physical Society (though these minutes are disappointingly short and formal). Even more valuable was their kind listing of all Rutherford's and Soddy's reports to that society. A photocopy of Professor Norman Shaw's article also came from McGill.

The letters from and to J. J. Thomson and Sir William Crookes are in CUL, but some are in the J. J. Thomson collection, namely those from Rutherford.

E. N. da C. Andrade's *Rutherford and the Nature of the Atom*, Doubleday, New York, 1964, has proved useful especially in explaining exponential laws. Alfred Romer's *The Discovery of Radioactivity and Transmutation*, Dover, New York, 1964, has useful linking material between some of the original papers and I have quoted from it once.

With this wealth of material Eve and Feather become less important, and the 1902 letter from Rutherford to his mother was first printed in the *Taranaki Herald*, having been provided by James Rutherford. It is also in Eve. J. J. Thomson's *Recollections and Reflections* was naturally used for the discovery of the electron. Although I have not quoted from Lucretius, I have based my opinions on R. E. Latham's version, Penguin Classics, 1951.

There is one extra important letter from Soddy (Bodleian, b. 189 237) which he wrote to Eve in April 1938, apparently for Eve's biography. I have not quoted from this but it forms general background, and in the same way I have drawn on impressions from Soddy's own book, *The Story of Atomic Energy*, Nova Atlantis, London, 1949.

The chapter ends with the quotation from Rutherford–Soddy "Radioactive Change", originally in *Philosophical Magazine*, May 1903, CPR p. 608.

6. *Life in North America*   pages 166–92

The major sources for Rutherford's scientific ideas in his later years at McGill are his own books: *Radioactivity*, Cambridge University Press, First Edition 1904, Second Edition 1905, and *Radioactive Transformations*, Constable, London, 1906.

As stated in the text, a certain amount of scholarly work has recently been published on this period of Rutherford's life. I have used and quoted from Dr Lewis Pyenson "The Incomplete Transmission of a European Image: Physics at Greater Buenos Aires and Montreal, 1890–1920",

Proc. Amer. Phil. Soc. Vol. 122, No. 2, April 1978, 92–114. From *Rutherford and Physics at the turn of the Century*, ed. Bunge and Shea, Dawson and Science History Publications, New York, 1979, which includes the papers from the Rutherford Symposium at McGill in 1977, I have used L. Badash "The Origins of Big Science: Rutherford at McGill", "Physics at McGill in Rutherford's Time", by John L. Heilbron, and Norman Feather, "Some Episodes of the Alpha-Particle Story, 1903–1977". Dr W. Bennet Lewis has kindly sent me a reprint of his Ontario Science Centre lecture of October 1979, "Nuclear Research in Canada from Rutherford to McGill to Candu".

The standard biographies, by Eve, Feather and Andrade, are all useful for this period, but earlier biographers did not have available the results of Professor Badash's work in the shape of *Rutherford and Boltwood: Letters on Radioactivity*, Yale University Press, 1969. Rutherford's letters to his mother are taken from Eve.

On the subject of Rutherford's lecturing I have drawn on many personal accounts given to me in interviews by people who knew him. But I have also quoted from Sir Henry Tizard's Memorial Lecture – reprinted in the *Journal of the Chemical Society*, 1946, pp. 980–986. Some unusual bias may have crept in here, for I feel that Rutherford's cavalier attitude towards elementary lectures and the élitism which Rutherford taught his own disciples persisted until the years after the Second World War in the Cavendish, and may well still persist today. Certainly the education offered to the student who was not going to be a brilliant physicist in the years 1945–1950, i.e. the instruction offered to me, was not of the highest class nor likely to arouse enthusiasm.

Hahn's extensive correspondence with Rutherford is almost entirely confined to matters scientific; his memories of McGill days were communicated to Eve in 1936 and are filed with similar material in CUL PA 312 and 313.

7. *Last Years in Canada*   pages 193–215

Rutherford's researches during his last years in Canada are covered in his book, *Radioactivity*, Cambridge University Press, Second Edition, 1905. His third book, *Radioactive Transformations*, Constable, London, 1906 (which is the printed version of his Silliman Lectures at Yale in 1905) also provides a major source for this chapter.

Radioactivity and Atomic Theory by Frederick Soddy, edited by T. J. Trenn, Taylor and Francis Ltd., London, 1975, which is a collection of Soddy's annual reports to the Chemical Society, shows how radioactivity research looked from a different angle.

For Rutherford's correspondence on his research, Lawrence Badash's work, *Rutherford and Boltwood: Letters on Radioactivity*, Yale University Press, 1969, is most useful. It is a great pity that Rutherford and Bumstead did not correspond more frequently, for their relationship, one may suspect, was livelier and more important in the long run.

G. M. Caroe's life of her father, *William Henry Bragg*, Cambridge

University Press, 1978, is important for the start of another of Rutherford's long friendships, but I believe some of the Bragg–Rutherford letters are quoted for the first time here. The letters of Erskine Murray on wireless patents and the Marconi Company are from CUL.

I have used Morris W. Travers' *Life of Sir William Ramsay*, Edward Arnold, London, 1956, to try to counterbalance the anti-Ramsay feeling which was so widespread among Rutherford's circle. But all the letters I quote from J.J. are in CUL, where Rutherford's replies to him are also to be found, among the Thomson papers.

Once again I am grateful to the McGill Archivist for the material mentioned in the Notes to Chapter 5.

8. *Starting in Manchester*   pages 216–37

With Rutherford's Manchester period we come to the time when there are a few living survivors who remember him – notably Lady Alice Bragg to whom I am most grateful for an interview in which she was able to give otherwise unavailable details of Rutherford's home life. This is also the time when Rutherford became widely-enough known to appear in works of a non-scientific character. The interesting fact that his relationship with A. J. Balfour started at this time emerges from *Arthur James Balfour* by Kenneth Young, Bell, London, 1963, and also from Chaim Weizmann's autobiography *Trial and Error*, Hamish Hamilton, London, 1949, which naturally includes Weizmann's own memories of Rutherford, Einstein and their mutual relations.

Rutherford's relations with Sir William Ramsay were in a very different tone and I have again used Travers' *Life of Sir William Ramsay* to try to balance the anti-Ramsay view that nearly every one of Rutherford's correspondents give. The correspondence with Sir Arthur Schuster, which I believe has never been used before, comes from the Royal Society Library.

9. *Science International*   pages 238–67

The bulk of the material in this chapter comes from the Rutherford correspondence in CUL, letters both outgoing and incoming. The authors or addresses have been named in all cases. The Hahn correspondence also comes from CUL (312, 313) and some additional material comes from the Soddy papers in the Bodleian Library.

I have made some use of the contributions of H. R. Robinson and E. N. da C. Andrade to the Rutherford Jubilee International Conference at Manchester University in September 1961. These appear in *Rutherford in Manchester*, edited by J. B. Birks, Heywood, London, 1962, and this collection now assumes greater value than the biographies of Rutherford by Eve, Feather and Andrade, though I have continued to use all of these to some small extent.

Marie Curie by Robert Reid, Collins, London, 1974, has much to say about Rutherford and the very different personality he could show when moved by sympathy.

10. The Atom   pages 268–307

Since this chapter covers Rutherford's greatest scientific discovery, and since many of those who worked with him at the time outlived him and were distinguished men at the time of the fiftieth anniversary of his discovery of the nucleus, there is a wealth of reminiscence available. *Rutherford in Manchester*, ed. J. B. Birks, Heywood, London, 1962, records the commemorative speeches at the Jubilee Celebrations of 1961, and also brings together much other material, reprinting some Rutherford Memorial Lectures, providing a bibliography of the papers published by the Manchester Physics Laboratory in the crucial years, giving some of his correspondence with Sir Arthur Schuster, etc. I have used particularly the contributions of H. R. Robinson in this chapter.

Norman Feather has been the most discerning of Rutherford's former pupils and memorialists. His contribution to the unofficial Royal Society's collection of memorial lectures, grouped under *Rutherford by those who knew him*, and originally published as "Rutherford–Faraday –Newton", *Notes and Records of the Royal Society of London*, Vol. 27, August 1972, is particularly helpful, and I have quoted briefly from it. Feather organised a display of original Rutherford materials for the Royal Society Centenary celebrations, and made the first proper search of the mass of correspondence and papers in CUL. He then found the marked reprint of Newton's *Principia* and the group of papers (CUL PA 194) marked "Theory of Structure of the Atom".

I have found "Rutherford and Recoil Atoms" by T. J. Trenn, *Historical Studies in Physical Sciences*, Vol. 6, 1975, p. 513, very useful.

I have quoted from J. L. Heilbron's *H. G. J. Moseley, the Life and Letters of an English Physicist, 1887–1915*, University of California Press, Los Angeles and London, 1974, both from the author's biographical sketch of Moseley and from the Moseley letters in this book, but I find it somewhat marred by an apparent antipathy to Rutherford.

I have also relied on Trenn in *ISIS*, Vol. 65, No. 226, March 1974, for his work on the Geiger–Marsden scattering paper and its origins. The Geiger–Marsden paper is "The Laws of Deflection of Alpha-Particles through large Angles". *Philosophical Magazine*, Vol. 25, 1913, 604–623. *Sir Ernest Marsden, 80th Birthday Book*, by A. H. and A. W. Read, Wellington, 1969, is a rare source with some valuable lights on the period, which I obtained through the kindness of Mrs A. B. Wood, who also provided a copy of her late husband's memories of Niels Bohr, which he compiled for Sir John Cockcroft's use. E. N. da C. Andrade, apart from academic memoirs, wrote a lively article for *New Scientist*, March 20th, 1958, entitled "When Studying Science was Fun", which also throws some less academic light on Rutherford's Manchester days.

Hans Geiger was the most devoted of Rutherford's disciples throughout his life, and he wrote a number of brief memoirs in German scientific magazines on such occasions as Rutherford's sixtieth birthday. These were collected by Eve and are to be found in the biographical material in CUL. Geiger also wrote a famous letter to *Nature* after Rutherford's death: *Nature*, Vol. 141, February 5th, 1938, p. 244.

All other quotations in this chapter come from Rutherford's correspondence in CUL, under names of correspondents.

Other material in this chapter from CUL includes Rutherford's brief start on autobiographical notes, PA 298 and laboratory notebooks for the period in PA 194, 175 and 178. In Eve's collection of material there is a note by Sir Max Perutz describing the history of the Braggs' discoveries in X-ray diffraction which I have used.

11. *The Atom in Action*   pages 308–38

As for the previous chapter, *Rutherford in Manchester*, edited by J. B. Birks, provides an important source of background material, especially in the form of the contributions by Niels Bohr, A. S. Russell and Sir Charles Darwin. Darwin published a number of other versions of the "Discovery of the Atomic Number", but there are no important variations between them. He also wrote a long and helpful letter to Eve describing the events of the "birth of the nucleus", which is in the Rutherford Papers, ADD 7653, in CUL.

I have quoted from the article, "The Genesis of the Bohr Atom" by J. L. Heilbron and T. S. Kuhn in *Historical Studies in Physical Sciences*, Vol. 1, 1969, p. 211. There are, however, passages in it which carry a sense of determination to denigrate Rutherford's role in the intellectual development of Bohr's ideas, and I cannot accept some of the authors' ideas where, by their own admission, they insist upon their own version of timing and development against Rosenfield, against the Manchester evidence and against Bohr's own memories, so that finally they are forced to adopt a psychological explanation for Bohr's telling them that their questions were "silly".

I have also used and quoted from T. S. Kuhn's *Black Body Theory and the Quantum Discontinuity*, Clarendon Press, Oxford, 1978, and I have looked briefly at the author's broader views on the sociology of science.

Soddy's annual reports to the Chemical Society, reprinted in *Radioactivity and Atomic Theory*, edited by T. J. Trenn, Taylor and Francis, London, 1975, provide an important running commentary on the progress of the rival theories through the years from 1910 to 1914.

All other quotations in this chapter come from Rutherford's correspondence in CUL, under the names of the correspondents, with the exception of letters in the Schuster collection in the Royal Society Library. The only Schuster letters that have been previously printed concern the appointment of Rutherford to Manchester, so these additional letters form a very valuable source, as Rutherford seems to have been unusually frank with his benefactor. For instance, after the meeting on the constitution of atoms at the Royal Society, which forms the conclusion of this chapter, Rutherford wrote to Schuster (and to von

Hevesy) that it had been "a great meeting" at which "the Rutherfordians celebrated the funeral of the old atom model of J. J. Thomson".

Other material from CUL includes a whole series of memories of Rutherford and praises of him from Niels Bohr in 312/14, 312/15, "Notes about my relations to Lord Rutherford" and 312/16.

Rutherford's manuscripts for his 1913 Royal Institution lectures provided some material, but possibly the most valuable item is the typescript of Rutherford's Presidential Address to the meeting of the Science Masters Association of January 2nd, 1923, titled "A Page of Scientific History" (CUL ADD 7653, PA 115). This was written within ten years of the revolution he and his colleagues wrought – time for memory to be still fresh and accurate while perspective has been achieved. It is also one of Rutherford's best and liveliest pieces of writing, over which he obviously took much trouble, and I have quoted from it extensively to give his own views of what happened and what was its significance.

12. *Rutherford at War*   pages 339–85

Very little has been written about Rutherford's work during the First World War, partly because, as has been explained in the text, the main outcome, the invention of ASDIC or sonar, was still an official secret at the time of the first biographies. Also, because this work is quite outside the mainstream of his scientific development in nuclear physics, it has been of little interest to the many academic scientists who have remembered him in memorial lectures or other reminiscences.

The general history of the BIR and the beginnings of the Naval Scientific Service are best given in "The A. B. Wood Memorial Number" of the *Journal of the Royal Naval Scientific Service*, Vol. 20, No. 4, July 1965, which reprinted four articles that Wood had written about the earliest years of the service. "Science and the Admiralty During World War I: The Case of the BIR" by Jack K. Gusewelle, which forms Chapter Seven of *Naval Warfare in the Twentieth Century*, ed. Gerald Jordan, Croom Helm, London, a collection of essays in honour of Arthur Marder, gives the story at the official level, and is highly critical of the Admiralty and the Royal Navy. The latest work on the development of ASDIC, and many other anti-submarine devices, is "Underwater Acoustics and the Royal Navy, 1893–1930" by W. D. Hackmann, *Annals of Science*, 36, (1979) 255–278. But no author has previously used either Rutherford's letters to Wood or the vital correspondence between Rutherford and Boyle in CUL. The claims in this chapter for Rutherford's previously unacknowledged leading role in the invention of sonar are fully supported by his laboratory notebooks – notably Numbers 13 to 17 – which again do not appear to have been consulted by previous authors. These notebooks show the vast amount of work Rutherford performed in the laboratory on "piezoelectriques", and even on the electrical amplifiers needed to make this system successful. The notes also show that he gave much consideration to the practical mounting of the quartz crystals in shipborne devices, including many sketches giving arrangements very similar to those that were later used in service. There is an interesting undated note in the correspondence files from the Duc de Broglie." Unfortunately, due to the broken nature of Rutherford's a specimen of the large French quartz crystals (and inviting him to his home) – and correspondingly the laboratory notebook for November 22nd, 1916 notes, "Testing of new 'piezo' brought over by Duc de Broglie". Unfortunately, due to the broken nature of Rutherford's wartime notes (which leads to even worse confusion in the next chapter) it is impossible to reconstruct the precise progress he was making.

The Public ·Records Office at Kew holds the official records of BIR under ADM 116/1430, including the Sothern Holland Report and the full version of Rutherford's report on his mission to the USA. Other BIR papers are in ADM 212/160, with details of the later work on supersonics, and ADM 212/159 is a useful list of BIR reports. But full copies of Rutherford's early "secret" reports on hydrophones, acoustics, etc., are in CUL in the Thomson papers under ADD 7654 and 7654 C42A and ADD 8243.

It has been made clear in the text that much new material has been found in Rutherford's letters to A. B. Wood, and I am most grateful to his widow, Mrs A. B. Wood, for letting me have these letters along with many other useful documents, such as the originals of the memoirs that Wood gave to Eve and to Chadwick. Mrs Wood kindly gave me her memories of life at the unhappy Hawkcraig Establishment.

Other personal memories of Hawkcraig are given by G. M. Caroe in her biography of her father, *William Henry Bragg* (CUP, 1978) and I have quoted from some of Bragg's letters that she has printed. But where Bragg wrote to Rutherford I have quoted direct from the letters in CUL. There are also some quotations from Kenneth Young's *Arthur James Balfour*, Bell, London, 1963. The letters from Rutherford to his wife, and the letter from his mother were printed by Eve, and all other quotations come from CUL.

13. *The Atom is Smashed*   pages 386–405

It is an interesting example of how the perspective of time changes the valuation we put upon past events that Eve, Rutherford's original biographer, writing in 1937, devoted less than a page of his book to the splitting of the atom, and even Feather, writing almost entirely about Rutherford's scientific work, but writing in 1940, accorded this piece of work only six pages.

Almost the whole of this chapter comes from the Rutherford Papers in CUL. The Hale lectures are in PA 108 and the notebooks relevant to 1914–1919 are NB 13 to NB 25. The matter of the shorthand writers is referred to in the letters to and from W. E. Hale.

All quotations from the final scientific papers are taken from CPR, and the passage from Marsden comes from *Rutherford at Manchester*, ed. Birks which gives the text of Sir Ernest Marsden's speech at the commemorative session.

14. *Cambridge and the Cavendish* pages 406–52

The manoeuvres to bring Rutherford to Cambridge can all be found in Rutherford's correspondence in CUL, but I have also added by using J. J. Thomson's *Recollections and Reflections*, Bell, London, 1936.

The bulk of the material for the whole of this chapter also comes from the Rutherford collection and from his published papers in CPR (Volume Three). In these years Rutherford was giving more and more public lectures, many of which are drafted in full in CUP; they give a clear picture of his slowly changing views and his steady struggle with the apparently insuperable problem of finding the structure of the nucleus – for instance his presidential address to the B.A. is in PA 134, as are his notes (on cards) for his addresses to the Royal Society. His address to the Franklin Institute of 1924 is in ADD 7653, PA 26, although this is also published in the *Journal of the Institute*, 198, 725–744. His speech to the Volta Centennial Conference is at the same reference in CUL, but is also in *Acti. Cong. Intern. die Fisica*, Como, 55–64.

Rutherford's opening moves at Cambridge – his "Report on the History and Needs of the Cavendish Laboratory", is PA 362. I am grateful to Lord Ashby for drawing my attention to Rutherford's leading role in the Cambridge controversies over degrees. All the material used here comes from the University Archives, principally the reports of the Board of Research Studies. *The Proceedings of the Second Congress of the Universities of the Empire, 1921*, ed. Alan Hill, G. Bell and Sons, London, 1921, has also been used.

The most important addition to the history of the Cavendish Laboratory and Cambridge physics of this period is provided in the long interviews recorded with Sir James Chadwick by Weiner of the American Institute for the History of Physics, and I am most grateful to that organisation for supplying transcript material. I have also used, and quoted from, Ronald W. Clark's *Einstein*, Hodder and Stoughton, London, 1979. For the brief comparison with Ernest Lawrence's Berkeley Laboratory, I have referred to *Cern Courier*, Volume 21, October 1981, where there is a long, unattributed article on the fiftieth anniversary of the laboratory.

The details of Rutherford's household arrangements are in CUL but were not formally catalogued when I was given access to them, since they had only just been found. Rutherford's notebooks relevant to this period run from NB 28 to NB 49 in CUL. The quotations from the Duc de Broglie's article are from *Nature*, 129, May 7th, 1932, p. 665.

15. *Politics and Power* pages 453–95

This chapter represents, I believe, the first study of Rutherford as a scientific administrator in quasi-government circles. He was remarkably discreet about this aspect of his life and obviously did not speak much about it to his scientific colleagues at Cambridge. Thus little of his correspondence deals with these matters, except for his letters to Marsden where he was able to discuss mutually interesting subjects with a man who had a similar standpoint. All letters quoted are to be found in CUL, filed under the name of the correspondent. The immediate post-war correspondence with Bumstead, Hales, Langmuir and Millikan about American scientific development all comes from this source and has not previously been reported. Among British correspondents Threlfall, J.J., Marsden and Merz are the most important for this period. The "Secret" memorandum of 1937 is also included under Merz's letter as M.101a. There are a group of letters to Walter Adams about the work of the Academic Assistance Council, but all other letters on this work come under the heading of the correspondents. Letters between J.J. and Merz and Threlfall in CUL ADD 7654 on the post-war development of naval science have been used, but Rutherford's letter to the Admiralty from the Lancashire Committee is filed as PA 332. I have again referred to Gusewelle in this section.

Rutherford's diary for 1926 when he was PRS is PA 303.

For Rutherford's work at the DSIR the best source is the Public Records Office at Kew under DSIR 1 (7) and (8) and (9) and DSIR 2 (68). The American Institute for the History of Physics interview with Chadwick gives useful additions, and there is further detail in *Tizard* by Ronald W. Clark, Methuen, London, 1965. *The Department of Scientific and Industrial Research*, by Sir Harry Melville, Allen and Unwin, London, 1962, the official history of DSIR, is not as helpful as might be expected, but has been quoted from.

I am particularly grateful to Mr Don Cawthron, of the Medical Research Council for calling my attention to the interesting Rutherford material in the Council's files. Correspondence on Dr R. H. T. P. Harris is 2110/1, and the problem of radium supplies to doctors and physicists is filed under 1074; but the most interesting part of this material is filed under 1776, the material about massive radium therapy which so well shows the working relationships of Fletcher, Rutherford, Frank Smith and others who managed business of all sorts at the Athenaeum.

The Written Archives Section of the British Broadcasting Corporation provided the material on Rutherford's work in and for the BBC, under the General Advisory Council, File 3 and R/51/339/1. The same source provided the material from the *Radio Times* of September 28th, 1923, and the script of Rutherford's National Lecture, published in *The Listener*, Vol. X, No. 249, October 18th, 1933.

The sources for Rutherford's work for German refugees are in his correspondence – letters to Walter Adams outstanding, and also some material from Eve, about the Albert Hall meeting. Rutherford's letter to *The Times* and the article "The Wandering Scholars" is CUL PA 335.

A Defence of Free Learning, by Lord (William) Beveridge, OUP, London, 1959, is a very good history of the movement and the people involved in it, and their motivations, while Max Born's autobiographical notes *My Life, Recollections of a Nobel Laureate*, Scribner, New York, 1979, gives a more personal view of some of these events and Rutherford's role. *The Division in British Medicine* by Frank Hogsbaum, Kogan Page, London, 1979, is a recent work which throws some unpleasant light on the sort of attitudes Rutherford had to face in trying to help refugees.

For the section on Rutherford and Defence Science in the 1930s I am indebted principally to the memories of Lord Snow, given in an interview, and to Ronald W. Clark, whose book, *Tizard*, has already been mentioned. That and *The Birth of the Bomb*, Phoenix House, London 1961, were kindly supplemented in personal conversation. *The Prof in Two Worlds, the Official Life of Prof. F. A. Lindemann (Viscount Cherwell)*, by the Earl of Birkenhead, Collins, London, 1961, has been the source of two short quotations, though the book is most remarkable for the almost complete absence of Rutherford from its pages, which perhaps means that the dislike of the men was mutual. Finally I have drawn from the Rutherford Memorial Lecture delivered to the Chemical Society on March 29th, 1939 by Sir Henry Tizard, and reprinted in the *Journal of the Chemical Society* in 1946, pp. 980–986.

16. *Kapitsa*   pages 496–537

The central core of this story comes from a file kept in CUL with the Rutherford papers but available only with the express permission of Academician P. Kapitsa. The material was kept in the Mond Laboratory and handed down from Cockcroft to Professor J. Allen, to Sir Lawrence Bragg, until the Cavendish finally deposited it with the University Library. I am most grateful to Academician Kapitsa for his permission to use this material and quote from it. It is easy to see that it was very "hot" in its day and I believe none of this material has ever been published before. Everything in this chapter that occurred historically after October 1934, that is the date of Kapitsa's detention in Russia, comes from this source.

Much of the material from the earlier part of the chapter also comes from Academician Kapitsa, by one route or another. Many of his own memories – and his views – are to be found in the collection of his articles and addresses published as *Experiment, Theory, Practice*, D. Reidel Publishing Company, Dordrecht, Holland, 1980. Possibly the most useful item is Professor Kapitsa's "Recollections of Lord Rutherford", the text of the address he gave to the Royal Society in 1966, *Proceedings of the Royal Society* A 294 123–137 (1966). Kapitsa's letters to his mother from Cambridge in his first days in England were originally printed in *Ernest Rutherford* by O. A. Staroselskaya-Nikitina – Esdatelstvo, Nayka, Moscow, 1967, from which I have had them translated. In a different translation some of them were also printed in *Peter Kapitsa; Life and Science*, ed. Albert Party, Macmillan, New York, 1968. I have used the *Great Soviet Encyclopedia* for background on some of the personalities that Kapitsa had to deal with in Moscow, notably V. I. Mezhlauk.

For the financing of Kapitsa's first large experiments – and the first warnings that he might one day return to Russia – the source is DSIR 3, 202–223, in PRO. I am most grateful to Professor J. Allen of St Andrew's University for guidance on the state of affairs at the Mond Laboratory after Kapitsa's departure. Finally, I have used C. P. Snow's essay on Rutherford in his *Variety of Men*, Penguin, London, 1969, and quoted from this source, and also the chapter by the same author, "Rutherford and the Cavendish", in *The Baldwin Age*, ed. John Raymond, Eyre and Spottiswoode, London, 1960. Lord Snow was a personal friend of Kapitsa and gave me further help in a personal interview.

17. *Final Triumphs*   pages 538–600

Naturally there are more people still alive who remember Rutherford in his last years, therefore this chapter is built more than any other on personal interviews with those who remember him as grandfather, friend, or Professor. The picture of Rutherford in the middle 1930s has emerged from the recollections of Professor Sir Harrie Massey, Professor Philip Dee, Professor T. E. Allibone, Dr J. A. Ratcliffe, Professor D. Shoenberg, Professor Sir Nevill Mott, Dr C. Kempson, and Sir Frederick Dainton. Most of the information from Cockcroft, Chadwick and Gamow also emerged in interview form, though their interviews were recorded for the Niels Bohr Archive of the American Center for the History of Physics, which has provided me with copies. In addition Professor E. T. S. Walton, Sir Mark Oliphant, Dr W. Bennet Lewis, and Professor J. Allen, have all kindly written to me with answers to my questions, and other information.

I have also used Sir James Chadwick's, "Some personal notes on the search for the Neutron", *Proc. Tenth Int. Cong. of the History of Science*, Ithaca, New York, 1962, pub. Paris, Herman, 1964 and "James Chadwick" by Sir Harrie Massey and Norman Feather, *Biog. Memoirs of Fellows of the Royal Society*, Vol. 22, November 1976. And Professor Allibone's recollections were supplemented by his address to the Royal Institution of London of October 29th, 1971, entitled "Rutherford: A Century of Nuclear Energy".

Formal scientific publications referred to are: J. Chadwick, "Possible existence of a neutron", *Nature*, No. 3252, Vol. 129, Feb. 27th, 1932, p. 312; J. Chadwick, "The existence of a neutron", *Proc. Royal Society*, A. 136, 1932, 692–708; Cockcroft and Walton, "The Disintegration of elements by high velocity protons" *Proc Royal Society* A. 137, 1932, 229–242. All references to Rutherford's own papers come from CPR.

By this stage of his fame Rutherford began to appear in more books on general subjects. I have used Lancelot Law Whyte, *Focus and Diversions*, Cresset Press, London, 1963; Gary Wersky, *The Visible College*, Allen Lane, London, 1978; Maurice Goldsmith, *Sage: A Life of J. D. Bernal*, Hutchinson, London, 1980; C. P. Snow's chapter on "Rutherford and the Cavendish" in *The Baldwin Age* ed. John Raymond, Eyre & Spottiswoode, London, 1963, and Snow's chapter on Rutherford in his *Variety of Men*, Penguin, London, 1969. Lord Snow also gave me an interview before his recent death.

Rutherford's own books of this period are *Radiations from Radioactive Substances*, by Sir Ernest Rutherford, James Chadwick and C. D. Ellis,

CUP, 1930, and *The Newer Alchemy*, by Lord Rutherford, CUP, 1937.

J. G. Crowther, the doyen of science journalists in Britain, wrote the official history of the Cavendish, *The Cavendish Laboratory 1874–1974*, Macmillan, London, 1974, and the important quotations from that work which I have used come from pp. 186 and 187; and from pp. 230–31.

I have also quoted from Crowther's chapter on Rutherford on his *British Scientists of the Twentieth Century*, Routledge and Kegan Paul, London, 1952. The recollections of that other pioneer of British science journalism, the late Lord Ritchie Calder, were given to me in an extremely pleasant personal interview in the House of Lords. I have also consulted the biographies of Eve and Feather again for this chapter.

More personal touches which can be found in print are in Sir Mark Oliphant, *Rutherford: Recollections of the Cambridge Days*, Elsevier, Amsterdam, 1972, and in the brief memoir by F. G. Mann, *Lord Rutherford on the Golf Course*, privately printed in Cambridge, 1976. I also received help in personal interviews with Lady Alice Bragg, Mrs Phyllida Cook, who took a large part in the upbringing of the Rutherford grandchildren, and from two of those grandchildren, Professor Peter Fowler and Dr Ruth Edwards. I found Professor S. Devons' memoir in *A Hundred Years of Cambridge Physics, 1874–1974*, ed. Dennis Moralee, published by the Cambridge University Physics Society, in CUL at 974.8.

All quotations from letters come from the Rutherford files in CUL where they can be found under the name of the correspondent. Virtually all the quotations from lectures come from the same source – PA 120 contains the Thomas Hawksley Lecture, the Boyle Lecture at Oxford, 1933, and the Mendeleev Centenary Lecture; PA 134 and PA 153 contain the notes for and the scripts of many other addresses and lectures. Rutherford's notebooks for the 1930s are PA 211 through to PA 236. The full version of his Norman Lockyer Lecture is CUL 9340 C.461.

Finally, I am again grateful to Dr W. Bennet Lewis for sending me a reprint of his interesting "Early detectors and counters" from *Nuclear Instruments and Methods*, 162 (1979) 9–14, North Holland Publishing, which reveals an unexpected facet of Cavendish technological development. I also found very interesting, and have quoted from Paul Forman, "Weimar Culture, Causality and Quantum Theory 1918–1927, Adaptation by German Physicists and Mathematicians to a Hostile Intellectual Environment", *Historical Studies in Physical Sciences*, Vol. 3, 1–115. I have to thank my daughter, Dr Clare Poulter, for pointing out this article and Gary Wersky's book.

Bibliography

ADAMS, J. B., "Four Generations of Nuclear Physicists", *Notes and Records of the Royal Society of London*, Vol. 27, Aug. 1972

ALLIBONE, T. E., *Rutherford, A Century of Nuclear Energy*, Royal Institution of London, Oct. 29, 1971

ANDRADE, E. N. da C., *Rutherford and the Nature of the Atom*, Doubleday, New York, 1964

ASHBY, Eric, *Community of Universities*, Cambridge University Press, Cambridge, 1963

BADASH, Lawrence, *Rutherford and Boltwood; Letters on Radioactivity*, Yale University Press, New Haven, 1969

BADASH, Lawrence, *Rutherford Correspondence Catalogue*, Centre for History of Physics, New York, 1974

BEISER, Arthur, *Concepts of Modern Physics*, McGraw Hill, Kogakusha, Tokyo, 1973

BEVERIDGE, Lord (William), *A Defence of Free Learning*, Oxford University Press, London, 1959

BIRKENHEAD, the Earl of, *The Prof. in Two Worlds – The Official Life of Prof. F. A. Lindemann, Viscount Cherwell*, Collins, London, 1961

BIRKS, J. B. (Ed.), *Rutherford at Manchester*, Heywood, London, 1962 (contains valuable contributions from Marsden, Darwin, Andrade, Bohr, Robinson, Russell and Blackett)

BLACKETT, P. M. S. (Lord), "Rutherford", *Notes and Records of the Royal Society of London*, Vol. 27, Aug. 1972

BLACKETT, P. M. S., "Rutherford Memorial Lecture", *Physical Society Year Book*, 1955

BOHR, Niels, "Rutherford Memorial Lecture", *Physical Society Year Book*, 1961

BOLTZ, C. L., *Ernest Rutherford (The Great Nobel Prizes)*, Heron Books, London, 1970

BORN, Max, *My Life: Recollections of a Nobel Laureate*, Taylor and Francis, London, 1978

BOWDEN, F. (Lord), "Professor the Lord Rutherford of Nelson, Christchurch's Most Famous Son", Unpublished lecture at Canterbury University, New Zealand, March 15th, 1979

BUNGE, M. and SHEA, W. R. (Eds), *Rutherford and Physics at the Turn of the Century*, Dawson and Science History Publications, New York, 1979 (contains valuable articles by Badash, Heilbron, Feather and Trenn)

CALDER, Ritchie, *Profiles of Science*, Allen and Unwin, London, 1951

CAROE, G. M., *William Henry Bragg, 1862–1942*, Cambridge University Press, Cambridge, 1978

A History of the Cavendish Laboratory, 1871–1910, Longman's Green and Co., London, 1910

CHADWICK, (Sir) J., *Radioactivity and Radioactive Substances*, Pitman, London, 1923

CHADWICK, (Sir) J. (Ed.), *The Collected Papers of Lord Rutherford*, 3 Vols. (CPR), Allen and Unwin, London, 1965

CLARK, Ronald W., *Einstein*, Hodder and Stoughton, London, 1979

CLARK, Ronald W., *Sir John Cockcroft*, Phoenix House, London, 1959

CLARK, Ronald W., *The Birth of the Bomb*, Phoenix House, London, 1961

CLARK, Ronald W., *Tizard*, Methuen, London, 1965

COCKCROFT, (Sir) J. D., "Niels Henrik David Bohr", *Biog. Memoirs of Fellows of the Royal Society*, Vol. 9, November 1963

COCKCROFT, (Sir) J. D., Rutherford Memorial Lecture, printed in *Rutherford by those who knew him*, Proc. Phys. Soc., 1943–51

CONN, G. K. T. and TURNER, H. D., *The Evolution of the Nuclear Atom*, Iliffe Press, London, 1965

COX, J., *Beyond the Atom*, Cambridge University Press, Cambridge, 1913

CROWTHER, J. G., *British Scientists of the Twentieth Century*, Routledge and Kegan Paul, London, 1952

CROWTHER, J. G., *The Cavendish Laboratory, 1874–1974*, Macmillan, London, 1974

CURIE, Eve, *Madame Curie*, Heinemann, London, 1941

DARWIN, C. G. (Sir), *The Discovery of Atomic Number*, Royal Institution of London, 1932

DAVIES, Paul, *Other Worlds*, Abacus, London, 1982

EVANS, Ivor B. N., *Man of Power: The Life Story of Baron Rutherford of Nelson*, Stanley Paul, London, 1939

EVE, A. S., *Rutherford*, Cambridge University Press, Cambridge, 1939

EVE, A. S. and CHADWICK, J., "Lord Rutherford", *Obituary Notices of the Royal Society of London*, No. 6, Vol. 2, January, 1938

FEATHER, Norman, *Lord Rutherford*, Blackie, Glasgow, 1940

FEATHER, Norman, "Rutherford–Faraday–Newton", *Notes and Records of the Royal Society of London*, Vol. 27, August 1972

FEYERABEND, Paul, *Against Method*, Verso, London, 1978

GLASSTONE, Samuel, *Sourcebook on Atomic Energy*, Macmillan, London, 1950

GOLDSMITH, Maurice, *Sage: A Life of J. D. Bernal*, Hutchinson, London, 1980

GOODCHILD, Peter, *J. Robert Oppenheimer*, BBC, London, 1980

GOWING, Margaret, *Britain and Atomic Energy*, Macmillan, London, 1964

GOWING, Margaret, *Independence and Deterrence*, 2 Vols, Macmillan, London, 1974

GOWING, Margaret, *Science and Politics*, eighth J. D. Bernal Lecture, Birkbeck College, London, 1977

HARTCUP, Guy, *The Challenge of War*, David and Charles, Newton Abbot, 1970

HEILBRON, J. L., *H. G. J. Moseley, The Life and Letters of an English Physicist, 1887–1915*, University of California Press, Los Angeles, 1974

HIGHT, James and CANDY, A. M. F., *A Short History of Canterbury College*, Whitcombe and Tombs, Auckland, 1927

HILL, Alex (Ed.), *Proceedings of the 2nd Congress of the Universities of the Empire*, Bell, London, 1921

HOFFMANN, Banesh, *Albert Einstein*, Paladin, St Albans, 1977

HOUGHTON, Walter E., *The Victorian Frame of Mind*, Yale University Press, New Haven, 1957

JAFFE, Bernard, *Moseley and the Numbering of the Elements*, Heinemann, London, 1972

JENKINSON, S. H., *New Zealanders and Science*, New Zealand Department of Internal Affairs, 1940

JORDAN, G. (Ed.), *Naval Warfare in the Twentieth Century, 1900–1945*, Croom Helm, London, 1977

JUNGK, Robert, *Brighter Than a Thousand Suns*, Gollancz, London, 1958

KAPITSA, P. L., "Recollections of Lord Rutherford", *Proc. Royal Society* A 294, 1966

KAPITSA, P. L., *Experiment, Theory, Practice*, Reidel, Dordrecht, 1980

KUHN, T. S., *Black Body Theory and Quantum Discontinuity 1894–1912*, Clarendon Press, Oxford, 1978

LARSEN, Egon, *The Cavendish Laboratory*, Ward, London, 1962

LATIL, Pierre de, *Enrico Fermi*, Souvenir Press, London, 1965

LEWIS, W. Bennett, "Some Recollections and Reflections on Rutherford", *Notes and Records of the Royal Society of London*, Vol. 27, August 1972

LODGE, (Sir) Oliver, *Man and the Universe*, 14th ed., Methuen, London, 1913

LOVELL, (Sir) Bernard, *The Story of Jodrell Bank*, Oxford University Press, London, 1968

LUCRETIUS, *On the Nature of the Universe* (trans. R. E. Latham), Penguin, London, 1951

MANN, F. G., *Lord Rutherford on the Golf Course*, Private, Cambridge, 1976

MARSDEN, (Sir) Ernest, *80th Birthday Book*, A. H. and A. W. Reed, Wellington, 1968

MARSDEN, (Sir) Ernest, "Rutherford Memorial Lecture", printed in *Rutherford by those who knew him*, Proc. Phys. Soc., 1943–51

MARSDEN, (Sir) Ernest, "Baron Rutherford of Nelson", *Trans. Royal Society of New Zealand*, Vol. 68, 1938

MASSEY, Sir Harrie and FEATHER, Norman, "James Chadwick", *Biog. Memoirs of Fellows of the Royal Society*, Vol. 22, November 1976

MASSEY, Sir Harrie, "Nuclear Physics Today and in Rutherford's

Day", *Notes and Records of the Royal Society of London*, Vol. 27, August 1972

MEDAWAR, P. B., *Induction and Intuition in Scientific Thought*, Methuen, London, 1970

MEDAWAR, P. B., *The Art of the Soluble*, Penguin, London, 1969

MELVILLE, (Sir) Harry, *The Department of Scientific and Industrial Research*, Allen and Unwin, London, 1962

MILLER, Harold, *New Zealand*, Hutchinson, London, 1957

MILNE, E. A., "Ralph Howard Fowler", *Obituary Notices of Fellows of the Royal Society*, Vol. 5, November 1945

MOORE, Ruth, *Niels Bohr*, Knopf, New York, 1966

MORALEE, Denis (Ed.), *A Hundred Years of Cambridge Physics, 1874–1974*, Cambridge University Physics Society, 1974

MOTT, (Sir) Nevill, "Rutherford and Theory", *Notes and Records of the Royal Society of London*, Vol. 27, August 1972

Nelson College Old Boys' Register, Nelson, New Zealand, 1956

OLIPHANT, (Sir) M. L., "Rutherford Memorial Lecture", printed in *Rutherford by those who knew him*, Proc. Phys. Soc., 1943–51

OLIPHANT, (Sir) M. L., *Rutherford – Recollections of the Cambridge Days*, Elsevier, Amsterdam, 1972

OLIPHANT, (Sir) M. L., "Some Personal Recollections of Rutherford, the Man", *Notes and Records of the Royal Society of London*, Vol. 27, August 1972

O'SHEA, P. P., "Ernest Rutherford, His Honours and Decorations", *Notes and Records of the Royal Society of London*, Vol. 27, August 1972

PAIS, Abraham, *Subtle is the Lord*, Oxford University Press, 1982

PARRY, Albert (Ed.), *Peter Kapitsa on Life and Science*, Macmillan, New York, 1968

POLKINGHORNE, J. C., *The Particle Play*, W. H. Freeman, Oxford, 1979

POPOVSKY, Mark, *Science in Chains*, Collins, London, 1980

RAYMOND, John (Ed.), *The Baldwin Age*, Eyre and Spottiswoode, London, 1960

REEVES, W. P., *The Long White Cloud*, 4th edn., Allen and Unwin, London, 1950

REID, Robert, *Marie Curie*, Collins, London, 1974

ROBINSON, H. R., "Rutherford Memorial Lecture", printed in *Rutherford by those who knew him*, Proc. Phys. Soc., 1943–51

ROMER, Alfred (Ed.), *The Discovery of Radioactivity and Transmutation*, Dover, New York, 1964

ROMER, Alfred, *The Restless Atom*, Doubleday, New York, 1960

ROWLAND, John, *Rutherford, Atom Pioneer*, Werner Laurie, London, 1955

RUSSELL, A. S., "Rutherford Memorial Lecture", printed in *Rutherford by those who knew him*, Proc. Phys. Soc., 1943–51

RUTHERFORD, E., *Radioactivity*, Cambridge University Press, Cambridge, 1904

Bibliography

RUTHERFORD, E., *Radioactive Transformations*, Constable, London, 1906

RUTHERFORD, E., *Radioactive Substances and Their Radiations*, Cambridge University Press, Cambridge, 1913

RUTHERFORD, Sir E. with CHADWICK, James and ELLIS, C. D., *Radiations from Radioactive Substances*, Cambridge University Press, Cambridge, 1930

RUTHERFORD, Lord, *The Newer Alchemy*, Cambridge University Press, Cambridge, 1937

SNOW, C. P., "He Started Something, Portrait of Lord Rutherford", Travel and Leisure Magazine *GO*, June 1951

SNOW, C. P., *The New Men*, Penguin, London, 1970

SNOW, C. P., *The Physicists*, Macmillan, London, 1981

SNOW, C. P., *Variety of Men*, Penguin, London, 1969

SODDY, Frederick, *Science and Life*, John Murray, London, 1920

SODDY, Frederick, *The Story of Atomic Energy*, Nova Atlantis, London, 1949

STAROSELSKAYA-NIKITINA, O. A., *Ernest Rutherford*, Nayka, Moscow, 1967

THOMSON, A. Landsborough, "Half a Century of Medical Research", Vol. II of *History of the Medical Research Council*, HMSO, London, 1960

THOMSON, (Sir) George P., *J. J. Thomson and the Cavendish Laboratory*, Nelson, London, 1964

THOMSON, Sir George, *The Inspiration of Science*, Oxford University Press, London, 1961

THOMSON, (Sir) J. J., *Recollections and Reflections*, Bell, London, 1936

TILDEN, (Sir) W. A., *Sir William Ramsay*, Macmillan, London, 1918

TIZARD, Sir Henry, "Rutherford Memorial Lecture", *Journal of the Chemical Society*, London, 1946

TRAVERS, Morris W., *A Life of Sir William Ramsay*, Arnold, London, 1956

TRENN, T. J., *Radioactivity and Atomic Theory by Frederick Soddy*, Taylor and Francis, London, 1975

TRENN, T. J., *The Self-Splitting Atom*, Taylor and Francis, London, 1977

TRICKER, R. A. R., *The Contributions of Faraday and Maxwell to Electrical Science*, Pergamon, Oxford, 1966

WEIZMANN, Chaim, *Trial and Error*, Hamish Hamilton, London, 1949

WERSKEY, Gary, *The Visible College*, Allen Lane, London, 1978

WESTCOTT, E. N., *David Harum*, Collins, London, 1911

WHYTE, Lancelot Law, *Focus and Diversions*, Cresset Press, London, 1963

WOOD, Alexander, *The Cavendish Laboratory*, Cambridge University Press, Cambridge, 1946

YOUNG, Kenneth, *Arthur James Balfour*, Bell, London, 1963

ZUKAV, Gary, *The Dancing Wu-Li Masters*, Fontana, London, 1980

INDEX

Compiled by Douglas Matthews

628

General Advisory Council, BBC, 469–71
George V, King, 265
Germany, 483–8
Gerrard, H., 358–60, 364–5, 370–1
Gibson, C. S., 485
Gill, Eric, 500
Gladstone, W. E., 19
Glasson, J. L., 547
Glazebrook, R. T., 77, 82–3, 169, 260–1, 344
Godlewski, Tadeusz, 183, 222, 344
Gollancz, Israel, 467
Goodlett, Brian L., 554
Göttingen University, 176
Gray, A., 60
Great Exhibition, London, 1851, 19; Scholarship, 59–62, 69, 85, 101, 119
Gregory, Sir Richard, 488, 523, 598
Grey, Sir Edward (later Viscount Grey of Fallodon), 250
Grier, Arthur G., 180–1
Griffiths, E. T., 169, 171
Groves, Gen. L. R., 599
Gurney, R. W., 559, 577

H-particles, 390, 392–8, 400–2, 404–5, 438–9, 447
Haast, Julius von, 21
Hahn, Otto: with ER at McGill, 183–5, 187; and radiochemistry, 197, 242; and thorium, 198, 282, 321; in Ramsay's lab., 199; on alpha-particles, 210; relations and correspondence with ER, 213, 231, 242, 252; and radium "family", 231, 242–3, 277; and Ramsay's emanation theory, 236; career, 242; ER visits, 247; and radium standard, 252–3; on Regener's scintillation experiments, 282; and atomic number, 322; and ER's knighthood, 340; post-war relations with ER, 430; and uranium fission, 583
Haldane, Richard Burdon, Viscount, 418
Hale, George Ellery, 432, 456–8
Hale, W. E., 344, 380–1
Hale lectures, Washington, 341, 387, 390
Halifax, Edward F. Lindley Wood, 1st Earl of, 485
Hall, Commodore, 383
Hallwachs, W. L. F., 253
Hankey, Sir Maurice (later Baron), 493–4
Harper's (magazine), 191, 207

Harty, Sir William Hamilton, 367
Hartley, Sir Harold, 492
Harwich: Parkeston Quay, 369, 374–6, 382, 455
Havelock, New Zealand, 29–31
Hawkcraig Experimental Establishment, Scotland, 349–50, 357–60, 365, 367, 369–70, 372–3
Haworth, Mary, 271
Hayles, W. H., 130, 538
Headlam, A. C., 219
Heath, Sir Frank, 438, 465, 473
Hector, James, 21
Heidelberg University, 186
Heidra (drifter), 360, 362, 369
Heisenberg, W., 303, 392, 578, 591, 593
helium, 164, 189, 195, 199; alpha-particles as, 205, 208–10, 246, 278–9, 284–5; and rare light gases, 396
helium 3, 447
Helmholtz, Hermann L. F. von, 74
Henderson, Arthur, 471
Henderson, W. C. Craig, 96
Henry, Joseph, 60, 92, 96
Henry Sidgwick Memorial Lectures, 573–4, 581
Hertz, H. R., 50, 56–7, 84, 109, 113, 144, 486
Hertz, Miss M., 486
Hess, Victor Francis, 344
Hevesy, George von: at Manchester, 275, 310; and Bohr, 327, 330, 336, 429; and ER's knighthood, 340; post-war relations with ER, 429; persecuted as Jew, 483; on losing assistants, 587; and ER's political views, 596
Hicks, W. M., 77
Hight, James, 48
Hill, A. V., 488, 490
Hitler, Adolf, 483–4, 503
Hobson, E. W., 418
Hochstetter, F. von, 21
Holland, Sir Reginald Sothern, 372, 382–4, 454
Hooker, Sir Joseph Dalton, 20
Hooker, Sir William, 20
Hoover, Herbert, 380
Hopkins, Sir Frederick Gowland, 420, 484, 515, 523, 545
Hopkinson, Alfred, 52–3, 222
Hopkinson, Bertram, 346
Houtermans, F. G., 486
Humphry, Sir George, 103–4
Hutton, Frederick W., 21
Hutton, Robert Salmon, 268
Huygens, Christiaan, 303
hydrogen, heavy, 567–9

hydrophones, 349, 351, 354, 357, 363, 365, 368, 374
Imperial College, London, 229
India, 596–8
Inskip, Sir Thomas, 491–2
Institut International de Physique, 430
Institution of Mechanical Engineers, 577
International Physics Conference, 1934, 563
International Physics Institute, 263
Ioffe, A., 496, 503, 514–16
ionium, 231, 277–8
ions, ionisation: nature of, 110, 112–14; of gases, 112–14, 117, 119–21, 125, 131–2, 136–8, 426; study of, 116–17; ER on mathematics of, 129; and alpha-rays, 132, 278–9; and excited radiation, 142; and cloud chamber, 426; and gamma-rays, 549
isotopes, 163, 197, 320–1, 324, 327, 342

Jackson, Sir Henry, 474
Japan, 241–2
Jeans, Sir James, 262, 334, 432–4, 467, 594, 597–8
Jeffries, Caroline (née Shuttleworth, then Thompson), 17–18
Jeffries, William, 18
Jellicoe, John Rushworth, 1st Earl, 367, 384
Jenkinson, S. H., 44
Jews: persecution of, 483–8, 543
Johnson, Samuel, 494
Joliot, Frédéric, 538, 548, 563, 573
Joliot-Curie, Irene, 538, 548–9, 551
Joliot-Curie Laboratory, Paris, 424, 538–9
Joly, J., 239
Joule, J. P., 178
Joynt, J. W., 36–7, 41
Jutland, Battle of (1916), 369

Kamenev, L. B., 508–9, 523
Kapitsa, Anna (Mrs Peter), 511, 516–18, 521–2, 526–9, 532, 534–6
Kapitsa, Peter L.: equipment transferred to USSR, 193, 529–35; work at Cavendish, 440, 496, 499–502, 506, 509–10, 545; and large machines, 449, 472–4, 504–7, 532, 554–5; ER supports, 489, 497, 506–7; relations with ER, 496–507,

631

635

638